Carbohydrate-Based Therapeutics

Carbohydrate-Based Therapeutics

Edited by Roberto Adamo and Luigi Lay

WILEY-VCH

Editors

Dr. Roberto Adamo
GlaxoSmithKline
Research Center
Via Fiorentina 1
53100 Siena
Italy

Prof. Luigi Lay
University of Milan
Department of Chemistry
Via Golgi 19
20133 Milan
Italy

Cover Images: © daniskg/Pixabay; © Linnas/Shutterstock; © sefa ozel/E+/Getty Images

Library of Congress Card No.: applied for

British Library Cataloguing-in-Publication Data:
A catalogue record for this book is available from the British Library.

Bibliographic information published by the Deutsche Nationalbibliothek
The Deutsche Nationalbibliothek lists this publication in the Deutsche Nationalbibliografie; detailed bibliographic data are available on the Internet at <http://dnb.d-nb.de>.

Print ISBN: 978-3-527-34870-1
ePDF ISBN: 978-3-527-83131-9
ePub ISBN: 978-3-527-83133-3
oBook ISBN: 978-3-527-83132-6

Typesetting: Straive, Chennai, India

In the memory of Veronica, wife and mother of my beloved children

Luigi Lay

In the memory of Raffaele, my father. Smell of lead and freshly printed books in his typography

Roberto Adamo

Contents

Foreword

Glycans are ubiquitous in nature, and in animal and human biology, they make up an average of 10% of cell membranes as proteoglycans, glycoproteins, and glycolipids and being attached to over 50% of proteins in the cell. Thus, it is not surprising that glycoconjugates have become an important facet of modern molecular medicine, with increasing recognition of the key roles they play in health and disease, ranging from nutrition, microbiomes, and natural aging processes to congenital disorders of glycosylation, inflammation, immune response, cancer, and microbial infections. Modern analytical, synthetic, and biochemical tools in the glycosciences have provided a deep structural and mechanistic insight into the complex functions of glycans, and this complexity can now be addressed by computational methods including modeling and machine learning. There are already small-molecule glycan derivatives in the clinic, such as the iminosugars miglustat and miglitol, but the most prominent success stories have been in biopharmaceuticals. Many of them are glycoconjugates, including carbohydrate-based vaccines against pneumonia and other infectious diseases, glycoproteins including hEPO and many therapeutic antibodies. This book provides the reader with an insight into some of the most topical applications of glycans in medicine.

The carbohydrate coats of microbes are highly diverse, and fungal and bacterial cell surface polysaccharides and viral glycoproteins present distinct microbe-specific targets for vaccination strategies. The chapter by Del Bino et al. reviews antifungal glycoconjugate vaccines in clinical development. Since polysaccharides are T-cell-independent antigens, long-lasting immune responses are achieved by conjugation of the polysaccharide epitope to carrier peptides and proteins. Complementary approaches include immunostimulatory adjuvants, and the use of the bacterial glycolipid Lipid A as adjuvants is discussed in the chapter by Shimoyama and Fukase. Given the potential toxic side effects of Lipid A, they discuss how an understanding of detailed structure–function relationships will be essential to develop effective analogues that are suitable in the clinic. Carbohydrate-based vaccines are relevant to many antimicrobial strategies, and Compostella et al. review the state-of-art of antibacterial vaccines, and Plata et al. discuss antiviral vaccines. With the current challenges presented to society by increasing resistance against many antimicrobial compounds, these vaccination strategies promise to be effective alternatives to fight infections both in humans and animals.

By understanding the biological function of cell-surface polysaccharides in animals and microbes, scientists can also use these glycans for biomedical applications. Two examples are discussed in the chapters by Russo et al. and by Cramer et al. Polysaccharides can be used as key components in tissue engineering, using their biocompatible and biodegradable properties as surface molecules. This has important applications in medical devices and regenerative medicine. Another interesting application of polysaccharides is in anti-adhesives for new therapeutic approaches to antimicrobials as discussed by Cramer et al.

In humans, glycan structures are very canonical with highly conserved biosynthetic pathways defining the different glycan structures such as lipids, *N*- and *O*-glycans, and proteoglycans. Diversity comes with misregulation of these pathways, resulting in over or under-expression of glycan motifs. We now understand that the resulting disorders of glycosylation can be either genetic or also observed in many diseases. Matassini et al. review therapeutic approaches to one class of congenital disorders of glycosylation, Lysosomal Storage Diseases, which are a group of metabolic disorders leading to accumulation of unmetabolized substrates in the lysosomes and subsequent cell dysfunction. Particularly promising therapeutic candidates are iminosugars, a class of small-molecule drugs initially designed as target enzyme transition state inhibitors.

Aberrant changes in glycosylation are also hallmarks of many diseases, in particular in diverse cancers. Duarte et al. review the analytical tools that allow for the identification of tumor-associated carbohydrate antigens in patient samples, resulting in new biomarkers for cancer progression. Andreana et al. review anti-cancer therapeutic monoclonal antibodies targeting glycans expressed on tumor cells. However, cancer is not the only disease hallmarked by mis-glycosylation and as one further example Matos et al. discuss carbohydrate-based therapeutic targets in Alzheimer's disease, including the role that dysregulation of *O*-GlcNAcs plays in formation of protein aggregates. Bisio et al. focus on the use of heparin-like glycosaminoglycans, as powerful drugs for prevention of thrombosis.

In conclusion, this book provides an authoritative introduction to the roles of glycan-based diagnostics and therapeutics and is written by leading experts in the field both from industry and academia. The book will be of great interest to glycoscientists by bringing together such a diverse set of topics. It should also convince a much broader community of scientists looking for new diagnostic tools and therapies in challenging diseases such as cancer that glycans have become an essential part of modern molecular medicine.

Sabine L. Flitsch
MIB & School of Chemistry,
The University of Manchester,
United Kingdom

Acknowledgments

We are glad that this project is finally coming to light. There is a plethora of books dealing with advancements in carbohydrate chemistry applied to biological challenges. However, our project aims at differing from all those publications as we wished to start from disease areas which need to be tackled in order to impact on health of millions of patients to highlight the modern tools that carbohydrate and glycoconjugate chemistry can nowadays offer.

We acknowledge all authors who kindly contributed to make this book possible, and Prof. S. Flitsch for providing an excellent preface to the chapters.

We are hopeful that this book can attract the interest of the broadly interdisciplinary scientific community around the glycoscience world. Most importantly, we hope it can be useful not only to senior scientists and specialized professionals from both industry and academy, but also to inspire students and young researchers who will form the next generation of glycoscientists.

Roberto Adamo and Luigi Lay

1

Antibacterial Carbohydrate Vaccines

Federica Compostella[1], Laura Morelli[1], and Luigi Lay[2]

[1]*University of Milan, Department of Medical Biotechnology and Translational Medicine, Via Saldini 50, 20133 Milano (MI), Italy*
[2]*University of Milan, Department of Chemistry, Via Golgi 19, 20133 Milano (MI), Italy*

1.1 Introduction

Despite the impressive advances of modern medicine and microbiological research, infectious and parasitic diseases are still a significant burden with a profound socio-economic impact worldwide, particularly in low- and middle-income countries (LMIC). Indeed, infectious diseases increase healthcare expenditures and decrease the country's economic growth, thereby representing a vital concern of the global economy to cope with outbreaks of novel pathogens, such as the SARS-CoV-2 pandemic. It should be noted that the impact of infectious diseases is not only confined to the healthcare system, but different cost categories in various sectors of the economy are involved. The production losses due to illness and premature death from the disease, or even broader economic effects such as those due to reduced trade and tourism are, indeed, additional factors that need to be considered.

In 1992, the World Bank and the World Health Organization (WHO) commissioned the first Global Burden of Disease (GBD) study. One of the main aims of this study was to quantify the burden of disease with a measure that could be used for cost-effectiveness analysis. The GBD study used a novel and single metric, the disability-adjusted life year (DALY), corresponding to the sum of the years of life lost to premature death (YLL) and the years lived with disability (YLD) for varying degrees of severity. DALY, therefore, represents a common metric for death and disability, disaggregating the contribution of comorbidity. The first GBD report covered eight geographic areas and five age groups, with estimates through 1990. Subsequent GBD studies progressively expanded the diseases and injuries number, the risk factors, and the geographic areas [1]. Besides injuries, causes of death and disability can be distinguished into communicable (infectious diseases, along with maternal, perinatal, and nutritional conditions) and noncommunicable (chronic) diseases. In

Carbohydrate-Based Therapeutics, First Edition. Edited by Roberto Adamo and Luigi Lay.

the GBD study 2001, the vast majority of DALYs caused worldwide by communicable diseases were due to infectious and parasitic diseases (21% out of 37%) [1]. The latest GBD report (2019) [2] reveals a substantial decline in the burden of communicable diseases and an improvement of the health of the world's population. For instance, global life expectancy at birth increased from 67.2 years in 2000 to 73.5 years in 2019 [3], even if the disability is becoming an increasingly large component of disease burden. However, the 2019 GBD study shows that the global burden of infectious and parasitic diseases still has a prominent place in terms of disability-adjusted life year's percentage (17% out of 26% total DALYs due to communicable diseases). In addition, another factor contributing to worsening the situation is the growing phenomenon of antimicrobial resistance (AMR), due to the constant emergence of antibiotic-resistant bacterial strains. It was in the 1990s when the first examples of AMR were documented [4, 5]. To ensure their survival, pathogens started to develop evolutionary defense mechanisms which progressively impaired the effectiveness of the antimicrobial agents, thus making the continuous development of new and more sophisticated antibiotics necessary. To date, AMR is one of the major global healthcare challenges, undermining the capacity to prevent and cure a number of infectious diseases that were once treatable. It has been estimated that AMR kills about 700 000 people each year worldwide, a number that, without urgent action, is expected to increase dramatically in the next decades, until reaching 10 million deaths per year by 2050 [6, 7].

The adoption of preventive strategies appears to be the most promising means to efficiently tackle both known and emerging infectious diseases. More than two centuries after the seminal experiments of Edward Jenner, vaccines have become one of the most powerful weapons of modern medicine in the fight against life-threatening infectious diseases, and capable of saving millions of lives, until being recognized by the WHO as "one of the most cost-effective ways to prevent disease" [8]. Indeed, records collected by the Centers for Disease Control and Prevention (CDC) since 1912 have shown a significant decrease in the number of reported cases of infectious diseases before and after the availability of a vaccine [9]. Particularly impressive is the impact of vaccination on life expectancy at birth: according to WHO, current immunization programs contributed to the substantial reduction in mortality of children under 5 years of age from 93 deaths per 1000 live births in 1990 to 39 deaths per 1000 live births in 2018 [10]. In addition, the vaccination practice is also a powerful tool to fight AMR [11]. Besides directly preventing the transmission of bacteria that are already resistant to antibiotic therapies, vaccination helps to reduce transmission of all types of infections, including viral infections, thus avoiding unnecessary or excessive antibiotic use that is a major cause of AMR.

1.1.1 A Brief History of Vaccines

Although the mass vaccination became a routine practice only in the twentieth century [12], it has had a tremendous impact on the health of the world's population. With the exception of safe water, no other intervention, not even antibiotics, has had such a major effect on mortality reduction and population growth. Vaccination, indeed, allowed to control 14 major diseases (smallpox, diphtheria, tetanus, yellow fever, pertussis, *Haemophilus influenzae* type b (Hib) disease, poliomyelitis, measles, mumps, rubella, typhoid fever, rabies, rotavirus, and hepatitis B), at least in parts of the world.

The word "vaccine" originates from the Latin term "variolae vaccinae" (smallpox), when Edward Jenner in 1798 observed that cowpox (a less virulent version of smallpox) could be used to prevent smallpox in humans. Contrary to the general belief, however, the first known real vaccination practice is owed not to a physician nor a scientist, but to a cattle breeder named Benjamin Jesty. Based on the observation that dairymaids seemed to be protected from smallpox after they had contracted cowpox, in 1774, he deliberately inoculated his wife and two children with cowpox to avoid a smallpox epidemic. His experiment was successful, as they were unaffected by the outbreak. Jesty never attempted to publicize his experiment or vaccinate anyone else. He nevertheless proved for the first time the key vaccination principle: inoculation with one moderately harmless disease (cowpox) could provide protection against another far more dangerous disease (smallpox). More than 20 years later, Edward Jenner demonstrated that cowpox could be passed directly from one person to another and that the stimulation of the immune system with a weaker version of the pathogen of interest can confer protection against the related disease [13, 14]. For this reason, despite Jesty's successful vaccination of his family, Jenner is considered the "father of modern human vaccination."

The second key milestone after Jenner's findings was the work of Louis Pasteur on the attenuation of the chicken cholera bacterium in the late 1870s, shortly thereafter followed by his research on anthrax bacillus. After intensive laboratory research, in 1880, Pasteur announced to the scientific community the vaccination against chicken cholera, while the first public controlled experiment of anthrax vaccination took place in 1881. Although both vaccines were not a success, Pasteur introduced the modern concept of vaccination, involving the creation of vaccines in the laboratory using the same agent that caused the disease. Pasteur's methodology paved the way for subsequent epochal developments. In the last decade of the nineteenth century, the novel advances in the newly born discipline called bacteriology provided a rationale for vaccine development based on techniques to inactivate whole bacteria, the discovery of bacterial toxins, and the production of antitoxins. It was realized that the immune system produced soluble substances (antibodies) capable of neutralizing toxins and stopping bacterial growth. Antitoxins and vaccines against rabies, diphtheria, tetanus, anthrax, cholera, plague, typhoid, tuberculosis, and more were developed through the 1930s, taking advantage of these new pieces of knowledge [15]. Likewise, the middle of the twentieth century was a very active time for vaccine research, fueled by the development of methods for growing viruses in the laboratory, which led to rapid discoveries and innovations, including the creation of vaccines for polio. Researchers targeted other common childhood diseases such as measles, mumps, and rubella, and vaccines for these diseases reduced the disease burden greatly. In the second half of the twentieth century, new and more sophisticated techniques were introduced and employed for vaccine development. Examples include the formulation of the first recombinant vaccine, licensed in 1986 against hepatitis B [16–19], and the recombinant, quadrivalent human papillomavirus (HPV) vaccine, licensed in the United States in 2006 and followed in 2014 by a nine-valent version [20]. In 1995, the first complete genome sequence of *H. influenzae* Rd was published [21], leading to a new breakthrough in vaccine research. The availability of bacterial strain genome sequences was successfully applied for the first time to identify vaccine candidates for Meningococcus serogroup B (MenB), thanks to the bioinformatic technique known as *reverse vaccinology* [22]. The sequencing of a MenB strain allowed to predict hundreds of specific protein antigens as possible vaccine candidates [23] and ultimately led to the licensure of the first vaccine against the MenB disease [24].

Today, vaccines can be classified into four different types:

- **Whole-cell killed or inactivated**: based on virulent microorganisms, no longer able to cause the disease, that have been killed by exposure to chemicals, heat, or radiation. Examples include the polio vaccine, hepatitis A vaccine, rabies vaccine, and some influenza vaccines.
- **Whole-cell attenuated**: contain live-attenuated microorganisms. Many of them are active viruses cultivated under conditions that disable their virulent properties, or closely related but less dangerous mutant strains that induce a broad immune response. Examples include the viral diseases yellow fever, measles, mumps, rubella, and the bacterial disease typhoid fever. Attenuated vaccines have some pros and cons. They typically provoke more durable immunological responses and are the preferred type for healthy adults, but they may not be safe in immunocompromised individuals and, on rare occasions, mutate to a virulent and disease-causing form.
- **Toxoid vaccines**: based on inactivated pathogenic toxins, but still retaining their immunizing capacity. Examples of toxoid-based vaccines include tetanus and diphtheria.
- **Subunit vaccines**: different from whole-cell-based vaccines, subunit vaccines contain only a specific and conserved microbial component of the microorganism. They can be obtained either by isolation and purification of the key antigens from the natural source (the pathogen) or by production of the antigen(s) by genetic engineering or chemical synthesis. Subunit vaccines do not contain "live" components of the pathogen; they cause only minor side-effects and are safer and more stable than other kinds of vaccines containing whole pathogens. Subunit vaccines can be divided into two groups:
  - *Protein-based vaccines* contain proteins present on the surface of the pathogen. An example is the subunit vaccine against Hepatitis B virus, composed of only the surface viral proteins, which were previously extracted from the blood serum of chronically infected patients and are now produced by recombination of the viral genes in yeast [19].
  - *PS-based vaccines* contain PSs either extracted from encapsulated bacteria (capsular polysaccharides (CPSs)) or portions of lipopolysaccharides (LPS, usually the O-antigen) of Gram-negative bacteria. However, as explained in more detail in Section 1.2, plain PSs are not able *per se* to induce B-cell-mediated immunological memory. The breakthrough in this field has been the introduction of *glycoconjugate vaccines*, obtained by chemical conjugation of pathogen-associated saccharide antigens to immunogenic proteins (e.g. toxins). The administration of glycoconjugate vaccines induces stronger activation of the immune system, resulting in persistent immunological memory and durable protection of the host.

An additional type of vaccine that has gained considerable attention in recent years, particularly following the COVID-19 pandemic, is mRNA vaccines [25]. Strictly speaking, they cannot be classified in any of the above-mentioned categories because they combine features of subunit vaccines and live-attenuated vectors. They are based on synthetic mRNA molecules that direct the production of the antigen that will generate an immune response and are able to mimic a viral infection eliciting both humoral and cellular immune responses. The interest in mRNA-based technology for the development of prophylactic vaccines against infectious disease stems from the potential to expedite vaccine development, to have improved safety and efficacy, have flexible production, to address maternal vaccination, and to tackle diseases that have not been possible to prevent with other approaches. However, this technology has been used to target mainly viral infections. There is only one study in the field of bacterial infections describing a self-amplifying mRNA vaccine to elicit immune responses against group A and group B streptococci [26], thus it is out of the scope of this chapter.

1.2 Carbohydrate-Based Vaccines

Carbohydrates are by far the most abundant organic molecules found in nature and are fundamental constituents of all living organisms. Besides their well-established structural and energetic (storage and production) functions, carbohydrates play a vital role in a great deal of biological and biochemical processes and are involved in cellular adhesion and differentiation, signal transmission, fetal development, fertilization, and all kinds of cellular recognition events. Both eukaryotic and prokaryotic cells expose on their surfaces a dense array of complex carbohydrates, mainly as components of glycoproteins, proteoglycans, and cell surface glycolipids, which are collectively referred to as *glycocalyx*. In mammalian cells, glycans can be attached not only to proteins and lipids on the cell surface, but also in the cytoplasm or even in the nucleus. The glycocalyx exerts a protective function from ionic and mechanical stress, preserving the integrity of the membrane and acting as a barrier from invading microorganisms. Furthermore, the glycocalyx is critical in determining the development of innate and adaptive immunity in response to pathogens or the growth and spread of cancer [27–29]. On the other hand, the glycocalyx of bacterial pathogens may include either CPSs of encapsulated bacteria, LPS of the outer membrane of Gram-negative microorganisms, or other lipid-linked or peptidoglycan-attached glycan chains. Bacterial surface PSs are key virulence factors. They can trigger bacterial adhesion and host cells infection, and interfere with innate immunity by preventing the activation of the alternative complement pathway. Overall, the bacterial glycocalyx plays many functions, regulating the interactions of the organism with its environment and allowing the bacterium to establish and maintain an infection. Additionally, since carbohydrates derive from posttranslational modifications, they remained structurally highly conserved throughout evolution and, in most cases, are uniquely associated to microbial species, even if some examples of bacterial surface carbohydrates mimicking host self-antigens have been reported [24]. It is, therefore, anything but surprising that these pathogenic glycans have proven to be attractive targets for vaccine development.

The first PS-based vaccine stemmed from the seminal findings of Avery and Heidelberger, who demonstrated the immunoreactivity of CPSs from the encapsulated bacterium *Streptococcus pneumoniae* (Sp) [30, 31]. Accordingly, a CPS-based vaccine targeting four relevant Sp serotypes was licensed in 1945 [32]. The advent of chemotherapeutics and antibiotics in later years, however, dampened the enthusiasm toward vaccines. It was a general belief that these new drugs could represent the ultimate solution to heal any kind of infectious disease, but the emergence and the constant increase of multidrug resistance phenomenon revealed the intrinsic limitations of antibiotic therapies and awakened a renewed interest in preventive strategies. In the following years, a large body of literature data highlighted the role of carbohydrate-specific antibodies in preventing microbial infections and provided a new impetus to the development of carbohydrate-based vaccines. These endeavors led to the approval of numerous CPS-based monovalent and multivalent vaccines against *Neisseria meningitidis* (MPSV4, introduced in 1978) [33], Sp (the current version including 23 out of approximately 100 known serotypes was launched by Merck and Co. in 1983 with the trade name PneumoVax) [34], Hib, and *Salmonella typhi* [15]. These PS-based vaccines were proven effective in preventing disease in healthy adults and older children. They appeared, however, to be poorly immunogenic in infants and young children (under two years of age), in the elderly, and in immunocompromised individuals. Even in adults, they induce only short-lasting antibody responses and fail to generate conventional B-cell-mediated immunological memory [35–38].

The limited clinical efficacy of PS vaccines is largely attributed to the T-cell-independent immune response they induce, which is typically triggered by repetitive polymeric antigens (see Section 1.2.1) [39, 40]. Once again, an old discovery was the key. In 1921, Landsteiner introduced the *hapten* concept, meaning any small organic molecule that alone does not elicit an immune response but is able to produce specific antibody responses when conjugated to an immunogenic protein [41]. A few years later, Avery and Goebel put in place this concept and chemically conjugated pneumococcal CPSs to proteins with the aim to enhance their immunogenicity [42]. This seminal experiment paved the way for the development of glycoconjugate vaccines, a revolutionary breakthrough in the field of vaccinology, although the first glycoconjugate vaccine, targeting Hib, was authorized only in 1987 [43, 44]. Subsequently, at least 12 more monovalent Hib conjugate vaccines have been licensed. Likewise, several formulations of glycoconjugate vaccines were licensed in the following years, targeting meningococcal [45] and pneumococcal disease (PCVs, pneumococcal conjugate vaccines). The first version of the latter was introduced in 2000 and contained seven serotypes (PCV7) [46]. Higher valent formulations (PCV15 and PCV20), however, have been authorized by the Food and Drug Administration (FDA) in 2021 to achieve broader serotype coverage [47, 48]. In the last few years, two *S. typhi* conjugate vaccines were also licensed [49, 50]. An updated picture of current glycoconjugate vaccines licensed by FDA and/or prequalified by WHO is given in Table 1.1. Table 1.2 reports the chemical structures of the repeating units (RUs) of the CPSs used in commercially available glycoconjugate vaccines.

The advent of glycoconjugate vaccines, capable of conferring long-term protective immunity even in high-risk groups, opened a new era of vaccinology. Glycoconjugate antigens raise an immune response improved in quality and quantity compared to plain PSs, as outlined in Section 1.3.1.

Table 1.1 List of glycoconjugate vaccines licensed by FDA[a] for use in the USA and/or prequalified by WHO[b] (content current as of 04 April 2022).

Pathogen	Serogroup *or* serotype	Pharmaceutical company[c] (country of manufacture)	Tradename (*if any*) (FDA Licensed, Prequalified by WHO)	Formulation type[d] (valency) (vaccine components)	Carrier protein[e]	Adjuvant	Saccharide antigen size[f]
Haemophilus influenzae	Type b	Merck Sharp & Dohme Corp. (USA)	Liquid PedvaxHIB (1990)	Monovalent	OMPC	Amorphous aluminum hydroxyphosphate sulfate	MEDIUM
	Type b	Sanofi Pasteur (France)	Act-HIB (1993, 1998) *used as vaccine component in Hexaxim® and Pentacel®*	Monovalent	TT	—	LONG
	Type b	Serum Institute of India Pvt. Ltd. (India)	*Haemophilus influenzae* type b conjugate vaccine (2008)	Monovalent	TT	*Information not available* (N/A)	N/A
	Type b	GlaxoSmithKline Biologicals SA (Belgium)	HIBERIX (2009)	Monovalent	TT	—	MEDIUM-LONG
	Type b	Centro de Ingenieria Genetica y Biotecnologia (Cuba)	Quimi-Hib (2010)	Monovalent	TT	Aluminum phosphate	SHORT
	Type b	Biological E. Limited (India)	ComBE Five (2011)	Combination (DTPw–HBV–Hib)	TT	Aluminum phosphate	N/A

(Continued)

Table 1.1 Continued

Pathogen	Serogroup *or* serotype	Pharmaceutical company[c] (country of manufacture)	Tradename (*if any*) (FDA Licensed, Prequalified by WHO)	Formulation type[d] (valency) (vaccine components)	Carrier protein[e]	Adjuvant	Saccharide antigen size[f]
	Type b	Panacea Biotec Ltd. (India)	Easyfive-TT (2013)	Combination (DTPw–HBV–Hib)	TT	Aluminum phosphate Gel	N/A
	Type b	Sanofi Pasteur (France)	Hexaxim (also known as Hexyon *or* Hexacima) (2014)	Combination (DTPa–HBV–IPV-Hib)	TT	Aluminum hydroxide	LONG
	Type b	Sanofi Healthcare India Private Limited (India)	Shan-5 (2014)	Combination (DTPw–HBV–Hib)	TT	N/A	N/A
	Type b	PT Bio Farma (Persero) (Indonesia)	Pentabio (2014)	Combination (DTPw–HBV–Hib)	TT	Aluminum phosphate	N/A
	Type b	LG Chem Ltd (Republic of Korea)	Eupenta (2016)	Combination (DTPw–HBV–Hib)	TT	Aluminum hydroxide	N/A
	Type b	MSP Vaccine Company (USA)	VAXELIS (2020)	Combination (DTPa–HBV–IPV–Hib)	OMPC	Aluminum salts	MEDIUM
	Type b	Sanofi Pasteur (Canada)	Pentacel (known in the UK and Canada as Pediacel) (2021)	Combination (DTPa–HBV–IPV–Hib)	TT	Aluminum phosphate	LONG

Neisseria meningitidis	A	Serum Institute of India Pvt. Ltd. (India)	MenAfriVac (2010)	Monovalent	TT	Aluminum phosphate	MEDIUM
	$ACYW_{135}$	Sanofi Pasteur Inc. (USA)	Menactra (2005, 2014)	Polyvalent (4-valent)	DT	—	MEDIUM-LONG
	$ACYW_{135}$	GlaxoSmithKline Biologicals SA (Belgium)	Menveo (2010, 2013)	Polyvalent (4-valent)	CRM_{197}	—	MEDIUM-LONG
	$ACYW_{135}$	Pfizer Europe MA EEIG (Belgium)	Nimerix (2016)	Polyvalent (4-valent)	TT	—	MEDIUM
	$ACYW_{135}$	Sanofi Pasteur Inc. (USA)	MenQuadfi (2020, 2022)	Polyvalent (4-valent)	TT	—	MEDIUM-LONG
Salmonella enterica	typhi Ty2 (Vi antigen)	Bharat Biotech International Ltd. (India)	Typbar-TCV (2017)	Monovalent	TT	—	N/A
	typhi Ty2 (Vi antigen)	Biological E. Limited (India)	TYPHIBEV (2020)	Monovalent	CRM_{197}	—	N/A
Streptococcus pneumoniae	1, 4, 5, 6B, 7F, 9V, 14, 18C, 19F, 23F	GlaxoSmithKline Biologicals SA (Belgium)	Synflorix (2009)	Polyvalent (10-valent)	NTHi PD, TT, and DT	Aluminum phosphate	MEDIUM
	1, 3, 4, 5, 6A, 6B, 7F, 9V, 14, 18C, 19A, 19F, 23F	Pfizer (USA)	Prevnar 13 (2010, 2010)	Polyvalent (13-valent)	CRM_{197}	Aluminum phosphate	LONG

(Continued)

Table 1.1 Continued

Pathogen	Serogroup *or* serotype	Pharmaceutical company[c] (country of manufacture)	Tradename (*if any*) (FDA Licensed, Prequalified by WHO)	Formulation type[d] (valency) (vaccine components)	Carrier protein[e]	Adjuvant	Saccharide antigen size[f]
	1, 5, 6A, 6B, 7F, 9V, 14, 19A, 19F and 23F	Serum Institute of India Pvt. Ltd. (India)	Pneumosil (2019)	Polyvalent (10-valent)	CRM_{197}	Aluminum phosphate	LONG
	1, 3, 4, 5, 6A, 6B, 7F, 9V, 14, 18C, 19A, 19F, 22F, 23F, 33F	Merck Sharp & Dohme Corp. (USA)	Vaxneuvance (2021)	Polyvalent (15-valent)	CRM_{197}	Aluminum phosphate	LONG
	1, 3, 4, 5, 6A, 6B, 7F, 8, 9V, 10A, 11A, 12F, 14, 15B, 18C, 19A, 19F, 22F, 23F, 33F	Pfizer (USA)	Prevnar 20/ Apexxnar (2021)	Polyvalent (20-valent)	CRM_{197}	Aluminum phosphate	LONG

a) https://www.fda.gov/vaccines-blood-biologics/vaccines/vaccines-licensed-use-united-states.
b) https://extranet.who.int/pqweb/vaccines/prequalified-vaccines.
c) MSP Vaccine Company is a U.S.-based joint-partnership between Merck and Sanofi Pasteur, Merck Sharp & Dohme Corp. is a U.S.-based subsidiary of Merck & Co.
d) DTPw–HBV–Hib = Diphtheria–Tetanus–Pertussis (whole cell), Hepatitis B Virus (rDNA), and *Haemophilus influenzae* type b conjugate vaccine (absorbed); DTPa–HBV–IPV–Hib = Diphtheria–Tetanus–Pertussis (acellular), Hepatitis B Virus, Poliovirus (Inactivated), and Haemophilus influenzae type b conjugate vaccine (absorbed); DTPa–IPV–Hib = Diphtheria–Tetanus–Pertussis (acellular), Poliovirus (Inactivated), and Haemophilus influenzae type b conjugate vaccine (absorbed).
e) OMPC = meningococcal outer membrane protein complex; TT = Tetanus Toxoid; DT = Diphtheria Toxoid; CRM_{197} = Cross Reacting Material 197; NTHi PD = Nontypeable *H. influenzae* Protein.
f) LONG = native PS; MEDIUM = depolymerized and/or sized-fractionated PS; SHORT = synthetic PS.

Table 1.2 Chemical structure of CPS repeating units within serogroups *or* types in licensed vaccines.

Pathogen	Serogroup *or* type	Repeating unit
Haemophilus influenzae	Type b	→3)-β-D-Rib*f*-(1→1)-D-Rib-ol-(5-OPO_3→
Neisseria meningitidis	A	→6)-α-D-Man*p*NAc(3/4OAc)-(1-OPO_3→
	C	→9)-α-D-Neu*p*5Ac(7/8OAc)-(2→
	W	→6)-α-D-Gal*p*-(1→4)-α-D-Neu*p*5Ac(7/9OAc)-(2→
	Y	→6)-α-D-Glc*p*-(1→4)-α-D-Neu*p*5Ac(7/9OAc)-(2→
Salmonella enterica	typhi Vi	→4)-α-D-Gal*p*NAcA(3OAc)-(1→
Streptococcus pneumoniae	1	→3)-α-D-AATGal*p*-(1→4)-α-D-Gal*p*A(2/3OAc)-(1→3)-α-D-Gal*p*A-(1→
	3	→3)-β-D-Glc*p*A-(1→4)-β-D-Glc*p*-(1→
	4	→3)-β-D-Man*p*NAc-(1→3)-α-L-Fuc*p*NAc-(1→3)-α-D-Gal*p*NAc-(1→4)-α-D-Gal*p*2,3(S)Pyr-(1→
	5	→4)-β-D-Glc*p*-(1→4)-[α-L-Pne*p*NAc-(1→2)-β-D-Glc*p*A-(1→3)]-α-L-Fuc*p*NAc-(1→3)-β-D-Sug*p*-(1→
	6A	→2)-α-D-Gal*p*-(1→3)-α-D-Glc*p*-(1→3)-α-L-Rha*p*-(1→3)-D-Rib-ol-(5→OPO_3→
	6B	→2)-α-D-Gal*p*-(1→3)-α-D-Glc*p*-(1→3)-α-L-Rha*p*-(1→4)-D-Rib-ol-(5→OPO_3→
	7F	→6)-[β-D-Gal*p*-(1→2)]-α-D-Gal*p*-(1→3)-β-L-Rha*p*(2OAc)-(1→4)-β-D-Glc*p*-(1→3)-[α-D-Glc*p*NAc-(1→2)-α-L-Rha*p*(1→4)]-β-D-Gal*p*NAc-(1→
	8	→4)-β-D-Glc*p*A-(1→4)-β-D-Glc*p*-(1→4)-α-D-Glc*p*-(1→4)-α-D-Gal*p*-(1→
	9V	→4)-α-D-Glc*p*(2/3OAc)-(1→4)-α-D-Glc*p*A(2/3OAc)-(1→3)-α-D-Gal*p*-(1→3)-β-D-Man*p*NAc(4/6OAc)-(1→4)-β-D-Glc*p*-(1→
	10A	→4)-β-D-Gal*p*NAc-(1→3)-α-D-Gal*p*-(1→2)-D-Rib-ol-(5→OPO_3→5)-β-D-Gal*f*-(1→3)-β-D-Gal*p*(1→
	11A	→3)-β-D-Gal*p*-(1→4)-β-D-Glc*p*-(1→6)-[Gro-(1→OPO_3→4)] -α-D-Glc*p*(3OAc)-(1→4)-α-D-Gal*p*(2OAc)-(1→
	12F	→4)-α-L-Fuc*p*NAc-(1→3)-β-D-Gal*p*NAc-(1→4)-β-D-Man*p*NAcA-(1→
	14	→4)-β-D-Glc*p*-(1→6)-[β-D-Gal*p*-(1→4)]-β-D-Glc*p*NAc-(1→3)-β-D-Gal*p*-(1→
	15B	→6)-[α-D-Gal*p*(2/3/4/6OAc)-(1→2)-[Gro-(2→OPO_3→3)]-β-D-Gal*p*-(1→2)]-β-D-Glc*p*NAc-(1→3)-β-D-Gal*p*-(1→4)-β-D-Glc*p*-(1→

(Continued)

Table 1.2 (Continued)

Pathogen	Serogroup *or* type	Repeating unit
	18C	→4)-β-D-Glc*p*-(1→4)-[α-D-Glc*p*(6OAc)-(1→2)] [Gro-(1→OPO_3→3)]-β-D-Gal*p*-(1→4)-α-D-Glc*p*-(1→3)-β-L-Rha*p*-(1→
	19A	→4)-β-D-Man*p*NAc-(1→4)-α-D-Glc*p*-(1→3)-α-L-Rha*p* -(1→OPO_3→
	19F	→4)-β-D-ManpNAc-(1→4)-α-D-Glc*p*-(1→2)-α-L-Rha*p* -(1→OPO_3→
	22F	→4)-β-D-Glc*p*A-(1→4)-[α-D-Glc*p*-(1→3)]-β-L-Rha*p*(2OAc)-(1→4)-α-D-Glc*p*-(1→3)-α-D-Gal*f*-(1→2)-α-L-Rha*p*-(1→
	23F	→4)-β-D-Glc*p*-(1→4)-[α-L-Rha*p*-(1→2)]-[Gro-(2→OPO_3→3)]-β-D-Gal*p*-(1→4)-β-L-Rha*p*-(1→
	33F	→3)-β-D-Gal*p*-(1→3)-[α-D-Gal*p*-(1→2)]-α-D-Gal*p*-(1→3)-β-D-Gal*f*-(1→3)-β-D-Glc*p*(1→5)-β-D-Gal*f*(2OAc)-(1→

Abbreviations: Glc, glucose; Gal, galactose; Neu5Ac, *N*-acetylneuraminic acid (sialic acid); Rha, rhamnose; GlcNAc, *N*-acetylglucosamine; GalNAc, *N*-acetylgalactosamine; FucNAc, *N*-acetylfucosamine; ManNAcA, *N*-acetylmannuronic acid; PneNAc, *N*-acetylpneumosamine (2-acetamido-2,6-dideoxytalose); GlcA, glucuronic acid; Gro, glycerol; Rib-ol, ribitol; Sug, 2-acetamido-2,6-deoxyhexose-4-ulose; AATGal, 2-acetamido-4-amino-2,4,6-trideoxy-galactose; 2,3(S)Pyr, *trans*-2,3-(S) cyclic pyruvate ketal modification at galactose; *p*, pyranose form; *f*, furanose form; OPO_3, phosphate group in phosphodiester linkages.

1.2.1 Mechanism of the Immune Response to Carbohydrate-Based Vaccines

Bacterial CPSs have been recognized as key virulence factors for more than a century and have been explored as potential vaccine antigens from the outset [51, 52]. The licensure of the previously mentioned CPS-based vaccines against *N. meningitidis*, *S. pneumoniae*, *H. influenzae* b, and *S. typhi* allowed an efficient control of the disease in adults and older children, especially for short-lasting exposures (travelers and soldiers in military campaigns). Plain PS-based vaccines, however, are unable to mount a protective immune response in the immature immune systems of newborns and young children (under two years of age), who are the major group at risk for these infections. Even in adults and adolescents, PS vaccines are not able to induce memory B cells (MBCs), avidity maturation, and antibody isotype switching from immunoglobulin M (IgM) to immunoglobulin G (IgG). Most of the antibodies produced, indeed, are low-affinity IgM, which are only poor activators of the complement system, a key arm of the innate immune system that enhances the humoral responses [53, 54]. Furthermore, immunization of adults and older children with PS vaccines leads to apoptosis of MBCs, thus reducing the response to subsequent administrations, a phenomenon usually referred to as hyporesponsiveness [55]. PSs are T-cell-independent type 2 (TI-2) immunogens, which are typically high-mass polymers with repeating structures of 5–10 nm. Due to their polymeric structure,

PSs activate B cells by cross-linking approximately 15–20 B-cell receptors (BCRs), triggering a series of protein phosphorylation steps which leads to an increase in free intracellular calcium (Figure 1.1a) [56]. The need to crosslink multiple BCRs to achieve B-cell activation is noteworthy, as it explains the reason why the immunogenicity of PS vaccines is size-dependent, with only high-molecular-weight antigens capable of inducing an effective immune response. Following stimulatory cytokines or co-stimulatory signals expressed by other cells of the immune system [57–59], B cells are finally activated. They mature into plasma cells and secrete antibodies

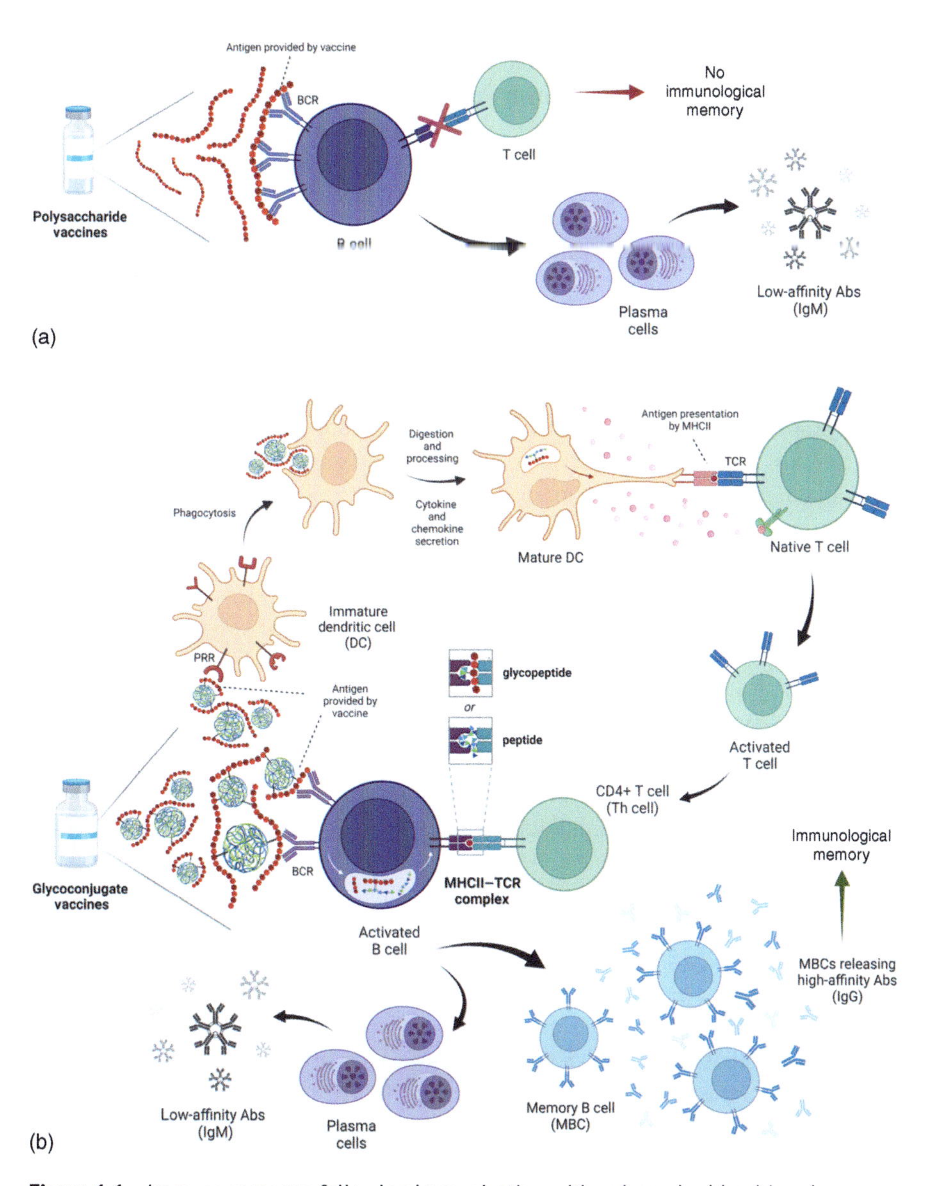

Figure 1.1 Immune response following immunization with polysaccharides (a) and glycoconjugates (b).

(mainly IgM) but since the entire process takes place without the direct involvement of T cells, the development of MBCs and the induction of immunological memory do not occur.

Unlike PSs, proteins and peptides are instead T-cell-dependent (TD) antigens, as they stimulate helper T lymphocytes to elicit an immune response leading to specific antibody response. TD antigens are immunogenic even in early childhood, the immune response induced can be boosted and enhanced by adjuvants, and it is characterized by an antibody class switch with the production of high-affinity and antigen-specific IgG. The hapten-carrier concept [60] underpinning the development of glycoconjugate vaccines is based on the fact that the covalently linked protein carrier confers T-cell-dependent properties to the glycoconjugate, which is eventually able to produce carbohydrate-specific MBCs. The molecular mechanism of the immune response to glycoconjugate vaccines has been thoroughly investigated [61, 62], albeit some steps are still unclear. After administration, the conjugate is taken up by antigen-presenting cells (APCs), mainly dendritic cells (DCs), but also macrophages and B cells can play the same role. Engulfment by APCs is promoted by stimulation of pathogen recognition receptors (PRRs), a large family of receptors expressed on APCs surface able to respond to a huge variety of pathogen-associated molecular patterns (PAMPs), structurally and chemically diverse compounds highly conserved in pathogens but absent in their multicellular host. PRRs stimulation creates the necessary pro-inflammatory context (expression of costimulatory molecules and secretion of soluble cytokines and chemokines), leading to full maturation of DCs, antigen uptake, and intracellular processing (Figure 1.1b). In particular, the glycoconjugate antigen is chopped through the intracellular endosomal compartments. While the saccharide portion is depolymerized by oxidative agents (reactive oxygen species, ROS, and reactive nitrogen species, RNS) [63, 64], the protein portion is processed by proteases into small peptides. Within a few days, mature DCs reach the draining lymph nodes, where peptide antigens are conveyed to the cell surface in association with the major histocompatibility complex class II (MHC-II) protein to be presented to T lymphocytes via interaction with the T-cell receptor (TCR). T lymphocytes are then activated as helper T cells (Th or $CD4^+$ T cells), which provide appropriate stimulatory signals to elicit a conventional TD immune response. Th cells prime the maturation process of resting B cells, driving their proliferation and differentiation into antibody-secreting (mainly IgM) plasma cells and MBCs, ensuring the establishment of the immunological memory. Contrary to plasma cells, MBCs survive for a long time in the body and respond rapidly to subsequent exposures of the same antigen by secreting high-affinity IgG antibodies.

Recent findings [65] provided more details on the mechanism of action of glycoconjugate vaccines, highlighting the crucial role of the formation of germinal centers (GCs) to elicit an immune response. GCs are sites in lymph nodes where mature B-cell proliferation and differentiation occur. The GC formation requires the presence of PS-specific B cells, which display the antibody on their surface as a BCR, the follicular helper T (Tfh) cells, which are able to recognize the protein carrier antigen, and the follicular dendritic cells (FDCs), highly specialized APCs, which contain and present the antigen to the B cells. The formation of GCs is an essential step during the immune response to an infection or after vaccination, since the affinity maturation process and the class switch from IgM to IgG occur in the GCs, specifically in two distinct regions called the light and dark zones.

Also, Kasper and coworkers suggested that not only peptides but also glycopeptide fragments are generated by glycoconjugate processing in APCs and presented to TCR in the context of MHC-II [66, 67]. According to this model, the lipophilic peptide portion of the glycopeptide antigen binds to MHC-II, whereas the hydrophilic carbohydrate portion is exposed to the TCR, leading to the generation of carbohydrate-specific Th, called T carb cells.

It should be emphasized that the different mechanisms of the immune response toward PS and glycoconjugate antigens also have practical consequences in vaccine design. Since crosslinking of surface immunoglobulin molecules on B cells is not required, glycoconjugate vaccines can also be produced from small saccharide chains obtained by size fractionation of native PSs or by chemical synthesis.

The development of glycoconjugate vaccines has been one of the greatest success stories of modern medicine [68, 69], as demonstrated by the drastic reduction of *S. pneumoniae*, *H. influenzae* b, and *N. meningitidis* infections in those countries where the corresponding conjugate vaccines have been introduced in routine vaccination programs. Overall, these vaccines have had a huge impact on global infant mortality and morbidity, saving millions of lives.

1.3 Components of Glycoconjugate Vaccines

A vaccine is a biological product able to induce an immune response that confers protection against an infection upon successive exposures to a pathogen [10]. The efficacy of protection is based on preexisting antibodies in the serum, which prevent the disease but not the infection [70]. Although most vaccines are based on TD protein antigens, PSs can also be used to induce protective immune responses, at least in an immunocompetent host. However, as explained in Section 1.2.1, the T-cell independence of PSs prevents the proliferation of MBCs and the occurrence of immunological memory, making PS-based vaccines ineffective for infants and young children, as well as for immunocompromised individuals. Protein–PS conjugate vaccines provided the solution to overcome these limitations, enabling the use of bacterial surface PSs as antigens to induce long-lasting protection from infectious diseases. Saccharide antigens employed for the construction of glycoconjugate vaccines are either derived from the pathogen or produced synthetically to mimic the components of the microorganism. They can be present in the form of microbial poly- or oligosaccharides (OSs) (the latter obtained by size-fractionation of the native polymer) or as synthetic low-molecular-weight compounds, covalently linked to a carrier protein. In addition, subunit glycoconjugate vaccines can be formulated with adjuvants to enhance the low immunogenicity of carbohydrate antigens.

Impressive progress has been made in the field of antibacterial vaccine development over the past two decades. Surprisingly, the important milestones that have outlined the history of antibacterial vaccines have been reached with an empirical approach [12]. The main reason is that only in recent times has immunology started to give a useful contribution to vaccine design. Recent advances in basic immunology are now unveiling the immunological principles that govern susceptibility to infections and protection of the host, ushering in the era of rational vaccine design. Indeed, progresses in vaccinology permitted us to realize that different, often interconnected, parameters can affect the immunogenicity of the glycoconjugate. Besides the choice of the T-cell helper protein, factors more related to the design of the saccharide antigen and its conjugation to the carrier protein have been recognized as important variables to determine immunogenicity. The elements to consider for the development of an efficient, immunogenic, and protective vaccine setting are now much clearer, and vaccinology is moving to a new era where the "vaccine" should be regarded as a pharmaceutical rather than as a biological product [71]. The following paragraphs of Section 1.3 report a more detailed analysis of each key component of a glycoconjugate vaccine, highlighting how they affect the efficacy of the construct and the robustness of the elicited immune response.

1.3.1 The Carbohydrate Antigen

Antibacterial glycoconjugate vaccines may contain full-length or size-reduced PSs, generally derived from CPS or the O-antigen portion of LPS. The PS portion of the construct is the key player in the glycoconjugate; indeed, the efficacy of the vaccine is measured by the magnitude and the quality of the immune response against the carbohydrate antigen. The size, and therefore the length, of the PS fragment is the principal determinant of its immunogenicity [72]. The glycan chain of a PS contains epitopes or antigenic determinants, also called glycoepitopes, that are the specific portions of the PS recognized by antibodies and responsible for inducing the antigen-specific immune response of the host. An epitope is commonly identified as a sequence of an average of six/seven contiguous residues (linear or branched), but for long saccharide chains, conformational epitopes are also formed, where discontinuous monosaccharide units closely organized in space are simultaneously engaged by the BCRs [73].

The lesson from traditional plain PS vaccines is that the large size of the PSs ensures cross-linking of multiple BCRs to elicit the TI immune response. The presence of repetitive antigenic determinants increases the avidity of antibody–antigen binding, thus facilitating the activation of B cells even when they express low-affinity antibodies [74]. On the other hand, the immunological mechanism of glycoconjugate vaccines does not depend exclusively on cross-linking of BCRs. In glycoconjugates, the simultaneous exposition of multiple copies of shorter PS fragments covalently attached to the carrier protein triggers the activation of the immune response. Yet, also in this case, a relationship between the length of the protein-conjugated PS fragment and the immunogenicity of the vaccine persists.

Consequently, significant efforts are aimed at determining the minimal OS length capable of maintaining the epitope determinants required to engage the antibody and elicit a T-cell-dependent protective immune response. However, the correlation between these two parameters depends on the type of glycoconjugate produced [72]. Typically, size-reduced PSs are obtained through chemical or mechanical fragmentation of the natural PSs isolated from cultures of pathogenic bacteria, followed by multiple purification and characterization steps to reduce glycan heterogeneity [75]. The use of more defined saccharide fragments facilitates the conjugation with the protein carrier and allows for improved batch-to-batch consistency of the final glycoconjugate. PS fragments are then conjugated to the protein either randomly, through the hydroxyl or carboxyl groups occurring along the saccharide chain, or via functional groups localized at the glycan end of the sugar chain (typically the anomeric carbon of the reducing end monosaccharide) [37]. The latter methodology generally includes the use of a linker to facilitate the glycan–protein coupling and to alleviate the steric hindrance between protein and saccharide (see Section 1.3.2). The two different strategies lead to heterogeneous cross-linked structures, usually involving multiple saccharide chains and protein molecules, and more defined structures with a radial exposition of saccharide chains end-terminal conjugated to a single protein molecule, respectively. The manufacture of random cross-linked conjugates uses PS fragments of about 100–300 kDa, while the selective attachment at the reducing end of the sugar is carried out on shorter fragments of 5–20 kDa, generally referred to as OSs [76]. Studies in the literature support the observation that higher immunogenicity of end-linked glycoconjugates correlates with the use of medium-size OSs, while longer PS fragments are preferable in obtaining randomly cross-linked effective glycoconjugates. As an example, a study in mice using end-linked conjugate preparation of *Salmonella enterica sv typhi* Vi antigen coupled to cross-reacting material 197 (CRM197) showed that shorter chain constructs (glycan size about 9.5 kDa) induced a more prolonged proliferation of Vi-specific B cells and a slower decline of Vi-specific IgG antibodies compared to their longer chain counterparts (glycan size about 165 kDa) [77]. On the other hand, a study on cross-linked glycoconjugates of *Francisella tularensis* OAgs coupled to tetanus toxoid (TT) evidenced that a genetically induced, very-high-molecular-size OAg (around 220 kDa) provided a marked increase in protection. The LPS-specific antibodies induced by this larger-sized O-antigen exhibit significantly enhanced relative affinity compared to low and native molecular weight preparations (25 and 80 kDa, respectively), albeit with comparable antibody titers [78]. This study supports the importance of conformational OAg epitopes in the protection induced by the vaccine.

The shape of the glycoconjugate and the way the glycoconjugates are presented to the immune system are critical and interconnected aspects of vaccine efficacy. In this regard, the carbohydrate-to-protein ratio is considered a significant parameter. Hib PS fragments with different chain lengths were coupled to diphtheria toxin (DT) carrier protein. The average degree of polymerization (avDP) was 8 or 20 monosaccharide RUs, and a maximum of three sugar moieties per protein molecule were conjugated [79]. The shorter avDP8 oligomer stimulated a poorer anticarbohydrate response than the avDP20, but immunogenicity comparable to the avDP20 oligomer was then obtained in a later study where four/five sugar fragments with avDP8 were coupled to TT, thus supporting the argument that a higher level of protein glycosylation could compensate the shorter chain length [80].

The evidence that even a short OS chain can induce the production of specific antibodies that can protect the host from a pathogen and current advances in OSs synthesis have opened a new era of vaccine research that aims to prepare the carbohydrate antigen by chemical synthesis [81]. Structurally defined synthetic OSs, designed based on molecular understanding of antigen–antibody interactions, offer a promising alternative for developing semisynthetic (containing fully synthetic carbohydrates conjugated to an immunogenic carrier protein) and fully synthetic glycoconjugate vaccines based on synthetic OS antigens conjugated to synthetic carrier molecules, such as glycolipids or peptides [82–84]. The access to the active components of the vaccines by chemical synthesis, made possible by the tremendous progresses and recent technological innovations in glycan assembly [85–95], would result in more reproducible and homogeneous vaccine preparations, characterized by robust biological properties and better safety profile. Accordingly, the manufacturing process of the glycoconjugates thus obtained would be facilitated and accelerated, enabling timely combat of emerging or antimicrobial-resistant infectious diseases and/or addressing currently unmet medical needs [71, 96].

One additional aspect to consider when investigating carbohydrate antigens for immunological applications is that the size and structure of the PS can be affected by the manufacturing process. The production of the PS is designed to improve the yields and consistency of the final product. For example, native CPSs are commonly isolated by bacterial culture. The yield (expressed in x mgl^{-1}, where x are mg of pure CPSs extracted from 1 l of culture broth) and the molecular weight of bacterial PS produced during fermentation are influenced by the genetic nature of the microorganism and the fermentation conditions (i.e. the carbohydrate source and the carbon/nitrogen ratio in the culture medium, pH, and temperature) [78, 97–99]. Other factors influence the consistency of the final products: the type of PS fragmentation (i.e. chemical or mechanical), the size-fractionation and purification processes, and the activation chemistry before conjugation. Due to the highly specific mechanism for immunological activation, it is important to reduce extensive chemical modifications of the native PS structure during the production of the glycoconjugates. For example, the loss of labile *O*-acetyl groups, which are present (often in random order) in different PSs, might impair the immunogenicity of a PS, leading to a less effective vaccine. In this regard, the meningococcal NeisVac-C vaccine, which uses de-O-acetylated MenC OSs [100], shows less efficacy than other licensed MenC glycoconjugate vaccines, like Menjugate, that are derived from acetylated OS. Also, the random periodate oxidation of the vicinal hydroxy groups of the PS units to generate multiple aldehyde groups to be reacted with the protein by reductive amination should be tightly controlled to affect only some monosaccharides in order to avoid extensive PS chemical modification and, consequently, the loss of immunogenicity.

1.3.2 Linkers for Carbohydrate–Protein Conjugation

In glycoconjugate vaccines, the covalent conjugation of the saccharide antigen to the carrier protein can be achieved either directly or through short linear spacer molecules, called linkers. The introduction of a linker facilitates the coupling of the glycan to the protein because the steric hindrance that may reduce the reactivity between two relatively large molecules is reduced. In addition, the presence of a linker affects the spatial orientation of the glycan epitopes around the protein and the way they are presented to the immune system. Indeed, the nature of the linker and the type of conjugation chemistry are other key factors in the development of a vaccine candidate. They are interconnected and have to be considered jointly because the conjugation chemistry should be based on a highly efficient chemoselective coupling step between two specific functional groups on the carbohydrate and the protein [61]. On one side, there are the chemically reactive sites of the carbohydrate moiety, which are mostly (i) aldehyde groups obtained by periodate oxidation of *cis*-diols in the sugar ring or at the glycerol moiety of sialic acid residues, (ii) carboxyl groups of sialic or uronic acid residues, (iii) cyanate esters generated by random functionalization of the sugar hydroxyls with cyanilating agents, and (iv) primary amino groups, generally obtained by an end terminal reductive amination of sized OSs. Bacterial OS fragments obtained by chemical synthesis are generally designed with short linkers at their reducing end bearing a terminal reactive group suitable for protein conjugation, such as an amine, a thiol, an alkyne, or an azide. On the other side, the common protein functional groups that are more suitable for chemical linkage to sugars are the carboxylic groups of aspartic and glutamic acid residues and the ε-amino group of lysines. These amino acid residues are generally well exposed on the surface of the protein, thanks to their hydrophilic character. Based on the functional groups that are present on the two coupling partners, the protein and the glycan, different types of linkers have been developed over the last 30 years [37].

Homobifunctional linkers contain the same reactive group at the two ends. The glycoconjugate construct is generally obtained through a two-step protocol, which involves the initial coupling of one of the two partners in the presence of an excess of the linker, followed by the conjugation of the activated intermediate to the second partner moiety. The Di-(*N*-succinimidyl)-adipate (DSA) and the dithio-propionates Dithio bis(succinimidylpropionate) (DSP) and Dithiobis(sulfosuccinimidylpropionate) (DTSSP) are examples of linkers containing two terminal carboxyls activated as *N*-hydroxysuccinimidyl (NHS) esters, which react with the amino groups of the lysine residues on the protein and of the sugar to give stable nonhydrolyzable amide bonds (Figure 1.2). Chemical conjugation through stable chemical bonds can also be obtained by exploiting the cystamine or ADH homobifunctional linkers that contain two amino or hydrazide functional groups at both ends, respectively (Figure 1.2). These are crosslinking reagents able to engage aspartic and glutamic residues on the protein in an EDC-mediated coupling, while they can give reductive amination with the formyl groups on the sugar or be employed in nucleophilic additions to cyanide groups to form

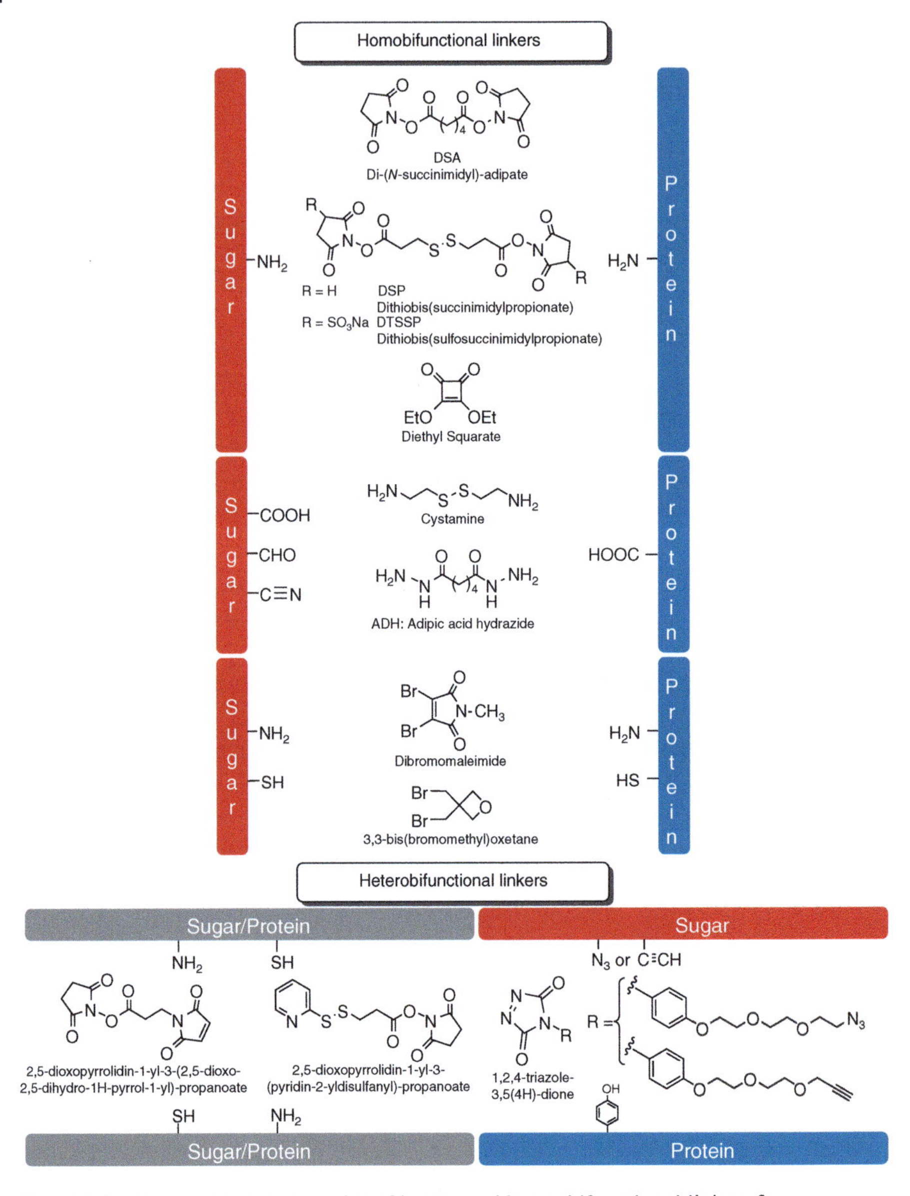

Figure 1.2 Representative examples of homo- and heterobifunctional linkers for glycoconjugation.

amidine bonds. All these homobifunctional linkers are soluble in water and ensure highly specific and easy conjugation protocols that have been exploited in the preparation of different glycoconjugate vaccines [101-103]. Diethyl squarate also reacts with amino groups on the sugar and the protein to give cross-linked coupling products [104]. Halogen-bearing homofunctional spacers, such as dibromomaleimide or 3,3-bis(bromomethyl)oxetane, are also available (Figure 1.2). They are characterized by a cyclic scaffold and by the presence of two halogens that can be displaced in a two-step protocol by nucleophiles such as amines or thiols either on the sugar or on the protein. While the protein conjugate with dibromomaleimide suffers from intrinsic instability, due to the reversible nature of the nucleophilic substitution reaction in a reducing environment, the cross-linked product with 3,3-bis(bromomethyl)oxetane gives a chemically defined and irreversible conjugate [105, 106].

The other class of spacer reagents is represented by heterobifunctional linkers that contain two different reactive end-groups connected by an organic spacer. The two ends of the linker target different functional groups in the reaction partners and are activated orthogonally for site-selective conjugations that require a two-step protocol. The family of heterobifunctional linkers is quite large due to the different combinations of functional groups that can be targeted in the protein and/or the sugar [37]. Among the most representative examples, it is worth mentioning the propanoate derivatives reported in Figure 1.2 that are designed to react with an amine group on one end and with a thiol on the other. Indeed, one of the two extremities of the spacer contains an activated ester, which is the first to be conjugated to the sugar/protein to give an amide, while at the other extremity is a specific group that can react with thiol-containing molecules to form a disulfide linkage or a thioether by Michael addition. In addition, heterobifunctional linkers that contain one functional group suitable for azide–alkyne Huisgen 1,3-dipolar cycloaddition (AAC) are gaining increasing prominence [107–109]. In this regard, triazole-3, 5-dione scaffolds can be easily substituted with alkylated phenols to obtain alkyne or azido-ending linkers suitable for AAC on one side and for ene-like reactions on the other.

An exhaustive picture of the type and chemistry of the linkers is out of the scope of the present account due to the great number of alternative options that have been recently developed. However, a general requirement in this field is that the linker should be easily introduced under mild conditions and should allow an efficient and selective coupling step that is easy to control and scale up. The use of a linker should allow the sugar and the protein to maintain the correct conformational arrangement to be recognized by the immune system, preserving sugar epitopes. In addition, several studies have demonstrated that a kind of intrinsic immunogenicity of the linker could be present, and any type of nonspecific response against the linker that drives the immune response away from the targeted hapten epitopes should be avoided [110, 111]. Data from the literature support that rigid and constrained spacers could induce a significantly high amount of undesirable antibodies, while these unwanted effects are reduced by the use of flexible alkyl linkers [112]. On the other hand, a study in mice with a GBS PS-based conjugate vaccine, where a GBS type II polysaccharide (PSII) was conjugated to a pilus protein, showed that the presence of moderate levels of antibodies against rigid triazole rings generated by click chemistry did not affect the anti-PS immune response [113]. In addition, it was reported that in some cases, the presence of the linker might divert the antibody response of the conjugate from the target sugar epitopes, or conversely, it can enhance the immunogenicity of carbohydrate antigens [101, 111, 114, 115]. Overall, the choice of the linker is case-specific and should be tailored to the particular glycoconjugate construct under development.

1.3.3 The Carrier Protein

The early discovery of Avery and Goebel that conjugation of PSs with proteins stimulates T-cell help and the production of IgG against encapsulated bacteria [42] set the stage for the development of the first glycoconjugate vaccine against *H. influenzae* [116], followed by all the other highly effective antibacterial vaccines based on the same technology. The protein antigen, linked to the saccharide, is the key agent responsible for the thymus-dependent (elicited by TD antigens) immune response and is the so-called T helper protein/peptide antigen. It generally contains multiple sites for saccharide attachment. Five carrier proteins have been developed and successfully used as T helper antigens in licensed glycoconjugate vaccines. TT and DT were the first TD antigens used for this scope. They were discovered at the beginning of the twentieth century by Gustav Ramon, who developed a method for inactivating the respective toxins using formaldehyde. TT and DT were initially selected as carriers for Hib conjugate vaccines due to their high safety profiles established over decades of vaccination against tetanus and diphtheria. Their use was then extended to other vaccines, such as Menactra®, the first licensed quadrivalent meningococcal vaccine [117, 118]. Later on, during mutagenesis studies of the phage containing the gene encoding DT [119], the CRM_{197} was identified. CRM_{197} and DT are structurally very similar [120] and differ only by a single amino acid substitution from glycine to glutamate in position 52 [121]. While they are immunochemically indistinguishable, the big advancement is that CRM_{197} is naturally non-toxic. For this reason, CRM_{197} was considered an ideal carrier for conjugate vaccines, and it is still currently used as a component of meningococcal and pneumococcal glycoconjugate vaccines [122]. It can be isolated from lysogens of *Corynebacterium diphtheriae* or produced by recombinant DNA techniques in heterologous organisms. The other two commonly used T helper antigens are the outer membrane protein complex (OMPC) of *N. meningitidis* serogroup B strain B11 and the outer membrane protein D (PD) derived from non-typeable *H. influenzae* (NTHi). They are components of vaccines against *H. influenzae* and pneumococcus [123–125].

The nature of the carrier protein is an important parameter that strongly affects the efficacy of glycoconjugate vaccines. Comparative studies on the clinical impact of the different carrier proteins used in licensed vaccines were not conclusive in proving the superiority of one protein over the other, due to the high number of variations in the formulation of different vaccines. However, some general hints for the selection and/or development of new carrier proteins can be defined. A good protein carrier should be safe and devoid of any toxic or enzymatic activity. In this regard, the production of carrier proteins with recombinant techniques is preferred over chemical detoxification, which can also result in extensive structure modification and heterogeneity. The maintenance of the protein effective T cell help should be balanced with eventual immune interferences caused by the carrier that diverts the immune system from the relevant saccharide target. Previous exposure to the carrier or carrier-related protein (i.e. coadministration of different vaccines) has been correlated with the carrier-induced epitope suppression, i.e. attenuation or even suppression of the antibody response against the glycoepitope. A high ratio by weight of the carrier protein with respect to the saccharide hapten in the vaccine formulation has also been associated to antigen hyporesponsiveness [126, 127]. In addition, a deep knowledge of the immunological and physicochemical characteristics of the protein is necessary to ensure a reproducible production of the glycoconjugate vaccine. The protein carrier should be soluble, stable, and contain a sufficient number of accessible amino acids for a controlled glycoconjugation. Lastly, the protein should be produced in high yield at a large scale and according to the cGMP protocols required by regulatory agencies.

Many other proteins have been tested in clinical and preclinical studies or are under development as alternatives to the currently used carriers. A larger range of options could reduce the problem of carrier-induced epitope suppression associated to the extensive use of current T helpers in vaccine manufacturing. A detailed picture of the research in this field has been recently reviewed [128, 129]. New carriers are derived from bacteria, like a recombinant nontoxic form of *Pseudomonas aeruginosa* exotoxin A (rEPA), which has been conjugated to PSs of *Staphylococcus aureus* types 5 and 8 [130], and the Vi antigen of *S. typhi* [131, 132]. Furthermore, most of the potential new carriers are produced through recombinant techniques with the purpose of reducing sophisticated culture conditions and developing solid protocols. For example, the tetanus toxin's native C-fragment (Hc) is a safe, low-cost, and highly immunogenic peptide with easy purification. It has been tested as a carrier for pneumococcal PSs (PS14 and PS23F) and has shown immunogenicity levels comparable to CRM_{197} and TT glycoconjugates [133].

The carrier protein may also provide protection against the pathogen from which it is derived. Protein D is one of the examples of proteins with dual role of carriers and antigens. It is expressed by NTHi and was selected as the carrier of 8 PSs in the 10-valent antipneumococcal protein D conjugate vaccine (PHiD-CV), in which the other two serotypes are conjugated to TT and DT, respectively, with the aim to provide protection against NTHi acute otitis media [134]. Other examples are *S. aureus* α toxin Hla and the *S. pneumoniae* protein PiuA, which were effective in animal models in inducing protection against *S. aureus* and pneumococcal infections [135, 136]. Another interesting approach regards a truncated version of the rotavirus spike protein VP8* that was covalently conjugated to Vi capsular PS of *S. typhi* to develop a bivalent vaccine. The Vi-ΔVP8* conjugate vaccine elicited high antibody titers and functional antibodies against *S. typhi* and rotavirus. This study showed that the conjugation of the antigenic peptide to the saccharide epitope enhanced the intrinsically low immunogenicity of the short ΔVP8 truncated peptide and allowed the switching to a specific TD response against the saccharide [137]. Due to their role as antigens eliciting functional antibodies, it is important to preserve the correct topography of the protein during the process of conjugation to avoid the masking of B-cell epitopes. In this regard, a significant example is the conjugate of the GBS type II (GBSII) PS with the antigenic GBS80 pilus protein, selected by genome-based reverse vaccinology. The GBSII PS was conjugated to the GBS80 protein by catalyst-free strain-promoted azide–alkyne cycloaddition (SPAAC) for glycan–protein coupling, exploiting a tyrosine-selective protein derivatization that enables targeting of predetermined less abundant and less exposed sites of the protein (tyrosine vs.

lysines), ensuring to preserve crucial protein epitopes. The GBSII-GBS80 construct was very effective in eliciting antibodies that recognized the glycan and the protein epitopes individually [113].

1.3.4 The Adjuvant

Adjuvants are immunostimulatory substances generally added to nonliving vaccines to enhance the immunogenicity of antigens. The goal of using adjuvants is to endow potentially weak antigens with the immunostimulatory capabilities of microbial pathogens. Adjuvants are in fact immune potentiators able to boost the immune system through a broad variety of molecular and cellular mechanisms of innate and adaptive immunity to provide a strong response even in less-responding individuals [138]. In addition, they are able to increase the stability and efficacy of the vaccine formulation [139]. Alum, one of the few adjuvants approved for human use, is the only adjuvant authorized for carbohydrate-based vaccines, but its beneficial effect in improving or amplifying the specific response to carbohydrate antigens cannot be generalized. Overall, it is well established that Alum is almost ineffective in plain PS vaccines, while it could act as an immune stimulator in glycoconjugate vaccines. Nevertheless, it is contained only in a few licensed antibacterial glycoconjugate vaccines, such as pneumococcal and anti-*H. influenzae* formulations. Alum is the adjuvant of choice for novel vaccines, thanks to its strong safety profile in all age groups [140]. It is a crystalline substance consisting of phosphate and/or hydroxide salts of aluminum. It induces an immune response by improving the attraction and uptake of the antigen from APCs and triggering innate immunity pathways. Glycoconjugates are often formulated with Alum in the form of Alhydrogel®, an aluminum hydroxide suspension. The vaccine formulation contains the glycoconjugate that is adsorbed at pH 5–7 onto the aluminum salt. Alhydrogel® particles possess a net positive charge at this pH and are well suited for the absorption of negatively charged antigens, like many saccharide epitopes that contain carboxylate or phosphate groups. The final product is stored as a lyophilized powder or as a suspension.

Although explored alternatives to the use of alum in vaccine formulation have been mostly unsuccessful so far, the last great advances in the understanding of the signaling pathways of innate immunity spurred and revitalized the search for new adjuvants [141]. New immune activators, including MF59 and AS03, the monophosphoryl lipid A (MPLA), and the CpG oligonucleotides, have been approved for human use and are active components of licensed vaccines for influenzae, herpes, and hepatitis B. MF59 and AS03 are two similar oil-in-water emulsions based on squalene [142, 143]. MPLA and CpG oligonucleotides are recognized by Toll-like receptors (TLRs), a group of PRRs expressed APCs surface. In particular, MPLA is a Toll-like receptor-4 (TLR4) agonist obtained by detoxification of LPS, while CpG oligonucleotides are characteristic nonmethylated oligonucleotide sequences recognized by TLR9 [144, 145]. Unfortunately, none of them were completely successful in enhancing the immune response in antibacterial conjugate vaccines. A human clinical trial in which the TLR9 agonist CpG 7909 has been co-administered with antipneumococcal PCV-7 vaccine to HIV-positive patients has shown a positive effect of the adjuvant in improving the titers and the persistence of the vaccine-specific IgG antibody response, but an increased number of mild side-reactions to PCV-7 vaccine have occurred [146]. In another clinical trial, the use of MPLA in co-administration with PCV-9 conjugate vaccine has not enhanced the production of IgG pneumococcal capsular PS antibodies with respect to the use of Alum in the control group [147]. The overall picture says that despite promising results in animal models, it is hard to translate preclinical data to clinic. The most probable explanation lies in the fact that the immunomodulatory effect of adjuvants is better in priming than in boosting the immune response. It is hence very difficult to enhance the immune response in humans that have already been primed or preexposed to the pathogen. Nevertheless, novel adjuvants capable of initiating and enhancing immune responses are under evaluation. The general strategy is to develop/identify agents able to target specific signaling activators of innate immunity (i.e. TLRs) with the purpose of delivering the adjuvant specifically to APCs and/or lymph nodes, thus avoiding unnecessary stimulation of other tissues. In this regard, a novel Alum-TLR7 adjuvant in which the TLR7 agonist SMIP7.10 is adsorbed to aluminum hydroxide has been proven in a mouse model to potentiate, right after the first immunization, the response to glycoconjugate antigens of *N. meningitidis* strains with respect to the addition of the sole aluminum hydroxide [148]. On the other side, a phase I randomized study in healthy adults who had received a MenC conjugate vaccine adjuvanted with a TLR7 agonist adsorbed on aluminum hydroxide (AS37) only highlighted a good safety profile of the new vaccine formulation. A comparison with the control group, vaccinated with a licensed MenC conjugate alum-adjuvanted vaccine, did not reveal any enhancement of the immune response [149]. Alternatively, the adjuvant can be directly covalently linked to the antigen to allow the simultaneous uptake of the two species by the specifically targeted APCs. This is the so-called self-adjuvanting approach, which is mainly explored in antitumor vaccines, even if some studies in animal models about the development of MPLA glycoconjugates have been reported [150]. An overall picture of the research in this context has been recently reviewed [38].

1.4 Technologies Employed for Production of Glycoconjugate Vaccines

As explained in Section 1.2, the T-cell independence of PS antigens calls for the development of glycoconjugate vaccines to improve the immunogenicity of saccharide haptens and ensure protective efficacy in infants and young children. However, despite the huge impact that conjugate vaccines have had on global health over the past 30 years, there are still some limitations to the use of this vaccination approach. For example, some immunogenicity issues still persist for certain groups at high risk, such as the elderly or immunocompromised individuals. Furthermore, continuous variations in global serotype distributions, as well as serotype replacement events, are reported [151, 152], and a constant check of the vaccine serotype coverage is required, often followed by the need to include new serogroups in already licensed vaccines [47, 48]. Additionally, the massive use of a limited number of protein carriers (TT, DT, and CRM_{197}) may cause the already-mentioned "carrier-induced epitope suppression," hence leading to reduced immunogenicity against the PS hapten [153].

Other relevant issues concern the complexity and manufacturing cost of the traditional, currently licensed glycoconjugate vaccines [154]. They are based on bacterial PSs originated by bacterial growth with an avDP; therefore, they have a heterogeneous composition with inevitable microbial contamination and batch-to-batch variability. Additionally, although the traditional technique is effective and safe, there are significant drawbacks related to the use of PSs derived from bacterial biomass, such as the production cost – mainly due to the need for multiple isolation and purification steps, quality control tests, and arduous physicochemical characterization of the final constructs – and the limited availability of B-cell antigen due to the generally low yields of isolated bacterial PSs. In this regard, innovative strategies are needed to avoid the hazardous handling of pathogens. Thus, there is an urgent demand for new solutions apt to obtain homogeneous and pure bacterial PSs in a safe and cost-effective manner to be used for the manufacture of improved vaccine constructs.

Various methodologies, based on advanced chemical and biochemical approaches, have been thoroughly investigated to overcome the intrinsic limitations of traditional glycoconjugate vaccines.

Chemical and (chemo-)enzymatic synthesis of carbohydrate antigens, glycoprotein bioengineering, use of glycoengineered vesicles, and glyconanoparticles are some of the innovative technologies aimed at optimizing traditional constructs and streamlining the manufacturing processes of glycoconjugate vaccines. In this section, we will describe the state of the art of the diverse technologies employed for the preparation of glycoconjugate vaccines (Figure 1.3), with a focus on the novel approaches explored for a faster production of vaccine constructs with a safer and more efficient profile.

1.4.1 Traditional Glycoconjugates

The traditional approach mainly consists of the conjugation of a PS or OS derived from cultivated pathogens to a carrier protein. As mentioned above, random conjugation leading to cross-linked architectures is usually employed for long sugar chains (PSs, more than 15 RUs), while end-linked chemistry is the preferred approach with shorter sugar chains (OSs, <20 RUs). Currently licensed vaccines are based on random conjugation chemistry (Figure 1.3a), and PSs are usually activated along the biopolymer chain mainly by oxidation of vicinal diols followed by reductive amination or by cyanylation followed by condensation. Periodate oxidation generates carbonyl groups along the chain (e.g. aldehydes from 1,2-diols of sialic acid residues). Oxidated PSs undergo either a direct reductive amination with ε-amino groups of protein lysines, or they are further converted with a bifunctional spacer (see Section 1.3.2) [155–158]. The cyanylation chemistry makes use of cyanogen bromide, recently replaced with the milder and safer 1-cyano-4-dimethylamminopyridinium

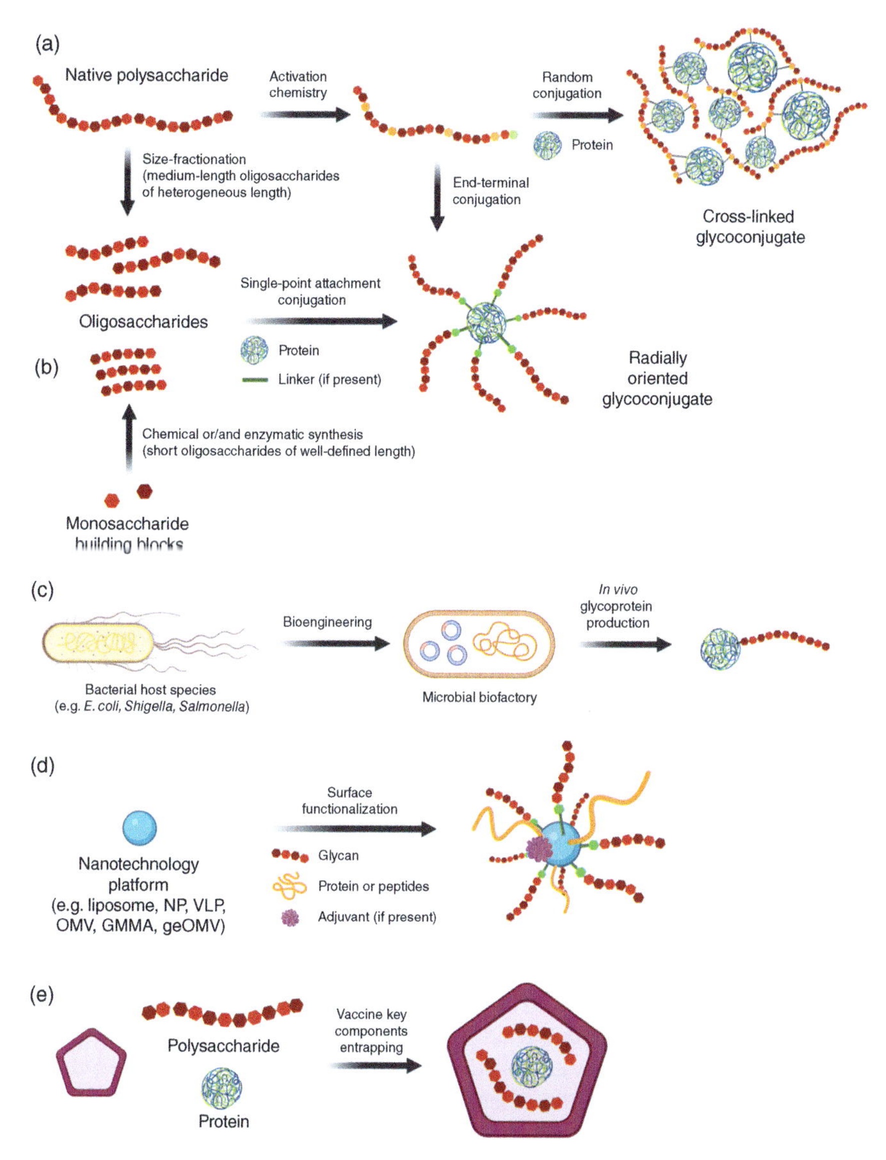

Figure 1.3 Summary of the technologies employed for glycoconjugate production: (a) traditional approaches, (b) chemical and/or enzymatic approach, (c) bioengineered synthesis, (d) nanotechnology-based approach, and (e) noncovalent constructs.

tetrafluoroborate (CDAP), to convert the carbohydrate hydroxyl groups into the corresponding cyanate esters, which are then conjugated with amine or hydrazide groups [159–161]. In selective end-terminal conjugation (Figure 1.3a), native CPSs should be fragmented via hydrolysis (acid or peroxide) and then size-modulated by ultrafiltration or chromatography techniques into shorter and better defined OSs. Shorter saccharides are selectively conjugated by direct reductive amination with ε-amino groups of protein lysine residues or further functionalized with a spacer prior to protein conjugation. For more details on conjugation techniques, we refer the reader to more specialized accounts [162].

Overall, traditional glycoconjugates prepared by random conjugation are obtained as rather heterogeneous mixtures of compounds with a variety of glycoforms containing a variable number of glycoepitopes and multiple glycosylation sites, making it very challenging to decipher their structure–immunogenicity relationship.

1.4.2 Glycoconjugates Based on Synthetic Carbohydrate Antigens

The use of chemical synthesis to obtain large amounts of saccharide antigens in a reproducible manner, often combined with site-selective protein conjugation chemistry, is considered a cutting-edge strategy for vaccine design [163]. In recent decades, improvements in various methodologies (e.g. automated solid-phase [164], one-pot [165, 166], or chemoenzymatic synthesis [167-169]) have fostered the use of synthetic procedures to attain OS fragments unachievable by microbial fermentation. In addition, chemical synthesis may provide analogs with improved stability [170, 171] or structural hybrids conceived to confer enhanced serotype coverage [172] [173]. Furthermore, synthetic saccharide fragments can be activated more efficiently to maximize the yield of protein glycoconjugation and achieve a controlled and reproducible carbohydrate-to-protein ratio. Compared to structures isolated from natural sources, synthetic OSs are highly pure, homogeneous, and well-defined, hence easier to characterize (Figure 1.3b). In addition, synthetic antigens may be helpful tools to detect the glycoepitopes within a PS chain [174–176].

The first and, so far, only licensed glycoconjugate vaccine based on a fully synthetic OS antigen was developed in 2004 by Vicente Verez-Bencomo and coworkers with the trade name Quimi Hib and is addressed to Hib [177]. The synthetic nature of the Quimi Hib vaccine was a revolutionary breakthrough, demonstrating that it is possible to produce a protective vaccine for human use by large-scale chemical synthesis under Good Manufacturing Practice (GMP) conditions.

Very recently, a semisynthetic glycoconjugate vaccine candidate targeting *Shigella flexneri* 2a has successfully completed preclinical studies and entered the first-in-human clinical trial in healthy adults [178–180]. *S. flexneri* 2a belongs to the *Shigella* species, and it is a major causative agent of shigellosis, a severe intestinal infectious disease considered a serious threat to human health and affecting children below five years of age.

Despite the great potential of synthetic carbohydrate antigens, there are crucial factors that still pose significant hurdles to their use in the construction of glycoconjugate vaccines. First, low-molecular-weight OSs, such as those typically achieved by chemical synthesis, are inherently poorly immunogenic. Moreover, the synthesis of complex carbohydrates is still a challenging process, which needs multistep procedures, including problematic and time-consuming purification of intermediates, leading to relatively low overall yield of the final product. The stereochemical control of the glycosylation steps may be difficult, and in particular the stereoselective formation of certain types of glycosidic linkages is still tricky (e.g. formation of 1,2-*cis*-glycosidic bonds). Optimization procedures are still needed to improve and simplify the large-scale chemical synthesis of complex PS fragments and to make synthetic carbohydrate antigens appealing enough for manufacture and industrial development.

1.4.2.1 Site-Selective Protein Conjugation

Over the last few years, numerous studies have highlighted that the type of chemistry used for conjugation may influence the efficacy and the immunogenicity of glycoconjugates. Accordingly, site-selective protein conjugation, intended as the preferential glycosylation of a specific set of amino acids over others of the same type present in the protein, has raised the interest of many researchers. In comparison to random approaches, glycoconjugates obtained by site-selective protein glycosylation should be characterized by higher homogeneity and easier physicochemical characterization, thereby accelerating the manufacture and regulatory processes.

Regioselective conjugation techniques comprise conjugation to specific protein amino acids, as well as unnatural residues artificially introduced by protein premodification, and *in vivo* production of glycoconjugates (glycoengineering, described in Section 1.4.4).

Site-selective conjugation of natural amino acids targets predominantly the ε-amino group of lysines, which are common amino acids highly exposed at the surface of the protein and easily accessible in water. Although Lys glycosylation is generally achieved with a random approach, a recent study shows that preferential conjugation of defined Lys residues over others can be attained [181]. In another example, the most solvent-exposed lysine residues of the protein flagellin (FliC) from *P. aeruginosa* were modified by the introduction of azide groups using a diazotransfer reaction, allowing the glycosylation of the chemically modified protein at predetermined sites [182]. On the other side, cysteine residues can also be reacted regioselectively with a variety of electrophiles and thiophilic agents with very high regioselectivity, thanks to the high nucleophilicity of the sulfhydryl group and to their relatively low abundancy within proteins [37]. In addition, cysteines can be delivered by reductive cleavage of disulfide bridges followed by stapling of the resulting thiol groups by alkylation with electrophilic agents [114]. Recently, a useful protein modification method targeting tyrosine residues has been developed and applied to the construction of glycoconjugate vaccine candidates. In particular, tyrosine residues can be reacted with 4-phenyl-1,2,4-triazoline-3,5-dione (PTAD) through an ene-like reaction. Using a properly modified PTAD, the protein is provided with a chemical handle regioselectively installed on a few Tyr residues and suitable for coupling with carbohydrate antigens [183].

On the other hand, unnatural amino acids (uAAs, i.e. not naturally encoded) suitable for site-selective conjugation can be installed into proteins by using molecular biology methods [184, 185]. Alternatively, proteins incorporating uAA can be expressed from *E. coli*-derived cell-free extracts [186, 187]. All these methods allow the introduction of diverse uAA at predetermined positions to enable regioselective protein glycosylation [188, 189]. Relevant examples include proteins modified with para-azido-L-phenylalanine [190], para-azidomethyl-L-phenylalanine [191], selenocysteine [192], and an uAA containing the SeH group instead of the SH of cysteine.

Incorporation of uAA-bearing bicyclononyne (BCN) and trans-cyclooctene (TCO) is also a very useful methodology to allow site-specific protein glycoconjugation by Huisgen cycloaddition and inverse electronic demand Diels–Alder [109, 193], respectively.

Many other protocols have been developed for chemical protein modifications, targeting both natural and uAAs, with the aim of obtaining homogeneous glycoproteins with a defined glycosylation pattern [194–198]. A comprehensive account of these methodologies is, however, beyond the scope of this chapter.

We believe that glycoconjugates obtained by site-selective protein modification with fully synthetic saccharide antigens represent the most promising candidates for a new generation of innovative, safer, and more effective carbohydrate-based vaccines.

1.4.3 Enzymatic and ChemoEnzymatic Approach

Enzymatic formation of glycosidic bonds is an attractive method to accomplish regio- and stereoselective linkages in high purity without applying a demanding protecting group strategy. Enzymatic synthesis occurs in aqueous solutions and can be considered a green approach since no toxic contaminants derive from experimental procedures. However, the exquisite substrate specificity still poses a limitation to the use of fully enzymatic synthesis of glycoproteins. Conversely, the combination of chemical and enzymatic methods has been recently developed for faster vaccine manufacture, taking advantage of the *in vitro* production of long PS chains by recombinant capsule polymerases. In principle, chemoenzymatic methods allow to reduce both synthetic and purification steps.

The chemoenzymatic synthesis of *N. meningitidis* serogroup C (MenC), W, and Y PSs was investigated [199, 200], and the specific enzymes needed for these processes have been identified. In particular, Vann and coworkers produced size-controlled MenC antigens using bacterial sialyl transferases [201]. They employed a recombinant sialyltransferase similar to the enzyme CSTII from *Campylobacter jejuni*, which recognizes lactosides as acceptors and is able to transfer multiple sialic acid residues. Lactosides functionalized at the reducing end with a short azido-spacer were sequentially sialylated using CMP-sialic acid as the donor. Polymer elongation was controlled using a CMP-9-deoxy-NeuNAc as process inhibitor. The OSs were then linked to TT protein, and the resulting glycoconjugates were able to elicit specific and protective anti-MenC antibody response [201, 202].

Recently, modern automated synthesis techniques have been employed in this area. In general, two approaches to immobilization are used to investigate the automated synthesis of PSs mediated by enzymes, where either the enzyme or the saccharide substrate is immobilized on the resin/support. The first example of chemical synthesis and enzymatic elongation combined by using an immobilized enzyme was addressed to the production of *N. meningitidis* serogroup X (MenX) OSs [203]. A mixture of the fully synthetic MenX trisaccharide acceptor [204] and the UDP-GlcNAc donor was eluted through a metal affinity column (HisTrap column, Ni Sepharose High-Performance column chromatography) and derivatized with a truncated MenX capsular polymerase (ΔN58ΔC99-CsxA), providing size-controlled

MenX fragments with an avDP of 12 units. Chemoenzymatic MenX-CRM_{197} conjugates were able to raise anti-MenX-protective antibodies. The enzyme is immobilized on the column and can only be removed at the end of the elongation process. A limitation of on-column chemoenzymatic synthesis is that pure OSs can be obtained only after careful purification of the eluates released by the column, which contain a mixture of PS fragments and other reagents. A recent methodology has been developed to overcome this issue. The new protocol has been applied to mammalian glycans and is based on the functionalization of sugar primers with sulfonate tags in order to facilitate the purification process by solid-phase extraction [205].

1.4.4 Bioengineered Glycoconjugates

In recent years, a new approach termed Protein Glycan Coupling Technology (PGCT) was developed, based on the use of a bacterial host species (in particular *E. coli*, *Shigella*, or *Salmonella*) as a biofactory for glycoprotein production (Figure 1.3c) [206]. In the last few years, PGCT has been finding application in the preparation of glycoconjugates for vaccine use. In PGCT, both the saccharide antigen and the carrier protein are biosynthesized and covalently coupled into the bacterial host species. This technique was first reported in the early 2000s, based on the remarkable discovery that the N-linked glycosylation system of *C. jejuni* can be functionally transferred into *E. coli* cells by the N-linking oligosaccharyltransferase (OTase) PglB [207]. In the following years, PGCT techniques employing both N-linking and O-linking OTases were developed and applied to the production of bacterial glycoconjugate vaccines. In particular, the *in vivo* production of glycoconjugates consists of three stages (Figure 1.4a): *glycan expression*, with a genetic locus (or loci) able to encode for glycan

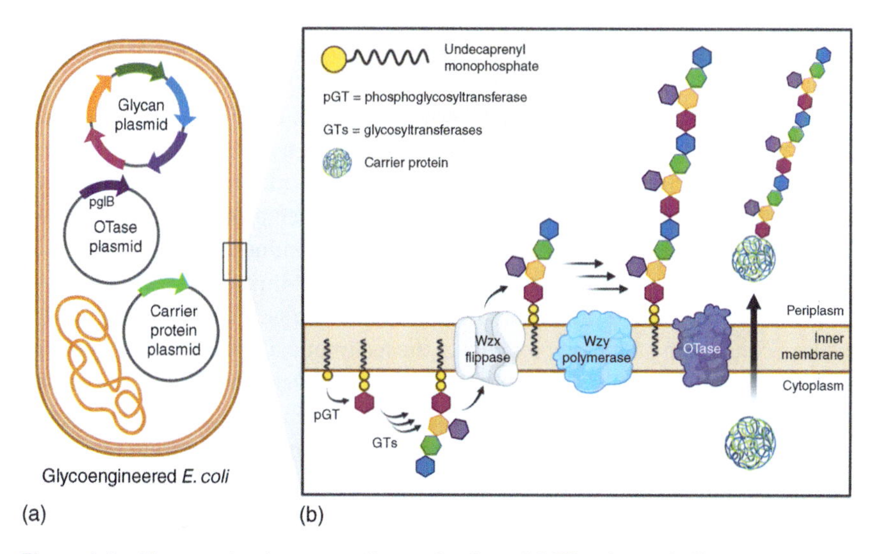

Figure 1.4 The protein glycan coupling technology (PGCT): schematic illustrations of (a) a glycoengineered *E. coli*, the preferred host for vaccine bioproduction, and (b) the biosynthetic Wzy-dependent pathway by which pure glycoconjugates are produced.

biosynthesis; *protein design and expression*, with a plasmid encoding a carrier protein; and *coupling* or *bioconjugation*, provided by an active OTase (N- or O-linking). In the cytosol (Figure 1.4b), a monosaccharide is firstly assembled on the undecaprenyl pyrophosphate (Und-PP) lipid carrier at the level of the inner membrane, and then the Und-PP-monosaccharide is sequentially elongated by glycosyltransferases (GTs).

Once glycan assembly is complete, the Und-PP-linked OS is flipped to the periplasm by a Wzx flippase, polymerized by a Wzy polymerase, and then transferred for final conjugation by the bacterial OTase complex to the acceptor protein. Protein glycosylation usually occurs in two ways, depending on the site for the attachment of the glycan: the Und-PP-linked OS is coupled to a side-chain amide nitrogen atom of asparagines within the Asp/Glu-X1-Asn-X2-Ser/Thr consensus sequence (N-linked, where X1 and X2 can be any amino acid except proline) or to a side-chain hydroxyl of a serine or threonine residue (O-linked).

The N-linking PglB system has been the most employed OTase to produce different bacterial glycoconjugates, such as *S. enterica*, *Shigella* species, *S. aureus* serotypes 5 and 8, *Francisella tularensis*, and extraintestinal pathogenic *E. coli* [208–211], but neither PglB nor PglL systems are able to transfer glycans with glucose at their reducing ends. On the contrary, the newly discovered PglS system possesses this ability, and it has been used for producing Group B *Streptococcus* glycoconjugates [212]. OTases can recognize various lipid-linked OSs and their target carrier proteins. CRM_{197} is a suitable carrier for bioconjugation, although recent PGCT-based vaccine candidates contain engineered *P. aeruginosa* exotoxin A (rEPA) carrier proteins [212] or other antigenic proteins derived from the same pathogen of the glycan hapten. The use of homologous proteins is optimal to confer protection against pathogens with multiple serotypes featuring many different PS structures. A significant example includes the PGCT-mediated conjugation of a *S. pneumoniae* protein antigen (PiuA) with different *S. pneumoniae* CPSs in *E. coli*. The obtained glycoconjugates were proven to stimulate an antibody response comparable to that generated by the commercial vaccine Prevnar-13 [136].

A limitation of the PglB-based PGCT platform is the requirement of a GlcNAc unit at the reducing end of the glycan hapten. The presence of a 2-acetamido group directly linked to the lipid carrier enables GTs and enzymes involved in sugar biosynthesis to transfer the glycan hapten onto the carrier protein [213]. A second structural limitation seems to be related to the sugar residues nearby the lipid carrier, since PglB OTase is unable to transfer an Und-PP-linked OS with a β-(1→4) linkage between the two terminal GlcNAc residues of the saccharide sequence [214]. However, considering the high potential of the technique, the mentioned shortcomings of the PglB-based PGCT platform are under optimization. Recent studies deal with the use of PglB orthologs as alternatives to cover a broader range of glycan structures and the search for other OTases able to recognize different protein sequons for site-selective conjugation. The OTase PglS, capable of transferring OSs containing glucose as the first residue of the growing sugar chain, has been used to develop the first vaccine candidate against hypervirulent *Klebsiella pneumoniae*, where glucose is the residue at the reducing end of K1 and K2 Klebsiella CPSs, the most virulent serotypes responsible for the majority of Klebsiella infections [215].

PGCT is a versatile platform and a promising alternative for the manufacture of a low-cost new generation of glycoconjugate vaccines, and many bioconjugated vaccine candidates are currently in different phases of clinical trials [216].

1.4.5 Nanotechnology-Based Glycoconjugate Vaccines

Various nanotechnology-based strategies have been explored to improve traditional glycoconjugate vaccines. The major advantage of nanotechnology tools is the ability to display carbohydrate antigens in a polyvalent fashion, thereby improving the instauration of multiple and simultaneous protein–glycan interactions, resulting in an increased binding affinity (Figure 1.3d) [217]. Multivalent presentation is particularly relevant for conjugates composed of low-molecular-weight OSs, typically characterized by low binding affinity. Nanotechnology-based vaccines have attracted much interest due to the high versatility of nanocarriers that can be easily functionalized with multicopies of ligands and, in addition, with more than one antigen (e.g. biantigenic NPs). For example, liposomes and gold nanoparticles (AuNPs) are biocompatible delivery systems used as multivalent scaffolds in recent novel antibacterial vaccine prototypes. The saccharide antigen is generally displayed on the surface of the nanoparticles but can also be encapsulated or associated with the nanomaterial.

In this regard, different nanostructured materials, allowing for modular functionalization with B- and T-cell antigens, have been explored and evaluated as carriers for a more efficient delivery of carbohydrate vaccine components, and they are briefly discussed in Sections 1.4.5.1 and 1.4.5.2.

1.4.5.1 Outer Membrane Vesicles (OMVs) and Generalized Modules for Membrane Antigens (GMMA)

Bacteria naturally release membrane vesicles (MVs) during growth through a "membrane blebbing" process in response to a specific external chemical stimulus. MVs are generally spherical lipid membrane nanoparticles with a diameter of 20–400 nm, and they are composed of microbial biomolecules (such as proteins, lipids, and nucleic acids) derived from the parental bacterial species, either Gram-negative [218] or Gram-positive [219]. MVs play different roles both in microbe–microbe interactions (e.g. biofilm formation, antibiotic resistance, nutrient acquisition, microbial defense) and microbe–host interactions (i.e. pathogenesis, communication) [220].

MVs were ignored and defined as insignificant for decades, but in the mid-1990s, it was found that outer membrane vesicles (OMVs) produced by Gram-negative bacteria, especially *P. aeruginosa*, mimic the bacterial surface, displaying various antigens (e.g. LPS, periplasmic proteins, outer membrane proteins, and lipoproteins). Since then, researchers have started to study OMVs in terms of their physiological and antigenic roles, and as delivery systems. Bacteria were genetically modified to increase blebbing and reduce the intrinsic toxicity of the lipid A portion [221], leading to Generalized Modules for Membrane Antigens (GMMAs) secretion [222]. GMMA combines the multivalent display of carbohydrates, favoring B-cell activation, with optimal size for immune stimulation. GMMA also works as self-adjuvants due to the presence of structures acting as agonists of TLRs 2 and 4.

GMMA were tested as novel tools for antibacterial vaccine design since their surface is naturally decorated with *O*-polysaccharide antigens. From a manufacturing point of view, only two steps of tangential flow filtration are needed to obtain GMMA of high purity. The simple and low-cost fabrication process makes these novel nanocarriers attractive for the production of vaccines addressed to developing countries [223]. In recent years, various GMMA-based vaccine candidates have been investigated and, in some cases, they have demonstrated greater serum bactericidal activity compared to classical protein-based glycoconjugates [224–227]. A novel GMMA-based vaccine candidate against *Shigella sonnei*, known as 1790GAHB, was evaluated in Kenya, a *Shigella*-endemic country, and tested in Phase 1 and 2 clinical trials. 1790GAHB was found to be highly immunogenic and well tolerated [228, 229]. Recently, OMVs were engineered in *E. coli* generating glycoengineered outer membrane vesicles (geOMVs). With this technique, both recombinant *O*-polysaccharide biosynthesis and vesciculation occur in the *E. coli* biofactory, and the surface of geOMVs can be decorated with heterologous polysaccharide antigens [230, 231].

Although OMV-based platforms are intriguing tools since they offer a simple and versatile method for low-cost production of antibacterial vaccines, further studies are still needed before they can be translated into viable and well-established vaccine settings approved for human use.

1.4.5.2 Gold Nanoparticles, Liposomes, and Virus-Like Particles

Vaccine protection is restricted to those serotypes or serogroups included in the formulations. However, the emergence of new epidemics requires novel vaccines with flexible composition that are easy to design and manufacture and available at affordable prices for a broader distribution worldwide. Carbohydrate-based vaccine candidates, which may meet these requirements, are based on single-point attachment chemistry, allowing short synthetic OSs to be displayed in a multivalent fashion on nanoparticles of differing conceptions such as AuNPs, liposomes, and virus-like particles (VLPs) [232] [6]. Peptides, oligonucleotides, and lipids are the proper immunogenic carriers for these types of nanoconstructs. Nanotechnology-based approaches exploit preferentially short synthetic glycans instead of long-chain native PSs because, in principle, there is no need for a high number of RUs to efficiently trigger TCR recognition. As mentioned above, the RU of PSs consists of one to a maximum of five (with a few cases extending beyond) monosaccharides, and the TCR-binding site can accommodate short sugar sequences (no more than four/six monosaccharides). The surface of the nanoconstructs allows for the presentation of the short synthetic sugar chains in a multivalent fashion. Another key advantage is that the number and type of ligand(s) loaded on the nanoconstruct can be controlled and modulated, so that NPs can be functionalized with more than one antigen (multiantigenic loading) with fine control of their stoichiometry.

AuNPs were the first nanosystems to be used as scaffolds for a fully synthetic pneumococcal vaccine candidate by Pènades and coworkers [233]. In this pioneering study, mice immunization showed that AuNPs bearing the synthetic RU of

S. pneumoniae serotype 14 (Sp14), together with $OVA_{323\text{-}339}$ peptide as T helper antigen, were able to elicit IgG titers even if in a lower amount than the positive control (Sp14-CRM_{197} conjugate). Despite this, AuNPs were proven to be very promising platforms for innovative vaccine development, and this study set the stage for further advancements. More recently, bi-antigenic AuNPs displaying synthetic Sp14 and Sp19F RUs combined on the same nanoparticle with $OVA_{323\text{-}339}$ peptide were able to trigger an anti-Sp14 immune response in mice comparable to licensed 13-valent pneumococcal vaccine [234].

Liposomes are nontoxic and attractive nanocarriers for multivalent antigen presentation. These phospholipid vesicles are safe and biocompatible nanotools for the co-delivery of the saccharide antigen(s) (encapsulated into the liposomal carrier) and a T helper entity (protein or antigenic peptide or natural killer T-cell adjuvant), which is exposed on the liposomal surface. In the last decade, liposomal encapsulation of polysaccharides (LEPS) has been widely investigated as an effective alternative strategy targeting pneumococcal diseases. Deng et al. demonstrated that a liposomal anti-Sp14 construct, made by a combination of saccharide and NKT cell antigens, is able to elicit a robust immune response comparable to the one obtained with the licensed vaccine [235]. This strategy was expanded to include multiple pneumococcal strains [236, 237, 238]. Jones and coworkers used the LEPS technique to successfully codeliver two pathogen-related proteins, GlpO (an α-glycerophosphate oxidase) and PncO (a bacteriocin ABC transporter transmembrane protein), together with 20 pneumococcal CPSs [237, 239]. This formulation was re-designed, increasing the CPS valency (up to 24 serotypes) and raising fourfold the dose of PncO as a result of the removal of GlpO protein antigen due to off-target immunogenicity associated with GlpO [238]. Indeed, a nonnegligible number of bacterial proteins in the human microbiome were found to share 50% or higher sequence identity to GlpO.

Mulard and co-workers have developed liposomes coated at their surface with two fully synthetic fragment mimics of *S. flexneri* 2a O-antigen (B-cell epitope) and a universal T helper peptide derived from influenza virus hemagglutinin ($HA_{307\text{-}319}$, known as PKY). The lipopeptide Pam_3CAG was used as adjuvant and anchor for the covalent linking of both sugar and peptide antigens to the liposomal surface [240]. These vesicles were able to induce a protective immune response in mice, demonstrating that liposomes are a promising and attractive alternative for the design of novel antibacterial vaccines.

VLPs are supramolecular constructs that mimic the architecture of a virus, preserving its conformational epitope but lacking the viral genome. VLPs are like empty shells able to induce a strong B-cell response by themselves, and they are, therefore, interesting platforms for glycoconjugate-based vaccine development. Bacteriophage Qβ VLPs, as an example, have been functionalized with short synthetic CPS fragments of *S. pneumoniae* serotypes 3 and 14. Qβ can be synthesized in *E. coli* as a self-assembled icosahedral particle and consists of 180 copies of a 132-amino acid monomeric protein [241]. VLP-pneumococcal prototypes were able to elicit protective, long-lived, and serotype-specific IgG antibodies with nM affinity in mice.

1.4.6 Nonprotein-Based Glycoconjugates

It is now a well-established principle that the intrinsically poor immunogenicity of saccharide haptens can be strongly enhanced by their covalent linking to an immunogenic carrier protein. Despite this, novel platforms based on the use of nonprotein carriers have been explored and proven useful for the development of semisynthetic carbohydrate-based vaccines.

The synthetic glycolipid antigen α-galactosyl ceramide (α-GalCer, KRN7000). α-GalCer is an activator of natural killer iNKT cells, which are a very abundant subpopulation of T cells, restricted by the nonpolymorphic CD1d molecule that is expressed by B cells. Recent immunization studies in mice reported that the coupling of *S. pneumoniae* PSs or OSs to α-GalCer gives conjugates able to induce high titers of class switched antibodies with high affinity and specificity for the PS used in the vaccination [235, 242, 243]. Indeed, the use of α-GalCer can offer an alternative approach to the development of a new type of self-adjuvanting glycovaccines where the TD response is evoked by the presence of the glycolipid.

The 1-*O*-dephosphorylated derivative of lipid A, known as MPLA, is a nontoxic glycolipid mimic of the hydrophobic anchor of bacterial LPS and a safe and effective vaccine adjuvant [144, 244]. Notably, Guo and coworkers showed that MPLA derivatives can be chemically linked to OS antigens and used as carrier molecules for the development of nonprotein-based semisynthetic antibacterial vaccines [150, 245, 246]. Short synthetic fragments of α-(2-9)-polysialic acid (the meningococcus C capsula) were conjugated to MPLA and investigated as antimeningococcal vaccine candidates [245]. Tri- and tetra-oligosialic acids linked to the glycolipid through a short spacer elicited a protective immune response comparable to the one induced by licensed protein-conjugate anti-MenC vaccine. In addition, in a recent work, an OS–MPLA conjugate was explored as an antituberculosis vaccine candidate. The authors extensively evaluated the adjuvant properties of MPLA carriers and the importance of the position of antigen conjugation to MPLA (1-MPLA vs. 6′-MPLA conjugates were compared) [246]. In this study, a 6′-amino-MPLA derivative was synthesized and coupled with the upstream tetrasaccharide fragment of lipoarabinomannan (LAM, a major virulence factor of *Mycobacterium tuberculosis*). LAM tetrasaccharide was also linked to 1-OH of MPLA, and the immunological activity of both prototypes was assessed. This study showed that the 6′-N-MPLA–LAM tetrasaccharide conjugate elicited higher IgG titers and is a better immunogen as compared to 1-O-MPLA–LAM conjugate.

1.4.7 Noncovalent Vaccines

In recent years, other noteworthy technologies have emerged for the development of innovative carbohydrate-based vaccines against infectious diseases. Although there are no examples reported in the literature supporting that the co-administration of both key antigenic components is effective [65, 66, 247], novel “covalent linkage free” macromolecular complexes have been investigated (Figure 1.3e). They include the Multiple Antigen Presenting System (MAPS) and the Protein Capsular Matrix Vaccines (PCMVs), with the aim of further simplifying vaccine manufacturing while preserving the ability to induce durable protection of the host.

Richard Malley and coworkers at Boston Children's Hospital were the first to develop the MAPS platform, where the covalent binding is replaced by an affinity-based coupling between a pathogen-specific protein genetically fused with rhizavidin and a biotinylated sugar antigen [248]. This strategy was first used to develop a pneumococcal vaccine candidate, ASP3772, including 24 pneumococcal CPSs and two pneumococcal proteins. The Phase 2 clinical trial recently conducted to assess the efficacy of the novel formulation in the elderly demonstrated a protective antibody response against both carbohydrate (all 24 PSs) and protein antigens [249]. These intriguing results open up the prospect of possible competitiveness with the well-established Prevnar multicomponent vaccine series (see Table 1.1), since MAPS platforms can optimize pneumococcal vaccine production, reducing production costs. Several MAPS platforms, conceived for including the most challenging bacterial pathogens, are currently under development by Affinivax Inc., a biopharmaceutical company in Cambridge, Massachusetts [250].

In 2015, Matrivax (a biotechnology company in Boston) [251] developed a cross-linked polymer matrix embedding the sugar antigen with the carrier protein in collaboration with John J. Mekalanos (Harvard Medical School) [252]. More recently, this "virtual conjugate" approach was found to be a viable alternative to classical conjugate vaccines to confer protection from typhoid fever [253]. Immunological data from a Phase 1 clinical study demonstrated that Typhax vaccine, made of Vi sugar antigen entrapped in a cross-linked CRM_{197} matrix, was able to induce IgG levels comparable to, and even better than, the commercial vaccine currently in use (Typhim Vi) [254].

1.5 Conclusion

Vaccination is a simple, safe, and effective way of protecting individuals against harmful, often life-threatening, infectious diseases, and there is no doubt that it has been one of the greatest revolutions of modern medicine. After the provision of adequate sanitation facilities and safe drinking water, the development of vaccines provided one of the most significant contributions to human health, leading to an increase in life expectancy from around 40 years up to almost double. Over the past century, vaccines were able to defeat devastating infectious diseases, in some cases until their complete eradication (smallpox), and for this reason, vaccination is considered by the WHO to be the most cost effective of all potential prevention strategies.

In this regard, carbohydrate-based vaccines hold a prominent role. They are classified as subunit vaccines based on the concept that only a component of the pathogen rather than the entire microorganism is sufficient to raise a specific immune response capable of inducing long-term protection of the host. The introduction of glycoconjugate vaccines represented a further fundamental advancement. They are based on CPSs or LPS OAg (for bacterial pathogens), as well as any other glycan uniquely expressed on the surface of the microorganism (viruses, fungi, parasites, or even tumor cells) and covalently linked to an immunogenic protein carrier. Glycoconjugate vaccines allow to overcome the intrinsic limitations of vaccines based on plain PSs that are poorly immunogenic in infants and young children (under two years of age), in the elderly, and in immunocompromised individuals due to the T-cell independence of unconjugated carbohydrates. Conversely, protein conjugation ensures a T-cell memory response and boosts effect of the vaccine, conferring persistent immunological memory and durable immune protection even in high-risk groups.

Interestingly, the fundamental milestones that marked the history and development of carbohydrate-based vaccines have been reached with an empirical approach. The tremendous advances in immunology and glycobiology of the last few decades, however, provided new insights to steer the immune response and to design innovative vaccine settings devoid of the side effects and drawbacks which typically distinguish the traditional, currently licensed glycoconjugates.

New technologies, including emerging techniques for the analysis of surface PSs structures and to decipher carbohydrate–protein interactions, became available and have been explored to produce in high yields more homogeneous glycoconjugates based on well-defined carbohydrate structures in order to simplify and accelerate the manufacture and regulatory processes. Notably, the use of synthetic chemistry is gaining an increasingly prominent role, fueled by the recent great progresses and achievements in the chemical synthesis of complex carbohydrates. The synthetic approach has indeed demonstrated that even short, but well-designed OSs can act as protective antigens. Accordingly, glycoconjugates obtained by site-selective protein modification with synthetic saccharide antigens are among the most promising candidates for a new generation of innovative, safer, and more effective carbohydrate-based vaccines. In addition, chemoenzymatic approaches, nanotechnology-based platforms, and glycoengineering techniques are highly promising outlooks to conceive novel vaccine candidates able to confer enhanced protection and broader serotype coverage.

Finally, a challenge for the near future is the development of novel glycoconjugates intended for infant vaccination to prevent future pandemic diseases. In particular, an additional effort is needed to develop economically viable technologies which will enable to introduce novel glycoconjugate vaccines (safer, easily scalable, flexible, and thermostable) at affordable prices into the market of LMIC.

Acknowledgments

We are grateful to COST Action CA18103 INNOGLY. Figures 1.1, 1.3, and 1.4 were created with BioRender.com.

References

1 Michaud, C.M. (2009). Global burden of infectious diseases. In: *Encyclopedia of Microbiology*, 3e (ed. M. Schaechter), 444–454. Oxford: Academic Press.

2 The, L. (2020). Global health: time for radical change? *The Lancet* 396 (10258): 1129.

3 Vos, T., Lim, S.S., Abbafati, C. et al. (2020). Global burden of 369 diseases and injuries in 204 countries and territories, 1990–2019: a systematic analysis for the Global Burden of Disease Study 2019. *The Lancet* 396 (10258): 1204–1222.

4 Mc Dermott, P.F., Walker, R.D., and White, D.G. (2003). Antimicrobials: modes of action and mechanisms of resistance. *International Journal of Toxicology* 22 (2): 135–143.

5 Kohanski, M.A., Dwyer, D.J., Hayete, B. et al. (2007). A common mechanism of cellular death induced by bactericidal antibiotics. *Cell* 130 (5): 797–810.

6 Morelli, L., Polito, L., Richichi, B., and Compostella, F. (2021). Glyconanoparticles as tools to prevent antimicrobial resistance. *Glycoconjugate Journal* 38 (4): 475–490.

7 GSK NRCM Fighting antimicrobial resistance with vaccines. https://www.nature.com/articles/d42473-021-00356-4#ref-CR1.

8 Organization WH Vaccines and immunization. https://www.who.int/health-topics/vaccines-and-immunization#tab=tab_1.

9 Ada, G. (2001). Vaccines and vaccination. *New England Journal of Medicine* 345 (14): 1042–1053.

10 Pollard, A.J. and Bijker, E.M. (2021). A guide to vaccinology: from basic principles to new developments. *Nature Reviews Immunology* 21 (2): 83–100.

11 Micoli, F., Bagnoli, F., Rappuoli, R., and Serruto, D. (2021). The role of vaccines in combatting antimicrobial resistance. *Nature Reviews Microbiology* 19 (5): 287–302.

12 Hilleman, M.R. (2000). Vaccines in historic evolution and perspective: a narrative of vaccine discoveries. *Vaccine* 18 (15): 1436–1447.

13 Jenner, E. (1802). An inquiry into the causes and effects of the variolae vaccinae: a disease discovered in some of the western counties of England, particularly Gloucestershire, and known by the name of the cow pox. Springfield [Mass.]: Re-printed for Dr. Samuel Cooley, by Ashley & Brewer. Cooley, Samuel.

14 Gross, C.P. and Sepkowitz, K.A. (1998). The myth of the medical breakthrough: smallpox, vaccination, and Jenner reconsidered. *International Journal of Infectious Diseases* 3 (1): 54–60.

15 Plotkin, S.A. and Plotkin, S.L. (2011). The development of vaccines: how the past led to the future. *Nature Reviews Microbiology* 9 (12): 889–893.

16 McAleer, W.J., Buynak, E.B., Maigetter, R.Z. et al. (1984). Human hepatitis B vaccine from recombinant yeast. *Nature* 307 (5947): 178–180.

17 Scolnick, E.M., McLean, A.A., West, D.J. et al. (1984). Clinical evaluation in healthy adults of a hepatitis B vaccine made by recombinant DNA. *JAMA* 251 (21): 2812–2815.

18 Michel, M.L., Pontisso, P., Sobczak, E. et al. (1984). Synthesis in animal cells of hepatitis B surface antigen particles carrying a receptor for polymerized human serum albumin. *Proceedings of the National Academy of Sciences* 81 (24): 7708–7712.

19 Emini, E.A., Ellis, R.W., Miller, W.J. et al. (1986). Production and immunological analysis of recombinant hepatitis B vaccine. *Journal of Infection* 13: 3–9.

20 McNeil, C. (2006). Who invented the VLP cervical cancer vaccines? *JNCI: Journal of the National Cancer Institute* 98 (7): 433.

21 Fleischmann Robert, D., Adams Mark, D., White, O. et al. (1995). Whole-genome random sequencing and assembly of *Haemophilus influenzae* Rd. *Science* 269 (5223): 496–512.

22 Rappuoli, R. (2000). Reverse vaccinology. *Current Opinion in Microbiology* 3 (5): 445–450.

23 Pizza, M., Scarlato, V., Masignani, V. et al. (2000). Identification of vaccine candidates against serogroup B meningococcus by whole-genome sequencing. *Science* 287 (5459): 1816–1820.

24 Giuliani Marzia, M., Adu-Bobie, J., Comanducci, M. et al. (2006). A universal vaccine for serogroup B meningococcus. *Proceedings of the National Academy of Sciences* 103 (29): 10834–10839.

25 Chaudhary, N., Weissman, D., and Whitehead, K.A. (2021). mRNA vaccines for infectious diseases: principles, delivery and clinical translation. *Nature Reviews Drug Discovery* 20: 817–838. https://doi.org/10.1038/s41573-021-00283-5.

26 Maruggi, G. et al.(2017). Immunogenicity and protective efficacy induced by self- amplifying mRNA vaccines encoding bacterial antigens. *Vaccine* 35: 361–368 https://doi.org/10.1016/j.vaccine.2016.11.040.

27 Rabinovich, G.A., van Kooyk, Y., and Cobb, B.A. (2012). Glycobiology of immune responses. *Annals of the New York Academy of Sciences* 1253 (1): 1–15.

28 Mahla, R., Reddy, C., Prasad, D., and Kumar, H. (2013). Sweeten PAMPs: role of sugar complexed PAMPs in innate immunity and vaccine biology. *Frontiers in Immunology* 4: 248.

29 Varki, A. (2017). Biological roles of glycans. *Glycobiology* 27 (1): 3–49.

30 Heidelberger, M. and Avery, O.T. (1923). The soluble specific substance of pneumococcus. *Journal of Experimental Medicine* 38 (1): 73–79.

31 Heidelberger, M. and Avery, O.T. (1924). The soluble specific substance of pneumococcus: second paper. *Journal of Experimental Medicine* 40 (3): 301–317.

32 MacLeod, C.M., Hodges, R.G., Heidelberger, M., and Bernhard, W.G. (1945). Prevention of pneumococcal pneumonia by immunization with specific capsular polysaccharides. *Journal of Experimental Medicine* 82 (6): 445–465.

33 Prevention C-CfDCa Prevention and control of meningococcal disease. Recommendations of the Advisory Committee on Immunization Practices (ACIP). https://www.cdc.gov/mmwr/preview/mmwrhtml/rr5407a1.htm.

34 Robbins, J.B., Austrian, R., Lee, C.J. et al. (1983). Considerations for formulating the second-generation pneumococcal capsular polysaccharide vaccine with emphasis on the cross-reactive types within groups. *The Journal of Infectious Diseases* 148 (6): 1136–1159.

35 González-Fernández, Á., Faro, J., and Fernández, C. (2008). Immune responses to polysaccharides: lessons from humans and mice. *Vaccine.* 26 (3): 292–300.
36 Segal, S. and Pollard, A.J. (2004). Vaccines against bacterial meningitis. *British Medical Bulletin.* 72 (1): 65–81.
37 Richichi, B., Stefanetti, G., Biagiotti, G., and Lay, L. (2021). Conjugation techniques and linker strategies for carbohydrate-based vaccines. In: *Comprehensive Glycoscience*, 2e (ed. J.J. Barchi), 676–705. Oxford: Elsevier.
38 Stefanetti, G., Borriello, F., Richichi, B. et al. (2022). Immunobiology of carbohydrates: implications for novel vaccine and adjuvant design against infectious diseases. *Frontiers in Cellular and Infection Microbiology* 11: 808005.
39 Mond, J.J., Lees, A., and Snapper, C.M. (1995). T cell-independent antigens type 2. *Annual Review of Immunology* 13 (1): 655–692.
40 Rappuoli, R. (2018). Glycoconjugate vaccines: principles and mechanisms. *Science Translational Medicine* 10 (456): eaat4615.
41 Landsteiner, K. (1921). Über heterogenetisches Antigen und Hapten. XV. Mitteilungen über Antigene. *Biochemische Zeitschrift* (Heterogeneous antigen and hapten. XV. Communication on antigens. Biochemical Journal) 119: 294–306.
42 Avery, O.T. and Goebel, W.F. (1929). Chemo-immunological studies on conjugated carbohydrate-proteins: II. Immunological specificity of synthetic sugar-protein antigens. *Journal of Experimental Medicine* 50 (4): 533–550.
43 Eskola, J., Peltola, H., Takala, A.K. et al. (1987). Efficacy of *Haemophilus influenzae* type b polysaccharide–diphtheria toxoid conjugate vaccine in infancy. *New England Journal of Medicine* 317 (12): 717–722.
44 Black, S.B., Shinefield, H.R., Fireman, B. et al. (1991). Efficacy in infancy of oligosaccharide conjugate *Haemophilus influenzae* type b (HbOC) vaccine in a United States population of 61 080 children. *The Pediatric Infectious Disease Journal* 10 (2): 97–104.
45 Bröker, M., Berti, F., and Costantino, P. (2016). Factors contributing to the immunogenicity of meningococcal conjugate vaccines. *Human Vaccines & Immunotherapeutics* 12 (7): 1808–1824.
46 Yildirim, I., Shea, K.M., and Pelton, S.I. (2015). Pneumococcal disease in the era of pneumococcal conjugate vaccine. *Infectious Disease Clinics of North America* 29 (4): 679–697.
47 Stacey, H.L., Rosen, J., Peterson, J.T. et al. (2019). Safety and immunogenicity of 15-valent pneumococcal conjugate vaccine (PCV-15) compared to PCV-13 in healthy older adults. *Human Vaccines & Immunotherapeutics* 15 (3): 530–539.
48 Hurley, D., Griffin, C., Young, M. Jr. et al. (2021). Safety, tolerability, and immunogenicity of a 20-valent pneumococcal conjugate vaccine (PCV20) in adults 60 to 64 years of age. *Clinical Infectious Diseases* 73 (7): e1489–e1497.
49 Mohan, V.K., Varanasi, V., Singh, A. et al. (2015). Safety and immunogenicity of a Vi polysaccharide–tetanus toxoid conjugate vaccine (Typbar-TCV) in healthy infants, children, and adults in typhoid endemic areas: a multicenter, 2-cohort, open-label, double-blind, randomized controlled phase 3 study. *Clinical Infectious Diseases* 61 (3): 393–402.

50 Micoli, F., Rondini, S., Pisoni, I. et al. (2011). Vi-CRM197 as a new conjugate vaccine against *Salmonella typhi*. *Vaccine* 29 (4): 712–720.

51 Blake, F.G. (1917). Methods for the determination of pneumococcus types. *Journal of Experimental Medicine* 26 (1): 67–80.

52 Wright, A., Parry Morgan, W., Colebrook, L., and Dodgson, R.W. (1914). Observations on prophylactic inoculation against pneumococcus infections. And on the results which have been achieved by it. *The Lancet* 183 (4715): 87–95.

53 Musher, D.M., Luchi, M.J., Watson, D.A. et al. (1990). Pneumococcal polysaccharide vaccine in young adults and older bronchitics: determination of IgG responses by ELISA and the effect of adsorption of serum with non-type-specific cell wall polysaccharide. *The Journal of Infectious Diseases* 161 (4): 728–735.

54 Lortan, J.E., Kaniuk, A.S., and Monteil, M.A. (1993). Relationship of in vitro phagocytosis of serotype 14 *Streptococcus pneumoniae* to specific class and IgG subclass antibody levels in healthy adults. *Clinical and Experimental Immunology* 91 (1): 54–57.

55 Richmond, P., Kaczmarski, E., Borrow, R. et al. (2000). Meningococcal C polysaccharide vaccine induces immunologic hyporesponsiveness in adults that is overcome by meningococcal C conjugate vaccine. *The Journal of Infectious Diseases* 181 (2): 761–764.

56 Snapper, C.M. and Mond, J.J. (1996). A model for induction of T cell-independent humoral immunity in response to polysaccharide antigens. *The Journal of Immunology* 157 (6): 2229.

57 Craxton, A., Magaletti, D., Ryan, E.J., and Clark, E.A. (2003). Macrophage- and dendritic cell—dependent regulation of human B-cell proliferation requires the TNF family ligand BAFF. *Blood* 101 (11): 4464–4471.

58 Balázs, M., Martin, F., Zhou, T., and Kearney, J.F. (2002). Blood dendritic cells interact with splenic marginal zone B cells to initiate T-independent immune responses. *Immunity* 17 (3): 341–352.

59 MacLennan, I.C.M. and Vinuesa, C.G. (2002). Dendritic cells, BAFF, and APRIL: innate players in adaptive antibody responses. *Immunity* 17 (3): 235–238.

60 Goebel, W.F. and Avery, O.T. (1929). Chemo-immunological studies on conjugated carbohydrate-proteins: I. The synthesis of p-aminophenol ß-glucoside, p-aminophenol ß-galactoside, and their coupling with serum globulin. *Journal of Experimental Medicine* 50 (4): 521–531.

61 Costantino, P., Rappuoli, R., and Berti, F. (2011). The design of semi-synthetic and synthetic glycoconjugate vaccines. *Expert Opinion on Drug Discovery* 6 (10): 1045–1066.

62 Jones, C. (2005). Vaccines based on the cell surface carbohydrates of pathogenic bacteria. *Anais da Academia Brasileira de Ciências* 77 (2): 293–324.

63 Cobb, B.A., Wang, Q., Tzianabos, A.O., and Kasper, D.L. (2004). Polysaccharide processing and presentation by the MHCII pathway. *Cell* 117 (5): 677–687.

64 Duan, J., Avci Fikri, Y., and Kasper, D.L. (2008). Microbial carbohydrate depolymerization by antigen-presenting cells: deamination prior to presentation by the MHCII pathway. *Proceedings of the National Academy of Sciences* 105 (13): 5183–5188.

65 Rappuoli, R., De Gregorio, E., and Costantino, P. (2019). On the mechanisms of conjugate vaccines. *Proceedings of the National Academy of Sciences* 116 (1): 14–16.

66 Avci, F.Y., Li, X., Tsuji, M., and Kasper, D.L. (2011). A mechanism for glycoconjugate vaccine activation of the adaptive immune system and its implications for vaccine design. *Nature Medicine* 17 (12): 1602–1609.

67 Avci, F.Y., Li, X., Tsuji, M., and Kasper, D.L. (2012). Isolation of carbohydrate-specific $CD4^+$ T cell clones from mice after stimulation by two model glycoconjugate vaccines. *Nature Protocols* 7 (12): 2180–2192.

68 Lesinski, G.B. and Julie Westerink, M.A. (2001). Vaccines against polysaccharide antigens. *Current Drug Targets - Infectious Disorders* 1 (3): 325–334.

69 Weintraub, A. (2003). Immunology of bacterial polysaccharide antigens. *Carbohydrate Research* 338 (23): 2539–2547.

70 Manz, R.A., Hauser, A.E., Hiepe, F., and Radbruch, A. (2004). Maintenance of serum antibody levels. *Annual Review of Immunology* 23 (1): 367–386.

71 Seeberger, P.H. (2021). Discovery of semi- and fully-synthetic carbohydrate vaccines against bacterial infections using a medicinal chemistry approach. *Chemical Reviews.* 121 (7): 3598–3626.

72 Anish, C., Beurret, M., and Poolman, J. (2021). Combined effects of glycan chain length and linkage type on the immunogenicity of glycoconjugate vaccines. *NPJ Vaccines.* 6 (1): 150.

73 Arnon, R. and Van Regenmortel, M.H.V. (1992). Structural basis of antigenic specificity and design of new vaccines. *The FASEB Journal.* 6 (14): 3265–3274.

74 Cavallari, M. and De Libero, G. (2017). From immunologically archaic to neoteric glycovaccines. *Vaccines.* 5 (1): 4.

75 Costantino, P., Norelli, F., Giannozzi, A. et al. (1999). Size fractionation of bacterial capsular polysaccharides for their use in conjugate vaccines. *Vaccine.* 17 (9): 1251–1263.

76 Hennessey, J.P., Costantino, P., Talaga, P. et al. (2018). Lessons learned and future challenges in the design and manufacture of glycoconjugate vaccines. In: *Carbohydrate-Based Vaccines: From Concept to Clinic. ACS Symposium Series. 1290* (ed. A.K. Prasad), 323–385. American Chemical Society.

77 Micoli, F., Bjarnarson Stefania, P., Arcuri, M. et al. (2020). Short Vi-polysaccharide abrogates T-independent immune response and hyporesponsiveness elicited by long Vi-CRM197 conjugate vaccine. *Proceedings of the National Academy of Sciences.* 117 (39): 24443–24449.

78 Stefanetti, G., Okan, N., Fink, A. et al. (2019). Glycoconjugate vaccine using a genetically modified O antigen induces protective antibodies to *Francisella tularensis. Proceedings of the National Academy of Sciences.* 116 (14): 7062–7070.

79 Anderson, P.W., Pichichero, M.E., Insel, R.A. et al. (1986). Vaccines consisting of periodate-cleaved oligosaccharides from the capsule of *Haemophilus influenzae* type b coupled to a protein carrier: structural and temporal requirements for priming in the human infant. *Journal of Immunology* (Baltimore, Md : 1950). 137 (4): 1181–1186.

80 Anderson, P.W., Pichichero, M.E., Stein, E.C. et al. (1989). Effect of oligosaccharide chain length, exposed terminal group, and hapten loading on the antibody response of human adults and infants to vaccines consisting of *Haemophilus influenzae* type b capsular antigen unterminally coupled to the diphtheria protein CRM197. *Journal of Immunology* (Baltimore, Md : 1950) 142 (7): 2464–2468.

81 Adamo, R., Nilo, A., Castagner, B. et al. (2013). Synthetically defined glycoprotein vaccines: current status and future directions. *Chemical Science.* 4 (8): 2995–3008.

82 Xin, H., Dziadek, S., Bundle David, R., and Cutler, J.E. (2008). Synthetic glycopeptide vaccines combining β-mannan and peptide epitopes induce protection against candidiasis. *Proceedings of the National Academy of Sciences.* 105 (36): 13526–13531.

83 Falugi, F., Petracca, R., Mariani, M. et al. (2001). Rationally designed strings of promiscuous $CD4^+$ T cell epitopes provide help to *Haemophilus influenzae* type b oligosaccharide: a model for new conjugate vaccines. *European Journal of Immunology.* 31 (12): 3816–3824.

84 Alexander, J., del Guercio, M.-F., Maewal, A. et al. (2000). Linear PADRE T helper epitope and carbohydrate B cell epitope conjugates induce specific high titer IgG antibody responses. *The Journal of Immunology.* 164 (3): 1625.

85 Morelli, L., Poletti, L., and Lay, L. (2011). Carbohydrates and immunology: synthetic oligosaccharide antigens for vaccine formulation. *European Journal of Organic Chemistry.* 2011 (29): 5723–5777.

86 Colombo, C., Pitirollo, O., and Lay, L. (2018). Recent advances in the synthesis of glycoconjugates for vaccine development. *Molecules.* 23 (7): 1712.

87 Seeberger, P.H. (2015). The logic of automated glycan assembly. *Accounts of Chemical Research.* 48 (5): 1450–1463.

88 Seeberger, P.H. and Haase, W.-C. (2000). Solid-phase oligosaccharide synthesis and combinatorial carbohydrate libraries. *Chemical Reviews.* 100 (12): 4349–4394.

89 Plante Obadiah, J., Palmacci Emma, R., and Seeberger, P.H. (2001). Automated solid-phase synthesis of oligosaccharides. *Science.* 291 (5508): 1523–1527.

90 Seeberger, P.H. (2008). Automated oligosaccharide synthesis. *Chemical Society Reviews.* 37 (1): 19–28.

91 Panza, M., Stine, K.J., and Demchenko, A.V. (2020). HPLC-assisted automated oligosaccharide synthesis: the implementation of the two-way split valve as a mode of complete automation. *Chemical Communications.* 56 (9): 1333–1336.

92 Kröck, L., Esposito, D., Castagner, B. et al. (2012). Streamlined access to conjugation-ready glycans by automated synthesis. *Chemical Science.* 3 (5): 1617–1622.

93 Panza, M., Pistorio, S.G., Stine, K.J., and Demchenko, A.V. (2018). Automated chemical oligosaccharide synthesis: novel approach to traditional challenges. *Chemical Reviews.* 118 (17): 8105–8150.

94 Wang, L.-X. and Huang, W. (2009). Enzymatic transglycosylation for glycoconjugate synthesis. *Current Opinion in Chemical Biology.* 13 (5): 592–600.

95 Yu, H., Huang, S., Chokhawala, H. et al. (2006). Highly efficient chemoenzymatic synthesis of naturally occurring and non-natural α-2,6-linked sialosides: A

P. damsela α-2,6-sialyltransferase with extremely flexible donor–substrate specificity. *Angewandte Chemie International Edition.* 45 (24): 3938–3944.

96 Zasłona, M.E., Downey, A.M., Seeberger, P.H., and Moscovitz, O. (2021). Semi- and fully synthetic carbohydrate vaccines against pathogenic bacteria: recent developments. *Biochemical Society Transactions.* 49 (5): 2411–2429.

97 Hegerle, N., Bose, J., Ramachandran, G. et al. (2018). Overexpression of O-polysaccharide chain length regulators in gram-negative bacteria using the Wzx-/Wzy-dependent pathway enhances production of defined modal length O-polysaccharide polymers for use as haptens in glycoconjugate vaccines. *Journal of Applied Microbiology.* 125 (2): 575–585.

98 Kalynych, S., Ruan, X., Valvano Miguel, A., and Cygler, M. (2011). Structure-guided investigation of lipopolysaccharide O-antigen chain length regulators reveals regions critical for modal length control. *Journal of Bacteriology.* 193 (15): 3710–3721.

99 Zeidan, A.A., Poulsen, V.K., Janzen, T. et al. (2017). Polysaccharide production by lactic acid bacteria: from genes to industrial applications. *FEMS Microbiology Reviews.* 41 (Supp_1): S168–S200.

100 Badahdah, A.-M., Rashid, H., and Khatami, A. (2016). Update on the use of meningococcal serogroup C CRM197-conjugate vaccine (Meningitec) against meningitis. *Expert Review of Vaccines.* 15 (1): 9–29.

101 Mawas, F., Niggemann, J., Jones, C. et al. (2002). Immunogenicity in a mouse model of a conjugate vaccine made with a synthetic single repeating unit of type 14 pneumococcal polysaccharide coupled to CRM197. *Infection and Immunity* 70 (9): 5107–5114.

102 Szu, S.C. (2013). Development of Vi conjugate – a new generation of typhoid vaccine. *Expert Review of Vaccines.* 12 (11): 1273–1286.

103 Szu, S.C., Bystricky, S., Hinojosa-Ahumada, M. et al. (1994). Synthesis and some immunologic properties of an O-acetyl pectin [poly(1-->4)-alpha-D-GalpA]-protein conjugate as a vaccine for typhoid fever. *Infection and Immunity* 62 (12): 5545–5549.

104 Safari, D., Dekker, H.A.T., Joosten, J.A.F. et al. (2008). Identification of the smallest structure capable of evoking opsonophagocytic antibodies against *Streptococcus pneumoniae* type 14. *Infection and Immunity* 76 (10): 4615–4623.

105 Boutureira, O., Martínez-Sáez, N., Brindle, K.M. et al. (2017). Site-selective modification of proteins with oxetanes. *Chemistry – A European Journal.* 23 (27): 6483–6489.

106 Lee, B., Sun, S., Jiménez-Moreno, E. et al. (2018). Site-selective installation of an electrophilic handle on proteins for bioconjugation. *Bioorganic & Medicinal Chemistry.* 26 (11): 3060–3064.

107 Jewett, J.C. and Bertozzi, C.R. (2010). Cu-free click cycloaddition reactions in chemical biology. *Chemical Society Reviews.* 39 (4): 1272–1279.

108 Lutz, J.-F. (2008). Copper-free azide–alkyne cycloadditions: new insights and perspectives. *Angewandte Chemie International Edition.* 47 (12): 2182–2184.

109 Blackman, M.L., Royzen, M., and Fox, J.M. (2008). Tetrazine ligation: fast bioconjugation based on inverse-electron-demand Diels–Alder reactivity. *Journal of the American Chemical Society.* 130 (41): 13518–13519.

110 Peeters, J.M., Hazendonk, T.G., Beuvery, E.C., and Tesser, G.I. (1989). Comparison of four bifunctional reagents for coupling peptides to proteins and the effect of the three moieties on the immunogenicity of the conjugates. *Journal of Immunological Methods.* 120 (1): 133–143.

111 Adamo, R., Hu, Q.-Y., Torosantucci, A. et al. (2014). Deciphering the structure–immunogenicity relationship of anti-Candida glycoconjugate vaccines. *Chemical Science.* 5 (11): 4302–4311.

112 Phalipon, A., Tanguy, M., Grandjean, C. et al. (2009). A synthetic carbohydrate-protein conjugate vaccine candidate against *Shigella flexneri* 2a infection. *The Journal of Immunology.* 182 (4): 2241.

113 Nilo, A., Morelli, L., Passalacqua, I. et al. (2015). Anti-group B streptococcus glycan-conjugate vaccines using pilus protein GBS80 as carrier and antigen: comparing lysine and tyrosine-directed conjugation. *ACS Chemical Biology.* 10 (7): 1737–1746.

114 Stefanetti, G., Hu, Q.-Y., Usera, A. et al. (2015). Sugar–protein connectivity impacts on the immunogenicity of site-selective *Salmonella* O-antigen glycoconjugate vaccines. *Angewandte Chemie International Edition.* 54 (45): 13198–13203.

115 Bartoloni, A., Norelli, F., Ceccarini, C. et al. (1995). Immunogenicity of meningococcal B polysaccharide conjugated to tetanus toxoid or CRM197 via adipic acid dihydrazide. *Vaccine.* 13 (5): 463–470.

116 Schneerson, R., Robbins, J.B., Barrera, O. et al. (1980). *Haemophilus influenzae* type B polysaccharide-protein conjugates: model for a new generation of capsular polysaccharide vaccines. *Progress in Clinical and Biological Research.* 47: 77–94.

117 Pizza, M., Bekkat-Berkani, R., and Rappuoli, R. (2020). Vaccines against meningococcal diseases. *Microorganisms.* 8 (10): 1521.

118 Dretler, A.W., Rouphael, N.G., and Stephens, D.S. (2018). Progress toward the global control of *Neisseria meningitidis*: 21st century vaccines, current guidelines, and challenges for future vaccine development. *Human Vaccines & Immunotherapeutics.* 14 (5): 1146–1160.

119 Uchida, T., Gill, D.M., and Pappenheimer, A.M. (1971). Mutation in the structural gene for diphtheria toxin carried by temperate phage β. *Nature New Biology.* 233 (35): 8–11.

120 Malito, E., Bursulaya, B., Chen, C. et al. (2012). Structural basis for lack of toxicity of the diphtheria toxin mutant CRM197. *Proceedings of the National Academy of Sciences.* 109 (14): 5229–5234.

121 Giannini, G., Rappuoli, R., and Ratti, G. (1984). The amino-acid sequence of two non-toxic mutants of diphtheria toxin: CRM45 and CRM197. *Nucleic Acids Research.* 12 (10): 4063–4069.

122 Shinefield, H.R. (2010). Overview of the development and current use of CRM197 conjugate vaccines for pediatric use. *Vaccine.* 28 (27): 4335–4339.

123 Donnelly, J.J., Deck, R.R., and Liu, M.A. (1990). Immunogenicity of a *Haemophilus influenzae* polysaccharide-*Neisseria meningitidis* outer membrane protein complex conjugate vaccine. *The Journal of Immunology.* 145 (9): 3071.

124 Prymula, R. and Schuerman, L. (2009). 10-valent pneumococcal nontypeable *Haemophilus influenzae* PD conjugate vaccine: Synflorix™. *Expert Review of Vaccines.* 8 (11): 1479–1500.

125 Forsgren, A. and Riesbeck, K. (2008). Protein D of *Haemophilus influenzae*: a protective nontypeable *H. influenzae* antigen and a carrier for pneumococcal conjugate vaccines. *Clinical Infectious Diseases.* 46 (5): 726–731.

126 Knuf, M., Kowalzik, F., and Kieninger, D. (2011). Comparative effects of carrier proteins on vaccine-induced immune response. *Vaccine.* 29 (31): 4881–4890.

127 Dagan, R., Eskola, J., Leclerc, C., and Leroy, O. (1998). Reduced response to multiple vaccines sharing common protein epitopes that are administered simultaneously to infants. *Infection and Immunity* 66 (5): 2093–2098.

128 Micoli, F., Adamo, R., and Costantino, P. (2018). Protein carriers for glycoconjugate vaccines: history, selection criteria, characterization and new trends. *Molecules.* 23 (6): 1451.

129 Bröker, M., Berti, F., Schneider, J., and Vojtek, I. (2017). Polysaccharide conjugate vaccine protein carriers as a "neglected valency" – potential and limitations. *Vaccine.* 35 (25): 3286–3294.

130 Fattom, A., Schneerson, R., Watson, D.C. et al. (1993). Laboratory and clinical evaluation of conjugate vaccines composed of *Staphylococcus aureus* type 5 and type 8 capsular polysaccharides bound to *Pseudomonas aeruginosa* recombinant exoprotein A. *Infection and Immunity* 61 (3): 1023–1032.

131 Burns, D.L., Kossaczka, Z., Lin Feng-Ying, C. et al. (1999). Safety and immunogenicity of Vi conjugate vaccines for typhoid fever in adults, teenagers, and 2- to 4-year-old children in Vietnam. *Infection and Immunity* 67 (11): 5806–5810.

132 Szu, S.C., Stone, A.L., Robbins, J.D. et al. (1987). Vi capsular polysaccharide-protein conjugates for prevention of typhoid fever. Preparation, characterization, and immunogenicity in laboratory animals. *Journal of Experimental Medicine.* 166 (5): 1510–1524.

133 Yu, R., Xu, J., Hu, T., and Chen, W. (2020). The pneumococcal polysaccharide-tetanus toxin native C-fragment conjugate vaccine: the carrier effect and immunogenicity. *Mediators of Inflammation.* 2020: 9596129.

134 Prymula, R., Peeters, P., Chrobok, V. et al. (2006). Pneumococcal capsular polysaccharides conjugated to protein D for prevention of acute otitis media caused by both Streptococcus pneumoniae and non-typable *Haemophilus influenzae*: a randomised double-blind efficacy study. *The Lancet.* 367 (9512): 740–748.

135 Wacker, M., Wang, L., Kowarik, M. et al. (2014). Prevention of *Staphylococcus aureus* infections by glycoprotein vaccines synthesized in *Escherichia coli*. *The Journal of Infectious Diseases.* 209 (10): 1551–1561.

136 Reglinski, M., Ercoli, G., Plumptre, C. et al. (2018). A recombinant conjugated pneumococcal vaccine that protects against murine infections with a similar efficacy to Prevnar-13. *NPJ Vaccines.* 3 (1): 53.

137 Park, W.-J., Yoon, Y.-K., Park, J.-S. et al. (2021). Rotavirus spike protein ΔVP8* as a novel carrier protein for conjugate vaccine platform with demonstrated antigenic potential for use as bivalent vaccine. *Scientific Reports.* 11 (1): 22037.

138 Bergmann-Leitner, E.S. and Leitner, W.W. (2014). Adjuvants in the driver's seat: how magnitude, type, fine specificity and longevity of immune responses are driven by distinct classes of immune potentiators. *Vaccines.* 2 (2): 1437–1441.

139 Petrovsky, N. and Aguilar, J.C. (2004). Vaccine adjuvants: current state and future trends. *Immunology & Cell Biology.* 82 (5): 488–496.

140 Kool, M., Fierens, K., and Lambrecht, B.N. (2012). Alum adjuvant: some of the tricks of the oldest adjuvant. *Journal of Medical Microbiology.* 61 (7): 927–934.

141 O'Hagan, D.T., Lodaya, R.N., and Lofano, G. (2020). The continued advance of vaccine adjuvants – 'we can work it out'. *Seminars in Immunology.* 50: 101426.

142 O'Hagan, D.T. (2007). MF59 is a safe and potent vaccine adjuvant that enhances protection against influenza virus infection. *Expert Review of Vaccines.* 6 (5): 699–710.

143 Garçon, N., Vaughn, D.W., and Didierlaurent, A.M. (2012). Development and evaluation of AS03, an adjuvant system containing α-tocopherol and squalene in an oil-in-water emulsion. *Expert Review of Vaccines.* 11 (3): 349–366.

144 Casella, C.R. and Mitchell, T.C. (2008). Putting endotoxin to work for us: monophosphoryl lipid A as a safe and effective vaccine adjuvant. *Cellular and Molecular Life Sciences.* 65 (20): 3231.

145 Klinman, D.M. (2006). Adjuvant activity of CpG oligodeoxynucleotides. *International Reviews of Immunology.* 25 (3, 4): 135–154.

146 Offersen, R., Melchjorsen, J., Paludan, S.R. et al. (2012). TLR9-adjuvanted pneumococcal conjugate vaccine induces antibody-independent memory responses in HIV-infected adults. *Human Vaccines & Immunotherapeutics.* 8 (8): 1042–1047.

147 Vernacchio, L., Bernstein, H., Pelton, S. et al. (2002). Effect of monophosphoryl lipid A (MPL®) on T-helper cells when administered as an adjuvant with pneumocococcal–CRM197 conjugate vaccine in healthy toddlers. *Vaccine.* 20 (31): 3658–3667.

148 Buonsanti, C., Balocchi, C., Harfouche, C. et al. (2016). Novel adjuvant Alum-TLR7 significantly potentiates immune response to glycoconjugate vaccines. *Scientific Reports.* 6 (1): 29063.

149 Gonzalez-Lopez, A., Oostendorp, J., Koernicke, T. et al. (2019). Adjuvant effect of TLR7 agonist adsorbed on aluminum hydroxide (AS37): a phase I randomized, dose escalation study of an AS37-adjuvanted meningococcal C conjugated vaccine. *Clinical Immunology.* 209: 108275.

150 Li, Q. and Guo, Z. (2018). Recent advances in toll like receptor-targeting glycoconjugate vaccines. *Molecules.* 23 (7): 1–24.

151 Geno, K.A., Gilbert Gwendolyn, L., Song Joon, Y. et al. (2015). Pneumococcal capsules and their types: past, present, and future. *Clinical Microbiology Reviews.* 28 (3): 871–899.

152 Ji, X., Yao, P.-P., Zhang, L.-Y. et al. (2017). Capsule switching of "*Neisseria meningitidis*" sequence type 7 serogroup A to serogroup X. *Journal of Infection.* 75 (6): 521–531.

153 Dagan, R., Poolman, J., and Siegrist, C.-A. (2010). Glycoconjugate vaccines and immune interference: a review. *Vaccine.* 28 (34): 5513–5523.

154 Plotkin, S., Robinson, J.M., Cunningham, G. et al. (2017). The complexity and cost of vaccine manufacturing – an overview. *Vaccine.* 35 (33): 4064–4071.

155 Wessels Michael, R., Paoletti Lawrence, C., Guttormsen, H.-K. et al. (1998). Structural properties of group B streptococcal type III polysaccharide conjugate vaccines that influence immunogenicity and efficacy. *Infection and Immunity* 66 (5): 2186–2192.

156 Turner, A.E.B., Gerson, J.E., So, H.Y. et al. (2017). Novel polysaccharide-protein conjugates provide an immunogenic 13-valent pneumococcal conjugate vaccine for *S. pneumoniae. Synthetic and Systems Biotechnology.* 2 (1): 49–58.

157 Bröker, M., Dull, P.M., Rappuoli, R., and Costantino, P. (2009). Chemistry of a new investigational quadrivalent meningococcal conjugate vaccine that is immunogenic at all ages. *Vaccine.* 27 (41): 5574–5580.

158 Zou, W. and Jennings, H.J. (2009). Preparation of glycoconjugate vaccines. In: *Carbohydrate-Based Vaccines and Immunotherapies* (ed. Z. Guo and G.-J. Boons), 55–88. Wiley.

159 Shafer, D.E., Toll, B., Schuman, R.F. et al. (2000). Activation of soluble polysaccharides with 1-cyano-4-dimethylaminopyridinium tetrafluoroborate (CDAP) for use in protein-polysaccharide conjugate vaccines and immunological reagents. II. Selective crosslinking of proteins to CDAP-activated polysaccharides. *Vaccine.* 18 (13): 1273–1281.

160 Lees, A., Nelson, B.L., and Mond, J.J. (1996). Activation of soluble polysaccharides with 1-cyano-4-dimethylaminopyridinium tetrafluoroborate for use in protein—polysaccharide conjugate vaccines and immunological reagents. *Vaccine.* 14 (3): 190–198.

161 Lees, A., Barr, J.F., and Gebretnsae, S. (2020). Activation of soluble polysaccharides with 1-cyano-4-dimethylaminopyridine tetrafluoroborate (CDAP) for use in protein–polysaccharide conjugate vaccines and immunological reagents. III optimization of CDAP activation. *Vaccines* 8 (4): 1–18.

162 Lu, L., Duong, V.T., Shalash, A.O. et al. (2021). Chemical conjugation strategies for the development of protein-based subunit nanovaccines. *Vaccines.* 9 (6): 1–24.

163 Anderluh, M., Berti, F., Bzducha-Wróbel, A. et al. (2021). Recent advances on smart glycoconjugate vaccines in infections and cancer. *The FEBS Journal.* 289 (14): 4251–4303.

164 Joseph, A.A., Pardo-Vargas, A., and Seeberger, P.H. (2020). Total synthesis of polysaccharides by automated glycan assembly. *Journal of the American Chemical Society.* 142 (19): 8561–8564.

165 Huang, X., Huang, L., Wang, H., and Ye, X.-S. (2004). Iterative one-pot synthesis of oligosaccharides. *Angewandte Chemie International Edition.* 43 (39): 5221–5224.

166 Wu, Y., Xiong, D.-C., Chen, S.-C. et al. (2017). Total synthesis of mycobacterial arabinogalactan containing 92 monosaccharide units. *Nature Communications.* 8 (1): 14851.

167 Fiebig, T., Litschko, C., Freiberger, F. et al. (2018). Efficient solid-phase synthesis of meningococcal capsular oligosaccharides enables simple and fast chemoenzymatic vaccine production. *Journal of Biological Chemistry.* 293 (3): 953–962.

168 Wang, Z., Chinoy Zoeisha, S., Ambre Shailesh, G. et al. (2013). A general strategy for the chemoenzymatic synthesis of asymmetrically branched N-glycans. *Science.* 341 (6144): 379–383.

169 Li, C. and Wang, L.-X. (2016). Endoglycosidases for the synthesis of polysaccharides and glycoconjugates. In: *Advances in Carbohydrate Chemistry and Biochemistry* (ed. D.C. Baker), 73–116. 73: Academic Press.

170 Enotarpi, J., Tontini, M., Balocchi, C. et al. (2020). A stabilized glycomimetic conjugate vaccine inducing protective antibodies against *Neisseria meningitidis* serogroup A. *Nature Communications.* 11 (1): 4434.

171 Gao, Q., Tontini, M., Brogioni, G. et al. (2013). Immunoactivity of protein conjugates of Carba analogues from *Neisseria meningitidis* A capsular polysaccharide. *ACS Chemical Biology.* 8 (11): 2561–2567.

172 Sanapala, S.R., Seco, B.M.S., Baek, J.Y. et al. (2020). Chimeric oligosaccharide conjugate induces opsonic antibodies against *Streptococcus pneumoniae* serotypes 19A and 19F. *Chemical Science.* 11 (28): 7401–7407.

173 Morelli, L., Lay, L., Santana-Mederos, D. et al. (2021). Glycan array evaluation of synthetic epitopes between the capsular polysaccharides from *Streptococcus pneumoniae* 19F and 19A. *ACS Chemical Biology.* 16 (9): 1671–1679.

174 Baek, J.Y., Geissner, A., Rathwell, D.C.K. et al. (2018). A modular synthetic route to size-defined immunogenic *Haemophilus influenzae* b antigens is key to the identification of an octasaccharide lead vaccine candidate. *Chemical Science.* 9 (5): 1279–1288.

175 Morelli, L., Fallarini, S., Lombardi, G. et al. (2018). Synthesis and biological evaluation of a trisaccharide repeating unit derivative of *Streptococcus pneumoniae* 19A capsular polysaccharide. *Bioorganic & Medicinal Chemistry.* 26 (21): 5682–5690.

176 Legnani, L., Ronchi, S., Fallarini, S. et al. (2009). Synthesis, molecular dynamics simulations, and biology of a carba-analogue of the trisaccharide repeating unit of Streptococcus pneumoniae19F capsular polysaccharide. *Organic & Biomolecular Chemistry.* 7 (21): 4428–4436.

177 Verez-Bencomo, V., Fernández-Santana, V., Hardy, E. et al. (2004). A synthetic conjugate polysaccharide vaccine against *Haemophilus influenzae* type b. *Science.* 305 (5683): 522–525.

178 van der Put, R.M.F., Kim, T.H., Guerreiro, C. et al. (2016). A synthetic carbohydrate conjugate vaccine candidate against *Shigellosis*: improved bioconjugation and impact of alum on immunogenicity. *Bioconjugate Chemistry.* 27 (4): 883–892.

179 Cohen, D., Atsmon, J., Artaud, C. et al. (2021). Safety and immunogenicity of a synthetic carbohydrate conjugate vaccine against *Shigella flexneri* 2a in healthy adult volunteers: a phase 1, dose-escalating, single-blind, randomised, placebo-controlled study. *The Lancet Infectious Diseases.* 21 (4): 546–558.

180 van der Put, R.M.F., Smitsman, C., de Haan, A. et al. (2022). The first-in-human synthetic glycan-based conjugate vaccine candidate against *Shigella*. *ACS Central Science.* 8 (4): 449–460.

181 Crotti, S., Zhai, H., Zhou, J. et al. (2014). Defined conjugation of glycans to the lysines of CRM197 guided by their reactivity mapping. *ChemBioChem.* 15 (6): 836–843.

182 Peng, C.-J., Chen, H.-L., Chiu, C.-H., and Fang, J.-M. (2018). Site-selective functionalization of flagellin by steric self-protection: a strategy to facilitate flagellin as a self-adjuvanting carrier in conjugate vaccine. *ChemBioChem.* 19 (8): 805–814.

183 Hu, Q.-Y., Allan, M., Adamo, R. et al. (2013). Synthesis of a well-defined glycoconjugate vaccine by a tyrosine-selective conjugation strategy. *Chemical Science.* 4 (10): 3827–3832.

184 Noren Christopher, J., Anthony-Cahill Spencer, J., Griffith Michael, C., and Schultz, P.G. (1989). A general method for site-specific incorporation of unnatural amino acids into proteins. *Science.* 244 (4901): 182–188.

185 Zhang, W.H., Otting, G., and Jackson, C.J. (2013). Protein engineering with unnatural amino acids. *Current Opinion in Structural Biology* 23 (4): 581–587.

186 Zimmerman, E.S., Heibeck, T.H., Gill, A. et al. (2014). Production of site-specific antibody–drug conjugates using optimized non-natural amino acids in a cell-free expression system. *Bioconjugate Chemistry.* 25 (2): 351–361.

187 Zawada, J.F., Yin, G., Steiner, A.R. et al. (2011). Microscale to manufacturing scale-up of cell-free cytokine production—a new approach for shortening protein production development timelines. *Biotechnology and Bioengineering.* 108 (7): 1570–1578.

188 Johnson, J.A., Lu, Y.Y., Van Deventer, J.A., and Tirrell, D.A. (2010). Residue-specific incorporation of non-canonical amino acids into proteins: recent developments and applications. *Current Opinion in Chemical Biology.* 14 (6): 774–780.

189 Wang, K., Sachdeva, A., Cox, D.J. et al. (2014). Optimized orthogonal translation of unnatural amino acids enables spontaneous protein double-labelling and FRET. *Nature Chemistry.* 6 (5): 393–403.

190 Goerke, A.R. and Swartz, J.R. (2009). High-level cell-free synthesis yields of proteins containing site-specific non-natural amino acids. *Biotechnology and Bioengineering.* 102 (2): 400–416.

191 Tookmanian, E.M., Fenlon, E.E., and Brewer, S.H. (2015). Synthesis and protein incorporation of azido-modified unnatural amino acids. *RSC Advances.* 5 (2): 1274–1281.

192 Li, X., Yang, J., and Rader, C. (2014). Antibody conjugation via one and two C-terminal selenocysteines. *Methods.* 65 (1): 133–138.

193 Machida, T., Lang, K., Xue, L. et al. (2015). Site-specific glycoconjugation of protein via bioorthogonal tetrazine cycloaddition with a genetically encoded trans-cyclooctene or bicyclononyne. *Bioconjugate Chemistry.* 26 (5): 802–806.

194 Gamblin, D.P., Scanlan, E.M., and Davis, B.G. (2009). Glycoprotein synthesis: an update. *Chemical Reviews.* 109 (1): 131–163.

195 Boutureira, O. and Bernardes, G.J.L. (2015). Advances in chemical protein modification. *Chemical Reviews.* 115 (5): 2174–2195.

196 Bernardes, G.J.L., Castagner, B., and Seeberger, P.H. (2009). Combined approaches to the synthesis and study of glycoproteins. *ACS Chemical Biology.* 4 (9): 703–713.

197 Chalker, J.M., Bernardes, G.J.L., and Davis, B.G. (2011). A "tag-and-modify" approach to site-selective protein modification. *Accounts of Chemical Research.* 44 (9): 730–741.

198 Takaoka, Y., Ojida, A., and Hamachi, I. (2013). Protein organic chemistry and applications for labeling and engineering in live-cell systems. *Angewandte Chemie International Edition.* 52 (15): 4088–4106.

199 Romanow, A., Haselhorst, T., Stummeyer, K. et al. (2013). Biochemical and biophysical characterization of the sialyl-/hexosyltransferase synthesizing the meningococcal serogroup W135 heteropolysaccharide capsule. *Journal of Biological Chemistry.* 288 (17): 11718–11730.

200 Romanow, A., Keys, T.G., Stummeyer, K. et al. (2014). Dissection of hexosyl- and sialyltransferase domains in the bifunctional capsule polymerases from *Neisseria meningitidis* W and Y defines a new sialyltransferase family. *Journal of Biological Chemistry.* 289 (49): 33945–33957.

201 Mosley, S.L., Rancy, P.C., Peterson, D.C. et al. (2010). Chemoenzymatic synthesis of conjugatable oligosialic acids. *Biocatalysis and Biotransformation.* 28 (1): 41–50.

202 McCarthy, P.C., Saksena, R., Peterson, D.C. et al. (2013). Chemoenzymatic synthesis of immunogenic meningococcal group C polysialic acid-tetanus Hc fragment glycoconjugates. *Glycoconjugate Journal.* 30 (9): 857–870.

203 Oldrini, D., Fiebig, T., Romano, M.R. et al. (2018). Combined chemical synthesis and tailored enzymatic elongation provide fully synthetic and conjugation-ready *Neisseria meningitidis* serogroup X vaccine antigens. *ACS Chemical Biology.* 13 (4): 984–994.

204 Morelli, L., Cancogni, D., Tontini, M. et al. (2014). Synthesis and immunological evaluation of protein conjugates of *Neisseria meningitidis* X capsular polysaccharide fragments. *Beilstein Journal of Organic Chemistry.* 10: 2367–2376.

205 Li, T., Liu, L., Wei, N. et al. (2019). An automated platform for the enzyme-mediated assembly of complex oligosaccharides. *Nature Chemistry.* 11 (3): 229–236.

206 Kay, E., Cuccui, J., and Wren, B.W. (2019). Recent advances in the production of recombinant glycoconjugate vaccines. *NPJ Vaccines.* 4 (1): 16.

207 Wacker, M., Linton, D., Hitchen Paul, G. et al. (2002). N-linked glycosylation in *Campylobacter jejuni* and its functional transfer into *E. coli*. *Science.* 298 (5599): 1790–1793.

208 Wetter, M., Goulding, D., Pickard, D. et al. (2012). Molecular characterization of the viaB locus encoding the biosynthetic machinery for Vi capsule formation in *Salmonella typhi*. *PLoS One* 7 (9): e45609.

209 Ihssen, J., Haas, J., Kowarik, M. et al. (2015). Increased efficiency of *Campylobacter jejuni* N-oligosaccharyltransferase PglB by structure-guided engineering. *Open Biology*. 5 (4): 140227.

210 Cuccui, J., Thomas, R.M., Moule, M.G. et al. (2013). Exploitation of bacterial N-linked glycosylation to develop a novel recombinant glycoconjugate vaccine against *Francisella tularensis*. *Open Biology*. 3 (5): 130002.

211 van den Dobbelsteen, G.P.J.M., Faé, K.C., Serroyen, J. et al. (2016). Immunogenicity and safety of a tetravalent *E. coli* O-antigen bioconjugate vaccine in animal models. *Vaccine*. 34 (35): 4152–4160.

212 Duke, J.A., Paschall, A.V., Robinson, L.S. et al. (2021). Development and immunogenicity of a prototype multivalent group B *Streptococcus* bioconjugate vaccine. *ACS Infectious Diseases*. 7 (11): 3111–3123.

213 Wacker, M., Feldman Mario, F., Callewaert, N. et al. (2006). Substrate specificity of bacterial oligosaccharyltransferase suggests a common transfer mechanism for the bacterial and eukaryotic systems. *Proceedings of the National Academy of Sciences*. 103 (18): 7088–7093.

214 Chen, M.M., Glover, K.J., and Imperiali, B. (2007). From peptide to protein: comparative analysis of the substrate specificity of N-linked glycosylation in *C. jejuni*. *Biochemistry*. 46 (18): 5579–5585.

215 Feldman Mario, F., Mayer Bridwell Anne, E., Scott Nichollas, E. et al. (2019). A promising bioconjugate vaccine against hypervirulent *Klebsiella pneumoniae*. *Proceedings of the National Academy of Sciences*. 116 (37): 18655–18663.

216 Huttner, A., Hatz, C., van den Dobbelsteen, G. et al. (2017). Safety, immunogenicity, and preliminary clinical efficacy of a vaccine against extraintestinal pathogenic *Escherichia coli* in women with a history of recurrent urinary tract infection: a randomised, single-blind, placebo-controlled phase 1b trial. *The Lancet Infectious Diseases*. 17 (5): 528–537.

217 Giuliani, M., Faroldi, F., Morelli, L. et al. (2019). Exploring calixarene-based clusters for efficient functional presentation of *Streptococcus pneumoniae* saccharides. *Bioorganic Chemistry*. 93: 103305.

218 Bayer, M.E. and Anderson, T.F. (1965). The surface structure of *Escherichia coli*. *Proceedings of the National Academy of Sciences*. 54 (6): 1592–1599.

219 Brown, L., Wolf, J.M., Prados-Rosales, R., and Casadevall, A. (2015). Through the wall: extracellular vesicles in Gram-positive bacteria, mycobacteria and fungi. *Nature Reviews Microbiology*. 13 (10): 620–630.

220 Caruana, J.C. and Walper, S.A. (2020). Bacterial membrane vesicles as mediators of microbe – microbe and microbe – host community interactions. *Frontiers in Microbiology*. 11: 1–24.

221 Rossi, O., Pesce, I., Giannelli, C. et al. (2014). Modulation of endotoxicity of *Shigella* generalized modules for membrane antigens (GMMA) by genetic lipid A modifications: relative activation of TLR4 And TLR2 pathways in different mutants. *Journal of Biological Chemistry*. 289 (36): 24922–24935.

222 Mancini, F., Micoli, F., Necchi, F. et al. (2021). GMMA-based vaccines: the known and the unknown. *Frontiers in Immunology* 12: 1–7.

223 Kis, Z., Shattock, R., Shah, N., and Kontoravdi, C. (2019). Emerging technologies for low-cost, rapid vaccine manufacture. *Biotechnology Journal.* 14 (1): 1800376.

224 Rossi, O., Caboni, M., Negrea, A. et al. (2016). Toll-like receptor activation by generalized modules for membrane antigens from lipid A mutants of *Salmonella enterica Serovars Typhimurium* and *Enteritidis. Clinical and Vaccine Immunology.* 23 (4): 304–314.

225 Schager Anna, E., Dominguez-Medina, C.C., Necchi, F. et al. (2018). IgG responses to porins and lipopolysaccharide within an outer membrane-based vaccine against nontyphoidal *Salmonella* develop at discordant rates. *mBio.* 9 (2): e02379–e02317.

226 De Benedetto, G., Alfini, R., Cescutti, P. et al. (2017). Characterization of O-antigen delivered by generalized modules for membrane antigens (GMMA) vaccine candidates against nontyphoidal *Salmonella. Vaccine.* 35 (3): 419–426.

227 Micoli, F., Rondini, S., Alfini, R. et al. (2018). Comparative immunogenicity and efficacy of equivalent outer membrane vesicle and glycoconjugate vaccines against nontyphoidal *Salmonella. Proceedings of the National Academy of Sciences.* 115 (41): 10428–10433.

228 Launay, O., Lewis, D.J.M., Anemona, A. et al. (2017). Safety profile and immunologic responses of a novel vaccine against *Shigella sonnei* administered intramuscularly, intradermally and intranasally: results from two parallel randomized phase 1 clinical studies in healthy adult volunteers in Europe. *eBioMedicine.* 22: 164–172.

229 Obiero, C.W., Ndiaye, A.G.W., Sciré, A.S. et al. (2017). A Phase 2a randomized study to evaluate the safety and immunogenicity of the 1790GAHB generalized modules for membrane antigen vaccine against *Shigella sonnei* administered intramuscularly to adults from a Shigellosis-endemic country. *Frontiers in Immunology.* 8: 1–11.

230 Price, N.L., Goyette-Desjardins, G., Nothaft, H. et al. (2016). Glycoengineered outer membrane vesicles: a novel platform for bacterial vaccines. *Scientific Reports.* 6 (1): 24931.

231 Chen, L., Valentine Jenny, L., Huang, C. Jr. et al. (2016). Outer membrane vesicles displaying engineered glycotopes elicit protective antibodies. *Proceedings of the National Academy of Sciences.* 113 (26): E3609–E3618.

232 Gregory, A., Williamson, D., and Titball, R. (2013). Vaccine delivery using nanoparticles. *Frontiers in Cellular and Infection Microbiology.* 3: 1–13.

233 Safari, D., Marradi, M., Chiodo, F. et al. (2012). Gold nanoparticles as carriers for a synthetic *Streptococcus pneumoniae* type 14 conjugate vaccine. *Nanomedicine.* 7 (5): 651–662.

234 Vetro, M., Safari, D., Fallarini, S. et al. (2016). Preparation and immunogenicity of gold glyco-nanoparticles as antipneumococcal vaccine model. *Nanomedicine.* 12 (1): 13–23.

235 Deng, S., Bai, L., Reboulet, R. et al. (2014). A peptide-free, liposome-based oligosaccharide vaccine, adjuvanted with a natural killer T cell antigen, generates robust antibody responses in vivo. *Chemical Science.* 5 (4): 1437–1441.

236 Bhalla, M., Nayerhoda, R., Tchalla, E.Y.I. et al. (2021). Liposomal encapsulation of polysaccharides (LEPS) as an effective vaccine strategy to protect aged hosts against *S. pneumoniae* infection. *Frontiers in Aging.* 2: 798868.

237 Jones Charles, H., Zhang, G., Nayerhoda, R. et al. (2017). Comprehensive vaccine design for commensal disease progression. *Science Advances.* 3 (10): e1701797.

238 Hill, A.B., Beitelshees, M., Nayerhoda, R. et al. (2018). Engineering a next-generation glycoconjugate-like *Streptococcus pneumoniae* vaccine. *ACS Infectious Diseases.* 4 (11): 1553–1563.

239 Li, Y., Hill, A., Beitelshees, M. et al. (2016). Directed vaccination against pneumococcal disease. *Proceedings of the National Academy of Sciences.* 113 (25): 6898–6903.

240 Said Hassane, F., Phalipon, A., Tanguy, M. et al. (2009). Rational design and immunogenicity of liposome-based diepitope constructs: application to synthetic oligosaccharides mimicking the *Shigella flexneri* 2a O-antigen. *Vaccine.* 27 (39): 5419–5426.

241 Polonskaya, Z., Deng, S., Sarkar, A. et al. (2017). T cells control the generation of nanomolar-affinity anti-glycan antibodies. *The Journal of Clinical Investigation.* 127 (4): 1491–1504.

242 Cavallari, M., Stallforth, P., Kalinichenko, A. et al. (2014). A semisynthetic carbohydrate-lipid vaccine that protects against *S. pneumoniae* in mice. *Nature Chemical Biology.* 10 (11): 950–956.

243 Bai, L., Deng, S., Reboulet, R. et al. (2013). Natural killer T (NKT)–B-cell interactions promote prolonged antibody responses and long-term memory to pneumococcal capsular polysaccharides. *Proceedings of the National Academy of Sciences.* 110 (40): 16097–16102.

244 Ulrich, J.T. and Myers, K.R. (1995). Monophosphoryl lipid A as an adjuvant. In: *Vaccine Design: The Subunit and Adjuvant Approach* (ed. M.F. Powell and M.J. Newman), 495–524. Boston, MA: Springer US.

245 Liao, G., Zhou, Z., Suryawanshi, S. et al. (2016). Fully synthetic self-adjuvanting α-2,9-oligosialic acid based conjugate vaccines against group C meningitis. *ACS Central Science.* 2 (4): 210–218.

246 Wang, L., Feng, S., Wang, S. et al. (2017). Synthesis and immunological comparison of differently linked lipoarabinomannan oligosaccharide–monophosphoryl lipid a conjugates as antituberculosis vaccines. *The Journal of Organic Chemistry.* 82 (23): 12085–12096.

247 Rappuoli, R. and De Gregorio, E. (2011). A sweet T cell response. *Nature Medicine.* 17 (12): 1551–1552.

248 Zhang, F., Lu, Y.-J., and Malley, R. (2013). Multiple antigen-presenting system (MAPS) to induce comprehensive B- and T-cell immunity. *Proceedings of the National Academy of Sciences.* 110 (33): 13564–13569.

249 ClinicalTrials.gov. A Single Ascending Dose Study in Adults (Stage 1) and Single Ascending Dose-Finding Study (Stage 2) in Elderly Subjects With ASP3772, A Pneumococcal Vaccine 2021. https://clinicaltrials.gov/ct2/show/NCT03803202.

250 Affinivax. https://affinivax.com/pipeline/overview/.

251 Matrivax. https://www.matrivax.com/

252 Thanawastien, A., Cartee Robert, T., Griffin Thomas, J. et al. (2015). Conjugate-like immunogens produced as protein capsular matrix vaccines. *Proceedings of the National Academy of Sciences.* 112 (10): E1143–E1151.

253 Cartee, R.T., Thanawastien, A., Griffin Iv, T.J. et al. (2020). A phase 1 randomized safety, reactogenicity, and immunogenicity study of Typhax: a novel protein capsular matrix vaccine candidate for the prevention of typhoid fever. *PLoS Neglected Tropical Diseases.* 14 (1): e0007912.

254 ClinicalTrials.gov. Safety and Immunogenicity of Typhax, a Typhoid Vaccine 2019. https://clinicaltrials.gov/ct2/show/NCT03926455.

2

Antifungal Glycoconjugate Vaccines

Linda Del Bino, Maria R. Romano, and Roberto Adamo

GSK, Via Fiorentina 1, Siena, 53100, Italy

2.1 Human Fungal Infections

Fungi are heterotrophic eukaryotes morphologically classified into yeast and filamentous forms. Most fungi are ubiquitous in the environment, and humans are exposed by inhaling spores or small yeast cells. Fungi are very proficient at responding to surrounding signals that promote their survival in several environments. As a result, they can interact with plants, animals, or humans in multiple ways, establishing symbiotic, commensal, latent, or pathogenic relationships. Out of hundreds of thousands of known fungal species in the world, only about 300 are human pathogens [1], with *Candida*, *Aspergillus*, *Cryptococcus*, and *Pneumocystis* spp. responsible for more than 90% of reported deaths due to fungal disease [2]. The manifestation of fungal infections can be mucocutaneous, mucosal, or tissue-invasive. The majority of fungal infections are opportunistic, since healthy people can mount an efficient immune response against them, and cause mainly mucosal or superficial infections. Advances in medicine and surgery over the past century have led to increased life expectancy, and many diseases previously considered to have a very poor prognosis can now be controlled in such a way that patients can live with them. This is the case for individuals with immunodeficiency due to chemotherapy, AIDS, diabetes, or organ transplant [3]. In this population, the number of high-risk groups exposed to invasive fungal infections (IFIs) has increased. Hospital-acquired fungal infections are less frequent than bacterial ones, but they account for higher mortality rates, longer hospitalization times, and increased healthcare costs. To date, concerns over IFIs are rising since they kill 1.5 million individuals annually with an unacceptable mortality rate, which for *Candida* has been estimated to be 27–55% [4]. On the other hand, mucosal fungi infections are common in non-immunocompromised subjects. They are generally not life-threatening; however, they are associated with high morbidity, socioeconomic impact, and low quality of life. The most common mucosal infection sites are the oral cavity and the genitourinary tract in apparently healthy people. Approximately two-thirds of all

Carbohydrate-Based Therapeutics, First Edition. Edited by Roberto Adamo and Luigi Lay.

women will experience an acute episode of candidal vaginitis at least once in their lives, and nearly 7% will develop recurrent vulvovaginal candidiasis (RVVC), which often needs chronic medical treatment [5]. The main therapeutic options for IFIs consist of a limited number of systemic drugs, whose antifungal activity comes together with severe adverse effects. Furthermore, new antimicrobial-resistant strains are emerging, enhancing the need to develop alternative treatments, especially for *Candida auris* and *Candida* spp. [6]. Particular concern is raised by the emergence of *C. auris* in health care settings due to its high resistance to drugs and capacity to spread from person to person, which increases the need of efficacious therapeutic measures [7] (Singh 2019 #728).

Immunoprophylaxis with antifungal vaccines represents an appealing therapeutic option, and, despite the fact that no licensed vaccines are currently on the market, a lot of work has been done on potential vaccine targets as well as on passive immunization with monoclonal antibodies (mAbs) against systemic mycosis.

In Table 2.1, the advantages of each approach and the corresponding drawbacks are reported:

Table 2.1 Main advantages and disadvantages of potential treatments for systemic mycosis.

	Antifungal drugs	**Glycoconjugate antifungal vaccines**	**mAbs targeting fungal glycans**
Advantages	• Only treatment currently available on the market • Low production cost • Easier to store and administer • Use in patients with underlying medical conditions causing immunodeficiency	• Potential to provide long-term immunity to systemic mycosis • Less expensive production compared to mAbs • Some fungal antigens could be used to produce a pan-fungal vaccine against multiple mycosis • Use to treat recurrent mucosal infections in immunocompetent patients • No selection of resistant strains	• Reduced toxicity risk • Immediate immunity is provided against systemic mycosis • Potentially efficient also in immunocompromised patients • Highly specific, so avoid selection of resistant strains • A Phase I clinical study has been completed
Disadvantages	• Few obsolete drugs are available • Severe adverse effects • Possible selection of resistant strains	• Only immunocompetent patients can mount an efficient response to fungal antigens • Proof of concept of their safety and immunogenicity in humans still missing • Weeks to months are needed to confer protection	• Highly specific, therefore, a precise diagnosis is required • Higher production cost compared to traditional drugs • More difficult to store and administer compared to traditional drugs

Source: Del Bino and Romano [8]/with permission of Elsevier.

Fungal cell wall (CW) is the outer component responsible for the initial recognition by the host immune system. Most fungal CWs are mainly composed of different polysaccharides, which are not present in humans and thus can be considered excellent targets for antifungal immunotherapy.

2.2 Immunity Against Fungal Pathogens

The interaction between fungal pathogens and the host immune system is a very complex mechanism involving both innate and adaptive immunity. Skin and epithelial surfaces are the first barrier against fungi, and, indeed, many fungal infections occur in patients in whom the integrity of the natural barriers is disrupted. Once the skin's physical barrier is passed, neutrophils, monocytes, macrophages, natural killer (NK) cells, and dendritic cells (DCs) sense the fungal pathogens, and innate and adaptive immune responses are both activated. The constitutive elements of innate immunity reside in the skin and the mucosal epithelial surfaces, where pattern recognition receptors (PRRs) of innate immune cells such as neutrophils and macrophages detect fungal pathogen-associated molecular patterns (PAMPs). The fungal CW is the main source of PAMPs recognized by PPRs in mammalian cells. This detection promotes the engulfment of fungal cells and their subsequent degradation within phagosomal compartments [9, 10]. At the same time, adaptive immunity is activated; for example, antigen-presenting cells like DCs prime T cells by presenting sampled antigens in association with Major Histocompatibility Complex Class II or Class I molecules, leading to the differentiation of $CD4^+$ or $CD8^+$ T cells, respectively [11]. The ability of DCs to recognize fungal antigens and activate adaptive T-cell immune responses makes them logical cellular targets for the development of fungal vaccination strategies [12]. T-cell immune responses, in particular Th1/IL12, are considered key for protective immunity to fungi, and a dominant Th1 cell response correlates with protective immunity against fungi and effective fungal vaccines [9]. Therefore, to achieve activation of adaptative immune responses, it is necessary to activate pathogen-detection mechanism of the innate immune system. Th17 cytokines have been shown to act as effector molecules during the immune response to fungal infections at the mucosal inflammation and seem to play a critical component of the protective host response to fungal infections. However, need to elicit Th17-driven response appears not to be crucial for developing an antifungal therapeutic. While antibody-mediated immunity has been considered for a long period of time less important in host defense against fungi, the advances in mAb technology have made it possible to elucidate their protective role, consisting of supporting infection clearance via opsonization or direct antifungal activity [13, 14]. Protective mAbs target protein as well as carbohydrate epitopes of fungi CW [15–17].

In addition, there is growing evidence that an efficacious antifungal therapy can be achieved by targeting CW components that exert critical functions in fungal CW structure and adherence to host cells [18].

2.3 Carbohydrate Antigens in Fungal Cell Wall

Carbohydrates dominate the CW of fungi, and, in the case of *Candida* and *Cryptococcus neoformans,* surface polysaccharides have been identified as involved in PAMP–pathogen recognition receptor (PRR) interactions initiating downstream immune responses [17].

Since polysaccharide vaccines act as T-cell-independent antigens, they do not generate a protective immune response in children under two years of age and do not induce immunological memory and high-affinity antibodies. The development of glycoconjugate vaccines, in which the polysaccharide antigen is covalently linked to a carrier protein, has made it possible to overcome this limit, creating T-cell-dependent antigens capable of inducing a potent and specific immune response and arousing protective immunological memory from infancy [19, 20]. Accordingly, to study their immunogenicity at preclinical level, several fungal carbohydrate antigens have been conjugated to selected protein carriers. Differently from bacterial glycans, which usually have a core repeating unit composed of one or a few monosaccharides, fungal polysaccharides show a higher level of complexity. Indeed, fungal CWs present a complex multilayered architecture where the inner skeleton is relatively conserved and composed mainly of chitin, chitosan, and glucan polysaccharides, while the outer layer presents highly variable specific polysaccharides and glycoproteins, often organized in an irregular structure [21] (Figure 2.1).

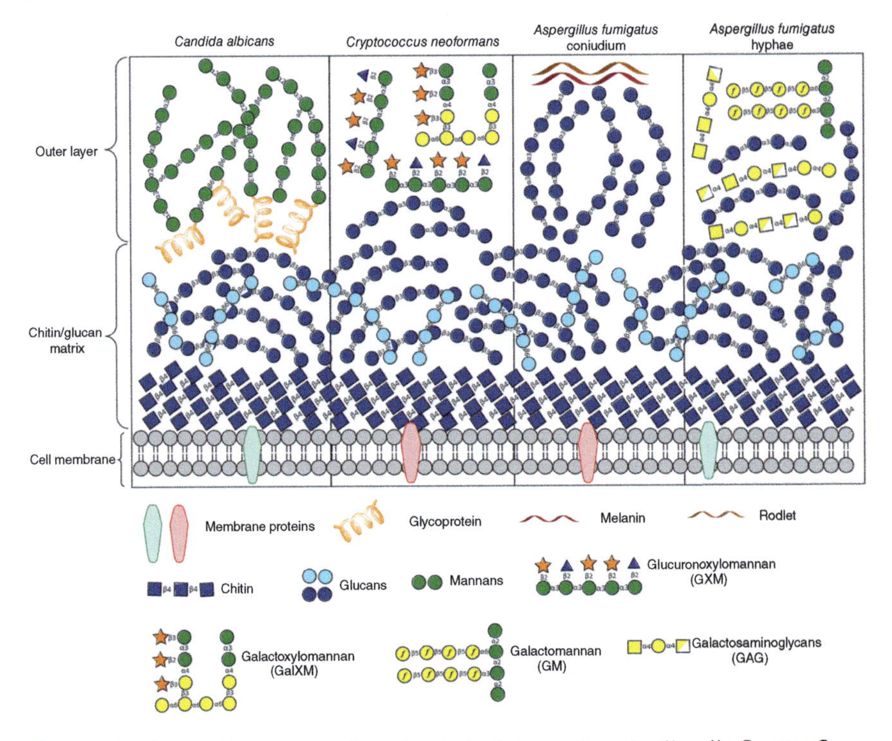

Figure 2.1 Schematic representation of carbohydrates in fungal cell walls. Source: Gow et al. [8]/with permission of Elsevier.

While bacterial vaccines have been successfully developed by conjugation of extracted polysaccharides to carrier proteins, this approach proves challenging for fungal sugars due to their complex organization and variable structure.

Given the challenge associated with fungal carbohydrate production, synthesis of oligosaccharide structures has appeared as an attractive alternative. Several synthetic glycoconjugate vaccines have been prepared and studied for the fungal pathogens causing most invasive infections in humans (*Candida*, *C. neoformans*). For *Aspergillus fumigatus* and *C. neoformans*, many efforts have been directed toward the synthesis of oligosaccharide components of the CW for structural studies aiming to identify the chemical features responsible for immunogenicity. Recent findings based on the use of synthetic oligosaccharides are summarized in this chapter.

2.4 Glycoconjugate Vaccines Against *Candida albicans/ Candida auris*

The outer core of *Candida albicans* CW is essentially composed of mannans, forming a network that functions as a scaffold for highly glycosylated proteins. This mannan polysaccharide is characterized by a set of different structural motifs, namely antigenic factors, distinguished by size, type of the glycoside bond, presence of phosphodiester linkages, branching points, etc., and its composition varies depending on species and strain of *Candida* microorganism. β-(1,3)-glucans, β-(1,6)-glucans, and chitin are instead present in the inner core of *Candida* CW. These polysaccharides are the main PAMPs that are recognized by PRRs.

Laminarin from the brown alga *Laminaria digitata* is composed of branched β-(1→3)-(1→6)-glucans and, due to its structural similarity with fungal sugars, has been considered a source of carbohydrates to develop vaccines against *Candida* infections [17, 22–29]. Other sources of β-glucans, such as the fully linear β-(1,3) glucan Curdlan (from *Alcaligenes faecalis* bacteria) or the β-(1,6) glucan Pustulan from *Umbilicaria papullosa*, have been exploited to determine the structural features needed for optimal immunogenicity. These studies indicated that the antibodies raised against β-(1,3) glucans were protective in a mouse model of systemic candidiasis, while structures containing β-(1,6) glucan side chains might induce non-protective antibodies [22, 23].

β-(1,3) glucan binds to the C-type lectin-like receptor Dectin-1, which is expressed on immune cells such as macrophages, neutrophils, and DCs [30, 31]. The binding residues have been identified in Trp221 and His223 [32]. The polysaccharide is structurally well organized [33], and its long form appears to assume a triple helical conformation, as it has been demonstrated by conformational nuclear magnetic resonance (NMR) studies [34] and also by combined static light scattering, dynamic light scattering, and atomic force microscopy[35]. However, a structurally disordered hexamer seems sufficient to bind to Dectin-1 [34].

The protective role of β-(1,3) glucans was thoroughly investigated by employing synthetic oligosaccharides conjugated with different conjugation chemistries to the carrier proteins in order to elucidate the optimal structure of the protective epitope and of the resulting glycoconjugate (saccharide/protein ratio, type of carrier, and conjugation site). By comparing a linear 15-mer with a branched 17-mer (**1**), it was observed that β-(1→6) branches need to be separated by at least six β-(1→3) residues for efficient antibody production [22].

Following this, a CRM_{197}-conjugate of a synthetic linear β-(1,3) glucan hexamer (**2**) was tested *in vivo* and showed to elicit in mice anti-β-glucan IgGs in a comparable manner to Lam-CRM_{197} [23]. This result well correlates with the findings that mAb 2G8, which is protective in mice challenged with *C. albicans,* mostly targets linear β-(1,3) glucan sequences. Subsequently, the impact of the conjugation site on CRM_{197} was further investigated by preparation of the linear β-(1,3) glucan hexamer conjugates through site-selective conjugation to tyrosines [24] via a newly developed 4-phenyl-1,2,4-triazoline-3,5-dione linker or controlled conjugation at the surface-exposed lysines [25] via active ester chemistry [36] (**4**). The resulting glycoconjugates were compared to those obtained through traditional random conjugation chemistry for CRM_{197} lysines [25]. All the constructs elicited a comparable level of IgGs [26, 37], and sera from immunized mice were able to inhibit the adhesion of *C. albicans* to human epithelial cells [25]. Therefore, it was concluded that four conjugation sites are sufficient to elicit a robust immune response in the animal model. The potential as an antigen of β-glucans was further confirmed by other investigators who conjugated a synthetic linear octasaccharide to the protein carrier Keyhole limpet hemocyanin (KLH) (**3**) [26]. The same team then investigated the immunogenicity of synthetic nonasaccharides of β-(1,3) glucans bearing β-(1,6) or β-(1,3) branches. Both conjugates showed similar immunological properties to the previously tested octasaccharide, and in addition to this, competitive ELISA experiments suggested that the majority of the induced antibodies were directed against the linear β-(1,3) backbone [37]. Overall, these studies suggested that β-(1,3) glucan oligomers could be promising candidates for antifungal vaccines and contributed to the development of a well-established synthetic route giving access to libraries of β-(1,3) glucan structures.

Due to its irregular structure, *Candida* mannan's immunogenic properties are difficult to reproduce, which makes its use problematic for the vaccine design. Therefore, the synthetic approach has been helpful in identifying the protective epitope to ensure the immunogenicity of the corresponding glycoconjugate vaccines. Short oligomers ranging from di- to hexasaccharide were synthesized in order to investigate the structure and size of the protective epitope based on the interaction with two protective mAbs generated by hybridoma techniques from mice immunized with a natural mannan–liposome preparation [16, 38]. Their findings indicated that a trisaccharide was recognized by both mAbs and, therefore, was a promising antigen for further vaccine development [39]. The synthetic β-(1,2)-mannan trisaccharide conjugated to TT (**5**) showed a robust secondary antibody response in rabbits but poor immunogenicity in mice [40, 41]. In order to improve the immunogenicity of the mannan trisaccharide, a synthetic conjugate was constructed using the 14-mer peptide Fba (**6**), deriving from a *C. albicans* CW protein and prepared through solid-phase peptide synthesis. After being tested in mice, the conjugate elicited a strong antibody response and offered protection against a lethal challenge of *C. albicans* [42, 43]. β-glucans act as immune potentiator molecules through Dectin-1 activation, and a hexamer has been shown to be sufficient to enhance the antibody response against protein antigens [44]. This feature has been harnessed to generate a tricomponent β-glucan and β-mannan conjugate, where the sugars were both conjugated to TT as carriers. This type of approach resulted in an enhanced antibody response to the β-mannan epitope of the conjugate while exploiting the proven targeting and antifungal response of the β-glucan conjugates [29]. Recently, a fully synthetic conjugate vaccine was constructed from a β-(1,2)-linked mannose trisaccharide conjugated to a T-cell peptide. This combined B- and T-cell epitope was synthesized and subsequently conjugated by click chemistry to an asymmetric dendrimer component bearing four copies of a β-(1,3)-linked hexaglucan DC epitope, obtaining a conjugate vaccine (**7**) that induced antibodies to all three epitopes of the fully synthetic construct. Unfortunately, the preparation of a synthetic dendrimer-based immunogen proved very challenging, discouraging the authors from continuing with further work on fully synthetic vaccines [45].

Other researchers focused their attention on the α-mannosides and studied the immunological properties of a series of synthetic oligo α-mannosides with and without branches conjugated to bovine serum albumin (BSA) via squarate chemistry. According to reported results, the presence of branches does not correlate with the induction of protective antibodies as measured by *in vitro* opsonophagocytic assay, while the length does not seem to play a major role, suggesting that oligomannoside structure more than their length might influence the quality of the antibody response and that the use of linear oligomannosides for vaccine design might be preferable [46].

Recent research demonstrated the immunobiological activity of synthetically prepared biotinylated α-mannooligosaccharides mimicking *Candida* antigenic factors. Macrophage exposure to a set of eight structurally different mannooligosaccharide conjugates induced the release of Th1, Th2, Th17, and Treg cytokine signature patterns, making these conjugates potential *in vitro* immunomodulative agents suitable for *in vitro Candida* diagnostics or prospectively for subcellular anti-*Candida* vaccine design [47] (Figure 2.2).

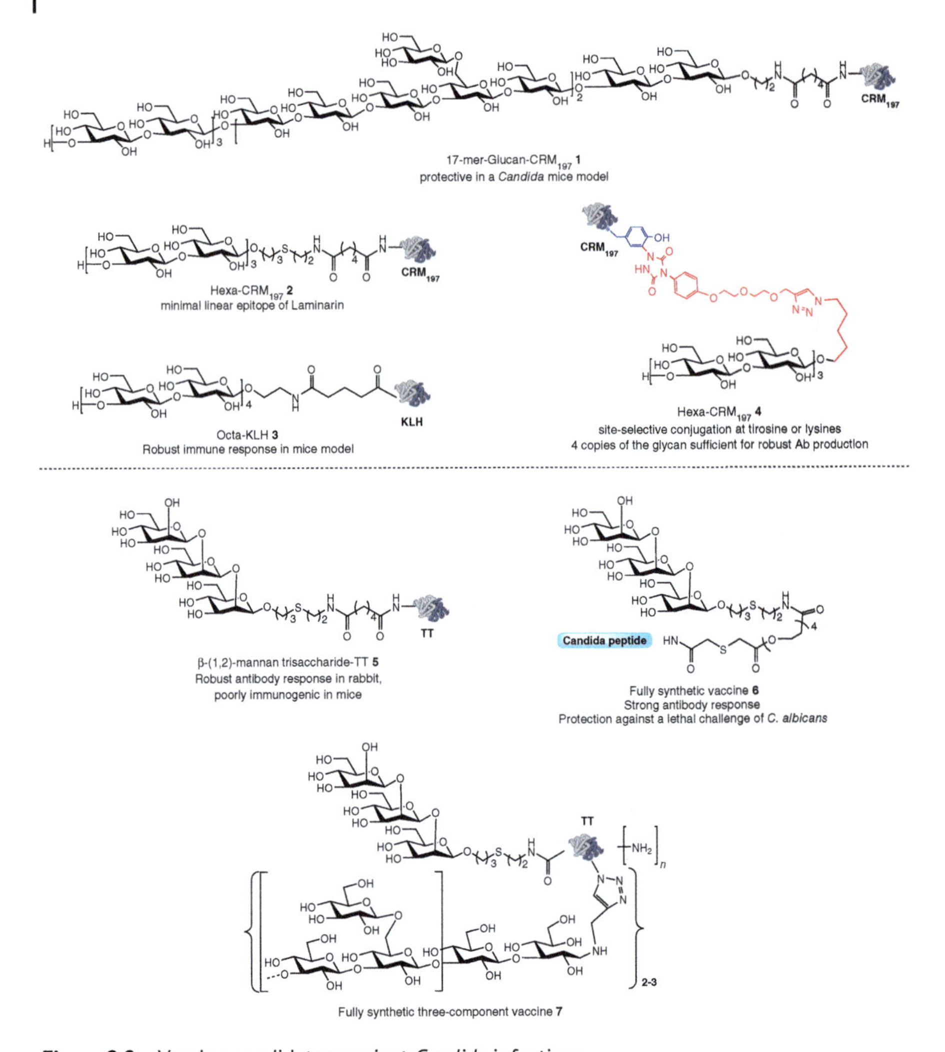

Figure 2.2 Vaccine candidates against *Candida* infections.

2.5 Glycoconjugate Vaccines Against *Cryptococcus neoformans*

The first fungal glycoconjugate vaccine was designed against *C. neoformans*, whose capsular polysaccharide is an important virulence factor and is composed mostly of glucuronoxylomannan (GXM), while galactoxylomannan (GalXM) and mannoproteins (MPs) are present to a lesser extent.

GXM has a very complex structure: it is a heteropolymer, not consisting of repeating units as bacterial polysaccharides but rather of six different chemotypes, occurring in various ratios depending on both strain and batch of the microbe and distributed among four serotypes: A, B, C, and D [48]. NMR experiments aiming to elucidate the structure of GXM concluded that this polysaccharide consisted of a linear α-(1,3) mannan trisaccharide backbone containing β-1,2 and β-1,4 xylose branches and a β-1,2 glucuronic acid branch attached at different mannoses of the repeating unit. Additional structural complexity and heterogeneity are introduced by O-acetylation, whose variability makes the identification of protective epitopes difficult.

GalXM represents about 7% of the capsular mass, and it is made of a α-1,6 galactan backbone with potentially four short oligosaccharide branch structures. Other carbohydrates present in the CW of *C. neoformans* are α-(1-3) glucans, which anchor the polysaccharide capsule to the CW, and β-(1-3) glucans.

Early attempts to make an anti-*Cryptococcal* glycoconjugate vaccine have been focused on GXM and resulted in a poorly characterized product, which showed immunogenicity in mice but did not give a protective antibody immune response [49]. The first evidence that GXM can be regarded as a target for immunotherapy was obtained using a natural polysaccharide conjugated with tetanus toxoid (GXM-TT conjugate **8**), which resulted in immunogenicity and protection in mice [50, 51]. The results of both active and passive protection experiments suggested that the presence of GXM-TT-elicited antibodies during the first —four to six weeks of infection was critical for the clearance of cryptococci from various organs, for limiting serum GXM titers from reaching immunosuppressive levels, and ultimately for survival [51, 52]. Thanks to the availability of anti-GXM mAbs isolated from mice infected with *C. neoformans* and mice immunized with GXM-TT, the presence of protective as well as nonprotective epitopes within GXM structure was demonstrated [53, 54].

Thus, further studies were directed toward determination of the protective epitope structure using synthetic oligosaccharides to be used in the conjugated vaccine. First, the synthetic heptasaccharide IX, representing the dominant putative epitope of *C. neoformans* serotype A GXM, was synthesized with different O-acetylation patterns and tested for binding against a library of seven mAbs. Both O-acetylated and nonacetylated forms of the heptasaccharide were strongly recognized by two mAbs (13F1 and 7B13). The mono-O-acetylated synthetic serotype A heptasaccharide was conjugated to human serum albumin (HSA) (**9**) and tested for immunogenicity in mice with or without Freund adjuvant. Only the conjugate administered with adjuvant was able to elicit anti-GXM antibodies, which unexpectedly recognized primarily and in an irregular manner the surface of serotype D and B strains in immunofluorescence experiments [55]. In a subsequent study, the ability of the antibodies elicited by the above conjugate to protect mice against challenge with serotype D *C. neoformans* was investigated, but unfortunately, no protection was observed [56].

To elucidate the oligosaccharide structure, which could be regarded as an efficient ligand for vaccine development, a synthetic glycan array containing immobilized oligosaccharides related to GXM fragments, ranging from di- to octadecasaccharides, was developed and the interactions of such oligosaccharides with available protective and nonprotective mAbs were investigated. The screening revealed that a serotype A decasaccharide was recognized by several neutralizing mAbs, making this structure a promising candidate for anti-*C. neoformans* conjugate vaccine development [57].

Another cell-wall polysaccharide that has been considered as vaccine antigen is GalXM, which was conjugated to BSA and to a protective antigen from *Bacillus anthracis* as protein carriers (**10**), demonstrating its effectiveness to induce robust

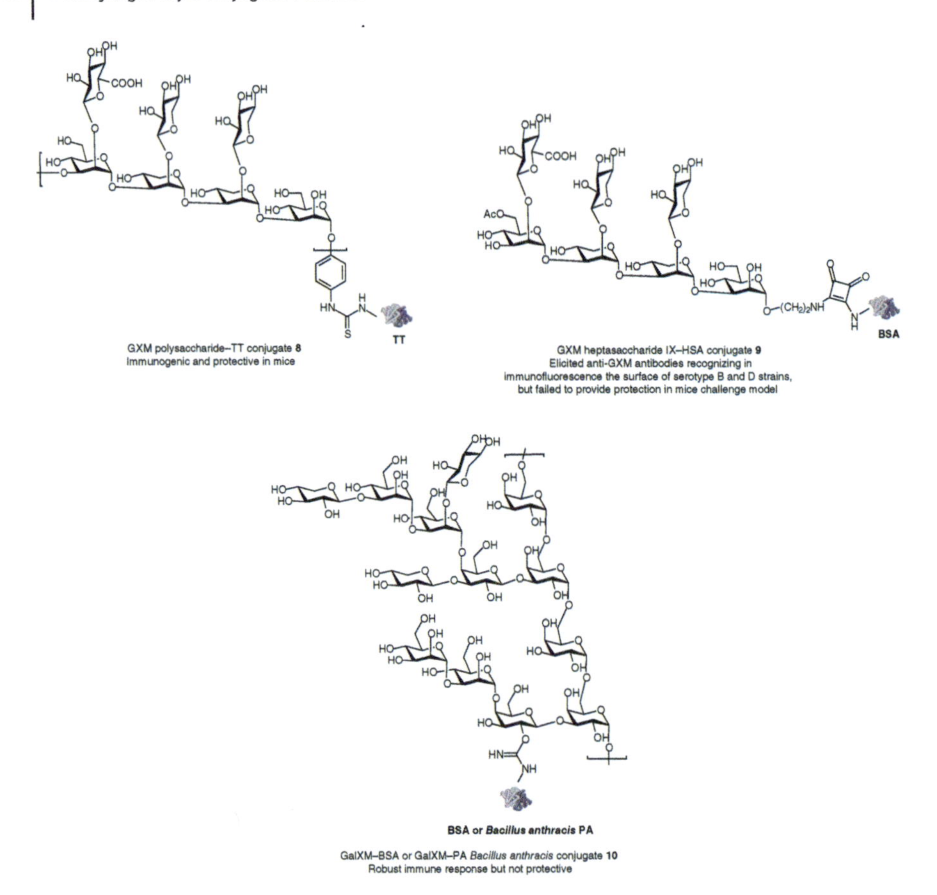

Figure 2.3 Vaccine candidates against *Cryptococcus neoformans.*

immune responses in the animal model, although the antibody responses were not protective [58, 59]. Further studies are, therefore, needed to clarify, possibly with the use of synthetic glycans as done for GXM, whether there are protective epitopes expressed on the polysaccharide (Figure 2.3).

2.6 Glycoconjugate Vaccines Against *Aspergillus fumigatus*

Similarly to other fungi, the CW of *A. fumigatus*, which is a main cause of pulmonary infections in immunocompromised patients, is dominated by carbohydrates, and its general composition not only comprises β-glucans crosslinked to chitin but also an extracellular matrix (ECM) composed of polysaccharides, mainly α-(1,3) glucan and galactosaminogalactan (GAG), creating a cell surface structure that is thought to be crucial to pathogenic expression. GAG consists of galactose (Gal), galactosamine (GalN), and *N*-acetylgalactosamine residues (GalNAc) [60].

The presence of α-(1→3)-glucans on their surface is shared by *Cryptococcus* and *Aspergilli*. However, in the first case, they anchor the polysaccharide capsule to the CW, whereas in *A. fumigatus*, α-(1→3)-glucans induce the aggregation of germinating fungal conidia [61].

The protective evidence of carbohydrate antigens from *Aspergillus* still needs to be deeply investigated. To address this, many research groups have recently focused on the synthesis of carbohydrate components of *A. fumigatus* to obtain synthetic oligosaccharides, which could aid to characterize structure–immunogenicity relationships of the cell-wall glycans. Nifantiev and coworkers synthesized, in 2015, an α-(1,3) pentaglucoside and then conjugated it with BSA (**11**) [62]. Immunization of mice with the BSA conjugate induced the generation of antibodies that recognize α-(1→3)-glucan on *A. fumigatus* CW and distinguish its morphotypes [62, 63]. Recently, the same group directed their synthetic efforts toward the preparation of oligo-α-(1,4) GalNs and their N-acetylated derivatives (**12**) [64–66]. After biotinylation, synthetic glycans were used as molecular probes in glycoarrays against sera from patients infected with pulmonary aspergillosis. Such investigations showed that human IgGs recognized both acetylated and non-acetylated oligo-GalNs with a degree of polymerization of at least 3. In the same direction, Codée, Barbero, and coworkers synthesized GAG structures (**13** and **14**), identifying a synthetic methodology for the assembly of GAG-oligomers capable of incorporating possible variations

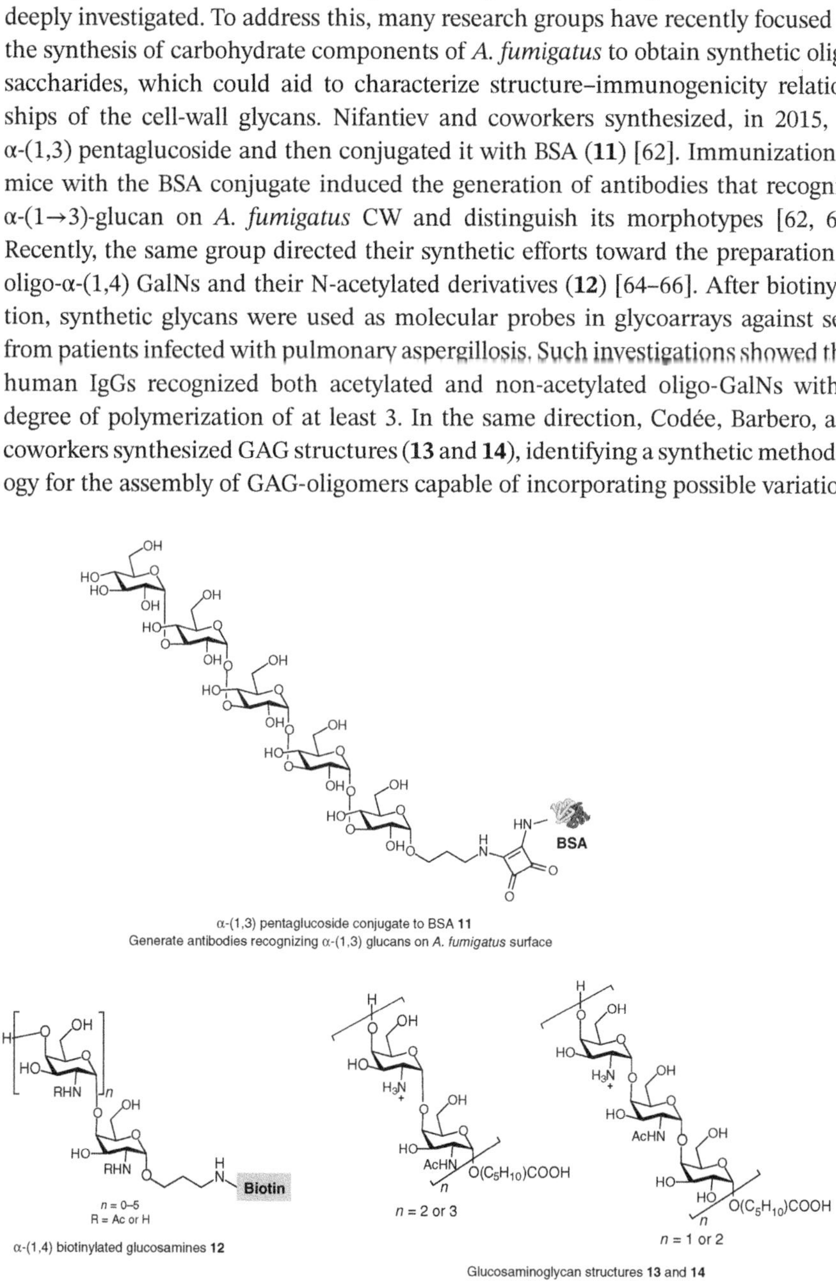

α-(1,3) pentaglucoside conjugate to BSA **11**
Generate antibodies recognizing α-(1,3) glucans on *A. fumigatus* surface

n = 0–5
R = Ac or H

α-(1,4) biotinylated glucosamines **12**

n = 2 or 3

n = 1 or 2

Glucosaminoglycan structures **13** and **14**

Figure 2.4 Vaccine candidates against *Aspergillus fumigatus*.

of the natural structures [67]. The conformation of the synthesized oligosaccharides was investigated by molecular dynamics and NMR experiments, and these glycans may find application in future binding studies to establish GAG epitopes, which can be used in the development of glycoconjugate vaccines against *A. fumigatus* (Figure 2.4).

2.7 Universal Fungal Polysaccharide Antigens

β-glucans are present in a conserved layer of the CW across many fungal species and, therefore, are an appealing target for the development of an effective "pan-fungal vaccine" [68]. The algae-derived sugar Laminarin conjugated to the nontoxic mutant of diphtheria toxin CRM_{197} has been shown to induce protection in a murine infection model against both systemic candidiasis and aspergillosis [17]. *Saccharomyces cerevisiae* β-glucans induced a protective response to coccidioidomycosis and aspergillosis [69–71] in a murine model, confirming the potential of these carbohydrate antigens as tools for pan-fungal vaccination.

Another potential target that has been studied for the potential development of a vaccine targeting multiple fungi is the poly-*N*-acetyl-1,6-glucosamine (PNAG), which has been detected on the surface of *C. albicans*. An anti-PNAG mAb, mAb F598, and polyclonal serum from mice immunized with a synthetic nonasaccharide, analog of deacetylated PNAG and conjugated to TT, provided protection in a mouse model of *C. albicans keratitis* [72] and were able to kill *A. fumigatus* and *Fusarium solani* in an opsonophagocytosis assay and to protect mice from *A. fumigatus keratitis* [73].

2.8 Conclusions and Future Prospects

IFIs are becoming an increasing threat in nosocomial settings due to increasing life expectancy, especially in the presence of underlying medical conditions such as cancer, HIV, or other immunosuppressive diseases. Current pharmaceutical treatments of fungal infections consist of a few old drugs with severe adverse events, while some fungal pathogens are starting to represent a concern for the emergence of antimicrobial-resistant strains. Preventive therapies such as vaccination, relying on the host immune response to fungal antigens, can represent an interesting alternative for IFIs in immunosuppressed patients and/or for recurrent mucosal infections in healthy people, such as vulvovaginal candidiasis. Polysaccharides are the main component of the fungal CW and are, therefore, regarded as potential vaccine antigens upon conjugation to carrier proteins, providing the T-cell help needed for the formation of memory B cells. However, the complexity of the fungal CW makes it very difficult to establish a correlation between oligosaccharide structures and immunogenicity. For this reason, many attempts have been made over the years to elucidate the glyco-epitopes expressed on the CW polysaccharides with the aim of designing glycoconjugate vaccines able to induce a robust and protective immune response. This task has been mainly in the hands of synthetic organic chemists, who have focused over the years on the preparation of glycans from three fungal pathogens (*Candida* spp., *C. neoformans*, and *A. fumigatus*). Synthetic well-defined oligosaccharide fragments conjugated to carrier proteins or peptides helped to highlight the saccharide structural requirements to induce functional antibodies in animal models. Moreover, the availability in the last few years of mAbs targeting fungal glycans has allowed the screening of libraries of oligosaccharide structures by glycoarray, helping to identify the length and the branches needed for epitope optimization. Synthetic glycoconjugate vaccines require the optimization of synthetic methods for complex oligosaccharide chain assembly. In particular, over the years, many research groups have developed efficient chemistries for the preparation of β-glucans and to achieve stereoselective 1,2-*cis* glycosylation, which are needed for the β-mannosylation essential to prepare *Candida mannans* and for the α-glucosylation for Aspergillus α-glucans [74]. Advances in synthetic methodologies together with the application of techniques that can map the interactions between oligosaccharides and proteins (glycoarray, surface plasmon resonance, NMR, and X-ray crystallography) can aid the rational structural design of modern and safe antifungal glycoconjugate vaccines. In addition to this, formulation technologies and adjuvants are also important factors in modulating the immune response.

References

1 Schmiedel, Y. and Zimmerli, S. (2016). *Swiss Medical Weekly* 146: w14281.
2 Brown, G.D., Denning, D.W., Gow, N.A.R. et al. (2012). *Science Translational Medicine* 4: 1–10.
3 Drummond, R.A. and Clark, C. (2019). *Pathogens* 8: 1–11.
4 Pfaller, M.A., Pappas, P.G., and Wingard, J.R. (2006). *Clinical Infectious Diseases* 43: 3–14.
5 Cassone, A. (2015). *BJOG An International Journal of Obstetrics and gynaecology* 122: 785–794.
6 Micoli, F., Costantino, P., and Adamo, R. (2018). *FEMS Microbiology Reviews* 42: 388–423.
7 Singh, S., Uppuluri, P., Mamouei, Z. et al. (2019). *PLoS Pathogens* 15: 1–25.
8 Del Bino, L. and Romano, M.R. (2020). *Drug Discovery Today: Technologies* 38: 45–55.
9 Romani, L. (2011). *Nature Reviews. Immunology* 11: 275–288.
10 Posch, W., Steger, M., Wilflingseder, D., and Lass-Flörl, C. (2017). *Expert Opinion on Biological Therapy* 17: 861–870.
11 Carvalho, A., Duarte-Oliveira, C., Gonçalves, S.M. et al. (2017). *Current Fungal Infection Reports* 11: 16–24.
12 Wan Tso, G.H., Reales-Calderon, J.A., and Pavelka, N. *Frontiers in Immunology* https://doi.org/10.3389/fimmu.2018.00897.

13 Arturo Casadevall, L.P. (2012). *Cell Host & Microbe* 11: 447–456.

14 Verma, A., Wüthrich, M., Deepe, G. et al. *Cold Spring Harbor Perspectives in Medicine* https://doi.org/10.1101/cshperspect.a019612.

15 Brena, S., Omaetxebarría, M.J., Elguezabal, N. et al. (2007). *Infection and Immunity* 75: 3680–3682.

16 Johnson, M.A., Cartmell, J., Weisser, N.E. et al. (2012). *The Journal of Biological Chemistry* 287: 18078–18090.

17 Torosantucci, A., Bromuro, C., Chiani, P. et al. (2005). *The Journal of Experimental Medicine* 202: 597–606.

18 Torosantucci, A., Chiani, P., Bromuro, C. et al. *PLoS One* https://doi.org/10.1371/journal.pone.0005392.

19 Pollard, A.J., Perrett, K.P., and Beverley, P.C. (2009). *Nature Reviews. Immunology* 9: 213–220.

20 Vella, M. and Pace, D. (2015). *Expert Opinion on Biological Therapy* 15: 529–546.

21 Gow, N.A.R., Latge, J.P., and Munro, C.A. (2017). *The Fungal Kingdom* 267–292.

22 Bromuro, C., Romano, M., Chiani, P. et al. (2010). *Vaccine* 28: 2615–2623.

23 Adamo, R., Tontini, M., Brogioni, G. et al. (2011). *Journal of Carbohydrate Chemistry* 30: 249–280.

24 Hu, Q.Y., Allan, M., Adamo, R. et al. (2013). *Chemical Science* 4: 3827–3832.

25 Adamo, R., Hu, Q.Y., Torosantucci, A. et al. (2014). *Chemical Science* 5: 4302–4311.

26 Guochao Liao, Z.G., Zhou, Z., Burgula, S. et al. (2015). *Bioconjugate Chemistry* 26: 466–476.

27 Paulovičová, E., Paulovičová, L., Pilišiová, R. et al. (2013). *FEMS Yeast Research* 13: 659–673.

28 Johnson, M.A. and Bundle, D.R. (2013). *Chemical Society Reviews* 42: 4327–4344.

29 Lipinski, T., Fitieh, A., Pierre, J.S. et al. (2013). *Journal of Immunology* 190: 4116–4128.

30 Brown, G.D., Taylor, P.R., Reid, D.M. et al. (2002). *Journal of Experimental Medicine* 196 (3): 407–412. https://doi.org/10.1084/jem.20020470.

31 Brown, G.D., Herre, J., Williams, D.L. et al. (2003). *The Journal of Experimental Medicine* 197: 1119–1124.

32 Adachi, Y., Ishii, T., Ikeda, Y. et al. (2004). *Infection and Immunity* 72: 4159–4171.

33 Synytsya, A. and Novak, M. (2014). *Annals of Translational Medicine* 2: 1–14.

34 Hanashima, S., Ikeda, A., and Tanaka, H. (2014). *Glycoconjugate Journal* 31: 199–207.

35 Xing Zheng, L.Z., Lu, F., and Xu, X. (2017). *Journal of Materials Chemistry B* 5: 5623–5631.

36 Crotti, S., Zhai, H., Zhou, J. et al. (2014). *ChemBioChem* 15: 836–843.

37 Guochao Liao, Z.G., Zhou, Z., Liao, J. et al. (2016). *ACS Infectious Diseases* 2: 123–131.

38 Han, Y., Riesselman, M.H., and Cutler, J.E. (2000). *Infection and Immunity* 68: 1649–1654.

39 Nitz, M., Ling, C.C., Otter, A. et al. (2002). *The Journal of Biological Chemistry* 277: 3440–3446.

40 Wu, X., Lipinski, T., Carrel, F.R. et al. (2007). *Organic & Biomolecular Chemistry* 5: 3477–3485.

41 Lipinski, T., Wu, X., Sadowska, J. et al. (2012). *Vaccine* 30: 6263–6269.

42 Xin, H., Dziadek, S., Bundle, D.R., and Cutler, J.E. (2008). *Proceedings of the National Academy of Sciences of the United States of America* 105: 13526–13531.

43 Xin, H., Cartmell, J., Bailey, J.J. et al. *PLoS One* https://doi.org/10.1371/journal.pone.0035106.

44 Donadei, A., Gallorini, S., Berti, F. et al. (2015). *Molecular Pharmaceutics* 12: 1662–1672. https://doi.org/10.1021/acs.molpharmaceut.5b00072.

45 Bundle, D.R., Paszkiewicz, E., Elsaidi, H.R.H. et al. *Molecules* https://doi.org/10.3390/molecules23081961.

46 Paulovičová, L., Paulovičová, E., and Bystrický, S. (2014). *Microbiology and Immunology* 58: 545–551.

47 Paulovičová, E., Paulovičová, L., Farkaš, P. et al. (2019). *Frontiers in Cellular and Infection Microbiology* 9: 1–14.

48 Cherniak, R., Valafar, H., Morris, L.C., and Valafar, F. (1998). *Clinical and Diagnostic Laboratory Immunology* 5: 146–159.

49 Goren, M.B. and Gardner, M. (1967). *Journal of Immunology* 98: 901–913.

50 Devi, S.J.N., Schneerson, R., Egan, W. et al. (1991). *Infection and Immunity* 59: 3700–3707.

51 Sj, D. (1996). *Vaccine* 14: 841–844.

52 Ueno, K., Yanagihara, N., Shimizu, K., and Miyazaki, Y. (2020). *Biological and Pharmaceutical Bulletin* 43: 230–239.

53 Mukherjee, J., Nussbaum, G., Scharff, M.D., and Casadevall, A. (1995). *The Journal of Experimental Medicine* 181: 405–409.

54 Mukherjee, J., Scharff, M.D., and Casadevall, A. (1992). *Infection and Immunity* 60: 4534–4541.

55 Oscarson, S., Alpe, M., Svahnberg, P. et al. (2005). *Vaccine* 23: 3961–3972.

56 Antonio Nakouzi, A.C., Zhang, T., and Oscarson, S. (2009). *Vaccine* 27: 3513–3518.

57 Guazzelli, L., Crawford, C.J., Ulc, R. et al. (2020). *Chemical Science* 11: 9209–9217.

58 De Jesus, M., Nicola, A.M., Rodrigues, M.L. et al. (2009). *Eukaryotic Cell* 8: 96–103.

59 Chow, S.K. and Casadevall, A. (2011). *Vaccine* 29: 1891–1898.

60 Yoshimi, A., Miyazawa, K., and Abe, K. (2016). *Bioscience, Biotechnology, and Biochemistry* 80: 1700–1711.

61 Fontaine, T., Beauvais, A., Loussert, C. et al. (2010). *Fungal Genetics and Biology* 47: 707–712.

62 Komarova, B.S., Orekhova, M.V., Tsvetkov, Y.E. et al. (2015). *Chemistry - A European Journal* 21: 1029–1035.

63 Strobl, S., Eckmair, B., Blaukopf, M. et al. (2020). *ACS Chemical Biology* 15: 369–377. https://doi.org/10.1021/acschembio.9b00794.

64 Krylov, V.B., Argunov, D.A., Solovev, A.S. et al. (2018). *Organic & Biomolecular Chemistry* 16: 1188–1199.

65 Kazakova, E.D., Yashunsky, D.V., Krylov, V.B. et al. (2020). *Journal of the American Chemical Society* 142: 1175–1179.

66 Sarah Sze Wah Wong, N.E.N., Krylov, V.B., Argunov, D.A. et al. (2020). *mSphere* 5: e00688–e00619.

67 Zhang, Y., Gómez-Redondo, M., Jiménez-Osés, G. et al. (2020). *Angewandte Chemie, International Edition* 59: 12746–12750.

68 Nicola, A.M., Albuquerque, P., Paes, H.C. et al. (2019). *Pharmacology & Therapeutics* 195: 21–38.

69 Clemons, K.V., Antonysamy, M.A., Danielson, M.E. et al. (2015). *Journal of Medical Microbiology* 64: 1237–1243.

70 Min Liu, D.A.S., Clemons, K.V., Bigos, M. et al. (2011). *Vaccine* 29: 1745–1753.

71 Clemons, K.V., Danielson, M.E., Michel, K.S. et al. (2014). *Journal of Medical Microbiology* 63: 1750–1759.

72 Cywes-Bentley, C., Skurnik, D., Zaidi, T. et al. (2013). *Proceedings of the National Academy of Sciences of the United States of America* 110: E2209–E2218.

73 Zhao, G., Zaidi, T.S., Bozkurt-Guzel, C. et al. (2016). *Investigative Ophthalmology and Visual Science* 57: 6797–6804.

74 Krylov, V.B. and Nifantiev, N.E. (2020). *Drug Discovery Today: Technologies* 35, 36: 35–43.

3

Carbohydrate-Based Antiviral Vaccines

Adrián Plata[1] and Alberto Fernández-Tejada[1,2]

[1] *CIC bioGUNE, Basque Research and Technology Alliance (BRTA), Chemical Immunology Lab, Biscay Science and Technology Park, Building 801A, Derio, Biscay 48160, Spain*
[2] *Ikerbasque, Basque Foundation for Science, Euskadi Plaza, 5, Bilbao, Biscay 48009, Spain*

3.1 Introduction

Carbohydrates play a critical role in numerous infections caused by viruses that are responsible for many diseases, including common cold, influenza [1], the more serious acquired immune deficiency syndrome (AIDS) [2], and different forms of the severe acute respiratory syndrome (SARS), best exemplified by the current, devastating coronavirus disease 2019 (COVID-19) pandemic due to the severe acute respiratory syndrome coronavirus-2 (SARS-CoV-2) virus [3]. In addition to infectious diseases caused by viral infections, some viruses are at the origin of several human cancers, most notably liver cancers resulting from chronic infections by hepatitis B and C viruses [4] and cervical cancers associated with long-lasting infection with the human papillomavirus (HPV) [5]. To fight against these serious diseases, the importance of safe, potent vaccines and therapeutic approaches is clear, contributing to the prevention and treatment of such viral infections for global health. A number of antiviral vaccines containing live-attenuated or inactivated viruses have been very effective in combating and even eradicating several viral infectious diseases in recent history, e.g. polio, measles, mumps, rabies, varicella, and smallpox [6]. However, this traditional approach has not been fully successful for some chronic and reemerging viral diseases, such as HIV, influenza, or hepatitis C. As such, the development of modern subunit vaccines based on purified and structurally defined immunogenic elements of a specific virus has become a preferred preventative strategy due to their improved safety and more precise immune targeting [6].

Carbohydrates are ubiquitous on many viral surface proteins and are crucially involved in viral pathobiology. Viral protein glycosylation plays pivotal functional roles in the infectious process [7], from initial adherence of the virus and tissue invasion to protection from the immune system, by mimicking the host-cell "self"

Carbohydrate-Based Therapeutics, First Edition. Edited by Roberto Adamo and Luigi Lay.

glycans, particularly *N*-glycans [8], by hijacking cellular glycosylation. Thus, in addition to the own viral genome information, the biosynthetic machinery and glycan processing events within the infected cell have important implications for shaping viral protein glycosylation. This provides further structural diversity for the virus beyond that arising from potential mutations occurring during virus evolution, impacting their virulence, infectivity, and immunogenicity [9].

Thus, realizing the importance of viral glycosylation and diversity in driving viral pathogenesis has provided fertile ground to exploit carbohydrates on the virus surface for the development of antiviral vaccines and therapeutic strategies using chemical approaches. In this chapter, we describe key and recent developments in synthetic carbohydrate-based vaccines against representative viral diseases (e.g. HIV, influenza, hepatitis, Ebola, and COVID-19), while also providing some examples of glycan-based immunoadjuvants and therapeutic agents.

3.2 Human Immunodeficiency Virus

Human immunodeficiency virus type 1 (HIV-1) is the causative agent of AIDS, a pandemic that has affected more than 76 million people since its onset in 1981, with over 33 million deaths (680 000 of them in 2020) because of AIDS-related diseases [2]. As such, the development of an effective and safe prophylactic vaccine against HIV-1 is of critical importance for global health. However, only a few clinical trials over the last decade have shown a positive outcome in terms of preventing HIV-1. The most successful results correspond to the RV144 efficacy trial, which used a replication-defective canarypox vector (ALVAC) together with the recombinant AIDSVAX B/E HIV-1 gp120 protein. Despite its promise, the protected efficacy of this vaccine was around 31%, mainly attributed to the synergistic contribution of both humoral and cellular immune responses [10, 11]. Therefore, the development of a successful HIV-1 vaccine still poses a significant scientific challenge, whereby induction of both neutralizing antibodies and T-cell responses should be ideal for optimal vaccine efficacy [12].

The HIV-1 virus surface is covered by a dense sugar coat, with the envelope glycoprotein (Env) spike being extensively glycosylated with host-synthesized carbohydrates that mask the protein antigens from immune recognition [13]. This, together with the high level of genetic diversity of the virus due to its high tendency to mutate, promotes viral escape from the host immune system. The HIV-1 Env is a trimer composed of three gp120–gp41 heterodimers consisting of the gp120 surface glycoprotein noncovalently associated with the gp41 transmembrane glycoprotein. Gp120 is heavily glycosylated with an extensive array of N-linked carbohydrates that constitute more than 50% of its total mass. These oligosaccharides form a dense glycan shield that covers the protein surface and contributes to immune evasion by hindering immune recognition of the underlying peptide epitopes by naturally induced broadly neutralizing antibodies (bnAbs) [14, 15]. Recent reports have shown that around 20% of HIV-1-infected individuals have circulating bnAbs, which are characterized by special, uncommon features (e.g. extensive somatic hypermutation, long heavy-chain third complementarity-determining regions [CDRH3], and/or self- or polyreactive nature) that make it difficult to induce such bnAbs by a designed HIV vaccine [16]. Notably, all bnAbs isolated so-far target-specific, conserved regions of vulnerability of the HIV-1 Env, particularly within established glycan/peptide domains corresponding to the gp120 variable loops 1/2 (V1V2) and 3 (V3), the CD4-binding site (CD4bs) on gp120, the bridging region between gp120 and gp41, and the gp41 membrane-proximal external region (MPER) [17]. Therefore, these bnAb epitopes represent promising targets for HIV-1 vaccine design, with the main goal of inducing such type of bnAb with the ability to neutralize multiple, diverse HIV-1 strains.

Given the key role of the HIV-1 glycans in viral transmission and infection as well as in masking the protein antigens for immune escape, the Env surface glycans are a primary target for the design of effective HIV-1 vaccines [13, 18]. In this context, the development of carbohydrate-based synthetic immunogens as minimal structural mimics of several bnAb epitopes has emerged as an important strategy in an attempt to develop vaccines capable of inducing bnAbs with safer and more precise immune targeting. This section describes the most significant milestones in the synthesis and immunological testing of glycan-based epitope mimics as promising targets for the development of bnAb-eliciting HIV-1 vaccine candidates [19, 20]. We summarize recent advances on carbohydrate-based minimal immunogen design for the induction of glycan-recognizing bnAbs that target three key domains on gp120: the outer domain high-mannose glycan cluster around N332 and the conserved glycopeptide-dependent epitopes in the V1V2 and V3 loops, respectively.

3.2.1 Vaccine Constructs Derived from gp120 High-Mannose *N*-Glycan Cluster

3.2.1.1 Surface Oligomannose Cluster-Targeting bnAb: 2G12 Antibody

The monoclonal antibody 2G12 was the first bnAb identified to bind the HIV-1 glycan shield. It was isolated from an HIV-1-positive patient and has been found to neutralize an array of HIV-1 virions [21] as well as to protect against simian-human immunodeficiency virus (SHIV) via passive immunization in macaques [22, 23]. As shown by epitope mapping studies, 2G12 recognizes a conserved high-mannose carbohydrate cluster on the gp120 surface that includes primarily *N*-glycans at the N295, N332, and N392 positions [24]. Moreover, the Manα1$\rightarrow$2Man disaccharide terminus was identified as a critical motif for binding [25]. The crystal structure of 2G12 showed an unusual assembly of two Fab (fragment antigen binding) regions into an interlocked V_H domain-swapped dimer, yielding an extended multivalent binding surface for the glycan cluster [26].

Additional binding analysis using well-defined carbohydrate antigens provided further details on glycan specificity. These studies showed that a $Man_9GlcNAc$ [27] or even a Man_4 structure mimicking the D1 arm of native $Man_9GlcNAc_2$ [28] presented the highest binding affinities with this bnAb [29–31], emphasizing the importance of the terminal Manα1→2Man moiety for 2G12 recognition and in agreement with previous crystallographic studies [26]. Based on these observations, several groups have focused on the development of different synthetic structures that can mimic the 2G12 epitope and have assessed their binding affinity *in vitro* (antigenicity) as well as their ability to elicit bnAbs *in vivo* (immunogenicity).

3.2.1.2 Synthesis and Immunological Evaluation of 2G12 Epitope Mimics

In 2004, Wang and Li synthesized several oligomannoside clusters by conjugating high-mannose glycans with cholic acid [32] and galactoside moieties [27] through a maleimide–thiol coupling reaction (Scheme 3.1a). Multivalent $Man_9GlcNAc_2$ structures (Figure 3.1a,b) presented relatively high binding affinity to 2G12, but they were not comparable to that of native gp120. In a subsequent study, the tetravalent galactoside-scaffolded moiety was conjugated through a maleimide-based linker to keyhole limpet hemocyanin (KLH, a carrier protein that promotes T-helper responses and multivalent antigen presentation) (Figure 3.1b) in order to evaluate its capacity to induce bnAbs in rabbits [33]. Unfortunately, most of the antibodies were raised against the maleimide linker, and only modest titers of carbohydrate-specific antibodies were elicited, which showed weak cross-reactivity against gp120 and no HIV-neutralizing activity.

Meanwhile, the Danishefsky laboratory performed the total synthesis of hybrid-type and $Man_9GlcNAc_2$ glycan structures and incorporated them into a gp120 peptide fragment via Lansbury aspartylation (Scheme 3.1b), obtaining fully synthetic N322-glycosylated gp120 (A316-R355) fragments [34, 35]. Although surface plasmon resonance (SPR) experiments showed weak binding of the compounds with 2G12 bnAb, subsequent dimerization of the structures through a disulfide bond (C331) resulted in enhanced antigenicity [36].

Collectively, the studies by the groups of Wang and Danishefsky suggested a crucial effect of glycan multivalency in binding the 2G12 bnAb. Therefore, later efforts have been focused on the design and synthesis of carbohydrate clusters by

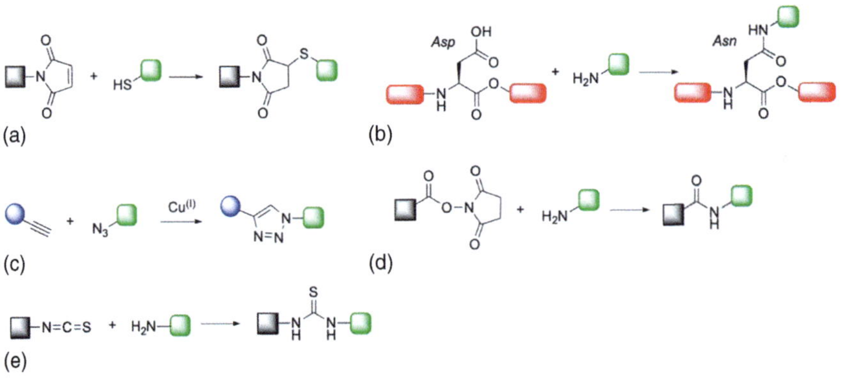

(a) (b) (c) (d) (e)

Scheme 3.1 Conjugation reactions and linker chemistry applied in the synthesis of 2G12 epitope mimics. (a) Thiol–maleimide coupling/thioether linkage, (b) Lansbury aspartylation/amide linkage, (c) Copper(I)-catalyzed alkyne–azide cycloaddition (CuAAC)/triazole linkage, (d)Amide linkage (via NHS ester), (e)Thiourea linkage.

Table 3.1 Significant examples of synthetic 2G12 epitope mimics.

Year(s)	Author(s)[a]	Coupling reaction/ linkage	Template/scaffold[b]	Reference(s)
2004–2006	Wang	Maleimide–thiol/ thioether	Cholic acid (**a**) and galactoside (**b**)	[27, 32, 33]
2004	Danishefsky	Lansbury aspartylation/amide	Gp120 peptide	[34–36]
2007	Wang	CuAAC/triazole	Cyclopeptide (**c**)	[37]
2007–2008	Danishefsky	Lansbury aspartylation/amide	Cyclopeptide (**d**)	[38, 39]
2008	Wong	CuAAC/triazole	Glycodendrimer (**e**)	[40]
2010	Costantino	NHS-based/amide	Glycodendrimer (**f**)	[41]
2008	Burton	Thiourea	BSA protein (**g**)	[42]
2019	Kosma	Thiourea	BSA protein (**h**)	[43]
2010	Finn, Burton	CuAAC/triazole	Qβ virus-like particle (**i**)	[44]
2010	Wilson, Davis	CuAAC/triazole	Qβ virus-like particle (**j**)	[45]
2011–2021	Krauss	CuAAC/triazole	DNA, peptide (**k**), and RNA	[46–52]

a) Corresponding author(s).
b) Structures indicated in Figure 3.1.

incorporating glycan moieties into different scaffolds using several conjugation strategies (Scheme 3.1). Table 3.1 and Figure 3.1 illustrate significant examples of synthetic 2G12 epitope mimics developed over the last 15 years.

The synthetic Man_4 (D1 arm of $Man_9GlcNAc_2$) tetravalent constructs (Figure 3.1c) synthesized by the Wang group using CuAAC (Scheme 3.1c) afforded relatively high 2G12 affinities, albeit decreased binding was observed for the fluorinated derivative [37]. In parallel, Danishefsky and coworkers prepared a range of $Man_9GlcNAc_2$ *N*-glycan clusters on a modular cyclic peptide (Figure 3.1d) via Lansbury aspartylation (Scheme 3.1b), with the divalent and trivalent structures showing significantly higher affinities compared to monovalent ones, which confirmed the importance of multivalent presentation. Subsequently, they attached the bivalent compound to the outer membrane protein complex (OMPC, an immunostimulatory carrier protein) via thiol–maleimide coupling (Scheme 3.1a) (~2000 glycopeptide monomers per conjugate) (Figure 3.1d) [38], in order to perform immunogenicity studies in guinea pigs and rhesus macaques. Although high levels of carbohydrate-specific antibodies were elicited in both species, they were not able to recognize a recombinant HIV gp120 precursor, thus failing to elicit a 2G12-like bnAb response [39]. These results show that despite their antigenicity, the synthetic high-mannose structures do not mimic the 2G12 antibody epitope realistically, and lack the ability to induce bnAbs, presumably due to a suboptimal oligosaccharide conformation/presentation necessary for efficient immune recognition.

Other multivalent structures, such as Wong's Man_9 glycodendron (Figure 3.1e) obtained via CuAAC (Scheme 3.1c) [40] or Costantino's Man_4- and Man_9-containing glycodendrimers (Figure 3.1f) synthesized using amide coupling (Scheme 3.1d) [41], both exhibited significant binding affinities with 2G12. The latter were linked to the CRM_{197} carrier protein via amide linkage (Scheme 3.1d) for subsequent immunization studies *in vivo* [41]. Analogously, Astronomo et al. synthesized a $(Man_4)_{14}$–BSA (bovine serum albumin) conjugate via a thiourea linkage (Scheme 3.1e) using BSA as a carrier protein for multimeric presentation (Figure 3.1g) and evaluated its immunogenicity in rabbits [42]. In both cases, carbohydrate-specific antibodies were generated, but the antisera were not cross-reactive to HIV-1 gp120 [41, 42]. Notably, Clark et al. showed that a bacterial lipooligosaccharide (LOS) derived from *Rhizobium radiobacter* Rv3 that included a Man4 D1-like arm was bound with

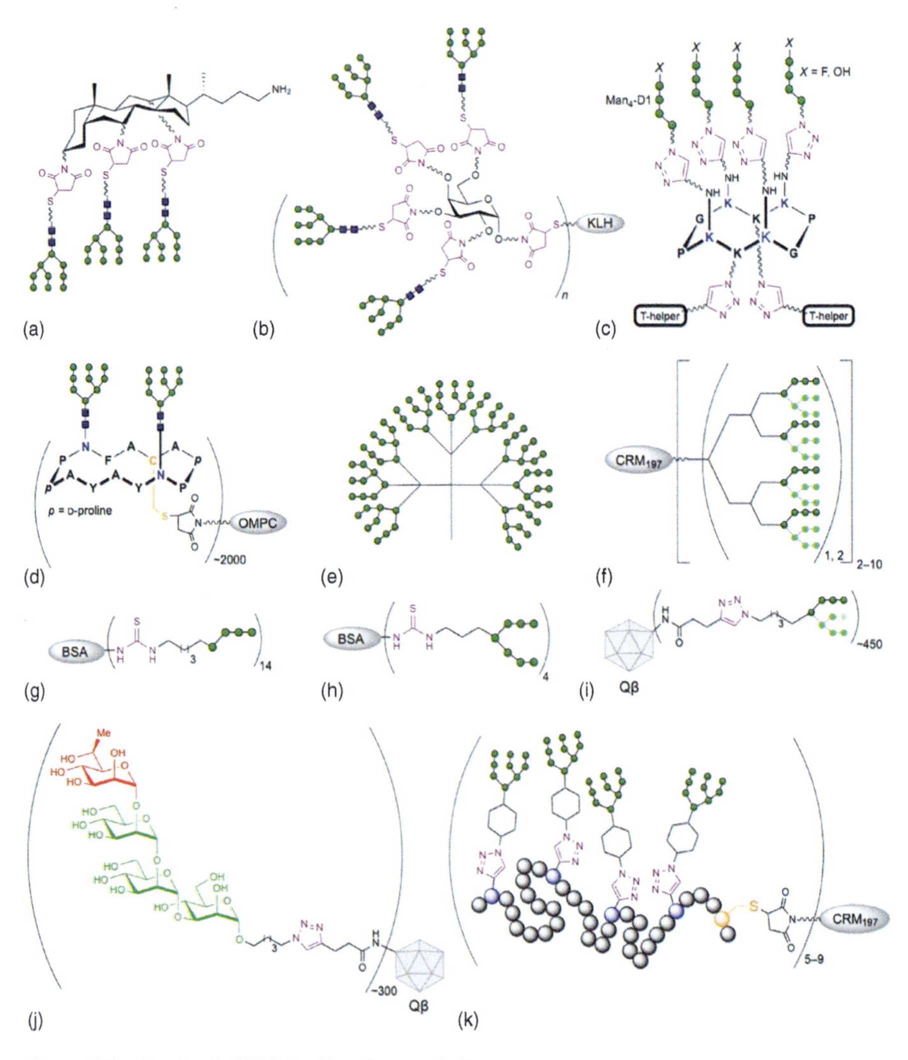

Figure 3.1 Synthetic 2G12 bnAb epitope mimics.

reasonable affinity by 2G12. Mouse immunization with heat-killed Rv3 bacteria elicited glycan-specific antibodies that recognized monomeric gp120 but could not neutralize HIV-1 virions [53]. Based on these findings and on the crystallographic structure of the 2G12/LOS complex [54], Kosma and coworkers conjugated several LOS-derived, penta- or heptamannose structures to BSA and assessed their antigenicity, with a thiourea-linked (Scheme 3.1e), β-anomeric Man_7 glycan (Figure 3.1h) showing the highest 2G12-affinity [43].

In other studies, Finn and collaborators investigated the use of virus-like particles (namely, bacteriophage Qβ), as scaffolds to facilitate a multivalent, ordered presentation of high-mannose glycans via triazole linkers (Scheme 3.1c) with the purpose of mimicking the oligomannose clustering on gp120. The highest 2G12 affinities were obtained with Qβ–Man_4, Qβ–Man_9, and especially with a mixture of Qβ–Man_8/Man_9 constructs (Figure 3.1i). The first two conjugate types (Qβ–Man_4 and Qβ–Man_9) induced mannose-specific antibodies in rabbits that recognized the respective glycans, but they did not cross-react with native gp120 or show HIV-1-neutralizing activity [44].

In a rationally designed strategy to enhance the immunogenicity of synthetic 2G12 epitope sugar mimics, Davis and coworkers synthesized a number of unnatural mannose-derived monosaccharides and their respective D1-arm tetrasaccharides for antigenicity and immunogenicity studies [45]. Interestingly, the nonself Man_4 glycan incorporating a terminal C-6-methylated mannose (see Scheme 3.2 for synthesis) showed the highest binding to 2G12, and its corresponding triazole-linked (Scheme 3.1c) Qβ conjugate (Figure 3.1j) was evaluated in rabbits. Although considerably increased titers of mannose-specific antibodies were induced, they were not cross-reactive with native gp120 and failed to neutralize HIV-1.

In the last decade, Krauss and coworkers have applied a directed evolution-based approach to the development of multivalent carbohydrate clusters as effective 2G12 epitope mimics. Following their early work with DNA-scaffolded Man_4 and Man_9

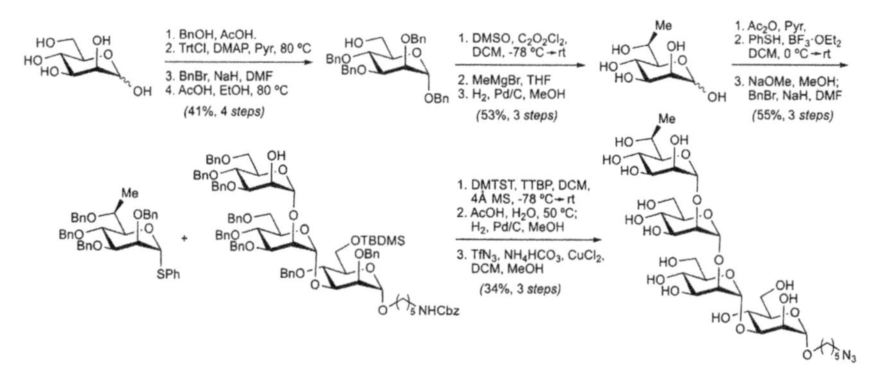

Scheme 3.2 Synthesis of unnatural C6-methylated mannose and assembly of a nonself tetrasaccharide D1-arm mimic.

glycoclusters [46, 47, 55], they developed a method to select Man_9-bearing multivalent glycopeptides as glycocluster scaffolds by combining mRNA presentation, incorporation of alkyne-containing unnatural amino acids, and subsequent glycan coupling via CuAAC (Scheme 3.1c) [48]. This strategy led to *in vitro* selected glycopeptides incorporating 3–5 oligosaccharides that showed binding affinities comparable with natural 2G12–gp120 interactions (in the picomolar to low nanomolar range), which represents the most antigenic 2G12 glycopeptide epitope mimic reported to date. The best Man_9-bearing peptides were conjugated to maleimide-functionalized CRM_{197} (Scheme 3.1a) (Figure 3.1k) [49] for immunological evaluation in rabbits [50]. While glycopeptide-reactive antibodies targeting the carbohydrate part were generated, low binding to the native-like soluble trimeric HIV Env protein (SOSIP) was observed (only in two cases), with negligible HIV-1-neutralizing activity. Moreover, the induced antibodies were found to bind mainly to the core mannoses rather than the Manα1→2Man termini recognized by 2G12, which might result from serum mannosidase trimming *in vivo* before immunogen presentation to B-cell receptors [50]. In a subsequent study to assess the effect of the vaccination regimen with a view to promote Manα1→2Man-specific antibodies, an evolved Man_4-glycopeptide immunogen was coadministered with the QS-21 adjuvant in liposomes using standard bolus dosing, an exponential series of mini doses, or continuous infusion [51]. The two latter regimens led to higher overall IgG titers to the glycopeptide, whereas bolus-immunized mice showed the strongest HIV Env-binding antibody response. Nonetheless, Manα1→2Man-binding antibodies were not induced in either case, suggesting that mannosidase activity might be saturated under the bolus immunization protocol, resulting in an increased presentation of intact Man_9 to B cells, albeit still insufficient to elicit antibodies to the Manα1→2Man motif. These findings highlight the need for an improved understanding of these biological processes in order to rationally develop optimal bnAb-eliciting HIV-1 vaccines. Recently, the Krauss group has also developed a novel directed evolution platform for the selection of stable 2′-fluoro-modified RNA-supported Man_9 glycoclusters that bind to 2G12 with low nanomolar affinities [52].

Taken together, while the synthetic glycoconjugates developed so far could induce oligomannose-specific antibodies with high 2G12-binding affinities (as described above), they have not proven to be effective HIV immunogens, failing to elicit 2G12-like antibodies that are cross-reactive with the native Env protein to lead to an HIV-1-neutralizing response. The lack of immunogenicity of these 2G12 epitope mimics can be attributed to several reasons. First and foremost, the conformation of the glycan and its presentation as part of the synthetic immunogen may be different from that of the natural gp120 protein, making it, therefore, unable to recapitulate the native spatial orientation in the Env spike. This may be due to their distinct, inherent physicochemical properties as well as the subsequent glycan processing *in vivo*. For instance, the inner $GlcNAc_2$ core that is missing in most of the constructs developed may have an important effect in defining the optimal carbohydrate orientation, which together with the high flexibility of the synthetic glycans may lead to unproductive immune recognition of irrelevant oligosaccharide conformations. This notion is supported by observations from Doms and coworkers, who used yeast-derived high-mannose glycoproteins as immunogens, presenting $Man_8GlcNAc_2$ glycans in a more dense, near-native form [56, 57]. Immunization in rabbits generated carbohydrate-specific antibodies that recognized gp120 and efficiently neutralized HIV-1 virions expressing high-mannose *N*-glycans but did not neutralize the wild-type virus. Second, considering the rare domain-exchange structure of 2G12, this class of bnAbs might be intrinsically difficult to induce, and some animal species may not have the ability to generate this unique and complex antibody type. Taking these concepts into account, a deeper knowledge of the key features of 2G12 bnAb evolution in HIV-1-infected individuals would be essential to design improved synthetic glycoconjugate immunogens that can elicit 2G12 bnAbs, leveraging a suitably devised vaccination approach.

3.2.2 Vaccine Constructs Derived from gp120 First and Second Variable Loops (V1V2)

3.2.2.1 V1V2-Targeting bnAbs

Since 2009, researchers have been continuously discovering new potent human bnAbs, which represents an important springboard toward the identification of new targets for HIV vaccine design. The PG9, PG16, CH01–04, and PGT141–145 antibodies were found to target the gp120 V1V2 apex of the HIV-1 Env trimer [58–63]. These bnAbs contain a long, anionic CDRH3 loop to penetrate the glycan shield and bind a quaternary motif within the first and second variable loops (V1V2). So far, synthetic glycan-based vaccine development has centered on the binding sites of PG9, PG16, and CH01 bnAbs, which include similar glycan-dependent conformational epitopes in the V1V2 region [59, 64]. Specifically, crystal structure studies of the complexes between PG9 and scaffolded V1V2 domains revealed that the antibody interacts with two high-mannose glycans at N160 and N156/N173 and a connected V1V2-peptide β-strand [65]. While the fine glycan specificities of the bnAb epitopes were yet uncertain, the available structural insights provided an important framework for the development of V1V2 carbohydrate-based immunogens as synthetic epitope mimics.

3.2.2.2 Synthetic V1V2 *N*-Glycopeptide Antigens as bnAb Epitope Mimics

Wang and coworkers designed and chemoenzymatically synthesized a number of gp120 V1V2 cyclic glycopeptides (V154–Y177) based on two HIV-1 strains with different glycosylation profiles, CAP45 (N156, N160) and ZM109 (N160, N173) (Scheme 3.3). Binding analysis by SPR and ELISA revealed that a $Man_5GlcNAc_2$ glycan at N160 was critical for recognition by PG9 and PG16, while the presence of an additional sialylated complex-type oligosaccharide at N156 or N173 further increased the binding affinity [66]. A more efficient chemoenzymatic approach was later developed for the site-selective glycosylation of the peptides with two distinct *N*-glycans by using orthogonally protected GlcNAc-Asn residues [67]. The important role of the sialylated *N*-glycan at the second glycosylation point was also corroborated by crystallographic studies with the PG16 bnAb [68]. Subsequently, Wu and coworkers synthesized unusual hybrid-type glycans bearing oligomannose and

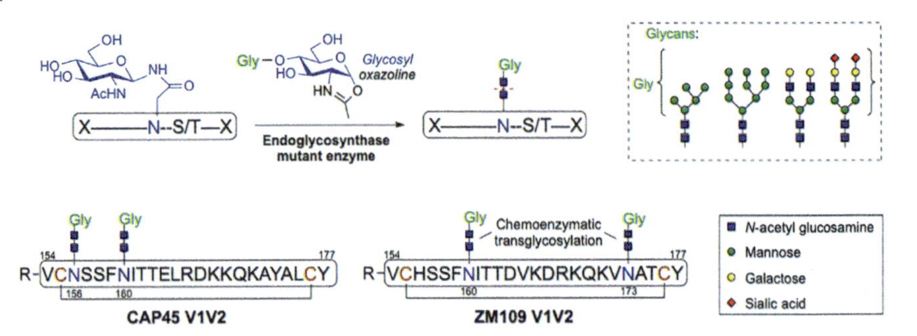

Scheme 3.3 Chemoenzymatic synthesis of V1V2 glycopeptides.

α-2,6-sialylated branches and analyzed their binding to PG9 and other bnAbs using glycan arrays. The high affinity obtained for these structures highlighted the critical role that the spatial distance between both glycan arms plays on antibody binding and provided uncommon glycans as potential epitope mimics for vaccine development [69, 70].

Meanwhile, Danishefsky and coworkers prepared several differently glycosylated gp120 V1V2 peptides based on the HIV-1 A244 strain (I148–I184) using chemical synthesis (Scheme 3.4). The corresponding glycosyl amines derived from Man_5-$GlcNAc_2$ and $Man_3GlcNAc_2$ were incorporated at the N156 and N160 residues of two individual peptides via Lansbury aspartylation, and the resulting fragments were coupled together in unprotected form by native chemical ligation (NCL). Binding studies confirmed the multivalent simultaneous interaction of PG9 with both the peptide backbone and the mannose-bearing N-linked oligosaccharides [71]. In a follow-up study, these glycopeptides were dimerized through a disulfide bond (C157), which resulted in even higher binding affinities to bnAbs (in the low

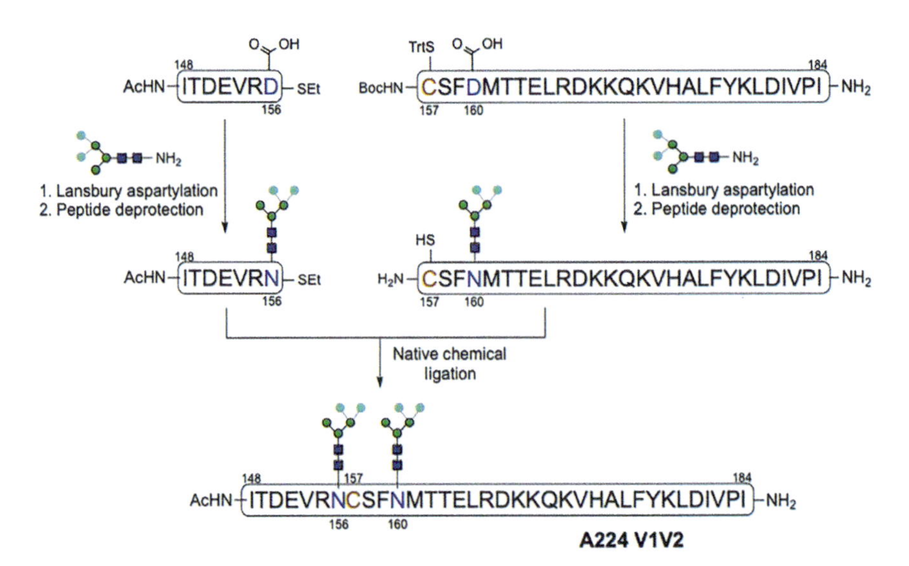

Scheme 3.4 Chemical synthesis of V1V2 glycopeptide.

nanomolar range) and their unmutated common ancestors [72]. In addition, circular dichroism experiments suggested that not only the Man_5/Man_3 *N*-glycans but also the disulfide bond-mediated dimerization may contribute to the more stable, β-stranded conformation necessary for bnAb binding. These results signal the promise of these rationally designed glycopeptide antigens for further development as potential synthetic immunogens to elicit V1V2-directed bnAbs.

3.2.3 Vaccine Constructs Derived from gp120 Third Variable Loops (V3)

3.2.3.1 V3-Targeting bnAbs

A majority of PGT bnAbs isolated from HIV-1-infected elite neutralizers were found to target epitopes on the V3 loop involving the N332 glycan and the V3 peptide backbone [60]. Further detailed characterization of the antibody-binding sites was obtained by crystallographic studies, which revealed an epitope formed by a V3 β-strand and 2 oligomannoses at N332 and N301 for PGT128 [73] and a preference toward complex-type glycans for PGT121 [74].

3.2.3.2 Synthetic Glycoconjugates and *N*-glycopeptides as V3-Directed bnAb Epitope Mimics

Kosma and collaborators observed that the previously mentioned β-linked Man_7-BSA conjugate (see Figure 3.1h) bound with high affinity to bnAbs of the PGT128-class and to their common germline precursor [75], likely because of the structural similarities of the bacterial LOS [54] and the PGT128 high-mannose glycan epitope [73]. In rat immunizations, this glycoconjugate induced reasonable levels of carbohydrate-specific IgM antibodies but low IgG titers, suggesting weak immunogenicity with B-cell activation in the absence of T-cell involvement. Interestingly, immune sera were cross-reactive with native gp120 and even exhibited neutralizing activity against some HIV-1 strains, presumably due to avid interactions with polymeric IgMs [75].

3.2.3.3 Synthetic V3 Glycopeptides as bnAb Epitope Mimics

In 2017, Wang and coworkers applied their glycosynthase-based chemoenzymatic strategy for the synthesis of a gp120 mini-V3 glycopeptide derived from the HIV-1 JR-FL strain (E292-N339). By using enzymatic transglycosylation and CuAAC cycloaddition, they prepared di- and trivalent constructs incorporating the high-mannose $Man_9GlcNAc_2$ glycan at N332, which were recognized by the PGT128- and PGT124-like bnAbs [76]. In a related study, the same group pinpointed the fine epitopes of some V3 bnAbs by exploiting differently glycosylated synthetic V3 glycopeptides (Figure 3.2a). Thus, PGT128 was found to exhibit binding affinity toward oligomannose glycopeptides with glycosylation-site flexibility (N301/N332), whereas the PG124 homolog recognized only the peptide having a high mannose at N332 and PGT121 required the presence at N301 of a sialylated complex-type oligosaccharide [77]. Later, they conjugated the high-mannose V3 glycopeptide (E293-N339) to a T-helper epitope from the tetanus toxoid (TT) carrier protein and

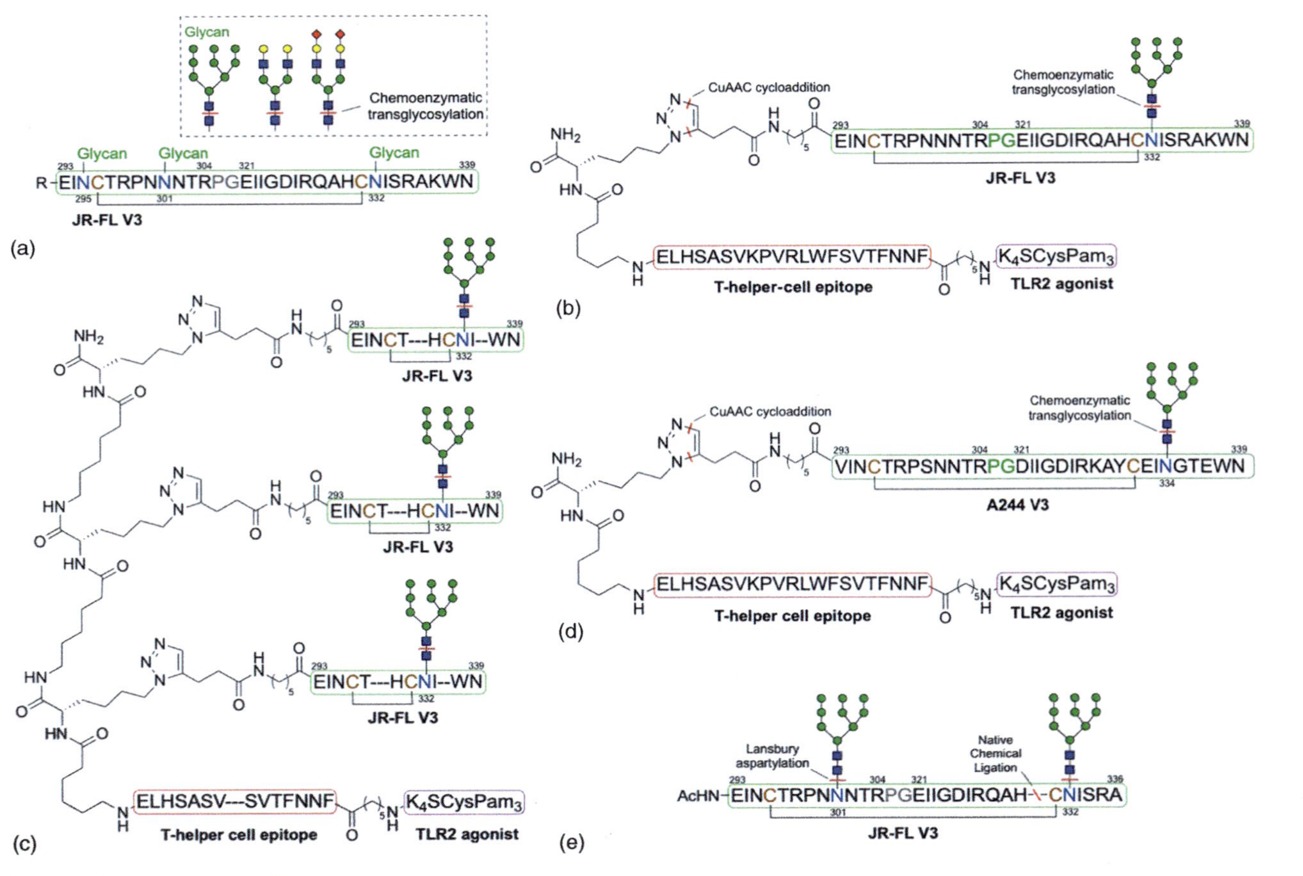

Figure 3.2 Synthetic V3-directed bnAb epitope mimics.

to the Pam_3CysSK_4 TLR2 ligand (as an adjuvant) (Figure 3.2b) [78]. In addition to this three-component "self-adjuvanting" construct, they also synthesized a trivalent analog presenting three copies of the V3 glycopeptide (Figure 3.2c) [79] as well as a monovalent variant with another V3 glycopeptide fragment derived from a different HIV-1 strain (A244) (Figure 3.2d) [80]. Rabbit vaccination studies with these structures showed induction of glycan-specific antibodies that cross-reacted with HIV-1 gp120/gp140 but did not neutralize HIV-1 virions [78–80].

Separately, Alam et al. designed and synthesized, through a two-step Lansbury aspartylation/NCL strategy, a minimal high-mannose V3 glycopeptide (Figure 3.2e) that was bound by PGT128 and PGT125 [81]. This Man_9 construct served to isolate V3 glycan bnAbs from an HIV-1-infected individual and elicited high-mannose-targeted antibodies in vaccinated rhesus macaques, thus mimicking the V3-glycan bnAb epitope. However, no HIV-1-neutralizing activity was observed. In a subsequent study, Seder and collaborators immunized nonhuman primates with a designed, dendrimer-based star nanoparticle system presenting several copies of a minimal synthetic immunogen consisting of a related V3 glycopeptide and a universal $CD4^+$ T-cell helper epitope (Pan DR-binding epitope, PADRE). Although high titers of V3-site-directed antibodies were generated, they showed weak affinity for native-like Env trimers and were not able to neutralize HIV-1 virions [82].

Despite important recent progress on the design and evaluation of minimal immunogens based on synthetic V3 *N*-glycopeptides, V3-targeted antibodies with broadly neutralizing activity have also not been generated. In part, this may be due to a different, irrelevant conformation/presentation of the V3 synthetic structures in comparison to that of the native Env glycoprotein epitopes, preventing elicitation of fully functional antibody responses. Moreover, in addition to the unique features of bnAbs, a potential reduction of the B-cell precursor pool because of immune tolerance could further limit the induction of these bnAbs by vaccination [16]. Preferably, an optimal immunogenic construct should activate these rare naive B cells in a selective manner, whereas further boost immunizations using rationally designed immunogens should ultimately produce bnAbs by driving B cells along desirable maturation pathways. With that objective in mind, the identification of clonally related bnAbs, a suitable B-cell-lineage design strategy, and fine structural determination of the epitopes recognized by intermediate B-cell receptors could yield critical insights for the development of effective minimal immunogens for HIV-1 [83].

3.3 Influenza A Virus

Influenza virus affects between 10% and 15% of the global population every year. In most cases, infection in healthy individuals results in a mild illness in the upper respiratory tract that does not require any type of surgery. However, it is estimated that between three and five million of these infections cause severe disease that progresses to the lower tract and viral pneumonia, resulting in up to 650 000 deaths each year, according to the World Health Organization [1]. Influenza viruses are formed into three different types: A, B, and C, albeit only A and B types seem to be pathogenic in humans. In fact, Influenza A viruses (IAVs) are responsible for the four pandemics that occurred in the last 100 years (1918, 1957, 1968, and 2009).

IAV is a highly mutable virus mostly associated with relatively mild diseases. However, IAV can be lethal to individuals with cardiac or pulmonary affections. On occasion, influenza viruses are transmitted from wild waterfowl to domestic poultry and are able to cause a human influenza pandemic [84]. IAV belongs to the *Orthomyxoviridae* family of enveloped, single-stranded RNA viruses with a genome that encodes several viral proteins, most notably hemagglutinin (HA) and neuraminidase (NA), and is classified based on 18 HA subtypes and 11 NA subtypes [85]. Both glycoproteins form the virus surface and are carbohydrate-recognizing proteins that function by recognizing sialic acid molecules on the host cell. HA binds sialylated receptors to enable attachment, whereas NA hydrolyzes sialic acid residues to help viral release and infection [86]. There exists a functional crosstalk between both proteins in viral attachment/release that depends on the HA glycans [87]. These carbohydrates play a role in immune evasion and constitute a dynamic glycan shield. While influenza virus vaccines exist, their efficacy is suboptimal due to mismatches between vaccines and circulating viral strains, requiring the development of a broadly protective vaccine to improve overall protection. Approaches are being attempted to attain such a universal vaccine aim to induce T-cell responses, or bnAbs, by targeting different internal or surface viral antigens, respectively [88]. The latter include the viral surface glycoproteins HA and NA, which are the main targets for the humoral immune response and the basis of currently available influenza vaccines, generating neutralizing antibodies.

3.3.1 Vaccine Constructs Based on Hemagglutinin (HA)

HA is the most abundant protein on the IAV surface. It plays a role in viral entry, is involved in receptor binding and membrane fusion, and is the major target of protective antibody responses upon infection or vaccination. HA is structured as a homotrimer on the surface of the virion, with each monomer comprising two subunits (HA1 and HA2) that originate from a distinct polypeptide precursor (HA0) [89]. The HA2 subunit presents a transmembrane region with a rather conserved cytoplasmatic stem that attaches the HA protein to the virion envelope. The receptor binding site is located in the more variable head domain of the HA1 subunit, which binds sialic acid residues on the host-cell surface, enabling viral entry [90]. Once the virion is internalized, a conformational change in HA exposes the N-terminus fusion peptide of the HA2 subunit stem, facilitating membrane fusion and release of the viral RNA into the cytoplasm of the host cell [91]. As the virion head domain contains the major immunodominant antigenic determinants, neutralizing antibodies against influenza are generally addressed to this region, interfering with HA binding to sialic acid and inhibiting its hemagglutination activity [92]. Thus, the hemagglutination inhibition test is used as a surrogate measurement to titrate the antibody response (anti-IAV-neutralizing antibodies) against influenza [93].

3.3.1.1 Hyperglycosylated HA Vaccines

In the design of vaccines directed to the more conserved HA stem domain, one strategy to increase stem-specific responses is to hyperglycosylate the variable HA1 head region in order to hide their immunodominant epitopes, thus directing the response to the stem [94]. Some studies showed that hyperglycosylated HA1 induced stronger anti-stem antibodies against the homologous H1 stem than wild-type HA [95]. The hyperglycosylated H1 also stimulated more cross-reactive antibodies to two heterologous H1 viruses and a heterosubtypic H5 virus. This strategy was also applied to the H5 stem, but it did not result in relevant antibody responses to other group 1 subtypes (H1, H3, and H9) [96]. Despite the increased anti-stem antibodies compared to wild-type HA, vaccination did not protect mice from critical influenza morbidity. Another example of this hyperglycosylation strategy has been used with the highly pathogenic avian influenza (HPAI) H5N1 viruses. The transmission capability of this virus from birds to humans has raised global concerns about a potential human pandemic. In these studies, hyperglycosylated HA vaccines were designed using N-linked glycan masking on highly variable sequences in the HA1 head domain [97]. Immunization with these hyperglycosylated HA DNA vaccines, followed by a flagellin-containing virus-like particle booster, was conducted in mice to evaluate neutralizing antibody responses against various clades of HPAI H5N1 viruses. However, no significant differences in anti-HA total Ig titers were found with these hyperglycosylated HA compared to the wild-type control.

A general approach to improve influenza vaccine's potency is the addition of adjuvants that increase antigen immunogenicity [98]. Thus, aluminum phosphate (alum) has been coadministered in HA-based DNA vaccines to enhance antibody production [99], and together with the oil-in-water MF59 and AS03, these are the three main adjuvants incorporated in licensed flu vaccines [100]. Other immunopotentiating substances, such as stimulatory glycolipids functioning as invariant natural killer T (iNKT) cell activators, have also been reported to exhibit adjuvant activities in protein and DNA vaccines [101].

3.3.1.2 α-Gal-Based Vaccine Constructs

An important stage for stimulating an adaptative immune response is the presentation of antigen fragments on the surface of antigen-presenting cells (APCs). α-Galactosylceramide (α-GalCer) was the first synthetic iNKT activator discovered, which was derived from a natural product extracted from marine sponges [102, 103]. Some research studies have included the α-GalCer glycolipid as a vaccine adjuvant against influenza [104, 105], while various delivery systems have also been developed (e.g. poly(lactic-*co*-glycolic acid) [PLGA] particles) to enhance its immunostimulatory properties and boost the immune response [106]. Moreover, recent developments in glycoconjugate vaccines are based on chemical combination of adjuvants (e.g. α-GalCer) and relevant carbohydrate antigens, which enables co-delivering of both vaccine components to the same immune cell for boosting the immune response [107].

This strategy of covalent conjugation was used by Anderson et al. for the development of a synthetic, influenza-targeting vaccine [108]. As opposed to antibodies

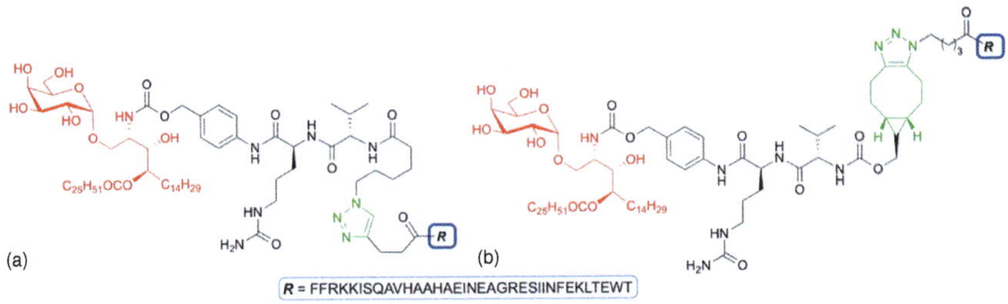

Figure 3.3 Antigen–α-GalCer prodrug conjugate vaccine against influenza challenge [108]. (a) CuAAC-coupled α-GalCer prodrug linker – SLP conjugate. (b) SPAAC-coupled α-GalCer prodrug linker – SLP conjugate. *R* represents the synthetic long peptide (SLP) containing an immunogenic sequence that involves the T-cell CD8$^+$ OVA_{257} (SIINFEKL) and CD4$^+$ OVA_{323} (ISQAVHAAHAEINEAGR) epitopes together with a protease cleavage sequence (FFRK).

binding cell surface HA, IAV-specific T cells recognize primarily conserved epitopes from internal viral proteins [109]. Thus, Painter and coworkers used click chemistry (CuAAC, Figure 3.3a; strain-promoted alkyne–azide cycloaddition SPAAC, Figure 3.3b) to link an α-GalCer prodrug derivative to a synthetic long peptide (SLP) from a virus-associated protein incorporating a well-known CD8$^+$ T-cell epitope from ovalbumin (OVA, see "R" substituent in Figure 3.3). *In vivo* studies in mice vaccinated with the SPAAC-coupled α-GalCer prodrug–SLP conjugate (Figure 3.3b) and challenged with a recombinant OVA-modified influenza virus showed induction of peptide-specific, memory T-cell responses that were protective against IAV infection [108].

In another approach, Galili and coworkers developed a carbohydrate-based method that leveraged the mechanism of antibody-dependent antigen uptake with a view to enhancing the immunogenicity of influenza vaccines [110]. Given the abundance of natural anti-Gal antibodies in humans, the authors incorporated a synthetic α-Gal epitope into the *N*-glycans of HA by applying a chemoenzymatic strategy that used α-1,3-galactosyltransferase (α-1,3GT) [111]. As such, modification of the virus *N*-glycans using recombinant α-1,3GT generated an influenza virus strain incorporating the α-Gal epitope. This engineered HA glycoprotein was bound by natural anti-Gal antibodies, leading to the formation of immune complexes, which results in targeting and uptake of the modified vaccine virus by APCs for stimulation of virus-specific T cells in the lymph nodes. In their study, mice vaccinated with this α-Gal-coated viral construct induced substantially increased antibody and cellular responses with higher protection than those immunized with the unmodified virus strain [110].

3.3.2 Vaccine Constructs Based on Neuraminidase (NA)

NA is the second most abundant glycoprotein on the viral surface and is implicated in viral delivery and propagation from infected cells. NA enables the release of lineage viruses from the host cell by cleaving terminal sialic acid from glycans both on the host cell and on the emerging virion [86]. Due to its role in the virus-replication cycle, NA has traditionally been a key target for antivirals or therapeutics based on inhibition of NA activity, including the commonly used oseltamivir (Tamiflu) and zanamivir (Relenza) [112]. Nonetheless, accumulated scientific evidence is leading to growing interest in considering NA as a target for vaccine design in the context of humoral immunity [113]. Thus, infection-induced human antibodies against NA were cross-reactive and able to inhibit the protein sialidase activity, blocking viral egress and providing protection from lethal influenza virus in mice [114, 115]. The major advantage of NA is its slower antigen evolution and the subsequent ability to induce longer lasting immunity and cross-protection than that offered by HA vaccines [116]. While the promise of natural NA-based immunity warrants further investigation into the inclusion of this glycoprotein in next-generation influenza vaccines targeting NA, there are still many knowledge gaps, including those regarding its immunogenicity, protection breadth, and mechanism, as well as the nature of the antigenic sites [117]. Notably, recent structural advances have yielded key insights into targeted epitopes and the basis of protection of anti-NA antibodies, providing templates for the rational design of NA-based vaccines and therapeutic agents [118, 119].

3.3.3 Acetalated Dextran as Adjuvant Carrier

Acetalated dextran (Ac-Dex) is a pH-responsive polysaccharide that can be readily synthesized from dextran through acetal formation with 2-methoxypropene. Ac-Dex is not soluble in water but is able to form microparticles loaded with different encapsulated cargoes by using emulsion techniques, releasing its content under acidic conditions [120]. It has been used as a vaccine carrier system to increase immune activation [121], whereby its microparticles can deliver the immunostimulatory agents (e.g. antigens and/or adjuvants) as a combined vaccine formulation, as shown against influenza and other infections [107].

This combination strategy was later used to apply the cyclic guanosine monophosphate–adenosine monophosphate (cGAMP)-encapsulated Ac-Dex together with soluble HA from the H1N1 subtype for anti-influenza vaccination [122]. This vaccine system induced a potent Th1-skewed neutralizing antibody response in mice, providing more than six-month protection from a lethal H1N1 challenge. In another example, Ainslie and coworkers co-formulated cGAMP and the ectodomain of the surface protein matrix 2 (M2e), both encapsulated within separate Ac-Dex particles, as an anti-influenza vaccine, which induced protective antibody and cellular immune responses [123].

3.3.4 Multivalent Constructs as Anti-Influenza Inhibitors

The multivalent interactions between HA and sialic acid residues on cell-surface receptors are the first step leading to viral internalization and consequent infection. Thus, for competitively blocking virus attachment to cells, synthetic multivalent glycoconjugates have been developed as inhibitors of influenza infection by using different scaffolds (e.g. dendrimers, proteins, and gold nanoparticles) decorated with numerous copies of sialic acid [124, 125].

Whitesides and coworkers introduced the first multivalent entry blockers in the 1990s. They developed α-sialoside–polyacrylamide copolymers that inhibited IAV adhesion to erythrocytes (measured as hemagglutination) more than 10 000 times more efficiently than its α-methyl sialoside monomer [126], in line with their greater binding affinity observed to the viral surface due to cooperative multivalent interactions [127]. Optimization studies led to high-affinity (in the nanomolar range) sialic acid-decorated polyacrylamide-based polymers [128] that, despite their considerable inhibitory potential, showed high cytotoxicity associated with their polyacrylamide backbone [129]. Additional designed multivalent conjugates consisted of polyamidoamine (PAMAM) dendrimers displaying sialyllactose, which showed *in vitro* micromolar inhibition and protected mice from H1N1 lethal challenge [130]. Haag and collaborators investigated other dendritic multivalent nanostructures based on chemical functionalization of gold nanoparticles [131, 132] and biocompatible polyglycerol nanogels (nPG) with sialic acid-terminated dendrons [133]. These carbohydrate nanosystems had high affinity for HA and showed 30% and 80% influenza infection inhibition, respectively. Another multivalent construct involved covalent conjugation of many copies of the anti-influenza drug zanamivir to a poly-L-glutamine scaffold, which led to zanamivir glycopolymers with enhanced potency (subnanomolar) that inhibited influenza viral fusion and release [134]. Overall, these examples highlight the importance of HA as a key target for further development of multivalent glycoconjugates as potential anti-influenza vaccines and drug candidates for the inhibition of influenza infection.

Among the options available to fight IAV, the most effective means to prevent influenza is through a universal vaccine that is broadly protective and does not need seasonal modification. With this aim, the development of such vaccine constructs should focus on conserved viral glycoproteins that are shared between virus strains and subtypes and can induce both antibody and cellular responses, providing effective cross-reactive immunity and prophylactic protection against IAV infection. Despite extensive research efforts in this direction by the scientific community, there is still a long pathway ahead to achieve more efficient and safer influenza vaccines, especially under the increasing threat of emerging global pandemics.

3.4 Hepatitis C Virus

Currently, hepatitis C virus (HCV) affects over 70 million people worldwide and has become the most important chronic liver disease, with the risk of developing into cirrhosis and hepatocellular carcinoma (HCC). HCV belongs to the *Flaviviridae* family and consists of an enveloped positive-sense single-stranded RNA virus that has six major genotypes and multiple subtypes [135, 136]. The HCV genome encodes one polyprotein precursor processed into three structural proteins (core protein and envelope glycoproteins E1 and E2) and seven nonstructural (NS) proteins [137]. The surface glycoproteins E1 and E2 contain around 5 and 11 potential N-glycosylation sites, respectively [138]. Most of them are rather conserved [139], suggesting that in addition to their role in protein structure and activity, they could serve as vaccine candidates. A heterodimer complex is formed by E1 and E2 on the viral particle, enabling HCV entry with involvement of E2 glycans [140]. Notably, the E2 glycoprotein includes the receptor-binding domain (RBD) that interacts with cell surface entry receptors [141, 142] and is also the major target for neutralizing antibodies [143]. Carbohydrates decorating HCV, especially the conserved E2 oligosaccharides, form a nonevolving glycan shield that protects underlying protein epitopes [144]. The E2 glycoprotein binds strongly to the lectin DC-SIGN (dendritic cell-specific intercellular adhesion molecule-grabbing nonintegrin), which may mediate viral infection in hepatocytes [145]. This is in line with the predominance of high-mannose glycans in this protein [146], which signals the potential of carbohydrate-binding agents (CBAs) as anti-HCV drug candidates that can disrupt glycan–protein interactions required for HCV entry and infection.

Thus, several mannose-specific CBAs targeting the oligomannose-containing E1E2 heterodimer have been developed that bound to envelope glycans and inhibited capture of HCV by DC-SIGN and viral entry [147]. Moreover, the lectin cyanovirin-N was also found to inhibit HCV infection, likely due to interaction with HCV envelope glycoproteins [148], highlighting the promise of CBAs as powerful lead drugs for blocking HCV at the cell entry phase.

In addition to enabling the screening of CBAs with anti-HCV activity, the abundance of conserved high-mannose glycans in the E1/E2 glycoproteins suggests the prospect of targeting these carbohydrate epitopes for the challenge of developing a so-far elusive HCV vaccine that can fight against the highly heterogeneous HCV genotypes. Nonetheless, some strategies based on recombinant HCV E2 protein have been investigated that can induce bnAbs against most HCV genotypes. For instance, an hyperglycosylation approach leading to an engineered subunit viral vaccine with distinct glycan patterns was employed to mask suboptimal epitopes associated with non-neutralizing antibodies, resulting in a more immunogenic HCV E2 protein that generated a more epitope-focused and protective neutralizing antibody response in mice [149].

3.5 Ebola Virus

Ebola virus (EBOV) is among the most lethal human pathogens, causing one of the deadliest infectious diseases in the world (Ebola hemorrhagic fever, or EBOV disease). It was first described in 1976 in Zaire (currently the Democratic Republic of the Congo), followed by various subsequent outbreaks, predominantly in Central Africa [150]. EBOV belongs to the Zaire ebolavirus species within the *Ebolavirus* genus of the family Filoviridae [151]. The last Ebola epidemic caused by EBOV in West Africa (2013–2016) reported over 28 000 cases of infected people and 11 000 deaths [152].

3.5.1 Glycoprotein-Based Vaccines

The envelope glycoprotein (GP) on the viral surface plays a key role in viral entry, being responsible for the interaction with host-cell receptors [153], and is the key target of neutralizing antibodies and protective immunity. It is highly immunogenic and has, therefore, been used as an immunogen in the development of Ebola vaccines. At present, two major vaccines exist against the Zaire ebolavirus species: the FDA-approved rVSV-ZEBOV (called Ervebo®) [154] and the two-component Ad26. ZEBOV/MVA-BN-Filo vaccine, authorized by the European Commission [155]. Mature GP is formed by GP1 and GP2, which are presented as trimers of disulfide-linked GP1–GP2 heterodimers. The GP1 subunit comprises two highly variable regions, the glycan cap and the mucin-like domain (MLD), that contain many N- and O-linked glycans [156]. Some of these *N*-glycans may serve as attachment points to host cells, for instance, via mannose-binding C-type lectins, which provides additional therapeutic opportunities (see below) [157]. Both heavily glycosylated domains can prevent neutralizing antibodies fr om binding the GP by protecting critical underlying epitopes [158]. Some research studies have investigated the influence of GP glycosylation, showing that mutation of two N-glycosylation sites on GP1 (388, 415 sites) may increase immunogenicity, presumably by exposing protective antibody epitopes, whereas deletion of the mucin region led to reduced protective efficacy in mice [159]. Because GP antibodies are mainly generated against the least conserved MLD, another study used a construct devoid of this domain for mouse vaccination, which resulted in the induction of cross-species immunity due to unmasking of more conserved regions by the immune system [160].

3.5.2 Monoclonal Antibodies and Carbohydrate Antiviral Agents as Therapeutics

Currently, there are two therapeutic options approved in 2020 by the FDA to treat Ebola, which are based on monoclonal antibodies that target the GP, blocking entry of the virus into the host cell [161, 162]. In addition to these antibody-based strategies, some small-molecule antivirals have been developed and tested in clinical trials, such as the broad-spectrum nucleoside analogs BCX4430 (Galidesivir) [163, 164] and Remdesivir [165], which act by inhibiting the viral RNA polymerase.

A general strategy exploited for the development of antiviral compounds involves the inhibition of viral transmission through blockade of DC-SIGN. This lectin interacts with highly mannosylated glycoproteins in a multivalent manner, including the EBOV GP [166], and is one of the most important pathogen-recognition receptors, being an important target for Ebola infection [167]. Consequently, a variety of multivalent carbohydrate structures have been developed with controlled glycan valency, size, and shape [168] that mimic the glycan presentation on the GP surface. Notably, some of these synthetic systems could block DC-SIGN and were able to inhibit infection by EBOV [169]. In a representative example, Davis and coworkers used a proteic platform (multimeric Qβ-based virus-like particles) for conjugation of mannose glycodendrons through CuAAC click chemistry, assembling multivalent glycodendrimeric nanoparticles that displayed >1600 monosaccharides in a homogeneous manner [170]. These glycoconjugates showed potent antiviral activity, preventing DC infection by pseudotype Ebola by competitively inhibiting the binding of DC-SIGN at low nanomolar/picomolar concentrations. Subsequently, Martin and collaborators efficiently synthesized a water-soluble multifullerene-based dendritic structure decorated with 120 mannose units using click chemistry (Figure 3.4), providing nanosized globular-shaped multivalent macromolecules that blocked DC-SIGN and inhibited EBOV infection in the subnanomolar range [171].

The high inhibitory potencies of these antiviral constructs open the door to investigating whether these multivalent systems could induce potent antibody responses *in vivo* that might potentially neutralize some of the relevant glycan–protein interactions operative at the virus–cell interface. Moreover, the modular nature and full chemical control enabled by this synthetic approach raise the possibility of creating dual/bifunctional structures that can incorporate immunogenic elements recognized by the immune system, leading to prospective novel molecular vaccines for inducing immune responses against EBOV.

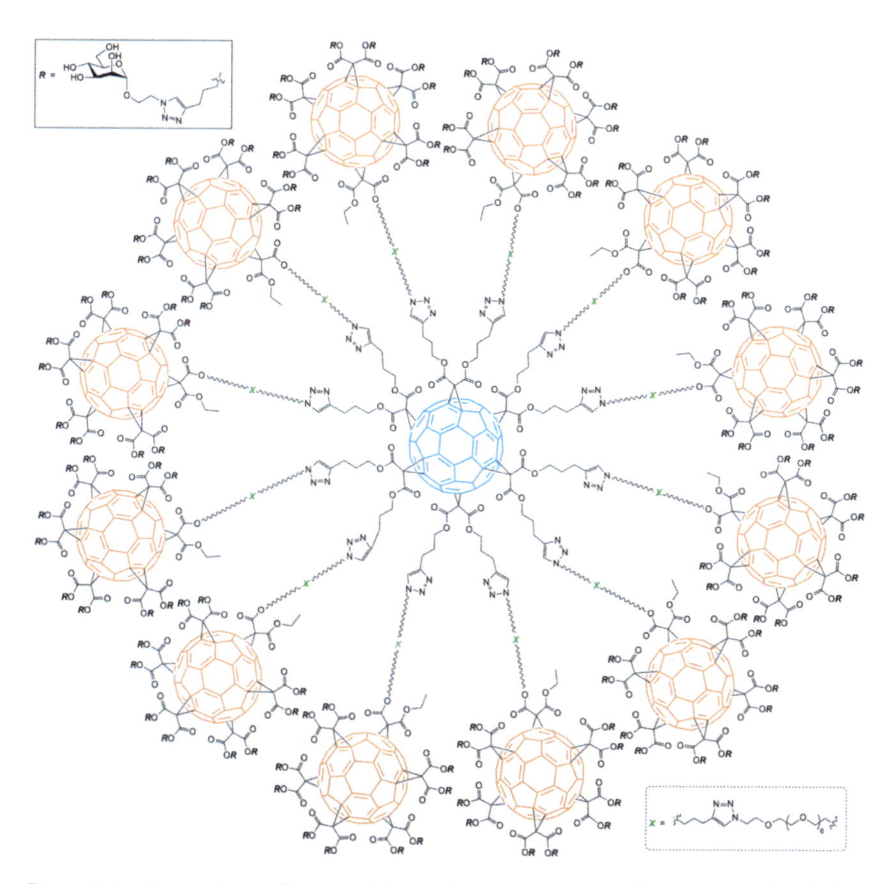

Figure 3.4 Glycofullerene "superballs" substituted with up to 120 mannose units [171].

3.6 SARS-CoV-2 Virus

COVID-19 has become a major pandemic and a global public threat since its first outbreak in China in December 2019. It is caused by a novel coronavirus called SARS-CoV-2 (WHO 2020) [172]. Coronaviruses are enveloped, single-stranded positive-sense RNA viruses that have a spike glycoprotein (S-protein) that plays a critical role in pathogenesis and in host immune response induction [173]. In addition to the spike glycoprotein, three other different proteins make up the SARS-CoV-2 structure: an envelope protein (E), membrane protein (M), and nucleocapsid protein (N) [174]. The SARS-CoV-2 spike has two functional subunits (i.e. S1 and S2) and contains 22 putative N-linked glycosylation sites and 4 potential O-linked glycosylation points [175, 176]. The function of the S1 subunit is associated with receptor binding, and the S2 subunit is responsible for membrane fusion [177]. As with many viral envelope glycoproteins, the carbohydrates around the spike form a SARS-CoV-2 glycan shield that helps evade the immune system. In addition to shielding, two *N*-glycans (N165 and N234) have been found to play a structural role in modulating the conformation of the RBD that binds to the angiotensin-converting enzyme (ACE2) host-cell receptor, priming the virus for infection [178]. Despite the structural similarity between SARS-CoV-2 and the first SARS-CoV virus, 5–6 key amino acid residues of the RBD implicated in ACE2 recognition are changed [179, 180], indicating distinct antigenic sites in the two viruses. The existence of different key epitopes points to the need to develop new therapeutic agents and vaccines that are specific for COVID-19 [181, 182]. In this context, the SARS-CoV-2 S protein serves as the main immunogen and a key target of neutralizing antibodies [183], representing an area of focus for vaccine development [184]. In particular, the spike glycan shield emerges, and its vulnerabilities [185] can be exploited for small-molecule drug development as well as for the design of potential carbohydrate-based vaccines to combat the COVID-19 pandemic [186, 187].

3.6.1 Prospective Vaccine Constructs Based on α-Gal Epitope

As mentioned in the end of Section 3.3.1.2 on influenza virus vaccines, endogenous human anti-Gal antibodies might also be exploited for increasing SARS-CoV-2 vaccine immunogenicity by glycoengineering α-Gal epitopes on inactivated SARS-CoV-2 or on S-protein subunit vaccines [188]. Presentation of these α-Gal epitopes would enable formation of anti-Gal/SARS-CoV-2$\alpha_{\text{-Gal}}$ or anti-Gal/S-protein$\alpha_{\text{-Gal}}$ immune complexes, resulting in higher APC targeting and uptake via specific receptor interactions and, as a result, enhanced immunogenicity and vaccine efficacy. This glycoengineering strategy would convert the native, protein-masking glycan shield into an α-Gal-modified carbohydrate coating that effectively directs the engineered vaccine to APCs via anti-Gal antibodies. Despite its promise, this α-Gal vaccine approach needs to be further demonstrated in a COVID-19 setting to fully assess its potential to boost anti-SARS-CoV-2 immune responses, both in terms of neutralizing antibody titers as well as T-cell immunity.

3.6.2 RBD-Based Constructs for Vaccine Development

The RBD glycoprotein within the S-protein trimer mediates viral infection by binding to ACE2 and represents a key target for the design of vaccines able to elicit neutralizing antibodies against SARS-CoV-2 [189]. In a notable example, a recombinant monomeric RBD vaccine adjuvanted with alum was shown to elicit a potent and protective antibody response in laboratory animals after *in vivo* SARS-CoV-2 challenge [190]. With a view to increasing immunogenicity and favoring presentation of the key RBD motif surface in search of enhanced neutralizing antibody induction, Vérez-Bencomo and coworkers chemically linked the recombinant RBD to the highly immunogenic protein carrier TT in a site-selective manner [191]. The resulting multivalent RBD–TT conjugates induced a robust IgG-neutralizing antibody and cellular response in mice when coadministered with alum, which prompted advancement of this conjugate vaccine candidate to clinical trials [192].

In an effort to guide the development of semisynthetic carbohydrate-based RBD vaccines, Wang and coworkers recently investigated the role of precise glycan structures on this domain (at T323, N331, and N343) by preparing homogeneous RBD glycoproteins using chemical synthesis and recombinant protein engineering [193]. Leveraging their homogenously glycosylated RBDs, they found no differences between various synthetic glycoforms in terms of ACE2 binding and revealed the influence of discrete RBD glycosylation on anti-SARS-CoV-2 RBD monoclonal antibody binding. In addition to deciphering carbohydrate structure–activity relationships, these studies have provided chemically defined glycosylated constructs as potential epitope mimics for further immunological evaluation *in vivo*, opening the door to the prospective rational development of synthetic glycan-based immunogens for future anti-SARS-CoV-2 vaccines and therapeutics.

3.6.3 Saponins as Carbohydrate-Based Adjuvant Candidates for COVID-19 Vaccines

The identification of novel vaccine adjuvants that can potentiate humoral, cellular, and memory immune responses to prevent COVID-19 infection is emerging as a frontline strategy in vaccine development against COVID as well as other infectious diseases [194, 195]. Triterpenoid and steroidal saponin natural products show antiviral activity against different viral groups. Aqueous extracts from the Chilean soapbark tree *Quillaja saponaria* Molina contain many physiologically active triterpenoid saponins, some of which show high adjuvant activity, potentiating the immune response against the coadministered antigen [196]. Purified saponin fractions (specifically QS-21), either alone or integrated as part of immunostimulating complexes (ISCOMs) and adjuvant systems, have been demonstrated to be powerful adjuvants inducing both antibody and cellular immunity in several vaccine clinical trials [197, 198]. AS01 is a potent adjuvant combination recently approved in herpes and malaria vaccines that includes two immunostimulants, 3-*O*-desacyl-4′-monophosphoryl lipid A (MPLA) and the naturally derived purified saponin fraction (QS-21), in a liposomal formulation [199]. Both immunopotentiating substances in this system appear to be critical for the stimulation and activation of antigen-specific cellular and humoral immune responses. The proprietary, saponin-based Matrix-M™ adjuvant has shown potent and well-tolerated immunostimulatory effects by inducing the influx of APCs into the site of injection and enhancing antigen presentation in the lymph nodes [200, 201]. Notably, Matrix-M has been licensed as part of Novavax recombinant nanoparticle subunit vaccine (NVAX-CoV2373) derived from the S-protein, acting by boosting B-cell and T-cell immunity for elicitation of more potent immune responses while also enabling dose-sparing [202]. NVX-CoV2373 has been successfully evaluated in clinical trials, demonstrating high vaccine efficacy against several SARS-CoV-2 variants with an acceptable safety profile [203, 204], which has led to its recent approval by the European Medicines Agency. This saponin-adjuvanted protein-based vaccine highlights the promise of saponin adjuvants and subunit vaccines to prevent not only traditional but also emerging viral diseases, including COVID-19 and potential new pandemics associated with further coronavirus infections.

3.7 Conclusions and Outlook

Viral glycosylation is a process mediated by the host-cell machinery that decorates the surface proteins of several pathogens with cell glycans, including the envelope glycoprotein (Env) of HIV-1, hemagglutinin glycoprotein (HA) of influenza virus, the envelope glycoproteins (E1 and E2) of HCV, the glycoprotein (GP) of EBOV, the coronavirus glycoprotein spike (S), and others. This protein glycosylation is characterized by high structural variation and is essential for the viral lifecycle and pathogenesis, playing a key role in specific viral functions such as host-cell attachment and entry, infectivity, and replication. Moreover, these viral carbohydrates form a glycan shield that has important implications for host immune responses to infection, protecting internal protein epitopes from immune recognition while also exposing potential targets for vaccine and drug development. For instance, the conserved oligomannose-type *N*-glycans of some envelope glycoproteins involved in trans-infection and immune evasion also serve as neutralizing antibody epitopes or binding receptors for CBAs to fight against viral infection.

Despite significant recent research on the development of carbohydrate vaccines and therapies against viral diseases, much further work is needed for the translation of additional glycan-based candidates into clinical applications in humans. Increased knowledge of viral glycobiology and site-specific protein glycosylation, as well as a better understanding of the immunological basis that drives viral diversity, will aid in guiding the rational design of next-generation vaccines and antiviral therapeutics in the future.

Acknowledgments

Funding from the European Research Council (ERC-2016-STG-716878 "ADJUVANT VACCINES") and the Spanish Ministry of Science and Innovation/State

Research Agency (MCIN/AEI) (CTQ2017-87530-R, RYC-2015-17888 to A.F.-T.; PRE2018-085772 to A.P.) is gratefully acknowledged. We thank Dr. Iñaki Bastida for assistance with the preparation of the manuscript. A F.-T. thanks Raquel Fernández for inspiration.

References

1 Liang, F., Guan, P., Wu, W. et al. (2018). A review of documents prepared by international organizations about influenza pandemics, including the 2009 pandemic: a bibliometric analysis. *BMC Infectious Diseases* 18: 383.
2 Eisinger, R.W. and Fauci, A.S. (2018). Ending the HIV/AIDS pandemic. *Emerging Infectious Diseases* 24 (3): 413–416.
3 Lai, C.C., Shih, T.P., Ko, W.C. et al. (2020). Severe acute respiratory syndrome coronavirus 2 (SARS-CoV-2) and coronavirus disease-2019 (COVID-19): the epidemic and the challenges. *International Journal of Antimicrobial Agents* 55 (3): 105924.
4 Ringelhan, M., McKeating, J.A., and Protzer, U. (2017). Viral hepatitis and liver cancer. *Philosophical Transactions of the Royal Society of London. Series B, Biological Sciences* 372 (1732): 20160274.
5 Crosbie, E.J., Einstein, M.H., Franceschi, S., and Kitchener, H.C. (2013). Human papillomavirus and cervical cancer. *Lancet* 382 (9895): 889–899.
6 Pollard, A.J. and Bijker, E.M. (2021). A guide to vaccinology: from basic principles to new developments. *Nature Reviews. Immunology* 21: 83–100.
7 Watanabe, Y., Bowden, T.A., Wilson, I.A., and Crispin, M. (2019). Exploitation of glycosylation in enveloped virus pathobiology. *Biochimica et Biophysica Acta, General Subjects* 1863 (10): 1480–1497.
8 Helenius, A. and Aebi, M. (2004). Roles of N-linked glycans in the endoplasmic reticulum. *Annual Review of Biochemistry* 73 (1): 1019–1049.
9 Vigerust, D.J. and Shepherd, V.L. (2007). Virus glycosylation: role in virulence and immune interactions. *Trends in Microbiology* 15: 211–218.
10 Rerks-Ngarm, S., Pitisuttithum, P., Nitayaphan, S. et al. (2009). Vaccination with ALVAC and AIDSVAX to prevent HIV-1 infection in Thailand. *The New England Journal of Medicine* 361 (23): 2209–2220.
11 Corey, L., Gilbert, P.B., Tomaras, G.D. et al. (2015). Immune correlates of vaccine protection against HIV-1 acquisition. *Science Translational Medicine* 7 (310): 310rv7.
12 Stephenson, K.E., Wagh, K., Korber, B., and Barouch, D.H. (2020). Vaccines and broadly neutralizing antibodies for HIV-1 prevention. *Annual Review of Immunology* 38: 673–703.
13 Behrens, A.-J., Seabright, G.E., and Crispin, M. (2017). Targeting glycans of HIV envelope glycoproteins for vaccine design. In: *Chemical Biology of Glycoproteins* (ed. Z. Tan and L.-X. Wang), 300–357. Cambridge: Royal Society of Chemistry.
14 Sok, D. and Burton, D.R. (2018). Recent progress in broadly neutralizing antibodies to HIV. *Nature Immunology* 19: 1179–1188.

15 Haynes, B.F., Burton, D.R., and Mascola, J.R. (2019). Multiple roles for HIV broadly neutralizing antibodies. *Science Translational Medicine* 11 (516): eaaz2686.

16 Haynes, B.F. and Verkoczy, L. (2014). Host controls of HIV neutralizing antibodies. *Science* 344 (6184): 588–589.

17 Walsh, S.R. and Seaman, M.S. (2021). Broadly neutralizing antibodies for HIV-1 prevention. *Frontiers in Immunology* 12: 712122.

18 Seabright, G.E., Doores, K.J., Burton, D.R., and Crispin, M. (2019). Protein and glycan mimicry in HIV vaccine design. *Journal of Molecular Biology* 431 (12): 2223–2247.

19 Bastida, I. and Fernández-Tejada, A. (2020). Synthetic carbohydrate-based HIV-1 vaccines. *Drug Discovery Today: Technologies* 35-36: 45–56.

20 Fernández-Tejada, A., Haynes, B.F., and Danishefsky, S.J. (2015). Designing synthetic vaccines for HIV. *Expert Review of Vaccines* 14 (6): 815–831.

21 Trkola, A., Purtscher, M., Muster, T. et al. (1996). Human monoclonal antibody 2G12 defines a distinctive neutralization epitope on the gp120 glycoprotein of human immunodeficiency virus type 1. *Journal of Virology* 70 (2): 1100–1108.

22 Mascola, J.R., Stiegler, G., Vancott, T.C. et al. (2000). Protection of macaques against vaginal transmission of a pathogenic HIV-1/SIV chimeric virus by passive infusion of neutralizing antibodies. *Nature Medicine* 6 (2): 207–210.

23 Hessell, A.J., Rakasz, E.G., Poignard, P. et al. (2009). Broadly neutralizing human anti-HIV antibody 2G12 Is effective in protection against mucosal SHIV challenge even at low serum neutralizing titers. *PLoS Pathogens* 5 (5): e1000433.

24 Sanders, R.W., Venturi, M., Schiffner, L. et al. (2002). The mannose-dependent epitope for neutralizing antibody 2G12 on human immunodeficiency virus type 1 glycoprotein gp120. *Journal of Virology* 76 (14): 7293–7305.

25 Scanlan, C.N., Pantophlet, R., Wormald, M.R. et al. (2002). The broadly neutralizing anti-human immunodeficiency virus type 1 antibody 2G12 recognizes a cluster of α1→2 mannose residues on the outer face of gp120. *Journal of Virology* 76 (14): 7306–7321.

26 Calarese, D.A., Scanlan, C.N., Zwick, M.B. et al. (2003). Antibody domain exchange is an immunological solution to carbohydrate cluster recognition. *Science* 300 (5628): 2065–2071.

27 Wang, L.-X., Ni, J., Singh, S., and Li, H. (2004). Binding of high-mannose-type oligosaccharides and synthetic oligomannose clusters to human antibody 2G12. *Chemistry & Biology* 11 (1): 127–134.

28 Lee, H.-K., Scanlan, C.N., Huang, C.-Y. et al. (2004). Reactivity-based one-pot synthesis of oligomannoses: defining antigens recognized by 2G12, a broadly neutralizing anti-HIV-1 antibody. *Angewandte Chemie, International Edition* 43 (8): 1000–1003.

29 Calarese, D.A., Lee, H.-K., Huang, C.-Y. et al. (2005). Dissection of the carbohydrate specificity of the broadly neutralizing anti-HIV-1 antibody 2G12. *Proceedings of the National Academy of Sciences of the United States of America* 102 (38): 13372–13377.

30 Adams, E.W., Ratner, D.M., Bokesch, H.R. et al. (2004). Oligosaccharide and glycoprotein microarrays as tools in HIV glycobiology. *Chemistry & Biology* 11 (6): 875–881.

31 Toonstra, C., Wu, L., Li, C. et al. (2018). Top-down chemoenzymatic approach to synthesizing diverse high-mannose *N*-glycans and related neoglycoproteins for carbohydrate microarray analysis. *Bioconjugate Chemistry* 29 (6): 1911–1921.

32 Li, H. and Wang, L.-X. (2004). Design and synthesis of a template-assembled oligomannose cluster as an epitope mimic for human HIV-neutralizing antibody 2G12. *Organic & Biomolecular Chemistry* 2 (4): 483–488.

33 Ni, J., Song, H., Wang, Y. et al. (2006). Toward a carbohydrate-based HIV-1 vaccine: synthesis and immunological studies of oligomannose-containing glycoconjugates. *Bioconjugate Chemistry* 17 (2): 493–500.

34 Mandal, M., Dudkin, V.Y., Geng, X., and Danishefsky, S.J. (2004). In pursuit of carbohydrate-based HIV vaccines, part 1: the total synthesis of hybrid-type gp120 fragments. *Angewandte Chemie, International Edition* 43 (19): 2557–2561.

35 Geng, X., Dudkin, V.Y., Mandal, M., and Danishefsky, S.J. (2004). In pursuit of carbohydrate-based HIV vaccines, part 2: the total synthesis of high-mannose-type gp120 fragments—evaluation of strategies directed to maximal convergence. *Angewandte Chemie, International Edition* 43 (19): 2562–2565.

36 Dudkin, V.Y., Orlova, M., Geng, X. et al. (2004). Toward fully synthetic carbohydrate-based HIV antigen design: on the critical role of bivalency. *Journal of the American Chemical Society* 126 (31): 9560–9562.

37 Wang, J., Li, H., Zou, G., and Wang, L.-X. (2007). Novel template-assembled oligosaccharide clusters as epitope mimics for HIV-neutralizing antibody 2G12. Design, synthesis, and antibody binding study. *Organic & Biomolecular Chemistry* 5 (10): 1529–1540.

38 Krauss, I.J., Joyce, J.G., Finnefrock, A.C. et al. (2007). Fully synthetic carbohydrate HIV antigens designed on the logic of the 2G12 antibody. *Journal of the American Chemical Society* 129 (36): 11042–11044.

39 Joyce, J.G., Krauss, I.J., Song, H.C. et al. (2008). An oligosaccharide-based HIV-1 2G12 mimotope vaccine induces carbohydrate-specific antibodies that fail to neutralize HIV-1 virions. *Proceedings of the National Academy of Sciences* 105 (41): 15684–15689.

40 Wang, S.-K., Liang, P.-H., Astronomo, R.D. et al. (2008). Targeting the carbohydrates on HIV-1: interaction of oligomannose dendrons with human monoclonal antibody 2G12 and DC-SIGN. *Proceedings of the National Academy of Sciences* 105 (10): 3690–3695.

41 Kabanova, A., Adamo, R., Proietti, D. et al. (2010). Preparation, characterization and immunogenicity of HIV-1 related high-mannose oligosaccharides-CRM197 glycoconjugates. *Glycoconjugate Journal* 27 (5): 501–513.

42 Astronomo, R.D., Lee, H.-K., Scanlan, C.N. et al. (2008). A glycoconjugate antigen based on the recognition motif of a broadly neutralizing human immunodeficiency virus antibody, 2G12, is immunogenic but elicits antibodies unable to bind to the self glycans of gp120. *Journal of Virology* 82 (13): 6359–6368.

43 Trattnig, N., Mayrhofer, P., Kunert, R. et al. (2019). Comparative antigenicity of thiourea and adipic amide linked neoglycoconjugates containing modified oligomannose epitopes for the carbohydrate-specific anti-HIV antibody 2G12. *Bioconjugate Chemistry* 30 (1): 70–82.

44 Astronomo, R.D., Kaltgrad, E., Udit, A.K. et al. (2010). Defining criteria for oligomannose immunogens for HIV using icosahedral virus capsid scaffolds. *Chemistry & Biology* 17 (4): 357–370.

45 Doores, K.J., Fulton, Z., Hong, V. et al. (2010). A nonself sugar mimic of the HIV glycan shield shows enhanced antigenicity. *Proceedings of the National Academy of Sciences* 107 (40): 17107–17112.

46 MacPherson, I.S., Temme, J.S., Habeshian, S. et al. (2011). Multivalent glycocluster design through directed evolution. *Angewandte Chemie, International Edition* 50 (47): 11238–11242.

47 Temme, J.S., MacPherson, I.S., DeCourcey, J.F., and Krauss, I.J. (2014). High temperature SELMA: evolution of DNA-supported oligomannose clusters which are tightly recognized by HIV bnAb 2G12. *Journal of the American Chemical Society* 136 (5): 1726–1729.

48 Horiya, S., Bailey, J.K., Temme, J.S. et al. (2014). Directed evolution of multivalent glycopeptides tightly recognized by HIV antibody 2G12. *Journal of the American Chemical Society* 136 (14): 5407–5415.

49 Bailey, J.K., Nguyen, D.N., Horiya, S., and Krauss, I.J. (2016). Synthesis of multivalent glycopeptide conjugates that mimic an HIV epitope. *Tetrahedron* 72 (40): 6091–6098.

50 Nguyen, D.N., Xu, B., Stanfield, R.L. et al. (2019). Oligomannose glycopeptide conjugates elicit antibodies targeting the glycan core rather than its extremities. *ACS Central Science* 5 (2): 237–249.

51 Nguyen, D.N., Redman, R.L., Horiya, S. et al. (2020). The impact of sustained immunization regimens on the antibody response to oligomannose glycans. *ACS Chemical Biology* 15 (3): 789–798.

52 Redman, R.L. and Krauss, I.J. (2021). Directed evolution of 2′-fluoro-modified, RNA-supported carbohydrate clusters that bind tightly to HIV antibody 2G12. *Journal of the American Chemical Society* 143: 35.

53 Clark, B.E., Auyeung, K., Fregolino, E. et al. (2012). A bacterial lipooligosaccharide that naturally mimics the epitope of the HIV-neutralizing antibody 2G12 as a template for vaccine design. *Chemistry & Biology* 19 (2): 254–263.

54 Stanfield, R.L., De Castro, C., Marzaioli, A.M. et al. (2015). Crystal structure of the HIV neutralizing antibody 2G12 in complex with a bacterial oligosaccharide analog of mammalian oligomannose. *Glycobiology* 25 (4): 412–419.

55 Temme, J.S., Drzyzga, M.G., MacPherson, I.S., and Krauss, I.J. (2013). Directed evolution of 2G12-targeted nonamannose glycoclusters by SELMA. *Chemistry - A European Journal* 19 (51): 17291–17295.

56 Agrawal-Gamse, C., Luallen, R.J., Liu, B. et al. (2011). Yeast-elicited cross-reactive antibodies to HIV Env glycans efficiently neutralize virions expressing exclusively high-mannose N-linked glycans. *Journal of Virology* 85 (1): 470–480.

57 Zhang, H., Fu, H., Luallen, R.J. et al. (2015). Antibodies elicited by yeast glycoproteins recognize HIV-1 virions and potently neutralize virions with high mannose *N*-glycans. *Vaccine* 33 (39): 5140–5147.

58 Walker, L.M., Phogat, S.K., Chan-Hui, P.-Y. et al. (2009). Broad and potent neutralizing antibodies from an African donor reveal a new HIV-1 vaccine target. *Science* 326 (5950): 285–289.

59 Bonsignori, M., Hwang, K.-K., Chen, X. et al. (2011). Analysis of a clonal lineage of HIV-1 envelope V2/V3 conformational epitope-specific broadly neutralizing antibodies and their inferred unmutated common ancestors. *Journal of Virology* 85 (19): 9998–10009.

60 Walker, L.M., Huber, M., Doores, K.J. et al. (2011). Broad neutralization coverage of HIV by multiple highly potent antibodies. *Nature* 477 (7365): 466–470.

61 Doria-Rose, N.A., Schramm, C.A., Gorman, J. et al. (2014). Developmental pathway for potent V1V2-directed HIV-neutralizing antibodies. *Nature* 509 (7498): 55–62.

62 Doria-Rose, N.A., Bhiman, J.N., Roark, R.S. et al. (2016). New member of the V1V2-directed CAP256-VRC26 lineage that shows increased breadth and exceptional potency. *Journal of Virology* 90 (1): 76–91.

63 Sok, D., van Gils, M.J., Pauthner, M. et al. (2014). Recombinant HIV envelope trimer selects for quaternary-dependent antibodies targeting the trimer apex. *Proceedings of the National Academy of Sciences* 111 (49): 17624–17629.

64 Doores, K.J. and Burton, D.R. (2010). Variable loop glycan dependency of the broad and potent HIV-1-neutralizing antibodies PG9 and PG16. *Journal of Virology* 84 (20): 10510–10521.

65 McLellan, J.S., Pancera, M., Carrico, C. et al. (2011). Structure of HIV-1 gp120 V1/V2 domain with broadly neutralizing antibody PG9. *Nature* 480 (7377): 336–343.

66 Amin, M.N., McLellan, J.S., Huang, W. et al. (2013). Synthetic glycopeptides reveal the glycan specificity of HIV-neutralizing antibodies. *Nature Chemical Biology* 9 (8): 521–526.

67 Toonstra, C., Amin, M.N., and Wang, L.-X. (2016). Site-selective chemoenzymatic glycosylation of an HIV-1 polypeptide antigen with two distinct *N*-glycans via an orthogonal protecting group strategy. *The Journal of Organic Chemistry* 81 (15): 6176–6185.

68 Pancera, M., Shahzad-ul-Hussan, S., Doria-Rose, N.A. et al. (2013). Structural basis for diverse *N*-glycan recognition by HIV-1–neutralizing V1–V2–directed antibody PG16. *Nature Structural & Molecular Biology* 20 (7): 804–813.

69 Shivatare, S.S., Chang, S.-H., Tsai, T.-I. et al. (2016). Modular synthesis of *N*-glycans and arrays for the hetero-ligand binding analysis of HIV antibodies. *Nature Chemistry* 8 (4): 338–346.

70 Shivatare, V.S., Shivatare, S.S., Lee, C.-C.D. et al. (2018). Unprecedented role of hybrid N- glycans as ligands for HIV-1 broadly neutralizing antibodies. *Journal of the American Chemical Society* 140 (15): 5202–5210.

71 Aussedat, B., Vohra, Y., Park, P.K. et al. (2013). Chemical synthesis of highly congested gp120 V1V2 *N*-glycopeptide antigens for potential HIV-1-directed vaccines. *Journal of the American Chemical Society* 135 (35): 13113–13120.

72 Alam, S.M., Dennison, S.M., Aussedat, B. et al. (2013). Recognition of synthetic glycopeptides by HIV-1 broadly neutralizing antibodies and their unmutated ancestors. *Proceedings of the National Academy of Sciences* 110 (45): 18214–18219.

73 Pejchal, R., Doores, K.J., Walker, L.M. et al. (2011). A potent and broad neutralizing antibody recognizes and penetrates the HIV glycan shield. *Science* 334 (6059): 1097–1103.

74 Mouquet, H., Scharf, L., Euler, Z. et al. (2012). Complex-type *N*-glycan recognition by potent broadly neutralizing HIV antibodies. *Proceedings of the National Academy of Sciences* 109 (47): E3268–E3277.

75 Pantophlet, R., Trattnig, N., Murrell, S. et al. (2017). Bacterially derived synthetic mimetics of mammalian oligomannose prime antibody responses that neutralize HIV infectivity. *Nature Communications* 8 (1): 1601.

76 Cai, H., Orwenyo, J., Guenaga, J. et al. (2017). Synthetic multivalent V3 glycopeptides display enhanced recognition by glycan-dependent HIV-1 broadly neutralizing antibodies. *Chemical Communications* 53 (39): 5453–5456.

77 Orwenyo, J., Cai, H., Giddens, J. et al. (2017). Systematic synthesis and binding study of HIV V3 glycopeptides reveal the fine epitopes of several broadly neutralizing antibodies. *ACS Chemical Biology* 12 (6): 1566–1575.

78 Cai, H., Orwenyo, J., Giddens, J.P. et al. (2017). Synthetic three-component HIV-1 V3 glycopeptide immunogens induce glycan-dependent antibody responses. *Cell Chemical Biology* 24 (12): 1513–1522.

79 Cai, H., Zhang, R., Orwenyo, J. et al. (2018). Multivalent antigen presentation enhances the immunogenicity of a synthetic three-component HIV-1 V3 glycopeptide vaccine. *ACS Central Science* 4 (5): 582–589.

80 Cai, H., Zhang, R.-S., Orwenyo, J. et al. (2018). Synthetic HIV V3 glycopeptide immunogen carrying a N334 *N*-glycan induces glycan-dependent antibodies with promiscuous site recognition. *Journal of Medicinal Chemistry* 61 (22): 10116–10125.

81 Alam, S.M., Aussedat, B., Vohra, Y. et al. (2017). Mimicry of an HIV broadly neutralizing antibody epitope with a synthetic glycopeptide. *Science Translational Medicine* 9 (381): eaai7521.

82 Francica, J.R., Laga, R., Lynn, G.M. et al. (2019). Star nanoparticles delivering HIV-1 peptide minimal immunogens elicit near-native envelope antibody responses in nonhuman primates. *PLoS Biology* 17 (6): e3000328.

83 Haynes, B.F., Kelsoe, G., Harrison, S.C., and Kepler, T.B. (2012). B-cell–lineage immunogen design in vaccine development with HIV-1 as a case study. *Nature Biotechnology* 30 (5): 423–433.

84 Alexander, D.J. (2000). A review of avian influenza in different bird species. *Veterinary Microbiology* 74: 3–13.

85 Zhuang, Q., Wang, S., Liu, S. et al. (2019). Diversity and distribution of type A influenza viruses: an updated panorama analysis based on protein sequences. *Virology Journal* 16 (1): 85.

86 Wagner, R., Matrosovich, M., and Klenk, H. (2002). Functional balance between haemagglutinin and neuraminidase in influenza virus infections. *Reviews in Medical Virology* 12: 159–166.

87 Klenk, H.-D., Wagner, R., Heuer, D., and Wolff, T. (2002). Importance of hemagglutinin glycosylation for the biological functions of influenza virus. *Virus Research* 82: 73–75.

88 De Jong, N.M.C., Aartse, A., Van Gils, M.J., and Eggink, D. (2020). Development of broadly reactive influenza vaccines by targeting the conserved regions of the hemagglutinin stem and head domains. *Expert Review of Vaccines* 19 (6): 563–577.

89 Wilson, I.A., Shekel, J.J., and Wiley, D.C. (1981). Structure of the haemagglutinin membrane glycoprotein of influenza virus at 3A° resolution. *Nature* 289: 366–373.

90 Steinhauer, D.A. (1999). Role of hemagglutinin cleavage for the pathogenicity of influenza virus. *Virology* 258 (1): 1–20.

91 Worch, R. (2014). Structural biology of the influenza virus fusion peptide. *Acta Biochimica Polonica* 61 (3): 421–426.

92 Bouvier, N.M. and Palese, P. (2008). The biology of influenza viruses. *Vaccine* 26: D49–D53.

93 Coudeville, L., Bailleux, F., Riche, B. et al. (2010). Relationship between haemagglutination-inhibiting antibody titres and clinical protection against influenza: development and application of a Bayesian random-effects model. *BMC Medical Research Methodology* 10: 18.

94 Bullard, B.L. and Weaver, E.A. (2021). Strategies targeting hemagglutinin as a universal influenza vaccine. *Vaccine* 9 (3): 257.

95 Eggink, D., Goff, P.H., and Palese, P. (2014). Guiding the immune response against influenza virus hemagglutinin toward the conserved stalk domain by hyperglycosylation of the globular head domain. *Journal of Virology* 88 (1): 699–704.

96 Lin, S.C., Liu, W.C., Jan, J.T., and Wu, S.C. (2014). Glycan masking of hemagglutinin for adenovirus vector and recombinant protein immunizations elicits broadly neutralizing antibodies against H5N1 avian influenza viruses. *PLoS One* 9 (3): e92822.

97 Lin, S.C., Lin, Y.F., Chong, P., and Wu, S.C. (2012). Broader neutralizing antibodies against H5N1 viruses using prime-boost immunization of hyperglycosylated hemagglutinin DNA and virus-like particles. *PLoS One* 7 (6): 1–8.

98 Nguyen, Q.T. and Choi, Y.K. (2021). Targeting antigens for universal influenza vaccine development. *Viruses* 13 (6): 973.

99 Ulmer, J.B., DeWitt, C.M., Chastain, M. et al. (1999). Enhancement of DNA vaccine potency using conventional aluminum adjuvants. *Vaccine* 18: 18–28.

100 Tregoning, J.S., Russell, R.F., and Kinnear, E. (2018). Adjuvanted influenza vaccines. *Human Vaccines & Immunotherapeutics* 14: 550–564.

101 Hung, J.T., Tsai, Y.C., Der Lin, W. et al. (2014). Potent adjuvant effects of novel NKT stimulatory glycolipids on hemagglutinin based DNA vaccine for H5N1 influenza virus. *Antiviral Research* 107 (1): 110–118.

102 Natori, T., Koezuka, Y., and Higa, T. (1993). Agelasphins, novel α-galactosylceramides from the marine sponge *Agelas mauritianus*. *Tetrahedron Letters* 34 (35): 5591–5592.

103 Morita, M., Motoki, K., Akimoto, K. et al. (1995). Structure-activity relationship of α-galactosylceramides against B16-bearing mice. *Journal of Medicinal Chemistry* 38 (12): 2176–2187.

104 Guillonneau, C., Mintern, J.D., Hubert, F.X. et al. (2009). Combined NKT cell activation and influenza virus vaccination boosts memory CTL generation and protective immunity. *Proceedings of the National Academy of Sciences of the United States of America* 106: 3330–3335.

105 Artiaga, B.L., Yang, G., Hackmann, T.J. et al. (2016). α-galactosylceramide protects swine against influenza infection when administered as a vaccine adjuvant. *Scientific Reports* 6: 23593.

106 Dölen, Y., Kreutz, M., Gileadi, U. et al. (2016). Co-delivery of PLGA encapsulated invariant NKT cell agonist with antigenic protein induce strong T cell-mediated antitumor immune responses. *Oncoimmunology* 5 (1): e1068493.

107 Lang, S. and Huang, X. (2020). Carbohydrate conjugates in vaccine developments. *Frontiers in Chemistry* 8 (April): 1–25.

108 Anderson, R.J., Li, J., Kedzierski, L. et al. (2017). Augmenting influenza-specific T cell memory generation with a natural killer T cell-dependent glycolipid-peptide vaccine. *ACS Chemical Biology* 12 (11): 2898–2905.

109 Wilkinson, T.M., Li, C.K., Chui, C.S. et al. (2012). Preexisting influenza-specific $CD4^+$ T cells correlate with disease protection against influenza challenge in humans. *Nature Medicine* 18: 274–280.

110 Abdel-Motal, U.M., Guay, H.M., Wigglesworth, K. et al. (2007). Immunogenicity of influenza virus vaccine is increased by anti-gal-mediated targeting to antigen-presenting cells. *Journal of Virology* 81 (17): 9131–9141.

111 Henion, T.R., Gerhard, W., Anaraki, F., and Galili, U. (1997). Synthesis of α-gal epitopes on influenza virus vaccines, by recombinant α1,3galactosyltransferase, enables the formation of immune complexes with the natural anti-Gal antibody. *Vaccine* 15 (11): 1174–1182.

112 von Itzstein, M. (2007). The war against influenza: discovery and development of sialidase inhibitors. *Nature Reviews. Drug Discovery* 6: 967–974.

113 Wohlbold, T.J. and Krammer, F. (2014). In the shadow of hemagglutinin: a growing interest in influenza viral neuraminidase and its role as a vaccine antigen. *Viruses* 6 (6): 2465–2494.

114 Chen, Y.Q., Wohlbold, T.J., Zheng, N.Y. et al. (2018). Influenza infection in humans induces broadly cross-reactive and protective neuraminidase-reactive antibodies. *Cell* 173 (2): 417–429.e10.

115 Gilchuk, I.M., Bangaru, S., Gilchuk, P. et al. (2019). Influenza H7N9 virus neuraminidase-specific human monoclonal antibodies inhibit viral egress and protect from lethal influenza infection in mice. *Cell Host & Microbe* 26 (6): 715–728.

116 Trombetta, C.M. and Montomoli, E. (2016). Influenza immunology evaluation and correlates of protection: a focus on vaccines. *Expert Review of Vaccines* 15 (8): 967–976.

117 Krammer, F., Fouchier, R.A.M., Eichelberger, M.C. et al. (2018). NAction! how can neuraminidase-based immunity contribute to better influenza virus vaccines? *mBio* 9 (2): e02332–e02317.

118 Wan, H., Yang, H., Shore, D.A. et al. (2015). Structural characterization of a protective epitope spanning A(H1N1)pdm09 influenza virus neuraminidase monomers. *Nature Communications* 6: 6114.

119 Zhu, X., Turner, H.L., Lang, S. et al. (2019). Structural basis of protection against H7N9 influenza virus by human anti-N9 neuraminidase antibodies. *Cell Host & Microbe* 26 (6): 729–738.e4.

120 Bachelder, E.M., Beaudette, T.T., Broaders, K.E. et al. (2008). Acetal-derivatized dextran: an acid-responsive biodegradable material for therapeutic applications. *Journal of the American Chemical Society* 130 (32): 10494–10495.

121 Bachelder, E.M., Beaudette, T.T., Broaders, K.E. et al. (2010). In vitro analysis of acetalated dextran microparticles as a potent delivery platform for vaccine adjuvants. *Molecular Pharmaceutics* 7 (3): 826–835.

122 Junkins, R.D., Gallovic, M.D., Johnson, B.M. et al. (2018). A robust microparticle platform for a STING-targeted adjuvant that enhances both humoral and cellular immunity during vaccination. *Journal of Controlled Release* 270: 1–13.

123 Chen, N., Johnson, M.M., Collier, M.A. et al. (2018). Tunable degradation of acetalated dextran microparticles enables controlled vaccine adjuvant and antigen delivery to modulate adaptive immune responses. *Journal of Controlled Release* 273: 147–159.

124 Matrosovich, M. and Klenk, H.D. (2003). Natural and synthetic sialic acid-containing inhibitors of influenza virus receptor binding. *Reviews in Medical Virology* 13 (2): 85–97.

125 Bhatia, S., Dimde, M., and Haag, R. (2014). Multivalent glycoconjugates as vaccines and potential drug candidates. *MedChemComm* 5 (7): 862–878.

126 Lees, W.J., Spaltenstein, A., Kingery-wood, J.E., and Whitesides, G.M. (1994). Polyacrylamides bearing pendant a-sialoside groups strongly inhibit agglutination of erythrocytes by influenza A virus: multivalency and steric stabilization of particulate biological system. *Journal of Medicinal Chemistry* 37 (20): 3419–3433.

127 Sigal, G.B., Mammen, M., Dahmann, G., and Whitesides, G.M. (1996). Polyacrylamides bearing pendant α-sialoside groups strongly inhibit agglutination of erythrocytes by influenza virus: the strong inhibition reflects enhanced binding through cooperative polyvalent interactions. *Journal of the American Chemical Society* 118: 3789–3800.

128 Mammen, M., Dahmann, G., and Whitesides, G.M. (1995). Effective inhibitors of hemagglutination by influenza virus synthesized from polymers having active ester groups. Insight into mechanism of inhibition. *Journal of Medicinal Chemistry* 38 (21): 4179–4190.

129 Gambaryan, A.S., Tuzikov, A.B., Chinarev, A.A. et al. (2002). Polymeric inhibitor of influenza virus attachment protects mice from experimental influenza infection. *Antiviral Research* 55 (1): 201–205.

130 Kwon, S.J., Na, D.H., Kwak, J.H. et al. (2017). Nanostructured glycan architecture is important in the inhibition of influenza A virus infection. *Nature Nanotechnology* 12 (1): 48–54.

131 Papp, I., Sieben, C., Ludwig, K. et al. (2010). Inhibition of influenza virus infection by multivalent sialic-acid- functionalized gold nanoparticles. *Small* 6 (24): 2900–2906.

132 Vonnemann, J., Sieben, C., Wolff, C. et al. (2014). Virus inhibition induced by polyvalent nanoparticles of different sizes. *Nanoscale* 6 (4): 2353–2360.

133 Papp, I., Sieben, C., Sisson, A.L. et al. (2011). Inhibition of influenza virus activity by multivalent glycoarchitectures with matched sizes. *ChemBioChem* 12 (6): 887–895.

134 Lee, C.M., Weight, A.K., Haldar, J. et al. (2012). Polymer-attached zanamivir inhibits synergistically both early and late stages of influenza virus infection. *Proceedings of the National Academy of Sciences of the United States of America* 109 (50): 20385–20390.

135 Lauer, G.M. and Walker, B.D. (2001). Hepatitis C virus infection. *The New England Journal of Medicine* 345 (1): 41–52.

136 The Polaris Observatory HCV Collaborators (2017). Global prevalence and genotype distribution of hepatitis C virus infection in 2015: a modelling study. *The Lancet Gastroenterology & Hepatology* 2 (3): 161–176.

137 Dubuisson, J. (2007). Hepatitis C virus proteins. *World Journal of Gastroenterology* 13 (17): 2406–2415.

138 Goffard, A. (2003). Glycosylation of hepatits C virus envelope proteins. *Biochemie* 85 (3, 4): 295–301.

139 Zhang, M., Gaschen, B., Blay, W. et al. (2004). Tracking global patterns of N-linked glycosylation site variation in highly variable viral glycoproteins: HIV, SIV, and HCV envelopes and influenza hemagglutinin. *Glycobiology* 14 (12): 1229–1246.

140 Op De Beeck, A., Cocquerel, L., and Dubuisson, J. (2001). Biogenesis of hepatitis C virus envelope glycoproteins. *The Journal of General Virology* 82: 2589–2595.

141 Pileri, P., Uematsu, Y., Campagnoli, S. et al. (1998). Binding of hepatitis C virus to CD81. *Science* 282 (5390): 938–941.

142 Scarselli, E., Ansuini, H., Cerino, R. et al. (2002). The human scavenger receptor class B type I is a novel candidate receptor for the hepatitis C virus. *The EMBO Journal* 21 (19): 5017–5025.

143 Ball, J.K., Tarr, A.W., and McKeating, J.A. (2014). The past, present and future of neutralizing antibodies for hepatitis C virus. *Antiviral Research* 105 (1): 100–111.

144 Helle, F., Goffard, A., Morel, V. et al. (2007). The neutralizing activity of anti-hepatitis C virus antibodies is modulated by specific glycans on the E2 envelope protein. *Journal of Virology* 81 (15): 8101–8111.

145 Cormier, E.G., Durso, R.J., Tsamis, F. et al. (2004). L-SIGN (CD209L) and DC-SIGN (CD209) mediate transinfection of liver cells by hepatitis C virus. *Proceedings of the National Academy of Sciences of the United States of America* 101 (39): 14067–14072.

146 Iacob, R.E., Perdivara, I., Przybylski, M., and Tomer, K.B. (2008). Mass spectrometric characterization of glycosylation of hepatitis C virus E2 envelope glycoprotein reveals extended microheterogeneity of *N*-glycans. *Journal of the American Society for Mass Spectrometry* 19 (3): 428–444.

147 Bertaux, C., Daelemans, D., Meertens, L. et al. (2007). Entry of hepatitis C virus and human immunodeficiency virus is selectively inhibited by carbohydrate-binding agents but not by polyanions. *Virology* 366 (1): 40–50.

148 Helle, F., Wychowski, C., Vu-Dac, N. et al. (2006). Cyanovirin-N inhibits hepatitis C virus entry by binding to envelope protein glycans. *The Journal of Biological Chemistry* 281 (35): 25177–25183.

149 Li, D., von Schaewen, M., Wang, X. et al. (2016). Altered glycosylation patterns increase immunogenicity of a subunit hepatitis C virus vaccine, inducing neutralizing antibodies which confer protection in mice. *Journal of Virology* 90 (23): 10486–10498.

150 Peters, C.J. and LeDuc, J.W. (1999). An introduction to Ebola: the virus and the disease. *The Journal of Infectious Diseases* 179 (Suppl 1): ix–xvi.

151 Kuhn, J.H., Becker, S., Ebihara, H. et al. (2010). Proposal for a revised taxonomy of the family Filoviridae: classification, names of taxa and viruses, and virus abbreviations. *Archives of Virology* 155 (12): 2083–2103.

152 WHO Ebola Response Team (2016). After Ebola in West Africa: unpredictable risks, preventable epidemics. *The New England Journal of Medicine* 375 (6): 587–596.

153 Yang, Z.Y., Delgado, R., Xu, L. et al. (1998). Distinct cellular interactions of secreted and transmembrane Ebola virus glycoproteins. *Science* 279 (5353): 1034–1037.

154 Medaglini, D. and Siegrist, C.A. (2017). Immunomonitoring of human responses to the rVSV-ZEBOV Ebola vaccine. *Current Opinion in Virology* 23: 88–94.

155 Agnandji, S.T. and Loembe, M.M. (2021). Ebola vaccines for mass immunisation in affected regions. *The Lancet Infectious Diseases* S1473-3099 (21): 00226–00227.

156 Lennemann, N.J., Rhein, B.A., Ndungo, E. et al. (2014). Comprehensive functional analysis of N-linked glycans on ebola virus GP1. *mBio* 5 (1): e00862–e00813.

157 Michelow, I.C., Lear, C., Scully, C. et al. (2011). High-dose mannose-binding lectin therapy for Ebola virus infection. *The Journal of Infectious Diseases* 203 (2): 175–179.

158 Yu, D.S., Weng, T.H., Wu, X.X. et al. (2017). The lifecycle of the Ebola virus in host cells. *Oncotarget* 8 (33): 55750–55759.

159 Dowling, W., Thompson, E., Badger, C. et al. (2007). Influences of glycosylation on antigenicity, immunogenicity, and protective efficacy of Ebola virus GP DNA vaccines. *Journal of Virology* 81 (4): 1821–1837.

160 Ou, W., Delisle, J., Jacques, J. et al. (2012). Induction of ebolavirus cross-species immunity using retrovirus-like particles bearing the Ebola virus glycoprotein lacking the mucin-like domain. *Virology Journal* 9: 1–13.

161 Tshiani Mbaya, O., Mukumbayi, P., and Mulangu, S. (2021). Review: insights on current FDA-approved monoclonal antibodies against Ebola virus infection. *Frontiers in Immunology* 12: 721328.

162 Iversen, P.L., Kane, C.D., Zeng, X. et al. (2020). Recent successes in therapeutics for Ebola virus disease: no time for complacency. *The Lancet Infectious Diseases* 20 (9): e231–e237.

163 Warren, T.K., Wells, J., Panchal, R.G. et al. (2014). Protection against filovirus diseases by a novel broad-spectrum nucleoside analogue BCX4430. *Nature* 508 (7496): 402–405.

164 Julander, J., Demarest, J., Taylor, R. et al. (2021). An update on the progress of galidesivir (BCX4430), a broad-spectrum antiviral. *Antiviral Research* 195: 105180.

165 Mulangu, S., Dodd, L.E., Davey, R.T. et al. (2019). A randomized, controlled trial of Ebola virus disease therapeutics. *The New England Journal of Medicine* 381: 2293–2303.

166 Simmons, G., Reeves, J.D., Grogan, C.C. et al. (2003). DC-SIGN and DC-SIGNR bind Ebola glycoproteins and enhance infection of macrophages and endothelial cells. *Virology* 305 (1): 115–123.

167 Baribaud, F., Doms, R.W., and Pöhlmann, S. (2002). The role of DC-SIGN and DC-SIGNR in HIV and Ebola virus infection: can potential therapeutics block virus transmission and dissemination? *Expert Opinion on Therapeutic Targets* 6 (4): 423–431.

168 Bhatia, S., Camacho, L.C., and Haag, R. (2016). Pathogen inhibition by multivalent ligand architectures. *Journal of the American Chemical Society* 138 (28): 8654–8666.

169 Illescas, B.M., Rojo, J., Delgado, R., and Martín, N. (2017). Multivalent glycosylated nanostructures to inhibit Ebola virus infection. *Journal of the American Chemical Society* 139 (17): 6018–6025.

170 Ribeiro-Viana, R., Sánchez-Navarro, M., Luczkowiak, J. et al. (2012). Virus-like glycodendrinanoparticles displaying quasi-equivalent nested polyvalency upon glycoprotein platforms potently block viral infection. *Nature Communications* 3: 1303.

171 Muñoz, A., Sigwalt, D., Illescas, B.M. et al. (2016). Synthesis of giant globular multivalent glycofullerenes as potent inhibitors in a model of Ebola virus infection. *Nature Chemistry* 8 (1): 50–57.

172 Zhou, P., Yang, X.L., Wang, X.G. et al. (2020). A pneumonia outbreak associated with a new coronavirus of probable bat origin. *Nature* 579 (7798): 270–273.

173 Du, L., He, Y., Zhou, Y. et al. (2009). The spike protein of SARS-CoV – a target for vaccine and therapeutic development. *Nature Reviews. Microbiology* 7 (3): 226–236.

174 Satarker, S. and Nampoothiri, M. (2020). Structural proteins in severe acute respiratory syndrome coronavirus-2. *Archives of Medical Research* 51 (6): 482–491.

175 Watanabe, Y., Allen, J.D., Wrapp, D. et al. (2020). Site-specific glycan analysis of the SARS-CoV-2 spike. *Science* 369 (6501): 330–333.

176 Shajahan, A., Supekar, N.T., Gleinich, A.S., and Azadi, P. (2020). Deducing the N- and O- glycosylation profile of the spike protein of novel coronavirus SARS-CoV-2. *Glycobiology* 30 (12): 981–988.

177 Huang, Y., Yang, C., Xu, X.F. et al. (2020). Structural and functional properties of SARS-CoV-2 spike protein: potential antivirus drug development for COVID-19. *Acta Pharmacologica Sinica* 41 (9): 1141–1149.

178 Casalino, L., Gaieb, Z., Goldsmith, J.A. et al. (2020). Beyond shielding: the roles of glycans in the SARS-CoV-2 spike protein. *ACS Central Science* 6 (10): 1722–1734.

179 Andersen, K.G., Rambaut, A., Lipkin, W.I. et al. (2020). The proximal origin of SARS-CoV-2. *Nature Medicine* 26: 450–452.

180 Lan, J., Ge, J., Yu, J. et al. (2020). Structure of the SARS-CoV-2 spike receptor-binding domain bound to the ACE2 receptor. *Nature* 581 (7807): 215–220.

181 Yi, C., Sun, X., Ye, J. et al. (2020). Key residues of the receptor binding motif in the spike protein of SARS-CoV-2 that interact with ACE2 and neutralizing antibodies. *Cellular & Molecular Immunology* 17 (6): 621–630.

182 Liu, C., Zhou, Q., Li, Y. et al. (2020). Research and development on therapeutic agents and vaccines for COVID-19 and related human coronavirus diseases. *ACS Central Science* 6 (3): 315–331.

183 Jiang, S., Hillyer, C., and Du, L. (2020). Neutralizing antibodies against SARS-CoV-2 and other human coronaviruses. *Trends in Immunology* 41 (5): 355–359.

184 Wang, D. (2020). Coronaviruses' sugar shields as vaccine candidates. *Current Trends in Immunology* 21: 17–23.

185 Watanabe, Y., Berndsen, Z.T., Raghwani, J. et al. (2020). Vulnerabilities in coronavirus glycan shields despite extensive glycosylation. *Nature Communications* 11 (1): 2688.

186 Duan, L., Zheng, Q., Zhang, H. et al. (2020). The SARS-CoV-2 spike glycoprotein biosynthesis, structure, function, and antigenicity: Implications for the design of spike-based vaccine immunogens. *Frontiers in Immunology* 11: 576622.

187 Kumbhar, P.S., Pandya, A.K., Manjappa, A.S. et al. (2020). Carbohydrates-based diagnosis, prophylaxis and treatment of infectious diseases: special emphasis on COVID-19. *Carbohydrate Polymer Technologies and Applications* 2: 100052.

188 Galili, U. (2020). Amplifying immunogenicity of prospective Covid-19 vaccines by glycoengineering the coronavirus glycan-shield to present α-gal epitopes. *Vaccine* 38 (42): 6487–6499.

189 Valdes-Balbin, Y., Santana-Mederos, D., Paquet, F. et al. (2021). Molecular aspects concerning the use of the SARS-CoV-2 receptor binding domain as a target for preventive vaccines. *ACS Central Science* 7 (5): 757–767.

190 Yang, J., Wang, W., Chen, Z. et al. (2020). A vaccine targeting the RBD of the S protein of SARS-CoV-2 induces protective immunity. *Nature* 586: 572–577.

191 Valdes-Balbin, Y., Santana-Mederos, D., Quintero, L. et al. (2021). SARS-CoV-2 RBD-tetanus toxoid conjugate vaccine induces a strong neutralizing immunity in preclinical studies. *ACS Chemical Biology* 16 (7): 1223–1233.

192 Toledo-Romaní, M.E., García-Carmenate, M., *et al.*, García-Rivera, D., Vérez-Bencomo, V., SOBERANA Phase 3 team. (2023). Safety and efficacy of the two doses conjugated protein-based SOBERANA-02 COVID-19 vaccine and of a heterologous three-dose combination with SOBERANA-Plus: a double-blind, randomised, placebo-controlled phase 3 clinical trial. The Lancet Regional Health – Americas, 18: 100423 doi.org/10.1101/2021.10.31.21265703.

193 Ye, F., Zhao, J., Xu, P. et al. (2021). Synthetic homogeneous glycoforms of the SARS-CoV-2 spike receptor-binding domain reveals different binding profiles of monoclonal antibodies. *Angewandte Chemie, International Edition* 60 (23): 12904–12910.

194 Pifferi, C., Fuentes, R., and Fernández-Tejada, A. (2021). Natural and synthetic carbohydrate-based vaccine adjuvants and their mechanisms of action. *Nature Reviews Chemistry* 5 (3): 197–216.

195 Schijns, V., Majhen, D., Van Der Ley, P. et al. (2021). Rational vaccine design in times of emerging diseases: the critical choices of immunological correlates of protection, vaccine antigen and immunomodulation. *Pharmaceutics* 13 (4): 501.

196 Kensil, C.R. (1996). Saponins as vaccine adjuvants. *Critical Reviews in Therapeutic Drug Carrier Systems* 13: 1–55.

197 Lacaille-Dubois, M.A. (2019). Updated insights into the mechanism of action and clinical profile of the immunoadjuvant QS-21: a review. *Phytomedicine* 60: 152905.

198 Wang, P. (2021). Natural and synthetic saponins as vaccine adjuvants. *Vaccine* 9 (3): 1–18.

199 Didierlaurent, A.M., Laupèze, B., Di Pasquale, A. et al. (2017). Adjuvant system AS01: helping to overcome the challenges of modern vaccines. *Expert Review of Vaccines* 16 (1): 55–63.

200 Bengtsson, K.L., Morein, B., and Osterhaus, A.D. (2011). ISCOM technology-based Matrix M™ adjuvant: success in future vaccines relies on formulation. *Expert Review of Vaccines* 10 (4): 401–403.

201 Bengtsson, K.L., Karlsson, K.H., Magnusson, S.E. et al. (2013). Matrix-M™ adjuvant: enhancing immune responses by "setting the stage" for the antigen. *Expert Review of Vaccines* 12 (8): 821–823.

202 Tian, J.H., Patel, N., Haupt, R. et al. (2021). SARS-CoV-2 spike glycoprotein vaccine candidate NVX-CoV2373 immunogenicity in baboons and protection in mice. *Nature Communications* 12 (1): 372.

203 Heath, P.T., Galiza, E.P., Baxter, D.N. et al. (2021). Safety and efficacy of NVX-CoV2373 Covid-19 vaccine. *The New England Journal of Medicine* 385 (13): 1172–1183.

204 Shinde, V., Bhikha, S., Hoosain, Z. et al. (2021). Efficacy of NVX-CoV2373 Covid-19 vaccine against the B.1.351 variant. *The New England Journal of Medicine* 384 (20): 1899–1909.

4

Bacterial Glycolipid Lipid As and Their Potential as Adjuvants

Atsushi Shimoyama and Koichi Fukase

Osaka University, Graduate School of Science, Department of Chemistry, 1-1 Machikaneyama, Toyonaka, Osaka 560-0043, Japan

4.1 Introduction

Bacterial components have long been known to regulate the immune system [1]. Tumor shrinkage due to bacterial infections has occasionally been reported for the past hundreds of years [2]. In 1893, Corey et al. first attempted to use immunotherapy for cancer with *Streptococcus pyogenes* and *Serratia marcescens*. The immunostimulatory effects of killed *Salmonella typhimurium* and *Mycobacterium tuberculosis* were confirmed in 1916 and 1924, respectively. These immunostimulatory effects, now widely known as the innate immune system, are triggered by recognizing molecular pattern characteristics of pathogens and microbes by various innate immune receptors in multicellular organisms. Because innate immune stimulants also activate acquired immune responses such as antigen–antibody interactions and cell-to-cell immunity, several studies have been conducted to develop innate immune stimulants as adjuvants [3], which are vaccine ingredients that enhance antibody production.

Lipopolysaccharide (LPS), a major glycoconjugate in the outer membrane of Gram-negative bacteria, is a well-known innate immune stimulator [4]. Lipid A, which is linked to the terminal of the polysaccharide part via the peculiar acidic sugar Kdo (2-keto-3-deoxy-D-mannooctanoic acid), is the active principal of LPS, and the chemical structure of an authorized *Escherichia coli* lipid A (**1**) is described in Figure 4.1. The recognition of LPS/lipid A by Toll-like receptor (TLR) 4/myeloid differentiation protein (MD)2 receptor induces various immune responses, including cytokine production, nitric oxide production, reactive oxygen species production, leukocyte migration, and lymphocyte activation, which trigger the host defense system against bacteria. LPS and lipid A are also extremely strong inflammatory agents and are known as endotoxins, which are the major contributors to sepsis and trigger serious systemic diseases that cause multiple organ failure, hypotension, and septic shock [4a]. Canonical *E. coli* LPS is highly toxic; therefore, its application as an

Carbohydrate-Based Therapeutics, First Edition. Edited by Roberto Adamo and Luigi Lay.

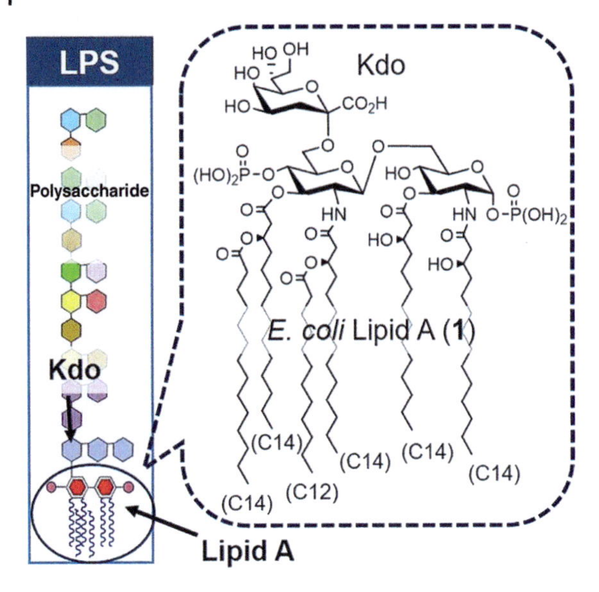

Figure 4.1 *E. coli* LPS and Kdo-lipid A.

adjuvant requires modification to attenuate any inflammatory effects and eliminate significant toxicity. Monophosphoryl lipid A (MPL) derivative, 3D-MPL (**2**) (Figure 4.2), has already been developed by GlaxoSmithKline (GSK) and approved as an adjuvant component [5]. Here, we introduce the structure–activity relationship of lipid A and the strategy for regulating the immune functions of lipid A for the development of lipid A as an adjuvant.

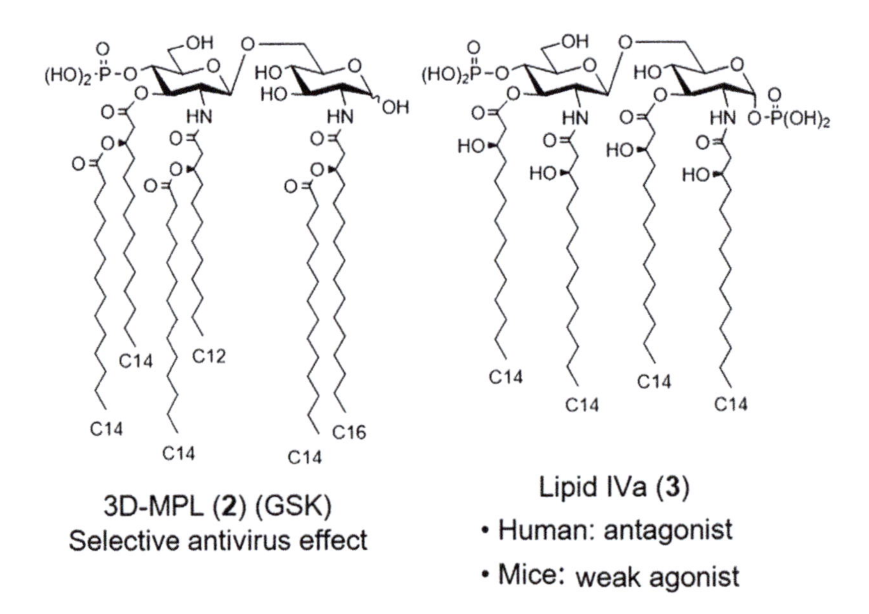

Figure 4.2 Chemical structure of 3D-MPL and lipid IVa.

4.2 Bacterial Glycolipid Lipid A: an Innate Immune Stimulant

In 1892, Pfeiffer (a disciple of the bacteriologist Koch) revealed that *Vibrio cholerae* produces two different toxic components: one is a heat-labile exotoxin and the other is a heat-stable endotoxin [6]. In 1945, Westphal (later the first director of the Max Planck Institute of Immunobiology and Epigenetics) reported that the active ingredient of endotoxin is LPS, the outer membrane component of Gram-negative bacteria [6]. In 1957, the LPS terminus-acylated disaccharide, glycolipid lipid A, was reported as the active core of LPS [6]. Shiba and Kusumoto began collaborating with a German research group and submitted the correct structure of *E. coli* lipid A (**1**) (Figure 4.1), and in 1985, they succeeded in the first total synthesis of *E. coli* lipid A (**1**), confirming that lipid A is the active core of endotoxin [7]. Simultaneously, Qureshi and Takayama identified the lipid A structure [8]. Shiba and Kusumoto also achieved lipid IVa (**3**) synthesis (Figure 4.2), a precursor of *E. coli* lipid A [1], and found that lipid IVa (**3**) has an immunostimulatory effect in mice but an antagonistic effect in humans [9]. The presence of an antagonist suggested the presence of receptors, which led to LPS receptor exploratory studies. In 1996, a breakthrough was achieved by Hoffmann, who revealed that the Toll gene, which regulates dorsoventral axis formation in Drosophila, is essential to the defense mechanism against the fungus, leading to the discovery of various innate immune receptors [10]. In 1997, TLRs were found to be human homologs of the Drosophila Toll protein [11], and Beutler identified TLR4 as an LPS receptor in 1998 [12]. To date, 10 types of TLRs (TLR1–10) in humans and 12 types (TLR1–9, TLR11–13) in mice have been identified.

TLRs are membrane glycoproteins containing a leucine-rich repeat motif in the ectodomain and a cytoplasmic signaling domain homologous to the interleukin 1 receptor (IL-1R), called the Toll/IL-1R (TIR) domain. TLR4 signaling is mediated via various adaptor molecules (MyD88, TRIF, TIRAP, and TRAM), including the TIR domain (Figure 4.3) [15]. MyD88-mediated signaling activates NF-κB, a transcription factor involved in inflammation, and induces the production of pro-inflammatory cytokines such as tumor necrosis factor (TNF)-α and IL-6. These inflammatory cytokines are produced as a protective response against infection. However, TRIF-mediated signaling leads to the activation of interferon (IFN) regulator 3 (IRF3) and induces the production of the antiviral cytokine type I IFN. Canonical *E. coli* LPS strongly activates both signals simultaneously (Figure 4.3), resulting in a massive inflammatory response leading to lethal toxicity. Therefore, lipid A attenuation and TLR4-signaling pathway regulation are essential for the development of lipid A-based adjuvants.

To achieve lipid A attenuation and TLR4-signaling regulation, it is essential to elucidate the molecular basis of lipid A recognition by TLR4. Miyake found that MD2, an accessory protein to TLR4, is essential for TLR4 signaling [16]. We synthesized a radiolabeled *E. coli* lipid A analog **4** (Figure 4.4), which Miyake used to elucidate the interaction between the TLR4/MD-2 complex and lipid A (Figure 4.4) [17]. Miyake also clarified that the species specificity of TLR4/MD-2 is due to differences in lipid A recognition by MD-2 [18]. Furthermore, X-ray crystallography revealed

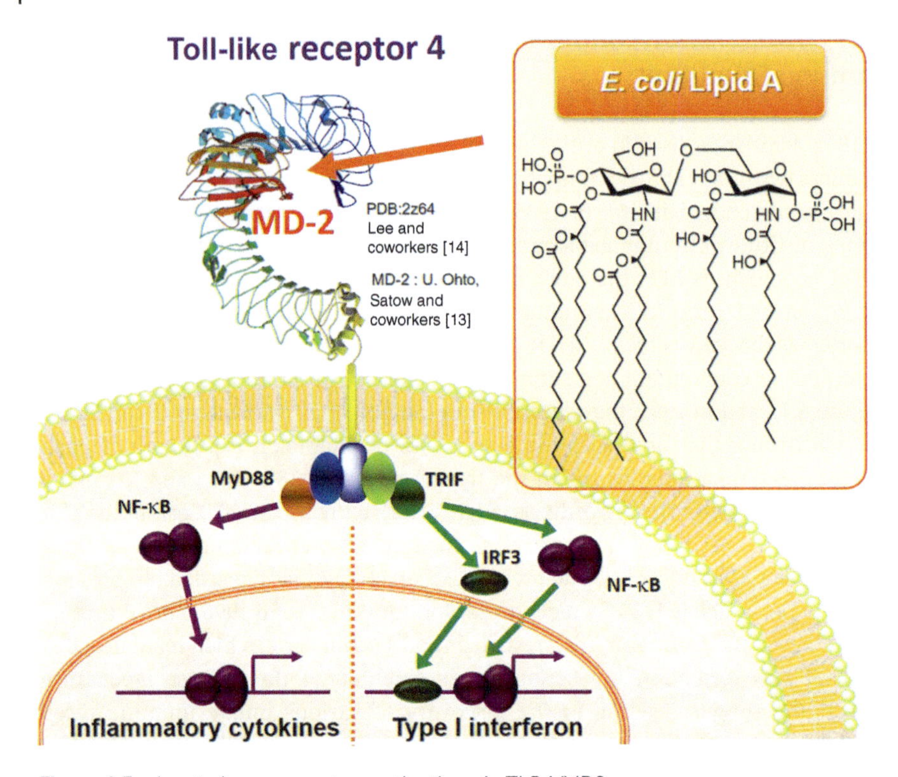

Figure 4.3 Innate immune system activation via TLR4/MD2.

the binding modes of TLR4/MD-2 to agonists and antagonists. The crystal structure of the complex of human MD-2 with lipid IVa (**3**) was revealed by Ohto and Satow [13], and the crystal structure of mouse TLR4/MD-2 complex with Eritoran (**5**) (Figure 4.5), a TLR4 antagonist developed by Eisai, was demonstrated by Lee [14]. In 2009, X-ray crystallography of the human TLR4/MD-2 complex with *E. coli* LPS was performed by Lee [19]. The results indicate that five of the six acyl chains of *E. coli* lipid A (**1**) are housed within the hydrophobic pocket of MD-2, while the remaining acyl chain interacts with the hydrophobic surface of the adjacent TLR4. These interactions trigger the dimerization of the TLR4/MD-2 complex to activate the immune response. For lipid IVa (**3**), an antagonist, the lipid A moiety binds to MD-2 in the form in which lipid A is rotated by 180° compared to the agonistic *E. coli* lipid A (**1**), and all acyl chains are placed within the MD-2 pocket, thus not causing TLR4/MD-2 dimerization. Ohto revealed the crystal structure of mouse TLR4/MD-2 with lipid IVa (**3**); incidentally, lipid IVa (**3**) acts as an antagonist in humans but as an agonist in mice [20]. In mice, three of the four acyl chains of lipid IVa (**3**) are housed within the MD-2 pocket, and the remaining one interacts with the hydrophobic surface of the adjacent TLR4, resulting in TLR4/MD-2 dimerization. These studies revealed that differences in the binding mode of lipid A to MD-2 significantly affect TLR4-mediated immune regulation.

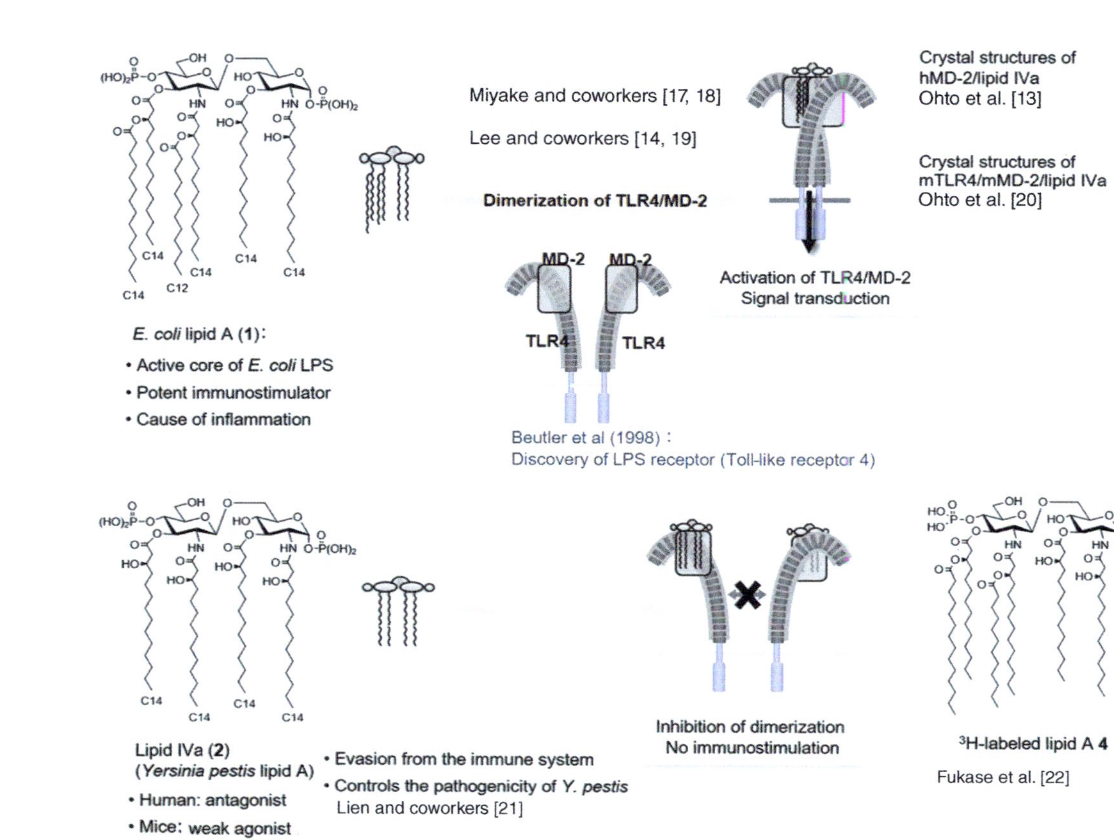

Figure 4.4 Molecular mechanism of TLR4/MD2 dimerization.

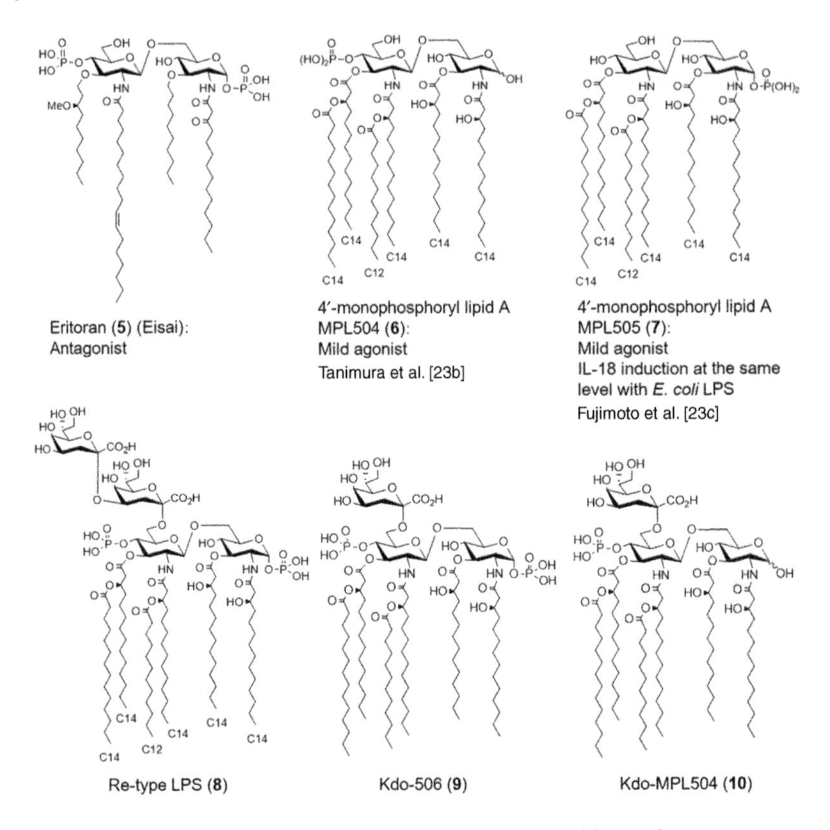

Figure 4.5 Chemical structures of various lipid As and lipid A analogs.

Previous structure–activity relationship studies [4a, 24] have revealed that agonistic and antagonistic effects can be controlled by the number of acyl chains and phosphate groups and the chain length (Figures 4.4 and 4.5). That is, hexa-acylated *E. coli* lipid A (**1**) is an agonist, whereas tetra-acylated lipid IVa (**3**) is an antagonist. MPL504 (**6**) (Figure 4.5), which is *E. coli* lipid A (**1**) without 1-phosphate, shows weaker IL-6-inducing activity than *E. coli* lipid A (**1**) [23]. MPL504 (**6**) is less dependent on CD14, a glycosylphosphatidylinositol-anchored receptor known to serve as a co-receptor for TLR4. Additionally, TLR4/MD2 dimerization in response to MPL is much lower than the response to *E. coli* LPS. MPL504 (**6**) has shown CD14-independent but MyD88-dependent TNFα-producing ability and TRIF-dependent CD86 upregulation and IFNβ-inducing ability [23b]. Similar to MPL504 (**6**), MPL505 (**7**), which lacks a 4′-phosphate, also exhibits mild immunomodulatory effects. However, while the ability of MPL504 (**6**) to induce IL-18 production is lower than that of *E. coli* LPS, MPL505 (**7**) exhibits the same IL-18 induction level as *E. coli* LPS [23c]. Therefore, these MPLs are expected to be developed as future adjuvants with different adjuvant effects. As described below, such structural modifications help regulate the effects of lipid A as a potential adjuvant.

LPS consists of an O-antigen polysaccharide part that is characteristic of each bacterial species, a core oligosaccharide part that has a high degree of commonality in chemical structure across bacterial species, and a lipid A component (Figure 4.1). We also investigated the structure–activity relationship of lipid A connected with the partial structure of the core oligosaccharide. There are R-mutant bacteria consisting of LPS lacking the O-antigen polysaccharide moiety, and the *E. coli* Re-mutant has a Re-LPS (**8**) consisting of lipid A linked to Kdo disaccharide. To elucidate the effect of Kdo on lipid A activity, we synthesized Re-LPS (**8**), Kdo-506 (**9**), and Kdo-MPL504 (**10**) (Figure 4.5), showing for the first time that Kdo enhances lipid A activity [25].

4.3 Vaccines Containing Natural LPS as Adjuvants

Vaccines are designed to attenuate or inactivate pathogens and their toxins. Most vaccines for bacteria contain natural bacterial components, and some of which would act as natural adjuvants. LPS is considered the main natural adjuvant in vaccines derived from Gram-negative bacteria. Here are some examples of vaccines that contain LPS.

4.3.1 Cholera Vaccines

Live-attenuated and inactivated whole-cell vaccines have been developed for cholera [26]. These vaccines include *V. cholerae* LPS and act as natural adjuvants.

An injectable whole-cell cholera vaccine was used in the 1960s, mainly in the USA and Japan, and was administered subcutaneously twice every five to seven days. The immune response rate was 50%, and protection continued for only six months. Side effects have been reported, and the World Health Organization recommended its discontinuation.

Dukoral®, comprising killed whole bacteria and recombinant cholera toxin B subunit, is an oral vaccine that was licensed in Sweden in 1991. This vaccine has an efficacy rate of 85–97% with few side effects. The duration of protection is approximately two to three years. It is licensed mainly in Europe, Canada, South Asia, and Latin America.

Live-attenuated oral cholera vaccines, Orochol® and Mutacol®, were developed using the *V. cholerae* Inaba strain, whose cholera toxin A subunit (toxic) was deleted. This vaccine is approved in the same country as Dukoral and has a similar effectivity rate and effectivity period. However, the production and sale of these vaccines have been discontinued.

4.3.2 *Salmonella enterica* Serovar Typhi Vaccines

Live-attenuated, inactivated whole-cell, and subunit vaccines have been developed for *Salmonella enterica* serovar Typhi [27].

Vivotif Berna®, derived from the attenuated typhoid Ty21a strain, is an oral live-attenuated vaccine developed in Switzerland. This vaccine is effective for more than two years and has few side effects. However, the administration of this vaccine to children under five years of age is not recommended. It is licensed in Africa, Europe, Asia, the USA, and South America.

Inactivated whole-cell vaccines (heat–phenol-inactivated or acetone-inactivated) were first developed in 1896. However, these have reported side effects, and most countries have withdrawn the use of these vaccines.

Typhim Vi® and Typherix®, using the virulence (Vi) capsular polysaccharide antigen purified from *Salmonella* Typhi, are subunit vaccines. Because the Vi capsular polysaccharide antigen was checked by the endotoxin test, LPS was omitted. This vaccine is effective for several years with a single injection. Its side effects are similar to those of whole-cell-inactivated vaccines but are relatively mild. It is licensed in Europe, Africa, Asia, Australia, and the USA.

4.3.3 Other Vaccines

Bexsero®, a vaccine against meningococcal group B (MenB), was developed by Novartis. It contains a meningococcal outer membrane vesicle, and LPS derived from meningococcal outer membrane vesicles can function as adjuvants in these vaccines [28]. In contrast, the MenB vaccine Trumenba® uses a recombinant lipoprotein containing the TLR2 ligand bound to the protein antigen [29].

For *Bordetella pertussis,* subunit and inactivated vaccines have been developed and are used as diphtheria, tetanus, and pertussis combination vaccine or diphtheria, pertussis, tetanus, and inactivated poliovirus combination vaccine. In most combination vaccines, the safer acellular pertussis vaccine is commonly used. However, a more effective whole-cell vaccine is still being used.

Similarly, for other Gram-negative bacteria, such as *Haemophilus influenzae* type b and *Neisseria meningitidis* (serogroups A, B, C, Y, and W-135), vaccines have also been developed, but all of them are capsular polysaccharide based vaccines; hence LPS was not included.

Therefore, vaccines that might contain LPS are widely used, and, in some cases, LPS may act as an adjuvant. In addition, it has been reported that some LPS could retain their immunostimulatory effects even when administered orally [30], in which case side effects would decrease compared to injected whole-cell vaccines with LPS.

4.4 LPS and Lipid A in the Environment or Fermented Foods as Adjuvants

Immunomodulatory functions of LPS and lipid A in the environment or fermented foods have been reported; thus, LPS and lipid A are attracting attention as safe adjuvant candidates.

The Gram-negative bacterium *Pantoea agglomerans*, which is widely present in the soil and plants such as wheat, rice, sweet potato, apple, and pear, has been detected during the fermentation process of rye bread [31]. *P. agglomerans* LPS exhibits immunostimulatory effects via oral administration [30]. *P. agglomerans* lipid A is a mixture of *E. coli* lipid A (**1**) and *Salmonella minnesota* lipid A (**11**) [32] (Figure 4.6), which are both agonists.

Kurozu (fermented black vinegar), an Asian fermented food, contains LPS derived from *Acetobacter* spp. This genus of Gram-negative bacteria is used in acetic acid fermentation. The chemical structure of *Acetobacter pasteurianus* LPS [35] and its lipid A **12** [33] (Figure 4.6) have been reported recently. Although *A. pasteurianus*

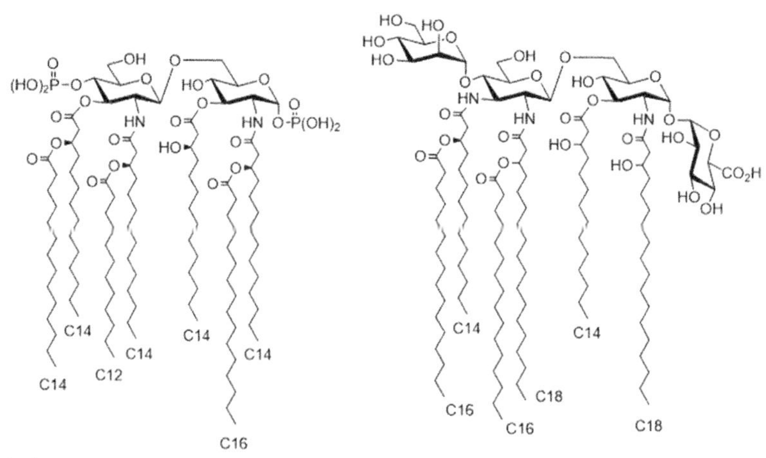

Salmonella minnesota lipid A (**11**)

Acetobacter pasteurianus lipid A (**12**)
Hashimoto et al. [33]

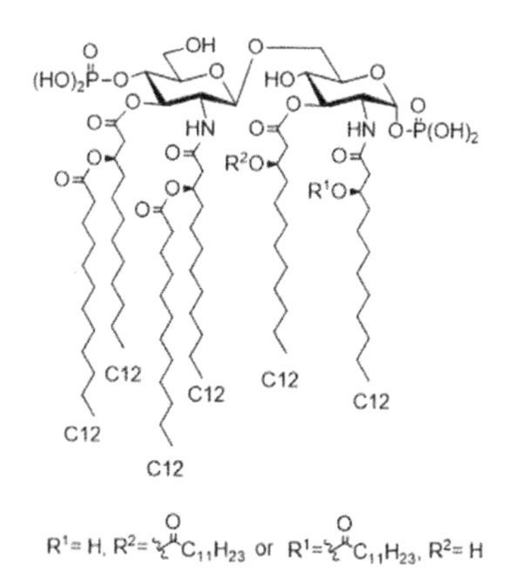

Acinetobacter lwoffii F78 lipid A (**13**)
Heine and coworkers [34]

Figure 4.6 Chemical structures of various lipid As.

LPS had weaker immunostimulating effects than *E. coli* LPS, *A. pasteurianus* LPS and lipid A are still expected to be novel adjuvants because of their safety as food-derived compounds.

The hygiene hypothesis states that exposure to environmental microbes during early childhood reduces the risk of developing allergic diseases. In the search for bacteria with an allergy-suppressing ability, *Acinetobacter lwoffii* F78 [34] was found in livestock feed. *A. lwoffii* LPS selectively induces T helper 1 (Th1) cell-derived cytokines, such as IL-12 and IFN-γ, which exhibit anti-allergic effects. *A. lwoffii* F78 LPS and its lipid A **13** (Figure 4.6) have potential as novel adjuvants.

4.5 Synthetic and Semisynthetic Lipid As as Adjuvants

3D-MPL (**2**) [5] (Figure 4.2), which has a 4′-monophosphate structure similar to MPL504 (**6**), was developed by GSK. By optimizing the lipid A structure, especially the acyl and phosphate groups, 3D-MPL (**2**) has been successfully attenuated and is currently being derivatized and produced from *S. minnesota* R595 LPS. 3D-MPL (**2**) selectively activates the TRIF-dependent pathway of the two signaling cascades downstream of TLR4/MD2 (Figure 4.3).

GSK has developed the liposome adjuvant AS01, a mixture of 3D-MPL (**2**), cholesterol, and QS21 (a saponin derived from the South American native tree Quillaja saponaria). AS01 was applied to the herpes zoster vaccine, composed of recombinant glycoprotein E of varicella-zoster virus and AS01.

Infectious sporozoites are injected into human blood via salivary glands during blood collection by *Anopheles* vector mosquitoes during infection with *Plasmodium falciparum*, the causative agent of malaria. Therefore, the development of vaccines targeting sporozoite surface proteins has been pursued. The recombinant protein RTS,S consists of a segment of a sporozoite protein and the hepatitis B virus (HBV) surface antigen, and the malaria vaccine candidate RTS,S/AS01 has been developed by GSK and is currently in phase III clinical trials. Adjuvant AS02, which consists of 3D-MPL (**2**), oil emulsion, squalene, and QS21, has also been developed by GSK, and the malaria vaccine RTS,S/AS02 is currently in phase III clinical trials.

GSK has also developed the adjuvant AS04, a mixture of 3D-MPL (**2**) and aluminum salts. Cell-mediated immune responses are induced by AS04, which exhibits antiviral effects. AS04 is practically used as an adjuvant for the human papillomavirus (HPV) vaccine Cervarix® and the HBV vaccine Fendrix®.

Furthermore, the MPL mimic RC-529 (**14**) (Figure 4.7) was approved as an adjuvant for the HBV vaccine in Argentina in 2003.

Lipid A adjuvants such as MPL can induce anti-inflammatory cytokines, including IL-10, while modulating the induction of inflammatory cytokines such as IL-6 by their chemical structural modifications [36]; therefore, lipid A adjuvants have a low risk of developing adjuvant-induced autoimmune diseases.

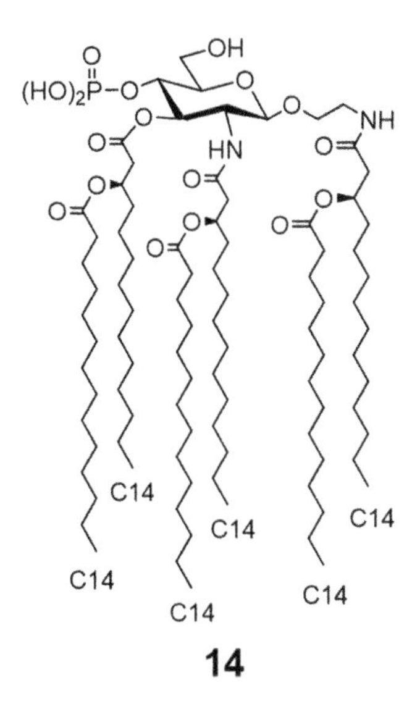

Figure 4.7 RC-529.

4.6 Developing Novel Lipid A Adjuvants

4.6.1 Parasitic Bacterial Lipid As

Recently, we undertook a lipid A-mediated host–bacterial chemical ecology study. Considering human symbiotic bacterial components as a pool of safe immunomodulators, we elucidated the immunomodulatory function of symbiotic bacterial lipid As, thereby investigating the lipid A-mediated chemical communication between host and symbiotic bacteria and developing their lipid As into safe and effective adjuvants [23c, 37].

Helicobacter pylori, which lives in the stomach, causes gastric ulcers, and *Porphyromonas gingivalis*, an oral bacterium, is a causative agent of periodontal disease. Extracted LPS from these parasitic bacteria has weak immunostimulatory effects and is associated with chronic inflammation and atherosclerosis [38]. *H. pylori* and *P. gingivalis* lipid As **15–21** are heterogeneous and have several different structures (Figure 4.8). The ability of these parasitic bacterial lipid As to regulate TLR4/MD2 was suggested to be a factor in the specific biological activity of the aforementioned parasitic bacterial LPS [39]. *E. coli* lipid A (**1**) (Figure 4.1) consists of six fatty chains (C12–C14), whereas *H. pylori* lipid As **15**, **16** have fewer (three to four) but longer (C16–C18) fatty chains. Regarding phosphate groups, *E. coli* lipid A (**1**) has two phosphate groups at the 1 and 4′-positions, whereas *H. pylori* lipid A has an MPL structure. That is, **15a, 16a** have a phosphate group only at the 1-position, and **15b, 16b** have an ethanolamine phosphate group only at the 1-position. *P. gingivalis*

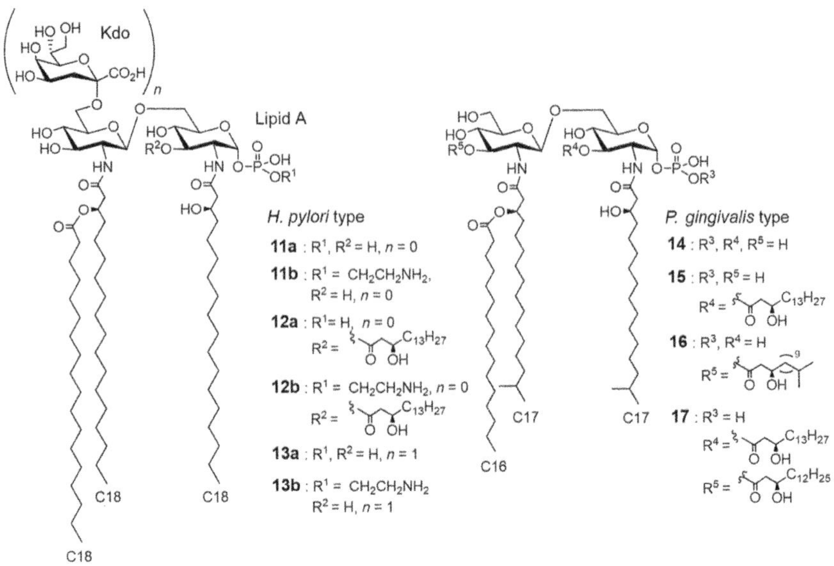

Figure 4.8 Chemical structures of parasitic bacterial LPS partial structures.

lipid As **18–21** have three to five fatty chains (C15–C17), including chains with terminal branches, and only the 1-position is phosphorylated. Parasitic bacterial lipid As **15–21** have the following common structural features: Compared to canonical *E. coli* lipid A (**1**), the fatty chains are longer and more diverse, and only 1-position is phosphorylated.

We have chemically synthesized these parasitic bacterial partial structures **15–21** comprehensively and evaluated their immunostimulatory functions (cytokine-inducing activities) in human peripheral whole blood. Their antagonistic effects on TLR4/MD-2 were evaluated by competition assays using *E. coli* LPS. Parasitic bacterial lipid As **15a**, **16a**, **18**, **19**, which have three to four fatty acid chains and one normal phosphate group, showed antagonistic activity in the induction of pro-inflammatory cytokines, such as IL-6 and TNF-α. In contrast, *H. pylori* lipid A **15b**, **16b** with three to four fatty acid chains and one ethanolamine phosphate group and *P. gingivalis* lipid A **20** with five fatty acid chains and one normal phosphate group showed IL-6- and TNF-α-inducing activity; however, the degree of activity was markedly lower than that of *E. coli* LPS. As shown in Figure 4.1, canonical *E. coli* lipid A (**1**) is linked to the polysaccharide part via Kdo and the immunostimulatory effect of lipid A is enhanced by the introduction of Kdo for *E. coli* lipid A [25]. In contrast, for *H. pylori* lipid A, **17a** with Kdo added to antagonist **15a** showed stronger antagonistic effects than **15a**, and **17b** with Kdo added to weak agonist **15b** switched to an antagonist [37a]. For *H. pylori* LPS, Kdo-lipid A, but not lipid A itself, was found to be the active principle. All parasitic bacterial lipid As **15-21** induced IL-12 and -18, which are involved in chronic inflammation, and **15a**, **16a**, **17-19** were found to selectively induce IL-12 and IL-18. Because the combination of IL-12 and IL-18 induces IFN-γ, which is involved in antitumor and anti-allergic responses, *H. pylori* lipid As, which selectively induce IL-12 and IL-18, are promising adjuvant candidates. LPS-mediated IL-18 induction was reported to be dependent on the

TRIF pathway [40]; however, a TRIF-independent pathway has also been reported [41]. The molecular mechanism of selective cytokine induction triggered by parasitic bacterial lipid As remains unclear. In 2014, it was reported that caspases 4, 5, and 11 are cytosolic LPS receptors [42]. Therefore, there may be a TLR4-independent pathway in humans via caspase 4 or 5 for caspase-1 activation, which is upstream of IL-18 induction.

4.7 Symbiotic Bacterial Lipid As

Parasitic bacterial lipid A research has suggested that parasitic bacteria evolve to escape the innate immune responses of the host, and their LPS/lipid A show antagonistic or extremely weak agonistic effects that favor infection of the host. Furthermore, these results indicate that parasitic bacteria might induce chronic inflammatory diseases while avoiding the bactericidal effects derived from acute inflammation (host immune response), suggesting that the activity of lipid A profoundly reflects the characteristics of the bacteria, i.e. that there is the presence of a lipid A-mediated bacterial–host chemical ecology (chemical communication between bacteria and host via lipid A). Thus, we hypothesized that symbiotic bacteria would have extremely low toxic immunomodulators and their components would be associated with maintaining homeostasis, and we chose symbiotic bacterial lipid A as a pool of safer immunomodulators.

Kiyono and Kunisawa revealed that the Gram-negative bacteria *Alcaligenes faecalis* inhabits gut-associated lymphoid tissues (GALT), Peyer's patches, which play an important role in the maintenance of homeostasis [43]. We hypothesized that *A. faecalis* lipid A has a homeostatic function, which is a key factor in establishing symbiotic relationships with the host, and could be applied as a safe, low-toxicity immunomodulator. Therefore, we performed purification, structural determination, and functional analysis of *A. faecalis* LPS. Canonical *E. coli* produce LPS consisting of tens to hundreds of sugar residues; however, some species produce lipooligosaccharides (LOS) with short sugar chains. We have revealed that *A. faecalis* produces a LOS consisting of a nonasaccharide (Figure 4.9) [37c].

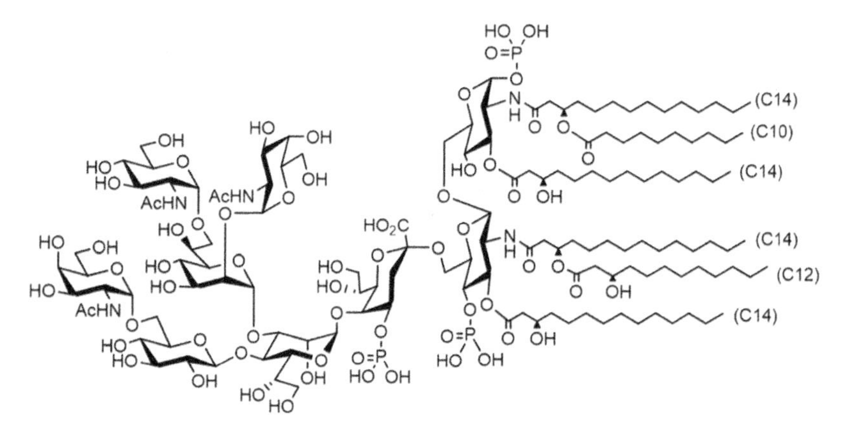

Figure 4.9 Chemical structures of *A. faecalis* LOS structures.

The extracted LOS fraction from *A. faecalis* significantly promoted IgA antibody production without toxicity and was comparable to that of toxic *E. coli* LPS, suggesting that the *A. faecalis* component is a promising safe adjuvant. Furthermore, the antibody production enhancement of the extracted *A. faecalis* LOS was TLR4-dependent, and it was suggested that lipid A is an adjuvant function core [44]. Structural analysis showed that *A. faecalis* lipid A was a mixture of **22–24** with different acyl chain patterns (Figure 4.10). Thus, we chemically synthesized **22–24** and

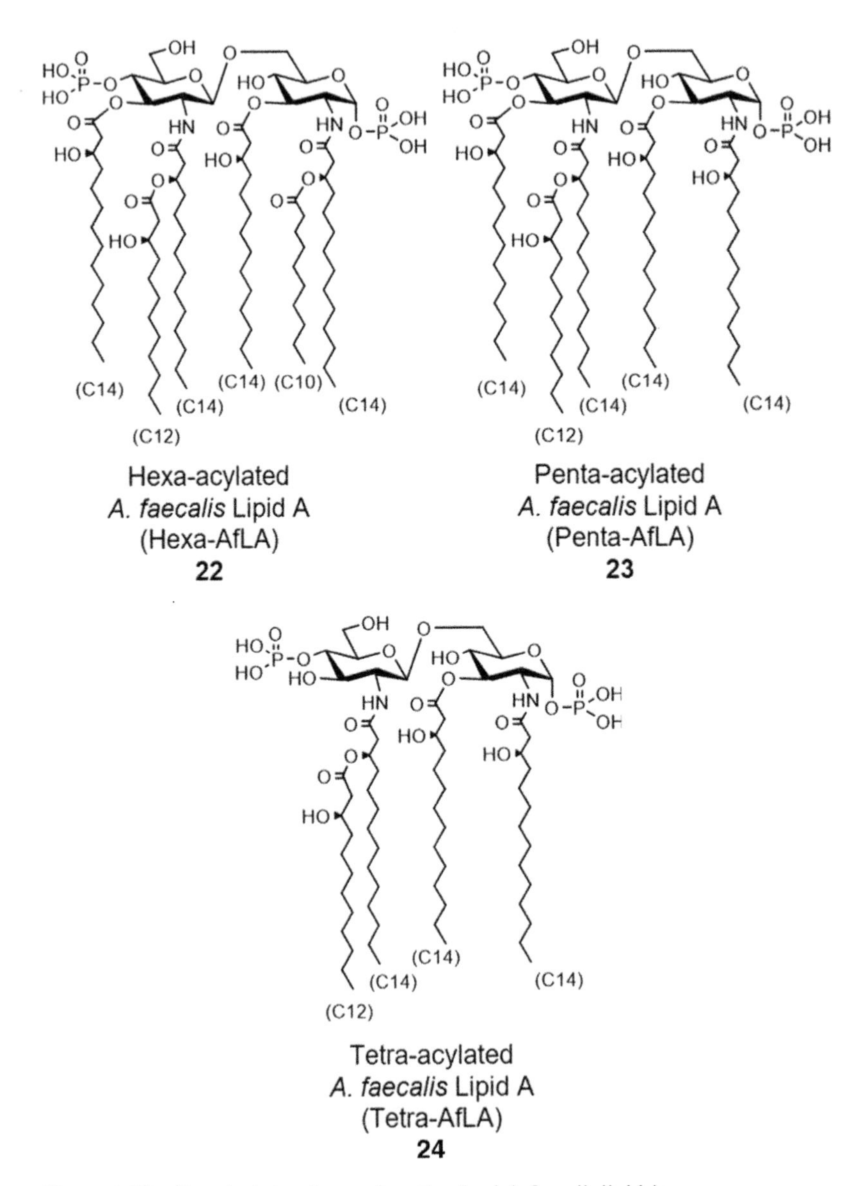

Figure 4.10 Chemical structures of synthesized *A. faecalis* lipid As.

evaluated their functions using human monocytic cells and found that only hexa-acylated *A. faecalis* lipid A **22** (**Hexa-AfLA**) showed immunostimulatory activity, which was almost identical to that of the extracted *A. faecalis* LOS, indicating that **Hexa-AfLA** is the active component of *A. faecalis* LOS [37c]. *In vivo* studies in mice confirmed that **Hexa-AfLA** has the same useful adjuvant effect (enhancement of antigen-specific IgA and IgG production and Th17-mediated protective immunity) as *A. faecalis* LOS without toxicity [45]. The efficacy of **Hexa-AfLA** as a safe nasal vaccine adjuvant has been demonstrated in *S. pneumoniae* infection models [45], making it an extremely promising adjuvant for vaccines against infectious diseases.

Hexa-AfLA from GALT resident *A. faecalis* can regulate the induction of IgA, which is responsible for maintaining the homeostasis of mucosal immunity, suggesting that **Hexa-AfLA** is a regulator of intestinal mucosal immunity. Based on bacterial–host chemical ecology research, we have succeeded in identifying key compounds for intestinal mucosal immunity by focusing on symbiotic bacteria inhabiting the GALT and immunoregulatory tissues in the gut and have found promising adjuvants that can safely regulate mucosal immunity.

4.8 Lipid A-Based Self-Adjuvanting Vaccines

The self-adjuvanting strategy promotes more efficient antibody production by complexing the antigen with the adjuvant. Recently, many studies on this strategy, especially those using lipopeptide adjuvants (Pam_3CSK_4 and TLR2 ligands), have been published [6, 46]. The antigen–adjuvant complex is actively taken up by dendritic cells via an innate immune ligand (adjuvant), which activates the immune system and induces cytokine production, resulting in efficient antibody production (Figure 4.11). The advantage of this strategy is that the antigen and adjuvant are taken up by the same dendritic cells and can trigger a specific immune response. It is also excellent in terms of quality retention and safety control because it is easy to obtain high-purity products. As for complexing methods between antigens and adjuvants, one is based on covalent bond formation [46], and the other is based on liposomes or self-aggregate formation [47]. In this section, covalent bond formation-type self-adjuvanting vaccines using lipid A are described.

Guo and coworkers synthesized an MPL-based self-adjuvanting vaccine, which is a covalently bound MPL adjuvant and antigen. They reported a complex MPL with GM3 [48] or α-2,9-oligosialic acid (meningococcal antigen) [49] (Figure 4.12a) and confirmed enhanced antibody production. Lewicky and Jiang synthesized an MPL mimic, RC-529 (**14**), conjugated with a Thomsen–Friedenreich antigen (a tumor-associated carbohydrate antigen) (Figure 4.12b) [50]. Codée and coworkers synthesized a lipid A mimic, CRX-527, conjugated with peptide antigen (Figure 4.12c), and T-cell immune responses against the antigen and specific killing of target cells expressing the antigen were observed [51]. Trumenba, the aforementioned vaccine against *N. meningitidis* group B, is a recombinant lipoprotein with TLR2-stimulating activity and is also a type of self-adjuvanting vaccine.

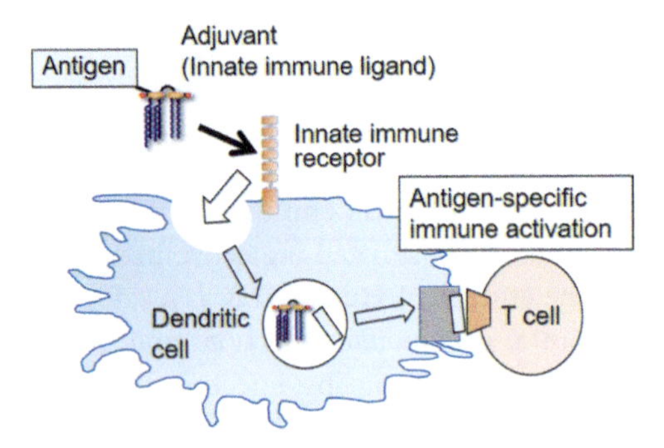

Figure 4.11 Self-adjuvanting strategy.

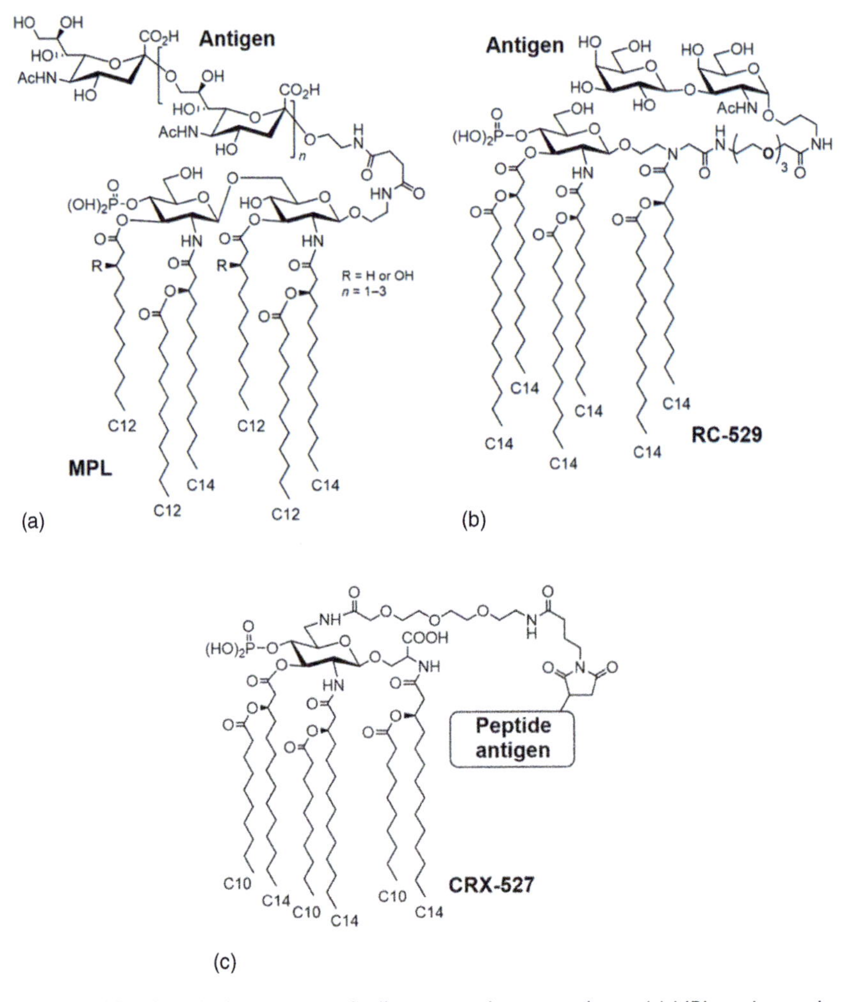

Figure 4.12 Chemical structures of adjuvant–antigen complexes. (a) MPL conjugated with α-2,9-oligosialic acid. (b) RC-529 conjugated with a Thomsen–Friedenreich antigen. (c) RX-527 conjugated with peptide antigen.

4.9 Conclusions

In this chapter, the structure–activity relationship of TLR4 ligands, especially of lipid As, and their potential as vaccine adjuvants have been discussed. Similar to the case of MPL and parasitic bacterial lipid As, structural modifications can regulate the activation of the TLR4/MD2 receptor and selective induction of intracellular signals. Therefore, cell-mediated, humoral, or mucosal immune responses can be controlled using a specific lipid A derivative. Various lipid A derivatives, mainly MPLs, are being developed as adjuvants, and next-generation safe adjuvants such as symbiotic bacterial lipid A also show great potential. Since the clinical use of lipid A-based adjuvants, including the AS series (GSK), has already been expanding in development as components of various novel vaccines such as anticancer vaccines and antiprotozoal vaccines, including antimalarial vaccines, the importance of lipid A adjuvants will increase in the future. The development of a self-adjuvanting strategy that can further enhance the function of lipid A adjuvants is also expected. In contrast, lipid A activity is significantly affected by subtle differences in its chemical structure, that is, the balance between the hydrophobic region formed by fatty acids and the hydrophilic region formed by sugar moieties, the number and position of phosphate groups, and the addition of Kdo. Hence, it is difficult to modify lipid A while retaining its immune function. Once a simple and universal lipid A modification method that can retain its function has been developed, it will be a breakthrough in developing innovative self-adjuvanting vaccines.

References

1 Kusumoto, S., Fukase, K., and Shiba, T. (2010). *Proceedings of the Japan Academy. Series B, Physical and Biological Sciences* 86: 322–337.

2 Wei, M.Q., Mengesha, A., Good, D., and Anne, J. (2008). *Cancer Letters* 259: 16–27.

3 Leroux-Roels, G. (2010). *Vaccine* 28 (Suppl 3): C25–C36.

4 (a) Molinaro, A., Holst, O., Di Lorenzo, F. et al. (2015). *Chemistry - A European Journal* 21: 500–519. (b) Di Lorenzo, F., Duda, K.A., Lanzetta, R. et al. (2022). *Chemical Reviews* 122 (20): 15767–15821.

5 Mata-Haro, V., Cekic, C., Martin, M. et al. (2007). *Science* 316: 1628–1632.

6 Rietschel, O.W.E.T. (1999). Endotoxin: historical perspectives. In: *Endotoxin in Health and Disease* (ed. H. Brade), 1–30. CRC Press.

7 (a) Imoto, M., Kusumoto, S., Shiba, T. et al. (1983). *Tetrahedron Letters* 24: 4017–4020. (b) Imoto, M., Kusumoto, S., Shiba, T. et al. (1985). *Tetrahedron Letters* 26: 907–908. (c) Imoto, M., Yoshimura, H., Shimamoto, T. et al. (1987). *Bulletin of the Chemical Society of Japan* 60: 2205–2214.

8 Takayama, K., Qureshi, N., and Mascagni, P. (1983). *The Journal of Biological Chemistry* 258: 12801–12803.

9 (a) Flad, H.D., Loppnow, H., Feist, W. et al. (1989). *Lymphokine Research* 8: 235–238. (b) Wang, M.H., Feist, W., Herzbeck, H. et al. (1990). *FEMS Microbiology Immunology* 2: 179–185.

10 Lemaitre, B., Nicolas, E., Michaut, L. et al. (1996). *Cell* 86: 973–983.

11 Medzhitov, R., Preston-Hurlburt, P., and Janeway, C.A. Jr. (1997). *Nature* 388: 394–397.

12 Poltorak, A., He, X., Smirnova, I. et al. (1998). *Science* 282: 2085–2088.

13 Ohto, U., Fukase, K., Miyake, K., and Satow, Y. (2007). *Science* 316: 1632–1634.

14 Kim, H.M., Park, B.S., Kim, J.I. et al. (2007). *Cell* 130: 906–917.

15 Kawai, T. and Akira, S. (2010). *Nature Immunology* 11: 373–384.

16 Shimazu, R., Akashi, S., Ogata, H. et al. (1999). *The Journal of Experimental Medicine* 189: 1777–1782.

17 Akashi, S., Saitoh, S., Wakabayashi, Y. et al. (2003). *The Journal of Experimental Medicine* 198: 1035–1042.

18 Akashi, S., Nagai, Y., Ogata, H. et al. (2001). *International Immunology* 13: 1595–1599.

19 Park, B.S., Song, D.H., Kim, H.M. et al. (2009). *Nature* 458: 1191–1195.

20 Ohto, U., Fukase, K., Miyake, K., and Shimizu, T. (2012). *Proceedings of the National Academy of Sciences of the United States of America* 109: 7421–7426.

21 Montminy, S.W., Khan, N., McGrath, S. et al. (2006). *Nat. Immunol.* 7 (10): 1066–1073.

22 Fukase, K., Kirikae, T., Kirikae, F. et al. (2001). *Bull. Chem. Soc. Jpn.* 74 (11): 2189–2197.

23 (a) Brade, L., Brandenburg, K., Kuhn, H.M. et al. (1987). *Infection and Immunity* 55: 2636–2644. (b) Tanimura, N., Saitoh, S., Ohto, U. et al. (2014). *International Immunology* 26: 307–314. (c) Fujimoto, Y., Shimoyama, A., Saeki, A. et al. (2013). *Molecular BioSystems* 9: 987–996.

24 (a) Fukase, K.F., Shimoyama, A., and Tanaka, K. (2012). *Journal of Synthetic Organic Chemistry, Japan* 70 (2): 113–130. (b) Kusumoto, S. and Fukase, K. (2006). *Chemical Record* 6: 333–343.

25 Yoshizaki, H., Fukuda, N., Sato, K. et al. (2001). *Angewandte Chemie International Edition* 40: 1475–1480.

26 Ryan, E.T. and Calderwood, S.B. (2000). *Clinical Infectious Diseases : An Official Publication of the Infectious Diseases Society of America* 31: 561–565.

27 Keystone, J. (1995). *The Canadian Journal of Infectious Diseases = Journal Canadien des Maladies Infectieuses* 6: 231.

28 Dowling, D.J., Sanders, H., Cheng, W.K. et al. (2016). *Frontiers in Immunology* 7: 562.

29 Luo, Y., Friese, O.V., Runnels, H.A. et al. (2016). *The AAPS Journal* 18: 1562–1575.

30 (a) Dutkiewicz, J., Mackiewicz, B., Lemieszek, M.K. et al. (2016). *Annals of Agricultural and Environmental Medicine* 23: 206–222. (b) Hebishima, T., Matsumoto, Y., Watanabe, G. et al. (2011). *Experimental Animals* 60: 101–109.

31 Kariluoto, S., Aittamaa, M., Korhola, M. et al. (2006). *International Journal of Food Microbiology* 106: 137–143.

32 Tsukioka, D., Nishizawa, T., Miyase, T. et al. (1997). *FEMS Microbiology Letters* 149: 239–244.

33 Hashimoto, M., Ozono, M., Furuyashiki, M. et al. (2016). *The Journal of Biological Chemistry* 291: 21184–21194.

34 Debarry, J., Hanuszkiewicz, A., Stein, K. et al. (2010). *Allergy* 65: 690–697.

35 Pallach, M., Di Lorenzo, F., Facchini, F.A. et al. (2018). *International Journal of Biological Macromolecules* 119: 1027–1035.

36 Martin, M., Michalek, S.M., and Katz, J. (2003). *Infection and Immunity* 71: 2498–2507.

37 (a) Shimoyama, A., Saeki, A., Tanimura, N. et al. (2011). *Chemistry - A European Journal* 17: 14464–14474. (b) Fujimoto, Y., Shimoyama, A., Suda, Y., and Fukase, K. (2012). *Carbohydrate Research* 356: 37–43. (c) Shimoyama, A., Di Lorenzo, F., Yamaura, H. et al. (2021). *Angewandte Chemie International Edition* 60 (18): 10023–10031.

38 (a) Hynes, S.O., Ferris, J.A., Szponar, B. et al. (2004). *Helicobacter* 9: 313–323. (b) Nielsen, H., Birkholz, S., Andersen, L.P., and Moran, A.P. (1994). *The Journal of Infectious Diseases* 170: 135–139. (c) Perez-Perez, G.I., Shepherd, V.L., Morrow, J.D., and Blaser, M.J. (1995). *Infection and Immunity* 63: 1183–1187. (d) Danesh, J., Wong, Y., Ward, M., and Muir, J. (1999). *Heart* 81: 245–247.

39 Triantafilou, M., Gamper, F.G., Lepper, P.M. et al. (2007). *Cellular Microbiology* 9: 2030–2039.

40 Imamura, M., Tsutsui, H., Yasuda, K. et al. (2009). *Journal of Hepatology* 51: 333–341.

41 Kanneganti, T.D., Lamkanfi, M., Kim, Y.G. et al. (2007). *Immunity* 26: 433–443.

42 Shi, J., Zhao, Y., Wang, Y. et al. (2014). *Nature* 514: 187–192.

43 (a) Obata, T., Goto, Y., Kunisawa, J. et al. (2010). *Proceedings of the National Academy of Sciences of the United States of America* 107: 7419–7424. (b) Fung, T.C., Bessman, N.J., Hepworth, M.R. et al. (2016). *Immunity* 44: 634–646. (c) Sonnenberg, G.F., Monticelli, L.A., Alenghat, T. et al. (2012). *Science* 336: 1321–1325.

44 Shibata, N., Kunisawa, J., Hosomi, K. et al. (2018). *Mucosal Immunology* 11: 693–702.

45 (a) Yoshii, K., Hosomi, K., Shimoyama, A. et al. (2020). *Microorganisms* 8. (b) Wang, Y., Hosomi, K., Shimoyama, A. et al. (2020). *Vaccines (Basel)* 8.

46 (a) Ingale, S., Wolfert, M.A., Gaekwad, J. et al. (2007). *Nature Chemical Biology* 3: 663–667. (b) Khan, S., Weterings, J.J., Britten, C.M. et al. (2009). *Molecular Immunology* 46: 1084–1091. (c) Kaiser, A., Gaidzik, N., Becker, T. et al. (2010). *Angewandte Chemie International Edition* 49: 3688–3692. (d) Wilkinson, B.L., Day, S., Malins, L.R. et al. (2011). *Angewandte Chemie International Edition* 50: 1635–1639. (e) Wilkinson, B.L., Day, S., Chapman, R. et al. (2012). *Chemistry – A European Journal* 18: 16540–16548. (f) Lakshminarayanan, V., Thompson, P., Wolfert, M.A. et al. (2012). *Proceedings of the National Academy of Sciences of the United States of America* 109: 261–266. (g) Cai, H., Chen, M.-S., Sun, Z.-Y. et al. (2013). *Angewandte Chemie International Edition* 52: 6106–6110. (h) Palitzsch, B., Hartmann, S., Stergiou, N. et al. (2014). *Angewandte Chemie International Edition*

53: 14245–14249. (i) Thompson, P., Lakshminarayanan, V., Supekar, N.T. et al. (2015). *Chemical Communications* 51: 10214–10217.

47 (a) Ingale, S., Wolfert, M.A., Buskas, T., and Boons, G.J. (2009). *ChemBioChem* 10: 455–463. (b) Aiga, T., Manabe, Y., Ito, K. et al. (2020). *Angewandte Chemie International Edition* 59: 17705–17711. (c) Skwarczynski, M., Zhao, G., Boer, J.C. et al. (2020). *Science Advances* 6: eaax2285.

48 Wang, Q., Zhou, Z., Tang, S., and Guo, Z. (2012). *ACS Chemical Biology* 7: 235–240.

49 Liao, G., Zhou, Z., Suryawanshi, S. et al. (2016). *ACS Central Science* 2: 210–218.

50 Lewicky, J.U., M., and Jiang, Z.H. (2016). *ChemistrySelect* 5: 906–910.

51 Reintjens, N.R.M., Tondini, E., de Jong, A.R. et al. (2020). *Journal of Medicinal Chemistry* 63: 11691–11706.

5

Antiadhesive Carbohydrates and Glycomimetics

Jonathan Cramer[1,2], *Lijuan Pang*[3], *and Beat Ernst*[1]

[1]*University of Basel, Department of Pharmaceutical Sciences, Pharmacenter, Klingelbergstrasse 50, Basel CH-4056, Switzerland*
[2]*Institute for Pharmaceutical and Medicinal Chemistry, Heinrich-Heine-University Düsseldorf, Universitätsstraße 1, Düsseldorf DE-40225, Germany*
[3]*WuXi AppTec UK LTD, 5 New Street Square, London EC4A 3TW, United Kingdom*

5.1 Introduction

Infectious diseases are still a major cause of death, disability, and social and economic disorder for millions of people throughout the world. Poverty, poor access to health care, human migration, emerging disease agents, and antibiotic resistance all contribute to the expanding impact of these illnesses [1]. Prevention and treatment strategies for infectious diseases are derived from a thorough understanding of the complex interactions between specific viral or bacterial pathogens and the human host.

Glycans are found on the surfaces of all bacteria and viruses, as well as on their hosts. Thus, a majority of interactions between microbial pathogens and their hosts are based on the interaction of carbohydrate epitopes on the one hand and glycan-binding receptors on the other hand [2]. This initially leads to the colonization of host epithelial surfaces, a prerequisite for spreading infection. In this chapter, the status of the development stage of *Antiadhesive Carbohydrates and Glycomimetics* is presented by means of some selected examples.

5.1.1 Carbohydrate–Protein Interactions in Viral Adhesion to Host Cells

The surfaces of viruses and host cells are densely covered with diverse glycans. Many viruses exploit carbohydrate–protein interactions for adhesion to host cells, initiation of virus internalization, and evasion of immune surveillance [3–5]. Whereas bacterial adhesion is most commonly mediated by the interaction of bacterial lectins with host-derived glycoproteins, viruses have a more diverse arsenal of adhesion mechanisms. Similar to bacteria, some viral pathogens display carbohydrate-binding proteins on their surface that can specifically recognize certain host glycans [5]. A well-studied example is the interaction of the hemagglutinin glycoprotein expressed on the surface of influenza

Carbohydrate-Based Therapeutics, First Edition. Edited by Roberto Adamo and Luigi Lay.

A viruses with sialic acid-containing host glycans. Furthermore, glycosaminoglycans, acidic linear polysaccharides that commonly decorate host cell surfaces, can serve as an initial attachment factor for many different viruses. Alternatively, endocytic carbohydrate-binding receptors on host cells can promote adhesion to viral envelope glycoproteins and subsequent internalization [3, 4]. An archetypical example of this process, the interaction of the myeloid C-type lectin DC-SIGN with viral glycoproteins, will be discussed in this book chapter. Many viruses employ different strategies for adhesion simultaneously. Besides mediating the interaction with sialic acid receptors, the influenza A hemagglutinin glycoprotein is also heavily modified with N-linked glycans. These structures are recognized by DC-SIGN and mediate a secondary, sialic acid-independent adhesion and entry mechanism into host cells [6].

5.1.2 Bacterial Adhesins and Antiadhesion Therapy

An essential step of bacterial infection and pathogenesis is the adherence of bacteria to cell surfaces of the host tissue, granting the bacteria substantial resistance to natural defense mechanisms, mechanical shear stress, and antibiotics [7]. Bacteria can express more than one type of adherence factors or "adhesins." Most of these adhesins are lectins that bind directly to cell-surface carbohydrate motifs on glycoproteins or glycosphingolipids via carbohydrate-recognition domains (CRDs) [8, 9]. Bacterial lectins commonly exist in the form of elongated, hair-like, multi-subunit protein appendages, known as fimbriae (hair) or pili (threads), protruding from the surface of bacteria [7, 8]. Although carbohydrate–lectin interactions are generally of low affinity, such pili structures provide a multivalent, Velcro-like binding to epithelial surfaces, hence facilitating bacterial survival and invasion [8, 10]. Therefore, antiadhesive agents that block bacterial adherence to host tissues may offer a novel strategy to combat infectious diseases.

Similar to animal lectins, bacterial lectins bind to terminal sugar residues or internal glycan sequences present in linear or branched oligosaccharide chains [11]. Since Sharon et al. first described bacterial surface lectins in 1970s, [12] researchers have identified a large fraction of the carbohydrate epitopes ("adhesin receptors") used by bacteria for colonization and entry into host tissues (Table 5.1) [31]. Although the natural carbohydrate epitopes show effectiveness in blocking microbial adhesion, their susceptibility to enzymatic degradation and undesirable pharmacokinetic properties hamper their clinical applications [32]. Based on resolved protein structures, structure-based rational design advanced the identification and optimization of antiadhesive glycomimetics, allowing improved metabolic stability, binding selectivity, and bioavailability [33–35]. In the era of increasing antimicrobial resistance, one exceptional advantage of antiadhesive therapeutics is that they do not kill or restrict the growth of the pathogens and are therefore less likely to promote antibacterial resistance. Additionally, antiadhesive agents could potentially reduce overuse of broad-spectrum antibiotics and thus prevent long-lasting detrimental effects on the healthy human microbiota. Furthermore, target-specific antiadhesion therapy makes precision antimicrobial treatment possible [36, 37]. The search for FimH and PA-IL/IIL inhibitors depicts representative examples of modern antiadhesive therapeutics.

5.1.3 Selected Examples

In this chapter on *Antiadhesive Carbohydrates and Glycomimetics*, we are forced to limit ourselves to a few selected examples. From a plethora of therapeutic targets for which the involved carbohydrate ligands and lectins have been elucidated, only the most prominent examples, namely DC-SIGN and the virulence factors FimH, PA-IL, and PA-IIL, were selected for discussion.

The C-type lectin receptor DC-SIGN is a pattern recognition receptor expressed on macrophages and dendritic cells (DCs). It has been identified as a promiscuous entry receptor for many pathogenic agents, including pandemic viruses such as SARS-CoV-2, ebola, and HIV [38]. The virulence factors FimH, PapG, PA-IL, and PA-IIL are expressed by pathogenic bacteria that represent an immediate or future threat to public health in the light of emerging antibiotic resistance.

FimH is one of the most studied adhesins expressed by uropathogenic *Escherichia coli* (UPEC) strains because it is a key determinant of urovirulence [39]. Urinary tract infections (UTIs) and catheter-associated urinary tract infections (CAUTIs) are becoming increasingly important threats to human health, and the antiadhesive strategy emerged as a relevant alternative therapeutic approach. Recently, the FimH antagonist GSK3882347 entered Phase I clinical trials in a collaboration between Fimbrion Therapeutics and GlaxoSmithKline.

Finally, *Pseudomonas aeruginosa* produces biofilms that can cause chronic opportunistic infections, which often cannot be treated effectively with traditional antibiotics [40]. Since *P. aeruginosa* is considered a model organism for the study of antibiotic-resistant bacteria, its virulence factors PA-IL and PA-IIL were extensively studied.

5.2 DC-SIGN-Mediated Viral Adhesion and Entry into Myeloid Cells

5.2.1 Introduction

C-type lectin receptors (CLRs), a class of proteins expressed on the membrane of myeloid cells such as DCs and macrophages, are often exploited as entry receptors by viral pathogens [38, 41]. Physiologically, CLRs recognize conserved carbohydrate epitopes on diverse pathogens and initiate tailored immune responses. However, some viruses have developed the ability to circumvent the physiological function of CLRs and infect myeloid cells themselves (*cis*-infection) or other cells under mediation of myeloid CLRs (*trans*-infections). The CLR DC-specific ICAM-3-grabbing nonintegrin (DC-SIGN, CD209) has been proven vulnerable to viral exploitation, most famously in the pathology of HIV infections [38, 41–43]. DC-SIGN-mediated adhesion and internalization of virus particles can result in trafficking to nonlysosomal compartments and virus persistence in a protected intracellular environment. Besides this, DC-SIGN has been demonstrated to serve as a receptor for many other viruses, such as dengue, zika, ebola, and coronaviruses [38]. Because of their

Table 5.1 Carbohydrate epitopes used by bacteria for colonization and entry in host tissues.

Pathogen	Adhesin	Binding epitope	References
Campylobacter jejuni	Flagella, LPS	Fucα(1-2)Galβ(1-4)GlcNAc	[13]
Escherichia coli	Type-1 fimbriae	Manα(1-3)Manα(1-6)Man	[14]
	P fimbriae	Galα(1-4)Gal	[15]
	S fimbriae	Neu5Acα(2-3)Galβ(1-4) GalNAc	[16]
	K99 fimbriae	Gangliosides GM3, Neu5Glcα(2-3)Galβ(1-4)Glc	[17]
	CFA1	AsialoGM1, Lewis A	[18, 19]
Haemophilus influenzae	HMW1 adhesin	Neu5Acα(2-3)Galβ(1-4) GlcNAc	[20]
Helicobacter pylori	BabA	Lewis B	[21]
	SabA	Sialyl Lewis X	[22]
Klebsiella pneumoniae	Type-1 fimbriae	Man	[23]
Mycobacterium tuberculosis	Heparin-binding hemag-glutinin adhesin (HBHA)	Heparan sulfate	[24]
Neisseria gonorrhoeae	Opa proteins	LacCer, Neu5Acα(2-3) Galβ(1-4)GlcNAc, syndecans, heparan sultate	[25]
Pseudomonas aeruginosa	PA-IL (LecA)	Galactosides	[26]
	PA-IIL (LecB)	Lewis A, Fuc	[27]
Salmonella typhimurium	Type-1 fimbriae	Man	[28]
Streptococcus pneumoniae	Carbohydrate-binding modules of β-galactosidase, BgaA	Lactose, *N*-acetyl-lactosamine, Neu5Acα(2-3) Gal	[29]
Streptococcus suis	SadP	Galα(1-4)Galα(1-4)Glc	[30]

potential to interfere with viral adhesion, carbohydrate-based molecules and glycomimetic drugs targeting DC-SIGN are of tremendous interest. This treatment strategy circumvents common resistance mechanisms through attenuation of virulence and provides a host-directed pharmacological response to not only established but also newly emerging infections with pandemic potential.

DC-SIGN is anchored to the cell membrane by a hydrophobic neck domain that induces tetramerization (Figure 5.1a). Carbohydrate ligands are bound to the CRD by a calcium ion acting as a cofactor in the primary binding site (Figure 5.1b).

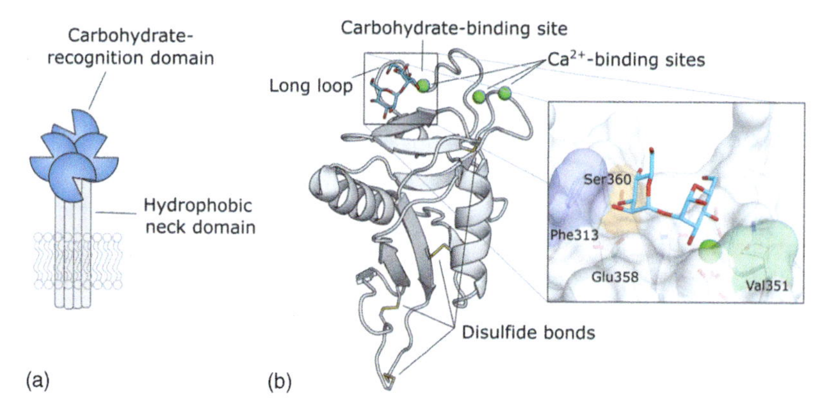

Figure 5.1 Structure of DC-SIGN. (a) Schematic depiction of DC-SIGN domain organization. (b) The carbohydrate-binding site of DC-SIGN in complex with α-1,2-mannobiose (PDB 2IT6). Secondary interaction sites commonly targeted by glycomimetics are highlighted.

DC-SIGN naturally binds to underprocessed high-mannose-type glycans (e.g. $Man_9GlcNAc_2$) that are abundantly presented on viral envelope glycoproteins. Fragments of this glycan as well as mannose itself also bind to DC-SIGN, albeit with lower binding affinity. In addition, DC-SIGN recognizes fucosylated glycans such as Lewis-type and ABO antigens. Canonically, mannose or fucose epitopes coordinate with the central calcium ion via their 3-OH and 4-OH groups [44, 45]. However, it has been demonstrated that mannose ligands are able to bind in a variety of transient binding modes, also employing other hydroxyl functions [46]. Various concepts have been employed to identify glycomimetic ligands that utilize secondary binding sites to achieve higher monovalent affinity toward DC-SIGN. Besides ligand-/structure-based design [47, 48], combinatorial [49, 50] and fragment-based approaches [51, 52] were successful. Important sites for additional secondary interactions in an extended binding site are highlighted in Figure 5.1: In the long loop, Val351 mediates binding of Lewis-type antigens to DC-SIGN [45, 53]. A hydrophobic subsite in the vicinity of Phe313 is an additional target for glycomimetics [52, 54]. This allosteric pocket, as well as several other distal areas, have been identified as binding sites for noncarbohydrate fragments. Finally, Glu358 and Ser360 have been shown to act as binding partners for positively charged residues [47]. In general, druggability of DC-SIGN and other CLRs has been soundly demonstrated by experimental and computational approaches [51].

Approaches for the development of carbohydrate-based inhibitors of viral attachment and entry can be categorized into two different groups. Firstly, natural mono- or oligosaccharide ligands of DC-SIGN have been utilized for the synthesis of various multivalent systems. A second approach relies on the design of carbohydrate derivatives or glycomimetics that surpass the affinity of natural ligands and can be employed as monovalent therapeutics or utilized for the construction of multivalent systems with improved affinity.

5.2.2 DC-SIGN Ligands Employing Natural Carbohydrate Epitopes

Section 5.2.2. gives an account of the multivalent scaffolds employed for the presentation of mannose- and fucose-based epitopes (Table 5.2), as well as general trends observed in several studies. It must be emphasized that the multitude of different assay formats used across the literature employing either recombinant protein or a cellular system renders a direct comparison of multivalent ligands impossible. In any case, numeric affinity values are mainly determined by specific assay setups and are difficult to compare. This problem could easily be avoided by including monovalent controls (e.g. methyl α-D-mannoside) in every study and reporting relative affinities compared to this control. Unfortunately, this easily implementable solution is only applied sporadically, and we would like to encourage this practice for future publications of affinity data for multivalent ligands in general.

Table 5.2 Overview over multivalent systems employed for DC-SIGN targeting.

Multivalent scaffold		Carbohydrate epitopes	References
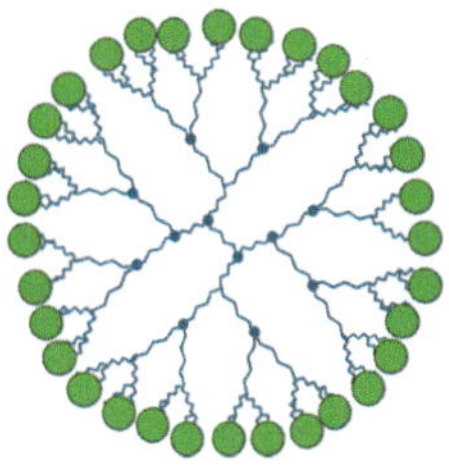	Dendrimers	Monosaccharides	[55, 56]
		α-1,2-mannobiose	[57]
		Glycan-derived oligosaccharides	[58, 59]
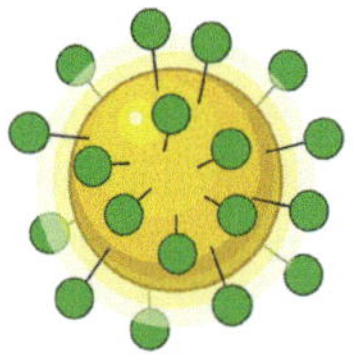	Nanoparticles	Monosaccharides	[60–62]
		α-1,2-mannobiose	[61–63]
		Glycan-derived oligosaccharides	[64, 65]
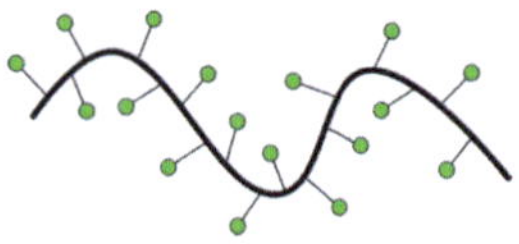	Polymers	Mannose	[66–69]
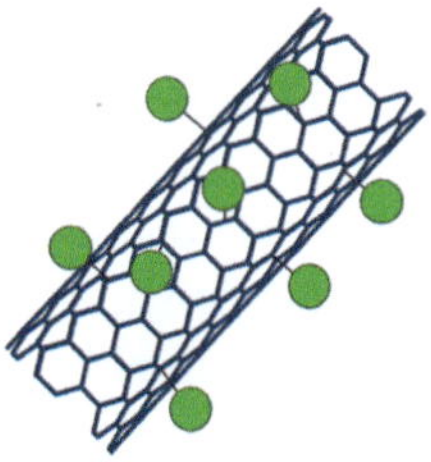	Other scaffolds	Monosaccharides	[70–73]

5.2.2.1 Dendrimers

Hyperbranched dendritic polymers and polyamido amine (PAMAM) dendrimers (Boltorn^{TM}) have been extensively used for the construction of multivalent DC-SIGN ligands [55–57]. As a general trend observed in these studies, only larger, high-valency compounds showed binding to DC-SIGN [55]. The most potent derivatives blocked DC-SIGN/gp120 interaction with a nanomolar IC_{50} in an ELISA assay, representing a 10^6-fold affinity enhancement compared with the monovalent mannose control [56]. It was also found that higher generation dendrimers, which were characterized by a larger particle size and wider glycan spacing, were more potent DC-SIGN binders [57]. Importantly, DC-SIGN binding and uptake of the dendrimers into DCs did not result in activation or maturation. In an impressive application, up to 360 α-1,2-mannobiose epitopes were displayed on [60]fullerene-based dendrimer scaffolds [58]. These particles were shown to inhibit the DC-SIGN-mediated infection of Jurkat DC-SIGN$^+$ cells with dengue and zika virus pseudotypes in picomolar concentration.

5.2.2.2 Nanoparticles

Gold nanoparticles (GNPs) decorated with linear and branched mannose oligosaccharides have been shown to inhibit the infection of human T cells in an HIV *trans*-infection assay in nano- to subnanomolar concentrations [59, 60]. Similarly, mannose- or α-1,2-mannobiose-functionalized GNPs inhibited ebola glycoprotein-driven infection of DC-SIGN$^+$ cells with up to 100 pM affinity [61]. Interestingly, glycan density on the GNPs, rather than the exact composition of the oligomannose ligand, determined the outcome of the experiment. Again, treatment of DCs with functionalized GNPs did not lead to maturation or induction of DC-SIGN-associated signaling, thus highlighting the pure targeting function of glyconanoparticles [62]. CdSe/ZnS quantum dots (QDs) functionalized with 369 carbohydrate epitopes potently bound to soluble recombinant DC-SIGN with an apparent K_D of 0.6 nM, an increase in binding affinity by a factor of 1.5×10^6 compared with monovalent analogs [63, 64]. Linker length was found to be a critical factor for the affinity of functionalized QDs. Long, more flexible PEG11 linkers resulted in weaker binding affinity compared with shorter PEG3 linkers. The authors hypothesized that the longer linker chains suffered from a more pronounced entropic penalty upon ligand binding, ultimately diminishing the overall free energy of binding. In an EBOVgp-pseudovirus infection assay, the QDs proved to be highly potent blockers of DC-SIGN-mediated infection, inhibiting virus internalization with an IC_{50} as low as 0.7 nM. Despite the high *in vitro* activity of CdSe/ZnS QDs, the authors acknowledge the severe drawbacks of the inherently cytotoxic nanoparticles for biological applications.

A notable feature of small, spherical nanoparticles is their efficient discrimination between DC-SIGN and the related receptor DC-SIGNR [61]. Despite an identical protein fold and high sequence similarity, these two proteins differ in the structural organization of their CRDs in their respective tetrameric assembly.

Whereas DC-SIGN aligns its CRDs vertically, orthogonal to the cell membrane ("closed flower" arrangement), DC-SIGNR orients its CRDs perpendicular to the plane of the cell membrane, thereby pointing each CRD in a different direction ("open flower" arrangement). As a result, DC-SIGN binds globular ligands with four CRDs simultaneously, capitalizing on additional multivalency effects and resulting in increased binding affinity. A single DC-SIGNR tetramer, however, binds globular ligands with a single CRD and, as a result, has lower affinity.

5.2.2.3 Polymers

Linear polymers, as well as nonlinear brush and bottlebrush polymers presenting mannose epitopes, have also been explored for their application toward DC-SIGN binding [65–67]. It was found that high valency was the main determinant for DC-SIGN affinity, whereas, unlike for other studied lectins, the degree of branching did not impact binding affinity significantly. When the polymer dextran was functionalized with mannose or *n*-heptyl α-D-mannoside (HM) [68], the polymer with the highest valency (902 HM epitopes) showed the highest binding affinity with an IC_{50} value in the subnanomolar range. This represents a 33-fold affinity enhancement per HM epitope compared with the monovalent control. The high affinity was corroborated in a trans-infection assay, where DC-SIGN-mediated trans-infection of permissive fibroblasts was inhibited with an IC_{50} of 20 pM, which is a 10^4-fold enhancement compared to the monovalent control methyl α-D-mannoside. Polymers based on a poly-L-lysine backbone can successfully inhibit viral adhesion to DC-SIGN-expressing cells with sub-nanomolar affinity [69]. It was shown that long polymers forming larger particles by aggregation were more potent inhibitors of glycoprotein binding compared to shorter, fully soluble analogs. A notable feature of poly-L-lysines is the fact that they are nonimmunogenic, biocompatible, and efficiently degraded by target cells, thus highlighting their potential for biological applications [69–72].

5.2.2.4 Other Multivalent Scaffolds

Besides the common multivalent carriers detailed above, a number of other systems have been used for targeting DC-SIGN. The most potent compounds from a study with calixarene and thiacalixarene scaffolds were able to inhibit *cis*-infection of $DC\text{-}SIGN^+$ cells in nanomolar concentrations [73, 74]. In another approach, dynamic micelles based on mono- and trivalent mannoside glycolipids have been investigated [75]. Finally, nanocarbon-based scaffolds carrying mannose dendrimers have been investigated for their ability to inhibit viral attachment to DC-SIGN-expressing cells [76]. In these experiments, single- and multiwall carbon nanotubes and nanohorns prevented DC-SIGN-mediated ebola virus infection at low concentrations down to $0.37\,\mu g\,ml^{-1}$.

5.2.3 DC-SIGN Ligands Employing Carbohydrate Derivatives or Glycomimetics

Besides improving the affinity of weak ligands by their multivalent presentation, the affinity of individual ligands can be enhanced by structural modifications of natural carbohydrates, enabling additional favorable interactions with the target.

Mitchell et al. reported DC-SIGN ligands based on C2-branched mannose derivatives obtained by a Kiliani ascension of fructose. In particular, 2-*C*-aminomethyl-D-mannose (**1**) stood out in a surface plasmon resonance (SPR) competition assay with a 48-fold improved binding affinity for DC-SIGN compared with mannose (from 17.1 to 0.35 mM) [77]. Glycomimetic DC-SIGN ligands identified from a library based on the shikimic acid core were reported by Garber et al. [49]. In a solid-phase fluorescence assay, the most potent hit **2** displayed an IC_{50} of 3.2 mM, which is about a fourfold increase compared to the positive control *N*-acetylmannosamine. When this molecule was subsequently incorporated into a linear polymer obtained by ring-opening metathesis, a multivalent ligand with micromolar affinity ($IC_{50} = 2.9\,\mu M$) was obtained. In an attempt to target two hydrophobic pockets adjacent to Phe313, Tomašić et al. introduced large hydrophobic substituents on a flexible bifurcated glycerol linker in the C1-position of α-D-mannose [48]. The most potent analog **3**, bearing two naphthyl substituents, showed an IC_{50} value of 40 μM in a competitive solid-phase immunoassay. Whereas molecular docking studies confirmed the intended binding mode targeting Phe313, molecular dynamics simulations starting from the predicted binding mode revealed residual flexibility of the ligand, allowing additional interactions for the hydrophobic substituents (Figure 5.2).

The Bernardi group pioneered the use of cyclohexane-based pseudo-dimannoside mimetics as DC-SIGN ligands. The glycomimetic scaffold **6** is accessible by oxidation of dimethyl (1*S*,2*S*)-cyclohex-4-ene-1,2-dicarboxylate (**4**) followed by selective

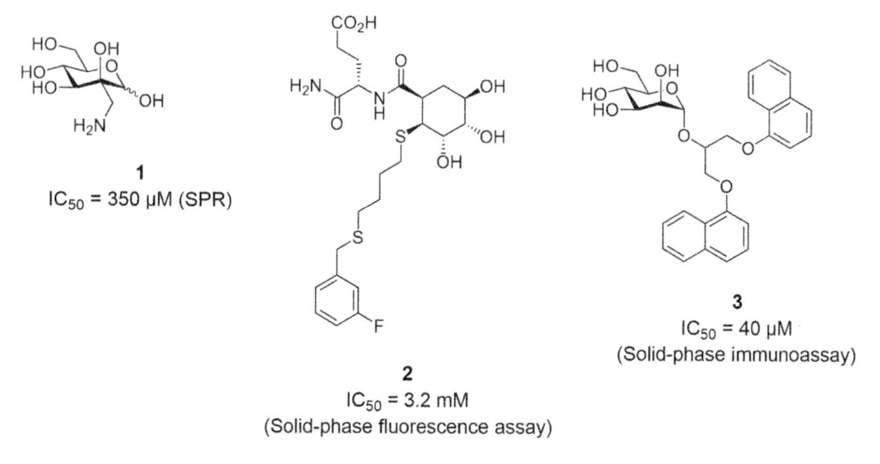

Figure 5.2 Monovalent carbohydrate derivatives and glycomimetics as DC-SIGN ligands.

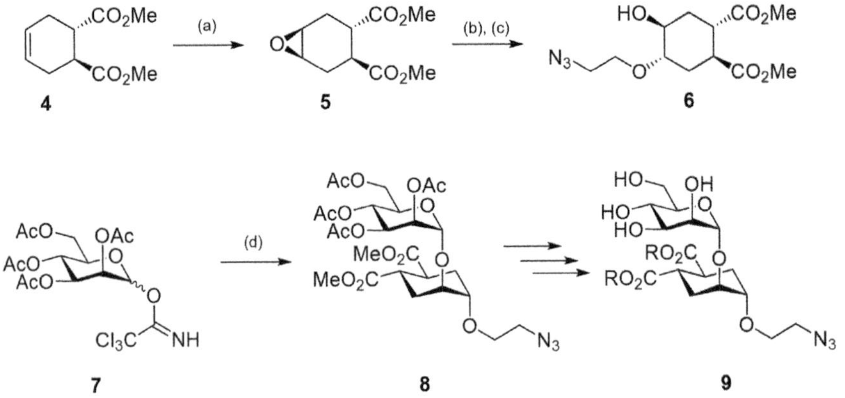

Scheme 5.1 Synthesis of α-1,2-mannobiose mimetics. (a) MCPBA, rt; (b) $Cu(OTf)_2$, 2-bromoethanol, rt; (c) NaN_3, 50 °C; and (d) TMSOTf, −20 °C.

alkoholysis of the resulting epoxide **5** [78]. This intermediate was then converted to dimannoside mimetics of the general formula **9** [79] (Scheme 5.1).

Since the first report of these glycomimetics, their binding affinity and selectivity toward the related CLR langerin, an anti-target in HIV therapy, have been continuously optimized. A key improvement was achieved by replacing the two carboxylic esters by benzylamides (**10**) [54, 80]. This enabled extended hydrophobic contacts to Val351 located in the long loop (see Figure 5.1), increasing potency to ca. $IC_{50} = 300\,\mu M$ in an SPR competition assay. The introduction of an amino substituent in position 6 of the mannose moiety yielded fully selective DC-SIGN ligand **11** with virtually no affinity toward langerin [81]. By virtual screening of a fragment library, an ammonium binding site was identified, which provided the incentive for the synthesis of a triazol library at the axial 2-position of the mannose moiety [47]. The best ligand **12** displayed an IC_{50} of 76 μM by SPR and a K_D of 52 μM by ITC (Figure 5.3).

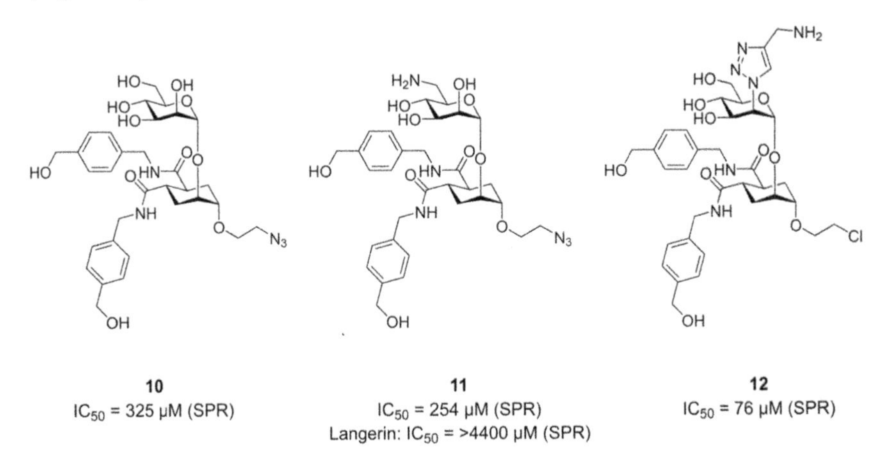

Figure 5.3 Representative structures of pseudo-dimannoside glycomimetics and their binding affinities to DC-SIGN.

Bernardi's DC-SIGN ligands were designed for multivalent applications. Grafting of pseudo-dimannoside **10** and a related pseudo-trisaccharide analog on a third-generation Boltorn dendrimer scaffold yielded multivalent ligands showing strong binding in an SPR competition experiment (IC_{50} = 1–2 μM) and efficient inhibition of DC-SIGN-mediated *cis*- (IC_{50} = 20 nM) and *trans*-infection (IC_{50} = 32–62 nM) in an Ebola virus infection model [82]. However, for high-valency derivatives, the multivalent presentation of glycomimetic **10** led to solubility issues [83]. When the same ligand **10** was presented on rigid molecular rods [84], affinity tended to increase with the length of the rod-like linker. The most potent compound, dubbed Polyman26 (**13**), reached the lower detection limit (IC_{50} = <5 μM) in a competitive SPR assay and efficiently prevented trans-infection of $CD4^+$ T cells in an HIV infection assay with an IC_{50} of 24 nM. Treatment of human immature monocyte-derived DCs with **13** induced the activation of immune responses with an elevated production of the chemokines CCL3, CCL4, and CCL5, as well as proinflammatory cytokines IL-1β, IL-6, and TNFα, highlighting the potential of DC-SIGN ligands for immunomodulatory applications [85]. The same compound has been investigated for its potential to block DC-SIGN-mediated trans-infection of $ACE2^+$ Vero E6 cells by SARS-CoV-2 pseudovirions. With an IC_{50} of 94 nM, the activity of **13** proved to be comparable to that in the HIV assay [86] (Figure 5.4).

Another class of mannose-based DC-SIGN ligands has been disclosed recently [87]. Supported by molecular docking experiments, a focused library of triazole glycomimetics with modifications in positions 2 and 6 was synthesized. Compounds with an aromatic substitution in position 2 were found to efficiently engage in a hydrophobic interaction with Val351. An extension of the anomeric position toward the ammonium binding pocket then generated **14**, which showed a K_D of 31 μM in an ITC assay. A multivalent poly-L-lysine polymer modified with a derivative of this glycomimetic ligand inhibited the *trans*-infection of susceptible Vero E6 cells by a SARS-CoV-2 pseudovirus with an IC_{50} of 4 nM (Figure 5.5).

5.2.4 Conclusion and Perspectives

Adhesion of pathogens to DC-SIGN and subsequent internalization is a conserved process that plays a role in a multitude of infectious diseases, predominantly of viral origin. Inhibition of DC-SIGN-mediated viral attachment to the cell surface with carbohydrate-derived therapeutics provides a host-directed pharmacological response, offering an attractive strategy for the development of broad-spectrum antivirals. Despite the affinity improvement from millimolar affinity for the monovalent natural ligand D-mannose to low-micromolar monovalent mimetics, the affinities are not yet sufficient to compete with the multivalent interaction between DC-SIGN clusters on the cellular surface and arrays of viral envelope glycoproteins. Thus far, only multivalent therapeutics have shown the potential for efficient competition in biologically relevant concentrations. Whereas multivalent compounds employing natural carbohydrate epitopes can reach impressive potency down to picomolar concentrations, an open question is the selectivity of these systems for DC-SIGN over other lectins binding mannose, such as Langerin and other CLRs or

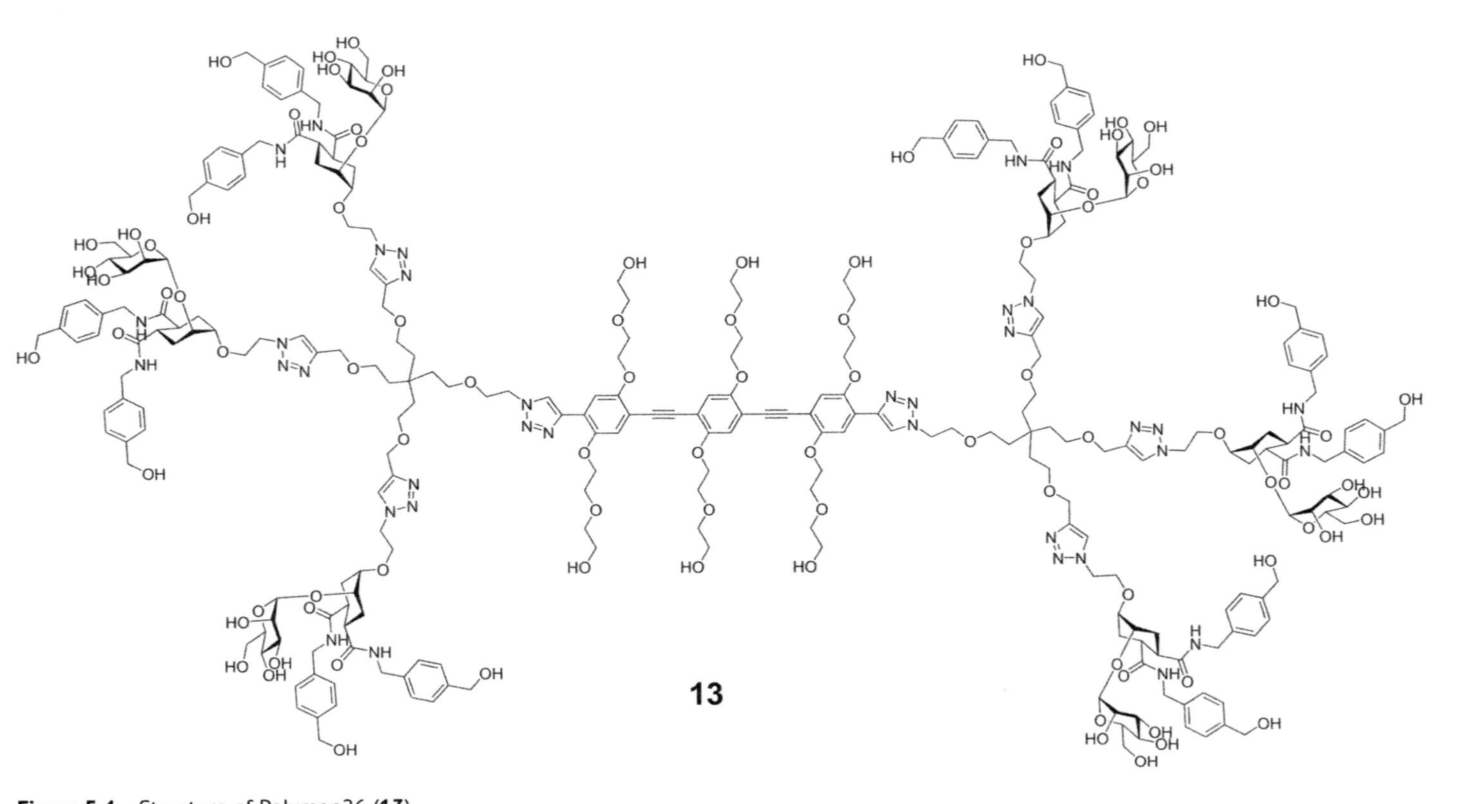

Figure 5.4 Structure of Polyman26 (**13**).

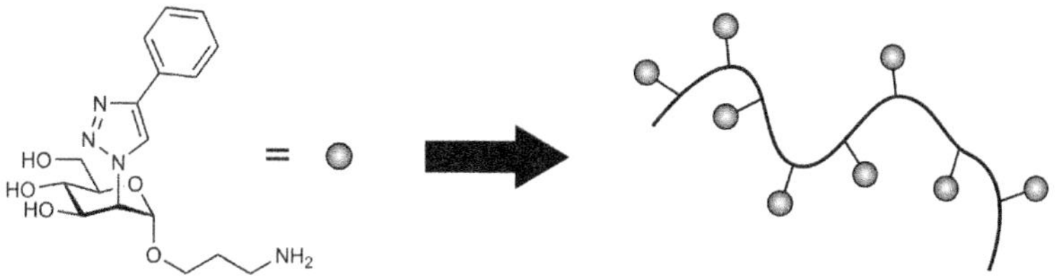

Figure 5.5 Structure of glycomimetic **14** and its multivalent presentation.

the mannose-binding lectin (MBL), which often also play a role in immune regulation. Here, the multivalent presentation of selective glycomimetics holds great potential to yield highly potent and selective therapeutics.

5.3 The Bacterial Adhesin FimH

5.3.1 UTIs and FimH

UTIs affect millions of people and account for significant morbidity and high medical costs worldwide [88, 89]. Statistics show that about 50% of women experience a UTI in their lives, and about 60% of them experience recurrent infections shortly after the preceding treatment [90, 91]. UTIs are primarily caused by Gram-negative UPEC, which constitute up to 90% of the diagnosed cases. UPEC infection involves a well-defined multistep cascade that has been demonstrated in mouse cystitis models and human UTIs [92]. At first, UPEC adhere with their type 1 fimbriae (pili) to mannosylated glycoprotein receptors, mainly uroplakin-Ia (UPIa), located on the surface of the urinary bladder mucosa [93]. This adhesion event, a prerequisite for bacterial invasion, prevents UPEC from being cleared by the shear stress of urine flow and triggers bacterial invasion into urothelial cells. After entering the host cells, UPEC starts replicating to form biofilm-like intracellular bacterial communities (IBC), protecting the bacteria from antibiotics treatment and host innate immune responses. Virtually all the clinical UPEC isolates express pili, which are uniformly distributed on their surface, amounting to 100–400 copies per cell [94–96]. Structurally, type 1 pili are 7 nm wide, several micrometers long, rod-like fibers. Assembled by the chaperone/usher pathway, the pilus rod is a right-handed helical structure composed of numerous immunoglobulin-like (Ig) FimA subunits, terminated by the fimbrial tip comprising FimF, FimG, and the lectin FimH.

5.3.2 FimH CRD

FimH (29 kDa) consists of two Ig-like domains: the N-terminal lectin domain ($FimH_{LD}$), which contains an α-D-mannose-specific carbohydrate-recognition domain (CRD), and the pilin domain ($FimH_{PD}$), which connects FimH to the pilus rod and regulates the switch between the low- and high-affinity states of the lectin

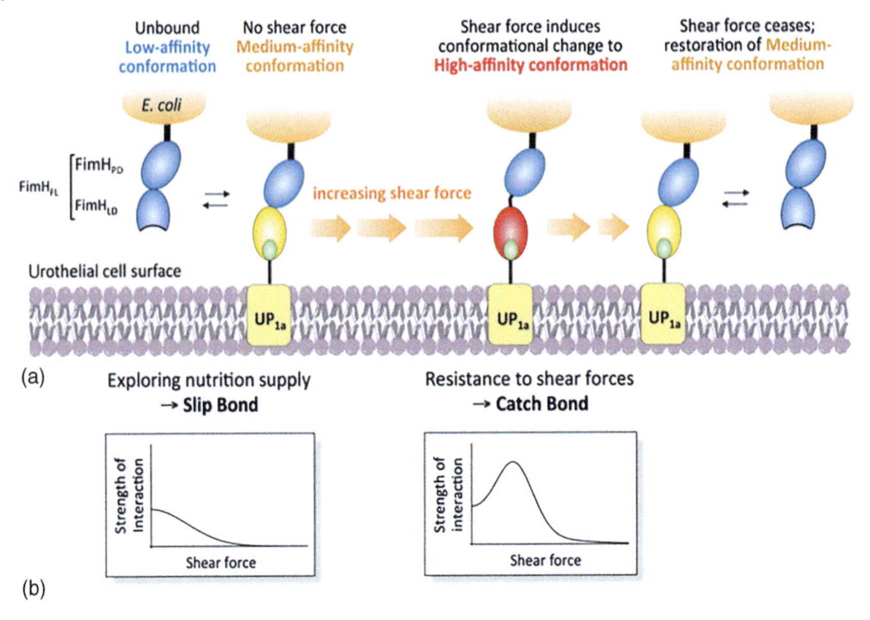

Figure 5.6 Schematic representation of FimH–uroplakin Ia (UP1a) interactions [97]. (a) Originally, bacteria were interested in undergoing only weak interactions with the host cells to retain mobility, a prerequisite to explore the urothelial surface for optimal nutrition supply. For this purpose, the equilibrium between the low-affinity conformation, characterized by an open binding pocket and intertwined domains (PDB ID: 4XOD), and medium-affinity conformation, characterized by well-defined binding pocket (PDB ID: 4XOE), is optimally suited. As urine flow arises, shear force induces a switch to high-affinity conformation with an approx. 2000-fold [98] improved affinity, allowing the bacteria to now withstand the shear force at least to a certain extent. In this conformation, the two domains are pulled apart (PDB ID: 4XOB). When shear force ceases, FimH restores the medium-affinity conformation. (b) When shear forces are low, bacteria can explore nutrition supply, and their pili are in the slip-bond state. Once shear forces increase, bacteria are protected from clearance, and the pili are in the catch-bound state. FL: FimH lectin, LD: lectin domain, PD: pilin domain. Reprinted with permission from Mayer, K., Eris, D., Schwardt, O. et al. (2017) Journal of Medicinal Chemistry 60: 5646–5662. Copyright 2017 American Chemical Society.

domain [96]. The two domains of FimH enable pathogens to induce a shift from low to high affinity (catch-bond, Figure 5.6) [97, 99]: Under tensile mechanical forces induced by urine flow, FimH forms stronger interactions with uroplakin located on the urothelial surface, preventing the elimination of bacteria by urination. More recently, Sauer et al. demonstrated that the domain-separated state of FimH resulted in a 2000-fold higher binding affinity compared to the domain-associated state of FimH [98]. The relatively weak affinity of FimH in the absence of shear force in turn enables bacterial motility on the urotheial surface, allowing rapid invasion of new tissue areas [100, 101]. Because the catch-bond effect is not observed with the FimH$_{LD}$ domain alone, full-length FimH serves as the best target for antiadhesive drug screening. The first crystal structure of the FimH lectin was determined in 1999 for FimC–FimH complexed with an oligomannoside ligand [102]. Since then, numerous structures of FimH (consisting of FimH$_{LD}$ and FimH$_{PD}$) alone or in complex with diverse mannoside ligands have been reported, greatly facilitating the

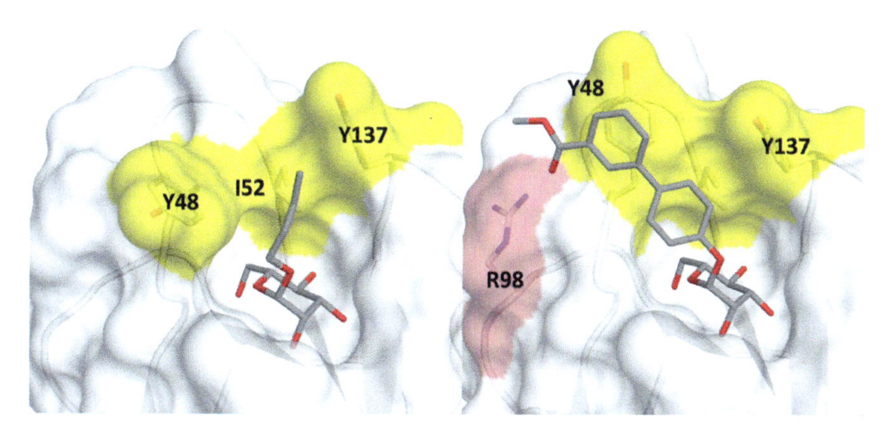

Figure 5.7 "Open" (left) and "closed" (right) conformation of the tyrosine gate. Left: crystal structure of *n*-butyl mannoside (PDB code: 1TR7) [104] bound to the FimH CRD, as a representative of the "open" conformation of tyrosine gate and "in"-docking mode of the ligand. Right: crystal structure of biphenyl mannoside bound to the FimH CRD (PDB code: 3MCY) [108], as a representative of the "closed" conformation of the tyrosine gate and "out"-docking mode of the ligand.

discovery of high-affinity FimH antagonists, such as the natural oligomannosides (Man3 and Man9) [98, 103] and glycomimetics of various structures [97, 104–107].

The main features of FimH CRD can be summarized as follows: (i) A deep and negatively charged pocket, the mannose moiety establishes direct hydrogen or water-mediated hydrogen bonds; (ii) the entrance of the binding site, referred to as the "tyrosine gate," formed by three hydrophobic amino acids (Tyr48, Ile52, and Tyr137), enables hydrophobic interactions with aliphatic/aromatic aglycones [35]; and (iii) the tyrosine gate can adopt two different conformations depending on the ligand structures (open and closed conformation, Figure 5.7) [103–105, 108].

5.3.3 FimH Antagonists

The different stages in the development of FimH antagonists are summarized in Figure 5.8. Already in 1979, Sharon and coworkers reported on the *in vivo* activity of methyl α-D-mannoside in a UTI mouse model [106]. In the following years, various structural modifications with the main goal of improving the affinity were explored. In 2005, Bouckaert et al. reported a series of alkyl α-D-mannosides (**1**, with $n = 0$–7) as potent FimH antagonists [104]. As they could show on the basis of the X-ray of *n*-butyl α-D-mannoside co-crystallized with $FimH_{LD}$, the potency is a result of van der Waals contacts of the alkyl aglycones with the so-called tyrosine gate, formed by Tyr48, Tyr137, and Ile52. The best representative of the alkyl mannoside series, *n*-heptyl α-D-mannoside, was used later on as reference compound. In addition to alkyl groups, aromatic aglycones were explored to reach the hydrophobic tyrosine gate. As early as the 1980s, Sharon and coworkers reported aromatic aglycones (→ **2**) to be able to enhance the binding affinity by a factor of approx. 600 compared to methyl α-D-mannoside [109]. Such findings were later rationalized with squaric acid

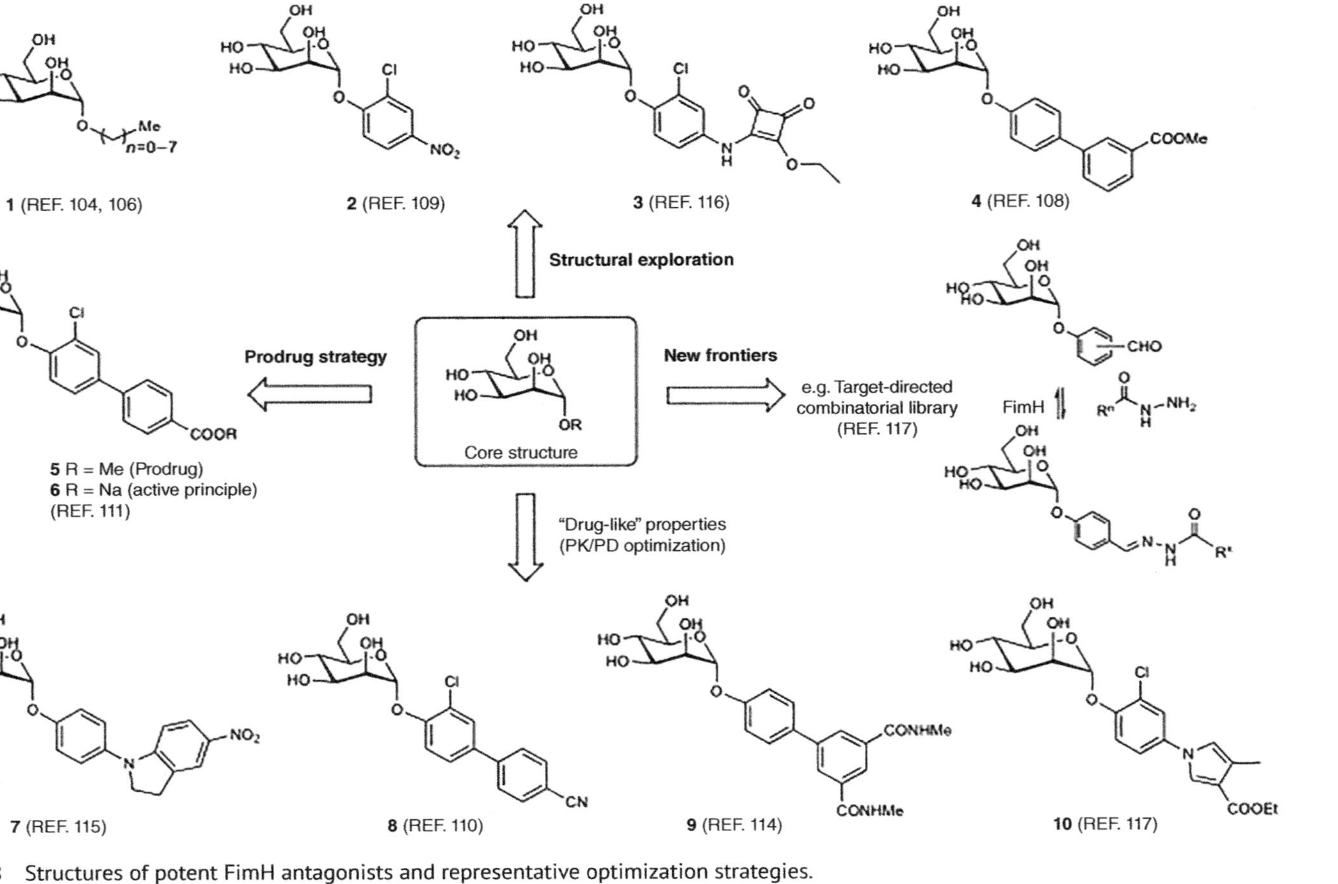

Figure 5.8 Structures of potent FimH antagonists and representative optimization strategies.

derivatives (→ **3**) and on the basis of various crystal structures (PDB code: 4AV5 [103], 3MCY [108], 4CST [110], and 4CSS [110]) with biphenyl mannosides (→ **4, 8**), indicating the positive effect of π–π stacking with the amino acid side chains of Tyr48 and Tyr137 of the "tyrosine gate." Furthermore, a favorable effect of *ortho*-substitution on the aromatic ring adjacent to the anomeric center of up to a factor 10 was reported by several groups (→ **2, 5,** and **6**) [109, 111].

Further studies on FimH antagonists with extended aromatic moieties lead to a series of modified biphenyl (→ **4-6, 8, 9**) [108, 110–114], indolinylphenyl (→ **7**) [115], squaric acid derivatives (→ **3**) [116], and pyrrolylphenyl (→ **10**) [117], all showing the affinities in low nanomolar range and oral availability.

For successful clinical applications of FimH antagonists, oral bioavailability and fast renal excretion for reaching the targets in the urinary tract are prerequisites. In 2010, Klein et al. reported both the *in vitro* and *in vivo* PK data of a series of biphenyl α-D-mannosides [111]. Because intestinal absorption and renal elimination are related to opposed properties, i.e. lipophilicity for intestinal absorption and hydrophilicity for renal elimination, a prodrug approach was applied to meet both conditions. For example, after high intestinal absorption, the ester (→ **5**) is hydrolzed by esterases in the enterocytes and in the liver ("first pass") releases the acid (→ **6**), which undergoes fast renal elimination to reach the target in the bladder, where it realizes the therapeutic effect. Later structural modifications on FimH antagonists, such as the indolinylphenyl derivative (→ **7**) [115] or bioisosteric replacement of the carboxylate [110], improved oral bioavailability *in vivo*. Among these modified structures, the indolinylphenyls showed a high therapeutic potential, resulting from optimized PK properties, and a substantial reduction of the dosage, i.e. a successful treatment of UTI with a low dosage of $1\,mg\,kg^{-1}$ without any additional administration of antibiotics. However, a major drawback of these indolinylphenyl antagonists is their low solubility, limiting their further *in vivo* applications. Therefore, a balance between solubility and permeability is another challenge for reaching oral bioavailability. Structural modifications, such as disruption of the molecular planarity and introduction of heteroatoms, provided promising solutions to fulfill this criterion [117]. Finally, FimH antagonist GSK3882347 (structure not disclosed) has already entered Phase I clinical study (NCT04488770) in a collaboration between Fimbrion Therapeutics and GlaxoSmithKline.

5.3.4 Conclusion and Perspectives

Diverse strategies, besides traditional chemical synthesis *in situ* target-directed dynamic combinatorial chemistry (Figure 5.8, New Frontiers) [107], were implemented in the search for high-potency FimH antagonists. Besides low-molecular-weight antagonists, carbohydrate-based clusters of mannosides [118–122] and carbohydrate dendrimers [123–126] were synthesized and extensively studied for FimH inhibition [127]. Although these multivalent mannosides have shown high potency, their sizes, polarity, and possible induction of gelation effects or hemagglutination *in vivo* make therapeutic application unlikely.

To summarize, *in vitro* and *in vivo* functionally active mannosides block UPEC adhesion through high-affinity binding to the mannose-binding site of FimH, thus preventing bacterial colonization on urinary tract surfaces. Over the last decade, our knowledge of FimH binding has been immensely expanded, thus facilitating the rational design of FimH antagonists with diverse structural complexity. Structural optimization for FimH antagonists has focused on improving pharmacokinetic and pharmacodynamic properties, aiming at *in vivo* potency and oral bioavailability. Ideally, after oral administration, the mannosides, which are characterized by acid stability in the stomach, will be quickly absorbed in the intestine and, once systemically available, will not be metabolized but rapidly eliminated by the kidney to reach their therapeutic target, the type 1 pili in the bladder and urinary tract. Encouragingly, some mouse models demonstrated that oral administration of the mannosides (→ **5**, **7**, **8**, and **9**) prevented UPEC colonization in the bladder in both acute and chronic UTIs [110, 111, 114, 115]. Additionally, biphenyl mannoside (→ **6**) has shown inhibitory potency on biofilm formation *in vitro* [114]. Recent research revealed that FimH adhesins act synergistically with PapG-II adhesins – another virulence factor prevalently expressed among the strains of UPEC causing pyelonephritis [128]. Therefore, in order to combat arising antibiotic resistance, combining both FimH and PapG antagonists could lead to a more effective treatment for UTIs [129, 130].

5.4 *Pseudomonas aeruginosa* Virulence Factors (PA-IL and PA-IIL)

5.4.1 Introduction

P. aeruginosa is an opportunistic Gram-negative pathogen that is part of the normal flora in healthy adults. However, it can become lethally pathogenic in immune-compromised patients [131]. *P. aeruginosa* is involved in both acute and chronic infections, especially in cystic fibrosis patients [131, 132]. The therapeutic options for these infections remain limited because this pathogen exhibits increasing resistance to many antibiotics [133]. *P. aeruginosa* utilizes lectins and adhesins, exposed on pili or flagella, for anchoring to the host cells. The soluble lectins PA-IL (or LecA) and PA-IIL (or LecB) are expressed by *P. aeruginosa*, specifically binding to galactosides and fucosides, respectively [27, 134, 135]. Both lectins are virulence factors under quorum-sensing control but are, by themselves, cytotoxic to primary epithelial cells [136]. Natively, PA-IL binds to α-galactosylated glycosphingolipids in lung epithelial cell membranes, while PA-IIL interacts with fucosylated and mannosylated epitopes but preferentially with Lewis^a oligosaccharides [137]. In a murine pneumonia model, both galactose and fucose reduced infection spread [138]. It was also reported that human milk oligosaccharides significantly prevent adhesion of *P. aeruginosa* to human respiratory epithelial cells [139]. In addition, a case report described that combination therapy of tobramycin with both D-galactose and L-fucose successfully cured an 18-month-old infant with systemic and pulmonary infections caused by tobramycin-resistant *P. aeruginosa* [140]. Given an increasing

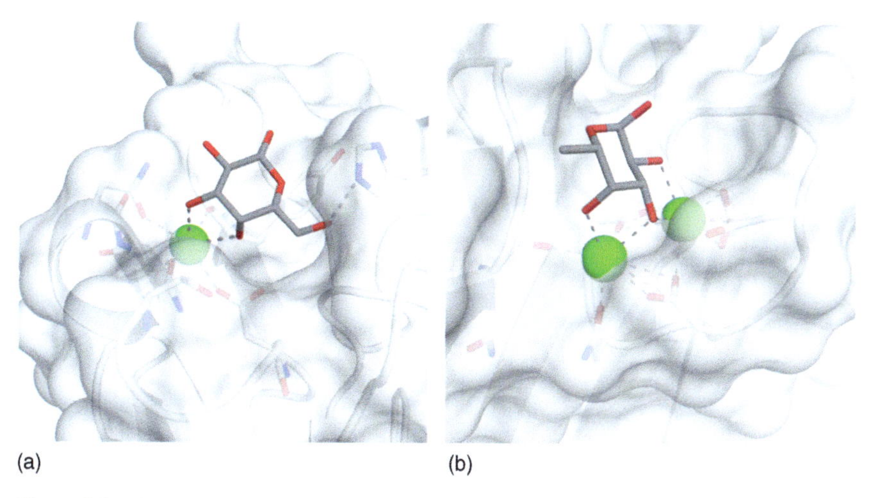

Figure 5.9 Crystal structures of PA-IL and PA-IIL in complex with D-galactose and L-fucose, respectively. (a) Binding sites of galactophilic lectin PA-IL complexed with D-galactose (PDB code: 1OKO). Source: Adapted from Cioci et al. [135]. (b) Binding sites of L-fucose-binding lectin PA-IIL complexed with L-fucose (PDB code: 1GZT). Source: Adapted from Mitchell et al. [27].

percentage of antibiotic resistance, antiadhesive strategy alone or in combination with antibiotics are expected to provide more effective clinical treatment to patients suffering *P. aeruginosa* infections [141, 142].

Both PA-IL and PA-IIL are tetrameric lectins that require Ca^{2+} ion for carbohydrate binding. PA-IL preferentially binds to terminal α-D-galactose in the presence of a bridging Ca^{2+} ion in the CRD with a K_D of 87.5 μM (Figure 5.9a) [135, 143], whereas PA-IIL binds unusually strongly (K_D: 625 nM) [144] to L-fucose with the involvement of two Ca^{2+} ions (Figure 5.9b) [27].

5.4.2 Mono- and Oligovalent Glycomimetic PL-Ligands

In nature, PA-IL displays high binding affinity toward α-linked galactosides, whereas studies with synthetic glycomimetics demonstrated its binding preference for β-aryl galactosides (Figures 5.10, 5.1 and 5.2) [145, 146]. Aromatic aglycones establish favorable contacts within the CRD, e.g. the 2-naphthyl moiety forms a CH–π "T-shape" interaction with His50, contributing to improved binding affinity [145].

When the aromatic aglycone of ligand **2** was presented tetravalently, even nanomolar affinity could be determined by ITC (→ **3**, Figure 5.10) [147]. Diversified multivalent scaffolds, such as tetravalent glycopeptide, glycoclusters of various sizes, fullerene, and gold nanoparticles, have been introduced and evaluated [34]. To date, a rationally designed divalent galactoside (→ **4**, Figure 5.10) with a K_D of 28 nM toward PA-IL has proven to be the ligand with the highest binding affinity so far [148]. Another 1,3-alternated galactosylated glycocluster (→ **5**, Figure 5.10) with a K_D of 176 nM against LecA has shown almost complete protection against

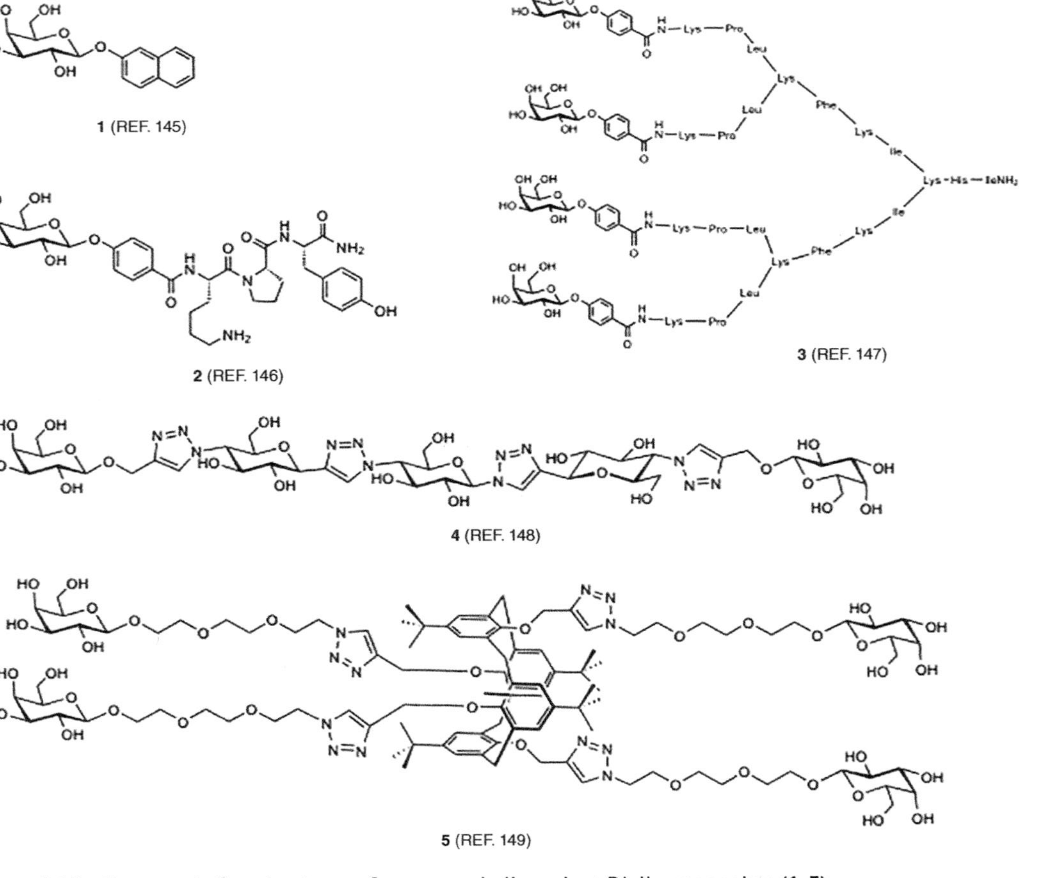

Figure 5.10 Representative structures of mono- and oligovalent PA-IL antagonists (**1-5**).

co-instillation of *P. aeruginosa* in a lung infection model in mice and therefore provides a promising drug candidate [149].

The proposed natural ligand of PA-IIL is the Lewis[a] trisaccharide, Galβ(1-3)[Fucα(1-4)]GlcNAc, which shows a dissociation constant of 210 nM [137]. To reduce the structural complexity of the trisaccharide, glycomimetics based on either Fuc alone or the Fucα(1-4)GlcNAc disaccharide were designed and synthesized (Figure 5.11) [150–160].

The crystal structure of PA-IIL in complex with sulfonamide **6a** (Figure 5.11) provides insight into binding details. A hydrogen bond of the sulfonamide with the carboxylate of Asp96 and lipophilic contacts with the protein surface represent the most

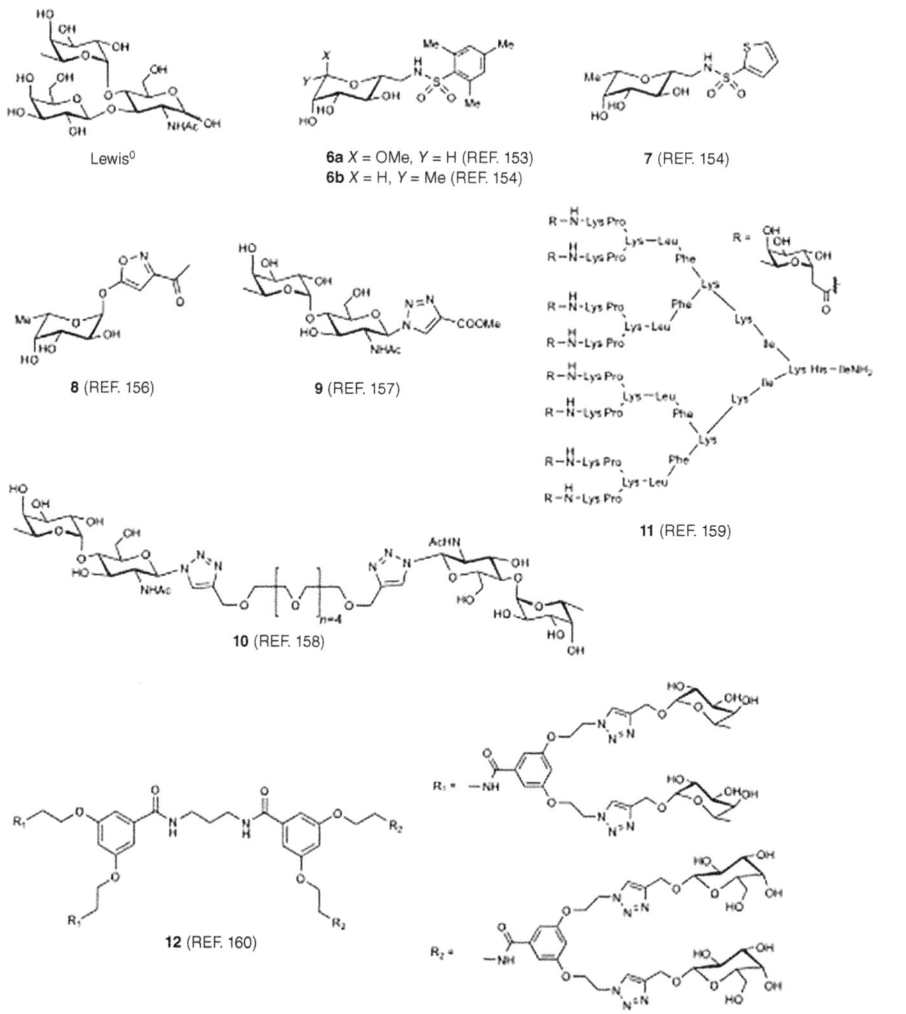

Figure 5.11 Representative structures of PA-IIL antagonists (**6-11**) and bifunctional glycodendrimer (**12**).

important contacts in this glycomimetic/lectin interaction [153]. Further structural optimization on the sulfonamide moiety resulted in two nanomolar glycomimetics (**6b** and **7**, Figure 5.11) with excellent receptor-binding kinetics and thermodynamic profiles [154]. Both **6b** and **7** efficiently blocked biofilm formation of *P. aeruginosa in vitro* and showed good oral bioavailability and pharmacokinetic properties *in vivo* [155]. The α-D-fucoside **8** bearing an isoxazol sulfonamide (Figure 5.11) showed a potency comparable to Lewis^a [156]. Additionally, partial structures of Lewis^a, such as derivative **9** (Figure 5.11) [157], also reached binding affinity in nanomolar range. ITC experiment with **9** revealed that increased entropy costs upon binding are probably related to increased flexibility of Fucα(1-4)GlcNAc compared to Lewis^a, which, however, is overcompensated by an enthalpy gain resulting from an extended hydrogen network. Finally, oligovalent ligands bearing Fucα(1-4)GlcNAc (→ **10**, Figure 5.11) [158] or L-Fuc (→ **11**, Figure 5.11) [159] exhibit increased binding activity toward PA-IIL; however, this effect is modest on a per saccharide basis.

Notably, heterobifunctional ligands presenting both D-Gal and L-Fuc in an oligovalent set-up (→ **12**, Figure 5.11) [160] showed efficacy in surgically stressed mice. Whereas 60% of the control group died within 48 hours after acute infection with *P. aeruginosa*, 100% of mice treated with **12** survived.

5.4.3 Conclusions and Perspectives

Studies on monovalent glycomimetics have revealed a series of high-affinity PA-IL and PA-IIL antagonists with low toxicity, good metabolic stability, and oral bioavailability. A notable feature of some candidates is their inhibitory potency against biofilm formation without affecting bacterial viability. Therefore, development of resistance toward these antibiofilm agents is unlikely, in contrast to traditional bactericidal or bacteriostatic antibiotics. Additionally, combining elements of both PA-IL and PA-IIL antagonists in one molecule represents a new, therapeutically valuable compound class for fighting *P. aeruginosa* infections. Future developments could include evaluation of antiadhesive therapeutics in a monotherapy treatment against biofilm-associated infections as well as their synergistic effects with antibiotics for eradication of bacteria outside biofilms [141, 142].

5.5 General Aspects

Two topics, namely resistance and affinity of carbohydrates, are of general importance and are therefore not presented in each chapter separately but in a general form in this last chapter.

Resistance. Drug resistance reduces the effectiveness of a medication, such as an antimicrobial or an antiviral, in treating a disease. The alarming increase in drug-resistant bacteria makes a search for novel anti-infective drugs mandatory [161].

It is well established that adhesion of enteric, oral, and respiratory bacteria is the initial step required for colonization and the subsequent development of disease. An attractive possibility to reduce these contacts is to provide agents that interfere with the ability of bacteria to adhere to tissues of their hosts. The validity of this approach has been demonstrated in experiments performed in a wide variety of animals, from mice to monkeys, and also in humans [10]. Because antiadhesive agents are not bactericidal, the propagation and spread of resistant strains are much less likely to occur after exposure to bactericidal agents, such as antibiotics. Inhibitor of adhesins, like FimH, PapG, or PA, involved in the first contact between bacteria and host cells [162]. Also, blockers of viral entry receptors like DC-SIGN are, therefore, regarded as a new means to fight infectious diseases.

Affinity of carbohydrate–lectin interactions. The interactions of carbohydrates with proteins are mainly mediated by the directional hydrogen bond contacts of hydroxyl groups. Thus, lectins have evolved to recognize the spatial arrangement of these functional groups on carbohydrate scaffolds. In a typically shallow and solvent-exposed lectin-binding pocket, highly mobile water molecules can easily assume all relevant interactions and functionally replace the carbohydrate ligand. In view of this, it is not self-evident why carbohydrate ligands bind to proteins with any measurable affinity at all. To compensate for the unfavorable consequences of highly polar interactions in solvent-exposed binding sites, carbohydrates and lectins utilize a set of physicochemical mechanisms to mitigate these unfavorable properties (cooperative desolvation and modulation of local dielectric properties) [163–165] or leverage other effects (hydrophobic CH–π interactions and preorganization) [166, 167] to enhance their binding affinity. However, due to competition with an omnipresent solvent, the resulting interactions typically remain weak on a monovalent level as experienced in case of DC-SIGN, PapG, or PA antagonists. High avidity in a biological context is achieved by expanding the monovalent contact to an oligo- or multivalent level [168]. In nature, interaction of multiple carbohydrate epitopes, e.g. on a glycoprotein, with a surface displaying an equally high number of receptors potentiates the weak monovalent interaction and gives thermodynamically complex systems with high overall avidity. Inspired by nature, medicinal carbohydrate chemistry applies comparable approaches to convert low-affinity natural carbohydrates or medium-affinity glycomimetics into high-affinity, therapeutically valuable compounds.

References

1 Holmes, K.K., Bertozzi, S., Bloom, B.R. et al. (2017). *Major Infectious Diseases: Key Messages from Disease Control Priorities*. Chapter 1, 3e (ed. K.K. Holmes, S. Bertozzi, and B.R. Bloom). Washington DC: International Bank for Reconstruction and Development.

2 Szymanski, C.M., Schnaar, R.L., and Aebi, M. (2015–2017). Chapter 42 – Bacterial and viral infections. In: *Essentials in Glycobiology*, 3e (ed. A. Varki, R.D. Cummings, J.D. Esko, et al.). NY: Cold Spring Harbor Laboratory Press.

3 Vigerust, D.J. and Shepherd, V.L. (2007). *Trends in Microbiology* 15: 211–218.

4 Watanabe, Y., Bowden, T.A., Wilson, I.A. et al. (1863). *Biochimica et Biophysica Acta, General Subjects* 2019: 1480–1497.
5 Raman, R., Tharakaraman, K., Sasisekharan, V. et al. (2016). *Current Opinion in Structural Biology* 40: 153–162.
6 Hillaire, M.L.B., Nieuwkoop, N.J., Boon, A.C.M. et al. (2013). *PLoS One* 8: e56164.
7 Pizarro-Cerdá, J. and Cossart, P. (2006). *Cell* 124: 715–727.
8 Sharon, N. (1987). *FEBS Letters* 217: 145–157.
9 Taylor, S.L., McGuckin, M.A., Wesselingh, S., and Rogers, G.B. (2018). *Trends in Microbiology* 26: 92–101.
10 Ofek, I., Hasty, D.L., and Sharon, N. (2003). *FEMS Immunology and Medical Microbiology* 38: 181–191.
11 Poole, J., Day, C.J., von Itzstein, M. et al. (2018). *Nature Reviews. Microbiology* 16: 440–452.
12 Sharon, N., Lis, H., and Lotan, R. (1974). *Coll. Int. CNRS* 221: 693–709.
13 Ruiz-Palacios, G.M., Cervantes, L.E., Ramos, P. et al. (2003). *The Journal of Biological Chemistry* 278: 14112–14120.
14 Wellens, A., Garofalo, C., Nguyen, H. et al. (2008). *PLoS One* 3: e2040.
15 Svenson, S.B., Hultberg, H., Källenius, G. et al. (1983). *Infection* 11: 61–67.
16 Parkkinen, J., Rogers, G.N., Korhonen, T. et al. (1986). *Infection and Immunity* 54: 37–42.
17 Lis, H. and Sharon, N. (1998). *Chemical Reviews* 98: 637–674.
18 Madhavan, T.P.V., Riches, J.D., Scanlon, M.J. et al. (2016). *Infection and Immunity* 84: 1642–1649.
19 Mottram, L., Liu, J., Chavan, S. et al. (2018). *Scientific Reports* 8: 11250.
20 St Geme, J.W. (1994). *Infection and Immunity* 62: 3881–3889.
21 Ilver, D., Arnqvist, A., Ogren, J. et al. (1998). *Science* 279: 373–377.
22 Pang, S.S., Nguyen, S.T., Perry, A.J. et al. (2014). *The Journal of Biological Chemistry* 289: 6332–6340.
23 Rosen, D.A., Pinkner, J.S., Walker, J.N. et al. (2008). *Infection and Immunity* 76: 3346–3356.
24 Pethe, K., Aumercier, M., Fort, E. et al. (2000). *The Journal of Biological Chemistry* 275: 14273–14280.
25 Dehio, C., Gray-Owen, S.D., and Meyer, T.F. (1998). *Trends in Microbiology* 6: 489–495.
26 Blanchard, B., Nurisso, A., Hollville, E. et al. (2008). *Journal of Molecular Biology* 383: 837–853.
27 Mitchell, E., Houles, C., Sudakevitz, D. et al. (2002). *Nature Structural Biology* 9: 918–921.
28 Kisiela, D., Laskowska, A., Sapeta, A. et al. (2006). *Microbiology* 152: 1337–1346.
29 Hobbs, J.K., Pluvinage, B., and Boraston, A.B. (2018). *FEBS Letters* 592: 3865–3897.
30 Kouki, A., Pieters, R.J., Nilsson, U.J. et al. (2013). *Biology (Basel)* 2: 918–935.
31 Sharon, N. (2006). *Biochimica et Biophysica Acta* 1760: 527–537.
32 Koropatkin, N.M., Cameron, E.A., and Martens, E.C. (2012). *Nature Reviews Microbiology* 10: 323–335.

33 Ernst, B. and Magnani, J.L. (2009). *Nature Reviews Drug Discovery* 8: 661–677.
34 Cecioni, S., Imberty, A., and Vidal, S. (2015). *Chemical Reviews* 115: 525–561.
35 Sattin, S. and Bernardi, A. (2016). *Trends in Biotechnology* 34: 483–495.
36 Lozupone, C.A., Stombaugh, J.I., Gordon, J.I. et al. (2012). *Nature* 489: 220–230.
37 Paharik, A.E., Schreibe, H.L., Spaulding, C.N. et al. (2017). *Genome Medicine* 9: 110.
38 Monteiro, J. and Lepenies, B. (2017). *Viruses* 9: 59.
39 Flores-Mireles, A.L., Walker, J.N., Caparon, M., and Hultgren, S.J. (2015). *Nature Reviews. Microbiology* 13: 269–284.
40 Ibrahin, D., Jabbour, J.-F., and Kanj, S.S. (2020). *Current Opinion in Infectious Diseases* 33: 464–473.
41 Bermejo-Jambrina, M., Eder, J., Helgers, L.C. et al. (2018). *Frontiers in Immunology* 9: 590.
42 Garcia-Vallejo, J.J. and van Kooyk, Y. (2015). *Immunity* 42: 983–985.
43 van Kooyk, Y. and Geijtenbeek, T.B.H. (2003). *Nature Reviews. Immunology* 3: 697–709.
44 Feinberg, H., Castelli, R., Drickamer, K. et al. (2007). *The Journal of Biological Chemistry* 282: 4202–4209.
45 Guo, Y., Feinberg, H., Conroy, E. et al. (2004). *Nature Structural & Molecular Biology* 11: 591–598.
46 Martínez, J.D., Valverde, P., Delgado, S. et al. (2019). *Molecules* 24: 2337.
47 Medve, L., Achilli, S., Guzman-Caldentey, J. et al. (2019). *Chemistry - A European Journal* 25: 14659–14668.
48 Tomašić, T., Hajšek, D., Švajger, U. et al. (2014). *European Journal of Medicinal Chemistry* 75: 308–326.
49 Garber, K.C.A., Wangkanont, K., Carlson, E.E. et al. (2010). *Chemical Communications* 46: 6747–6749.
50 Ng, S., Bennett, N.J., Schulze, J. et al. (2018). *Bioorganic & Medicinal Chemistry* 26: 5368–5377.
51 Aretz, J., Wamhoff, E.-C., Hanske, J. et al. (2014). *Frontiers in Immunology* 5: 323.
52 Aretz, J., Baukmann, H., Shanina, E. et al. (2017). *Angewandte Chemie International Edition* 56: 7292–7296.
53 Valverde, P., Delgado, S., Martínez, J.D. et al. (2019). *ACS Chemical Biology* 14: 1660–1671.
54 Obermajer, N., Sattin, S., Colombo, C. et al. (2011). *Molecular Diversity* 15: 347–360.
55 Tabarani, G., Reina, J.J., Ebel, C. et al. (2006). *FEBS Letters* 580: 2402–2408.
56 Wang, S.-K., Liang, P.-H., Astronomo, R.D. et al. (2008). *Proceedings of the National Academy of Sciences of the United States of America* 105: 3690–3695.
57 Garcia-Vallejo, J.J., Koning, N., Ambrosini, M. et al. (2013). *International Immunology* 25: 221–233.
58 Ramos-Soriano, J., Reina, J.J., Illescas, B.M. et al. (2019). *Journal of the American Chemical Society* 141: 15403–15412.
59 Martínez-Ávila, O., Bedoya, L.M., Marradi, M. et al. (2009). *ChemBioChem* 10: 1806–1809.

60 Di Gianvincenzo, P., Chiodo, F., Marradi, M. et al. (2012). *Methods in Enzymology* 509: 21–40.
61 Budhadev, D., Poole, E., Nehlmeier, I. et al. (2020). *Journal of the American Chemical Society* 142: 18022–18034.
62 Arosio, D., Chiodo, F., Reina, J.J. et al. (2014). *Bioconjugate Chemistry* 25: 2244–2251.
63 Guo, Y., Sakonsinsiri, C., Nehlmeier, I. et al. (2016). *Angewandte Chemie International Edition* 55: 4738–4742.
64 Guo, Y., Nehlmeier, I., Poole, E. et al. (2017). *Journal of the American Chemical Society* 139: 11833–11844.
65 Shamout, F., Monaco, A., Yilmaz, G. et al. (2020). *Macromolecular Rapid Communications* 41: e1900459.
66 Beyer, V.P., Monaco, A., Napier, R. et al. (2020). *Biomacromolecules* 21: 2298–2308.
67 Becer, C.R., Gibson, M.I., Geng, J. et al. (2010). *Journal of the American Chemical Society* 132: 15130–15132.
68 Brument, S., Cheneau, C., Brissonnet, Y. et al. (2017). *Organic & Biomolecular Chemistry* 15: 7660–7671.
69 Cramer, J., Aliu, B., Jiang, X. et al. (2021). *ChemMedChem* 16: 2345–2353.
70 Rai, R., Alwani, S., and Badea, I. (2019). *Polymers (Basel)* 11: 745.
71 Herrendorff, R., Hänggi, P., Pfister, H. et al. (2017). *Proceedings of the National Academy of Sciences* 114: E3689–E3698.
72 Aliu, B., Demeestere, D., Seydoux, E. et al. (2020). *Journal of Neurochemistry* 154: 486–501.
73 Morbioli, I., Porkolab, V., Magini, A. et al. (2017). *Carbohydrate Research* 453, 454: 36–43.
74 Taouai, M., Porkolab, V., Chakroun, K. et al. (2019). *Bioconjugate Chemistry* 30: 1114–1126.
75 Schaeffer, E., Dehuyser, L., Sigwalt, D. et al. (2013). *Bioconjugate Chemistry* 24: 1813–1823.
76 Rodríguez-Pérez, L., Ramos-Soriano, J., Pérez-Sánchez, A. et al. (2018). *Journal of the American Chemical Society* 140: 9891–9898.
77 Mitchell, D.A., Jones, N.A., Hunter, S.J. et al. (2007). *Tetrahedron: Asymmetry* 18: 1502–1510.
78 Bernardi, A., Arosio, D., Manzoni, L. et al. (2001). *The Journal of Organic Chemistry* 66: 6209–6216.
79 Reina, J.J., Sattin, S., Invernizzi, D. et al. (2007). *ChemMedChem* 2: 1030–1036.
80 Varga, N., Sutkeviciute, I., Guzzi, C. et al. (2013). *Chemistry - A European Journal* 19: 4786–4797.
81 Porkolab, V., Chabrol, E., Varga, N. et al. (2018). *ACS Chemical Biology* 13: 600–608.
82 Luczkowiak, J., Sattin, S., Sutkevičiute, I. et al. (2011). *Bioconjugate Chemistry* 22: 1354–1365.
83 Varga, N., Sutkeviciute, I., Ribeiro-Viana, R. et al. (2014). *Biomaterials* 35: 4175–4184.

84 Ordanini, S., Varga, N., Porkolab, V. et al. (2015). *Chemical Communications* 51: 3816–3819.

85 Berzi, A., Ordanini, S., Joosten, B. et al. (2016). *Scientific Reports* 6: 35373.

86 Thépaut, M., Luczkowiak, J., Vivès, C. et al. (2021). *PLoS Pathogens* 17: e1009576.

87 Cramer, J., Lakkaichi, A., Aliu, B. et al. (2021). *Journal of the American Chemical Society* 143: 17465–17478.

88 Fihn, S.D. (2003). *The New England Journal of Medicine* 349: 259–266.

89 McLellan, L.K. and Hunstad, D.A. (2016). *Trends in Molecular Medicine* 22: 946–957.

90 Foxman, B., Somsel, P., Tallman, P. et al. (2001). *Journal of Clinical Epidemiology* 54: 710–718.

91 Hooton, T.M., Scholes, D., Hughes, J.P. et al. (1996). *The New England Journal of Medicine* 335: 468–474.

92 Cegelski, L., Marshall, G.R., Eldridge, G.R., and Hultgren, S.J. (2008). *Nature Reviews. Microbiology* 6: 17–27.

93 Capitani, G., Eidam, O., Glockshuber, R., and Grütter, M.G. (2006). *Microbes and Infection* 8: 2284–2290.

94 Schilling, J.D., Mulvey, M.A., and Hultgren, S.J. (2001). *The Journal of Infectious Diseases* 183: S36–S40.

95 Hahn, E., Wild, P., Hermanns, U. et al. (2002). *Journal of Molecular Biology* 323: 845–857.

96 Waksman, G. and Hultgren, S.J. (2009). *Nature Reviews. Microbiology* 7: 765–774.

97 Sauer, M.M., Jakob, R.P., Eras, J. et al. (2016). *Nature Communications* 7: 10738.

98 Sauer, M.M., Jakob, R.P., Luber, T. et al. (2019). *Journal of the American Chemical Society* 141: 936–944.

99 Thomas, W.E., Trintchina, E., Forero, M. et al. (2002). *Cell* 109: 913–923.

100 Aprikian, P., Tchesnokova, V., Kidd, B. et al. (2007). *The Journal of Biological Chemistry* 282: 23437–23446.

101 Le Trong, I., Aprikian, P., Kidd, B.A. et al. (2010). *Cell* 141: 645–655.

102 Choudhury, D., Thompson, A., Stojanoff, V. et al. (1999). *Science* 285: 1061–1066.

103 Wellens, A., Garofalo, C., Nguyen, H. et al. (2008). *PLoS One* 3: e2040.

104 Bouckaert, J., Berglund, J., Schembri, M. et al. (2005). *Molecular Microbiology* 55: 441–455.

105 Wellens, A., Lahmann, M., Touaibia, M. et al. (2012). *Biochemistry* 51: 4790–4799.

106 Aronson, M., Medalia, O., Schori, L. et al. (1979). *The Journal of Infectious Diseases* 139: 329–332.

107 Frei, P., Pang, L., Silbermann, M. et al. (2017). *Chemistry - A European Journal* 23: 11570–11577.

108 Han, Z.F., Pinkner, J.S., Ford, B. et al. (2010). *Journal of Medicinal Chemistry* 53: 4779–4792.

109 Firon, N., Ashkenazi, S., Mirelman, D. et al. (1987). *Infection and Immunity* 55: 472–476.

110 Kleeb, S., Pang, L., Mayer, K. et al. (2015). *Journal of Medicinal Chemistry* 58: 2221–2239.

111 Klein, T., Abgottspon, D., Wittwer, M. et al. (2010). *Journal of Medicinal Chemistry* 53: 8627–8641.

112 Pang, L.J., Kleeb, S., Lemme, K. et al. (2012). *ChemMedChem* 7: 1404–1422.

113 Han, Z.F., Pinkner, J.S., Ford, B. et al. (2012). *Journal of Medicinal Chemistry* 55: 3945–3959.

114 Cusumano, C.K., Pinkner, J.S., Han, Z. et al. (2011). *Science Translational Medicine* 3: 109–115.

115 Jiang, X.H., Abgottspon, D., Kleeb, S. et al. (2012). *Journal of Medicinal Chemistry* 55: 4700–4713.

116 Sperling, O., Fuchs, A., and Lindhorst, T.K. (2006). *Organic & Biomolecular Chemistry* 4: 3913–3922.

117 Pang, L., Bezençon, J., Kleeb, S. et al. (2017). FimH antagonists – solubility vs. permeability. In: *Carbohydrate Chemistry: Volume 42* (ed. A.P. Rauter, T. Lindhorst, and Y. Queneau), 248–273. Cambridge, UK: The Royal Society of Chemistry.

118 Touaibia, M., Shiao, T.C., Papadopoulos, A. et al. (2007). *Chemical Communications* 4: 380–382.

119 Sperling, O., Dubber, M., and Lindhorst, T.K. (2007). *Carbohydrate Research* 342: 696–703.

120 Almant, M., Moreau, V., Kovensky, J. et al. (2011). *Chemistry - A European Journal* 17: 10029–10038.

121 Ortega-Caballero, F., Gimenez-Martinez, J.J., and Vargas-Berenguel, A. (2003). *Org. Lett* 5: 2389–2392.

122 Bouckaert, J., Li, Z., Xavier, C. et al. (2013). *Chemistry - A European Journal* 19: 7847–7855.

123 Lindhorst, T.K., Kieburg, C., and Krallmann-Wenzel, U. (1998). *Glycoconjugate Journal* 15: 605–613.

124 Touaibia, M., Wellens, A., Shiao, T.C. et al. (2007). *ChemMedChem* 2: 1190–1201.

125 Nierengarten, I., Buffet, L., Holler, M. et al. (2013). *Tetrahedron Letters* 54: 2398–2402.

126 Durka, M., Buffet, K., Iehl, J. et al. (2011). *Chemical Communications* 47: 1321–1323.

127 Lindhorst, T.K. (2015). Small molecule ligands for bacterial lectins: letters of an antiadhesive glycopolymer code. In: *RSC Polymer Chemistry Series No. 15. Glycopolymer Code: Synthesis of Glycopolymers and Their Applications* (ed. C.R. Becer and L. Hartmann), 1–16. Cambridge, UK: The Royal Society of Chemistry.

128 Tseng, C.C., Lin, W.H., Wu, A.B. et al. (2020). *Journal of Microbiology, Immunology, and Infection* https://doi.org/10.1016/j.jmii.2020.09.001.

129 Sarshar, M., Behzadi, P., Ambrosi, C. et al. (2020). *Antibiotics (Basel)* 9: 397.

130 Ribić, R., Meštrović, T., Neuberg, M., and Kozina, G. (2019). *Medical Hypotheses* 124: 17–20.

131 de Bentzmann, S. and Plesiat, P. (2011). *Environmental Microbiology* 13: 1655–1665.

132 Hogardt, M. and Heesemann, J. (2013). *Current Topics in Microbiology and Immunology* 358: 91–118.

133 Lambert, P.A. (2002). *Journal of the Royal Society of Medicine* 95 (Suppl 41): 22–26.

134 Imberty, A., Wimmerova, M., Mitchell, E.P., and Gilboa-Garber, N. (2004). *Microbes and Infection* 6: 222–229.

135 Cioci, G., Mitchell, E.P., Gautier, C. et al. (2003). *FEBS Letters* 555: 297–301.

136 Winzer, K., Falconer, C., Garber, N.C. et al. (2000). *Journal of Bacteriology* 182: 6401–6411.

137 Perret, S., Sabin, C., Dumon, C. et al. (2005). *The Biochemical Journal* 389: 325–332.

138 Chemani, C., Imberty, A., de Bentzmann, S. et al. (2009). *Infection and Immunity* 77: 2065–2075.

139 Weichert, S., Jennewein, S., Hüfner, E. et al. (2013). *Nutrition Research* 33: 831–838.

140 von Bismarck, P., Schneppenheim, R., and Schumacher, U. (2001). *Klinische Pädiatrie* 213: 285–287.

141 Wagner, S., Sommer, R., Hinsberger, S. et al. (2016). *Journal of Medicinal Chemistry* 59: 5929–5969.

142 Meiers, J., Zahorska, E., Röhrig, T. et al. (2020). *Journal of Medicinal Chemistry* 63: 11707–11724.

143 Joachim, I., Rikker, S., Hauck, D. et al. (2016). *Organic & Biomolecular Chemistry* 14: 7933–7948.

144 Mitchell, E.P., Sabin, C., Snajdrová, L. et al. (2005). *Proteins* 58: 735–746.

145 Kadam, R.U., Garg, D., Schwartz, J. et al. (2013). *ACS Chemical Biology* 8: 1925–1930.

146 Kadam, R.U., Bergmann, M., Garg, D. et al. (2013). *Chemistry* 19: 17054–17063.

147 Kadam, R.U., Bergmann, M., Hurley, M. et al. (2011). *Angewandte Chemie International Edition* 50: 10631–10635.

148 Pertici, F., de Mol, N.J., Kemmink, J., and Pieters, R.J. (2013). *Chemistry* 19: 16923–16927.

149 Boukerb, A.M., Rousset, A., Galanos, N. et al. (2014). *Journal of Medicinal Chemistry* 57: 10275–10289.

150 Sommer, R., Wagner, S., Varrot, A. et al. (2016). *Chemical Science* 7: 4990–5001.

151 Sommer, R., Exner, T.E., and Titz, A. (2014). *PLoS One* 9: e112822.

152 Sommer, R., Hauck, D., Varrot, A. et al. (2015). *ChemistryOpen* 4: 756–767.

153 Hauck, D., Joachim, I., Frommeyer, B. et al. (2013). *ACS Chemical Biology* 8: 1775–1784.

154 Sommer, R., Wagner, S., Rox, K. et al. (2018). *Journal of the American Chemical Society* 140: 2537–2545.

155 Sommer, R., Rox, K., Wagner, S. et al. (2019). *Journal of Medicinal Chemistry* 62: 9201–9216.

156 Imberty, A., Chabre, Y.M., and Roy, R. (2008). *Chemistry* 14: 7490–7499.

157 Marotte, K., Sabin, C., Préville, C. et al. (2007). *ChemMedChem* 2: 1328–1338.

158 Marotte, K., Préville, C., Sabin, C. et al. (2007). *Organic & Biomolecular Chemistry* 5: 2953–2961.

159 Kolomiets, E., Swiderska, M.A., Kadam, R.U. et al. (2009). *ChemMedChem* 4: 562–569.

160 Magnani, J.L., Patton, J.T., Sarkar, A.K. US Patent 7517980B2, filled 08 August 8 2006 and issued 14 April 2009.

161 Ventola, C.L. (2015). *Pharmacology and Therapeutics* 40: 277–283.

162 Cusumano, Z.T., Klein, R.D., and Hultgren, S.J. (2016). *Microbiology Sprectrum* 4: 1–31.

163 Cramer, J., Sager, C.P., and Ernst, B. (2019). *Journal of Medicinal Chemistry* 62: 8915–8930.

164 Cramer, J., Jiang, X., Schönemann, W. et al. (2020). *RSC Chemical Biology* 1: 281–287.

165 Sager, C.P., Eriş, D., Smieško, M. et al. (2017). *Beilstein Journal of Organic Chemistry* 13: 2584–2595.

166 Hudson, K.L., Bartlett, G.J., Diehl, R.C. et al. (2015). *Journal of the American Chemical Society* 137: 15152–15160.

167 Binder, F.P.C., Lemme, K., Preston, R.C. et al. (2012). *Angewandte Chemie International Edition* 51: 7327–7331.

168 Kiessling, L.L., Young, T., Gruber, T.D. et al. (2008). Multivalency in protein–carbohydrate recognition. In: *Glycoscience* (ed. B.O. Fraser-Reid, K. Tatsuta, and J. Thiem), 2483–2523. Berlin/Heidelberg: Springer.

6

Targeting Carbohydrates in Cancer – Analytical and Biotechnological Tools

Henrique O. Duarte[1,2], *Joana Gomes*[1,2], *and Celso A. Reis*[1,2,3,4]

[1] *Universidade do Porto, i3S – Instituto de Investigação e Inovação em Saúde, Rua Alfredo Allen n°208, 4200-135, Porto, Portugal*
[2] *IPATIMUP – Instituto de Patologia e Imunologia Molecular da Universidade do Porto, Rua Júlio Amaral de Carvalho n°45, 4200-135, Porto, Portugal*
[3] *Universidade do Porto, ICBAS – Instituto de Ciências Biomédicas Abel Salazar, Porto, Portugal*
[4] *University of Porto, FMUP – Faculty of Medicine, Alameda Prof. Hernâni Monteiro, 4200-139, Porto, Portugal*

6.1 Aberrant Protein Glycosylation in Cancer

Glycosylation is defined as the enzymatic assembly of complex carbohydrate chains, or glycans, from simple monosaccharide sugar building blocks, and their covalent attachment to a diverse range of macromolecules to form an ensemble of distinct types of glycoconjugates, which, as a whole, constitute the cellular glycome [1]. Depending on the nature of their nonglycan component, glycoconjugates can be grouped into glycosphingolipids, proteoglycans, and glycoproteins. Glycoproteins constitute the main focus of this chapter. Although numerous cytoplasmic proteins represent eligible targets for dynamic glycosylation, the majority of a cell's glycan repertoire decorates both secreted and membrane-bound macromolecules [2]. The oligosaccharidic component of cell surface glycoproteins, which faces the extracellular space, forms an electron-dense layer known as the glycocalyx, which, under homeostasis, actively regulates a plethora of biological processes occurring inside and around a cell. Indeed, this layer of glycoconjugates constitutes a vital interface between a cell and the surrounding microenvironment, which includes not only neighboring cells but also noncellular components, such as elements of the extracellular matrix (ECM). This privileged localization grants the cellular glycome significant control over key biological processes, including proliferation, differentiation, motility, cytoskeletal rearrangements, inter- and extracellular communication, and neoplastic transformation [3].

The major glycan signatures that are significantly enriched in tumor cells and that have been mechanistically linked to malignant cellular transformation include:

Carbohydrate-Based Therapeutics, First Edition. Edited by Roberto Adamo and Luigi Lay.

highly branched *N*-glycan chains [4], highly fucosylated and sialylated glycans (including Lewis antigens) [5, 6], extended lactosamine polymers [7], and short prematurely terminated *O*-glycan structures (such as Tn, sialyl Tn [STn], and T antigens) [8]. The upregulation of specific aberrant carbohydrate antigens is accompanied by the concomitant reduction of specific homeostatic signatures, including bisected *N*-glycans. Aberrant glycan traits actively tune key malignant properties of proteins participating in central cellular processes driving oncogenic transformation, including mitogenic signaling, cell adhesion, motility, and invasion; metabolic regulation; interaction with cellular and acellular components of the immune system; angiogenic growth; apoptosis evasion; and acquisition of molecular resistance to targeted therapeutic agents [1, 9–11].

Given that tumor-associated carbohydrate antigens (TACAs) have their expression highly restricted to neoplastic tissues and play an undeniable role in the governing of malignant cell behavior, they have emerged as valuable theranostic tools in the clinical oncology field, either as robust and specific biomarkers for disease detection and monitoring, or as promising, yet still underexplored candidates for therapeutic targeting [10, 12–14]. Such efforts require the thorough characterization of aberrant tumor-specific glycan alterations in patient neoplastic lesions as well as the mechanistic dissection of their oncogenic role through the establishment of robust *in vitro* and *in vivo* models of disease. However, several structural and biological features of glycans pose unique challenges to their study. Firstly, despite the limited number of monosaccharides from which complex carbohydrates can be generated, glycans exhibit remarkable structural diversity stemming from: the specific composition and sequence of their constituent building blocks; the precise anomeric configuration and position of glycosidic linkages; variable degrees of branching and extension; and their potential to undergo further structural modifications, including sulfation, phosphorylation, and acetylation. Secondly, as glycosylation reflects the post-translational modification of multiple proteins, the accurate identification and tissue mapping of specific glycan epitopes through the use of traditional immunohistochemical approaches based on monoclonal antibodies (mAbs) and glycan-binding proteins (GBPs), including lectins of mammalian or plant origin, can be difficult to achieve. Furthermore, the expression of a single carbohydrate product results from the coordinated expression, activity, and localization of multiple isoenzymes, often showing partially redundant and overlapping specificities. This significantly increases the biological complexity required from genome-edited *in vitro* and *in vivo* models for the dissection of glycan biosynthetic pathways. Moreover, the cellular glycosylation landscape is highly dynamic and sensitive to spatiotemporal regulation, which further compromises the translational value of data retrieved from simpler glycoengineered cell and animal models. In addition, since identical monosaccharide compositions often reflect distinct tridimensional carbohydrate structures with distinct functional attributes, the development of analytical tools capable of retrieving isomeric linkage information has become fundamental for glycan structural characterization. Finally, although information on the glycan site occupancy and microheterogeneity of a given glycoprotein is essential to define its biological role, it may prove difficult to determine experimentally. Indeed, linking a particular glycan structure to a defined biological function remains a challenging task.

This chapter will discuss how recently developed analytical and biotechnological tools have significantly contributed to overcome the challenges posed by the unique features of glycans and their intricate biosynthetic pathways, thus supporting the structural and functional characterization of protein glycosylation in the context of human neoplastic transformation and the successful establishment of glycan-based biomarkers and therapeutic targets for the clinical management of cancer patients (Figure 6.1).

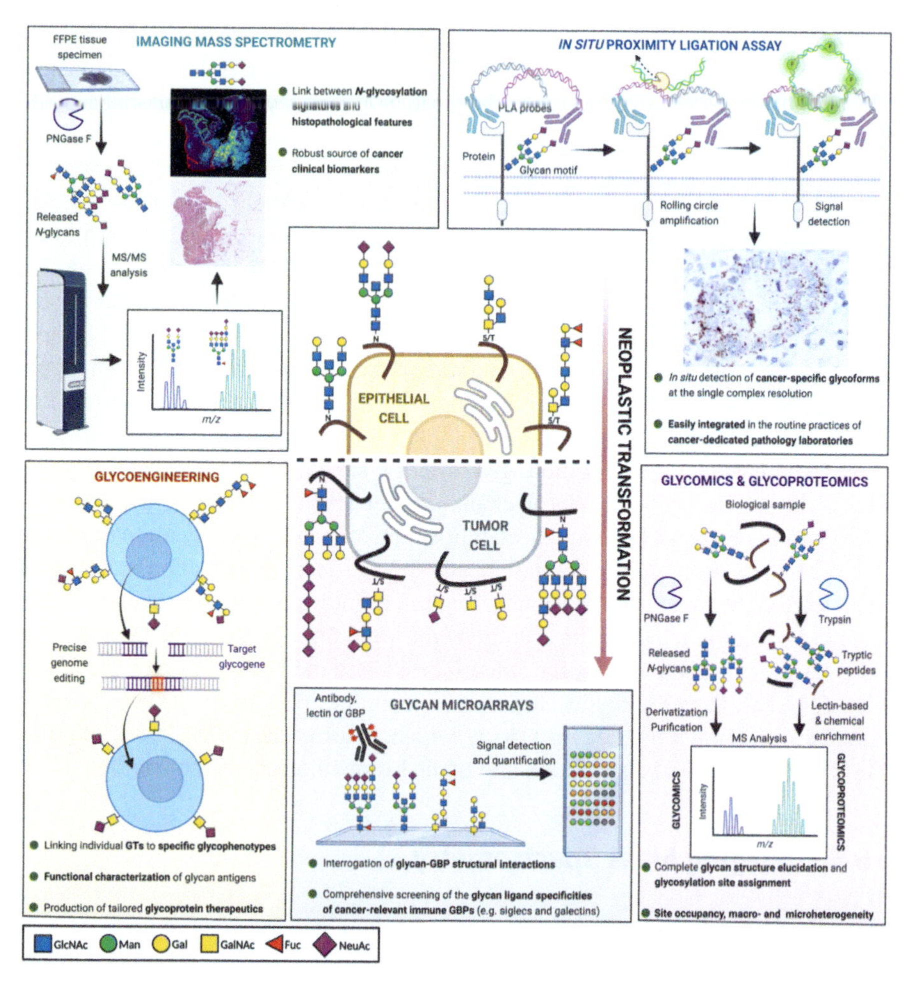

Figure 6.1 Biotechnological and analytical tools for the identification, functional characterization, and clinical application of glycan-based cancer biomarkers; FFPE - formalin-fixed paraffin-embedded; MS - mass spectrometry; GT - glycosyltransferase; GBP - glycan-binding protein.

6.2 Detection and Mapping of Carbohydrate-Based Antigens in Human Neoplastic Tissues

Over the past decades, numerous studies have sought to comprehensively characterize both the nature and impact of glycosylation alterations occurring within malignant cells and tissues (reviewed in [9]). In particular, the advent of hybridoma technology just under 50 years ago has supported the generation of a virtually unlimited catalog of highly specific mAbs for the reliable detection of any type of biological antigen, including glycans, and has, therefore, revolutionized the field of cancer biomarker discovery [15]. Indeed, the development of highly specific mAbs targeting TACAs, such as the prematurely truncated *O*-glycan determinant STn, allowed the unprecedented disclosure of their tissue-based cancer-specific expression pattern [16–18]. In the hybridoma system, activated immunoglobulin-secreting B cells, previously challenged with an isolated and structurally defined antigen of interest (ideally), are fused with myeloma cells, generating a hybrid, isogenic, and immortalized cell line with antibody-producing capacity [19]. Generally, mAb-based tissue mapping of entirely peptidic epitopes with a known sequence can be performed with reasonable certainty, as the specificity of the used mAb can be accurately determined. However, the signals produced by carbohydrate-binding mAbs often reflect the expression of a variety of distinct protein carriers, which hinders the precise identification of the target glycan epitope [20]. The same limitation applies to other glycan-recognizing molecules whose carbohydrate specificity is not fully determined, including polyclonal antibody mixtures and other GBPs. The poor affinity of lectins, usually in the low micromolar range, frequently limits their application for histochemical staining purposes [1, 21]. Indeed, these glycan-binding molecules do not provide insights on complete monosaccharide compositions, glycan tridimensional conformation, glycosylation site occupancy, microheterogeneity, or the precise location of glycan chains on the backbone of specific glycoproteins. Such affinity reagents facilitate the identification of broader glycosylation traits (e.g. *N*-glycan branching, core fucosylation, and linkage-specific sialylation) rather than elucidating the exact tridimensional structure of individual glycan species. In fact, such information may only be accurately retrieved through the implementation of increasingly complex mass spectrometry (MS)-based methods. However, the majority of glycan-directed analytical techniques, including MS-based workflows, require analyte extraction from the target tissue, which, in turn, leads to the loss of information on glycan spatial distribution and tissue histopathological architecture.

6.3 Imaging Mass Spectrometry

The recent development and maturation of imaging mass spectrometry (IMS) technology has significantly contributed to the circumvention of such technical limitations since it does not rely on target-specific reagents but rather on direct molecular measurements [22]. Indeed, this technique has been used to directly characterize the *N*-glycosylation profile of neoplastic tissue sections by generating two- and three-dimensional (3D) molecular maps of hundreds of distinct glycan species across a wide mass range while also providing information on analyte relative abundance and on-tissue spatial distribution, which can be directly linked to the histopathological data from the same clinical specimen [22–29]. The IMS technology provides the mass accuracy and chemical specificity of MS-based detection and supports further on-tissue tandem MS fragmentation (e.g. collision-induced fragmentation) to achieve exact structure identification while still preserving the spatial distribution of individual analytes and the histopathological landscape of the target tissue. The implementation of IMS workflows in the cancer research field is of particular relevance since most clinically approved cancer biomarkers are either glycoproteins or carbohydrate antigens.

The typical IMS glycomic workflow requires the total release of asparagine-linked *N*-glycans through the surface digestion of the target tissue section with Peptide-*N*-glycosidase F (PNGase F), an endoglycosidase that efficiently hydrolyzes the amide bond linking the innermost *N*-acetylglucosamine (GlcNAc) of the *N*-glycan core to the asparagine's side chain of the protein's peptidic backbone, and the subsequent MS-based identification and relative quantification of the released carbohydrate species. Possible additives in the enzyme's storage buffer, including glycerol or detergents, may cause ion suppression during the ionization process, which diminishes the quality of the retrieved spectra [30, 31]. IMS-based analysis of biological samples can be performed using one of several MS ionization techniques, which offer complementary capabilities regarding both spatial resolution and the mass range of the target analytes. The matrix-assisted laser desorption/ionization (MALDI)-IMS has become increasingly popular in the analysis of glycans due to its high sensitivity and wide mass range. The preparation of tissue sections involves the automated and uniform coating of the target sample with an energy-absorbing matrix for efficient and homogeneous analyte ionization. Conveniently, IMS workflows are compatible with on-tissue sialic acid (Neu5Ac) chemical derivatization, which allows for linkage-specific discrimination of sialylated glycan species, and with positive glycan labeling, which significantly improves signal-to-noise ratios [29, 32]. Generated molecular maps of analyte spatial distribution and relative abundance can then be overlapped with brightfield optical images, such as the corresponding hematoxylin/eosin (H&E) staining of the same tissue section. The obtained structural and semiquantitative data can then be allocated to well-defined histological regions (e.g. nontransformed adjacent mucosa, immune infiltrate, necrosis, and tumor regions) based on a pathologist's annotation, making the IMS technology particularly valuable in the study of solid tumors. Furthermore, subsequent off-tissue extraction and MS-based fragmentation of released *N*-glycan species allow for unequivocal structural identification. Typically, the MALDI-IMS analysis of one cancer tissue section allows the reliable detection of 40–60 distinct glycan structures [33].

This methodology is highly versatile, and its robustness has been validated across various types of clinical samples, including formalin-fixed paraffin-embedded (FFPE) tissue blocks and fresh frozen tissue specimens [23, 24, 34–36]. The incorporation of multiple individual FFPE tissue specimens in the tissue microarray (TMA) format reduces intersample technical variability and further allows the IMS-based multiplexed analysis of larger cohorts of clinical samples, which is of extreme relevance to the cancer biomarker discovery field, particularly when conducting retrospective studies [36]. Moreover, the proven applicability of IMS in the analysis of FFPE tissues is of extreme significance since these samples can be easily archived at room temperature for several years in tissue banks and biorepositories and are more widely available than cryopreserved clinical specimens. Additionally, by directly linking detailed and spatially resolved structural data to well-defined histopathological regions at the individual sample scale, IMS-based analysis of whole tissue sections provides invaluable molecular insights on intratumor heterogeneity. This is of particular relevance when defining tumor margins and interfaces. IMS analysis of neoplastic tissues has clearly demonstrated the expression of tumor-associated molecular signatures in apparently healthy histological regions. The comprehensive comparative characterization and relative quantification of glycan species from nontransformed adjacent mucosa, premalignant lesions, and fully transformed neoplastic regions may lead to the identification of robust glycan-based biomarkers of malignant transformation capable of accurately discriminating patient clinical outcomes and tumor subtypes. The MALDI-IMS technology has been successfully used to illustrate the astonishing differences in the glycosylation patterns between corresponding healthy and malignant tissues across multiple epithelial cancers, including prostate, pancreatic, ovarian, gastric, and hepatocellular carcinoma, as well as myxoid liposarcoma [24–28, 33, 35, 37]. IMS technology has thus emerged as a novel source of robust cancer-specific biomarkers, either as individual glycan masses or as more complex panels of combined mass spectra from multiple glycan species, for unequivocal tissue region identification.

Recently optimized methods are capable of the simultaneous multimodal acquisition of MALDI-IMS spectra derived from both *N*-glycans and proteolytic peptides from the same tissue section [38]. The spatial distribution map of identified peptidic sequences can then be combined with the corresponding *N*-glycan map for the identification of overlapping regions. Subsequent bioinformatic analysis may lead to the identification of glycoprotein candidates, and their spatial distribution can be further assigned to well-defined histological regions. Such studies may provide mechanistic insights on the functional roles played by particular glycan signatures in malignant cell transformation, in particular through the identification of proteins modified with specific aberrant glycan determinants. Furthermore, the combination of *N*-glycan- and protein-derived structural data may provide novel combinatorial sets of cancer-specific biomarkers with improved sensitivity and specificity.

6.4 *In Situ* Proximity Ligation Assay

The unbiased and comprehensive identification of glycoproteins for biomarker discovery purposes through on-tissue IMS analysis can be laborious and time-consuming. Moreover, glycoprotein validation through direct immunoprecipitation experiments from cryopreserved clinical specimens can be technically challenging due to heterogeneous or reduced expression of the target protein and the requirement of considerable amounts of starting frozen material. Furthermore, although methodologies based on genetically modified cell models represent powerful tools to dissect the interactome of specific proteins, they are not compatible with patient samples. On the other hand, traditional immuno- or lectin-based histochemical staining methods remain limited to the detection and tissue mapping of individual proteins and carbohydrate signatures while overlooking biologically significant molecular interactions. In addition, such techniques do not allow the unequivocal identification of the protein carriers that are modified with a given type of glycan structure. To circumvent such technical limitations, numerous studies have instead pursued the direct on-tissue validation of specific protein glycoform candidates, initially identified in *in vitro* models of malignant transformation, using large cohorts of tumor clinical samples. As a result, several oncogenic proteins modified with specific glycosylation signatures may emerge as more sensitive and specific biomarkers of human malignancy when compared to more classical biomarkers targeting fully peptidic or carbohydrate-based epitopes.

The *in situ* proximity ligation assay (PLA) allows the on-tissue detection, imaging, and relative quantification of a plethora of cellular events at single-molecule resolution, including protein–protein interactions, protein translation, and degradation, as well as multiple post-translational modifications such as protein phosphorylation and glycosylation. This assay relies on the dual binding of highly specific affinity reagents, such as primary antibodies or lectins, for the *in situ* detection of the molecular proximity between two prespecified target epitopes on either FFPE or cryopreserved tissue specimens [39, 40]. In this method, a pair of affinity reagents labeled with single-stranded DNA molecules, also termed oligonucleotide proximity/detection probes, are bound to their respective target epitopes in a whole tissue section. If the epitopes recognized by both antibodies/lectins are found in close molecular proximity (10–40 nm), a proper detection complex is formed, which allows for the antibody-bound oligonucleotide molecules to be enzymatically joined into a circular DNA strand through rolling-circle amplification. The resulting DNA ligation product can then serve as a template for PCR-based multimeric signal amplification. The final amplification product can then be hybridized with fluorescent or chromogenic oligonucleotide strands (detection probes) for microscopic visualization and counting of individual spots at single complex resolution. The requirement for two proximal recognition reactions ensures highly selective mapping of interacting complexes since individual probe binding is insufficient to produce visible detection signals. In the typical approach to identify proteins carrying specific glycosylation traits, one antibody binds to a peptidic epitope within the target protein, and a second glycan-binding affinity reagent (mAb or lectin) targets the carbohydrate motif. Conveniently, the *in situ* PLA method can be performed using labeled secondary affinity reagents, avoiding the need for the conjugation of numerous pairs of primary antibodies or lectins. The PLA methodology is highly versatile, as a myriad of affinity reagents can be readily and easily converted into proximity probes. This is of particular significance when studying protein-specific glycosylation reactions, as it allows the use of carbohydrate-binding lectins for the detection of the target glycoprotein's glycan component. The successful implementation of the PLA system requires several methodological considerations, such as the optimization of histochemical staining protocols for each individual target antigen and possible incompatibilities with antigen retrieval steps. Furthermore, for the same target epitope, multiple affinity reagents may have to be tested since the oligonucleotide conjugation reaction may sterically hinder their antigen-binding capacity. Although the exact minimal distance required for the generation of a positive PLA signal has yet to be fully investigated, such values may be estimated and optimized based on the dimensions of the selected affinity reagents and lengths of the oligonucleotide strands. Despite its high selectivity and nanometric resolution, *in situ* PLA provides only indirect evidence on molecular interactions or protein modifications [41].

Over the past decade, several studies have successfully used the PLA technology to validate, in whole tissue tumor clinical samples, the modification of several cancer-associated proteins with aberrant glycan antigens, aiming at the identification of clinically relevant diagnostic and prognostic biomarker candidates. Such is the example of the MUC2 intestinal mucin and the CD44 co-receptor, both disclosed by fluorescence-based PLA as carriers of the short-truncated *O*-glycan epitope STn in advanced gastric adenocarcinomas [42–44]. The PLA-based association between the sialyl Lewis x (sLe^X) tetrasaccharide and several membrane-anchored proteins in gastric cancer tissues, including the RON receptor tyrosine kinase (RTK) and the carcinoembryonic antigen (CEA) adhesion molecule, has further unveiled the active role played by aberrant glycans in tuning the malignant features of oncogenic receptors [45, 46]. Of note, the expression of sLe^X-containing CEA proteoforms was associated with worse patient clinical outcome, portraying the PLA technology as a valuable source of prognostic markers. Recently, the E-cadherin cellular adhesion molecule has been validated as a molecular carrier of highly branched *N*-glycan chains by brightfield PLA analysis of advanced gastric adenocarcinoma tissue sections, highlighting this particular E-cadherin glycoform as a robust predictive biomarker of patient dismal prognosis [47]. In addition, the use of PLA technology has demonstrated the modification of the oncogenic ErbB2 RTK, which currently remains one of the few actionable therapeutic targets in the gastric cancer setting, with α2,6-linked Neu5Ac moieties in whole tissue sections of intestinal-type gastric carcinomas [48]. *In situ* PLA analysis has also provided valuable insights on the glycosylation status of mucin receptors required for *Helicobacter pylori* (*H. pylori*) adhesion to the gastric epithelium of glycoengineered mice models [49]. The combination of glycan metabolic labeling with the PLA-based analysis of protein–carbohydrate interactions has allowed the disclosure of the mechanistic contribution of specific glycan traits (e.g. sialylation and fucosylation) to various functional aspects of cancer-relevant proteins at the subcellular scale, including the turnover and ligand-induced dimerization of oncogenic cell surface receptors [50].

Importantly, the *in situ* PLA technology holds tremendous translational potential as it may be readily incorporated into the routine workflows of cancer-dedicated pathology laboratories as a robust source of novel glycoprotein-based biomarkers bearing both diagnostic and prognostic utility, thus improving patient stratification and clinical management.

6.5 Glycan Microarrays

Although the exact dimension of the cellular glycome remains a matter of debate, it is estimated to be in the range of 100 000–500 000 unique glycan structures [51]. Such structural diversity has created the need for molecular tools capable of linking the structure of a given glycan to its biological function. The development of glycan microarrays, just under 20 years ago, has propelled the high-throughput systematic interrogation of glycan-based molecular interactions [52–55]. By facilitating the fast and highly reproducible screening of a great number of glycan epitopes and compositions, this technology has allowed the unprecedented elucidation of the carbohydrate-binding specificity of a multitude of pathogens, whole cells, and GBPs. Indeed, the use of glycan microarrays for the comprehensive validation of the glycan ligand repertoire of several mAbs and plant-derived lectins has solidified their applicability as invaluable research tools to address the role of glycans in biological systems [54, 56–58]. The Consortium for Functional Glycomics (CFG) has used glycan microarray technology to disclose the detailed specificity of over 100 plant lectins (http://www.functionalglycomics.org/) [59]. Furthermore, the binding specificity and affinity data provided by glycan microarrays will pave the way for the rational design of carbohydrate-based therapeutic agents. Printed carbohydrate microarrays enable the simultaneous analysis of thousands of binding events between a single target analyte (e.g. lectin, soluble ligand, and mAb) and a miniaturized catalog containing hundreds of spatially defined glycan species immobilized onto a solid phase in a covalent or noncovalent manner [54, 60]. Moreover, the condensation of hundreds of different ligands into a miniaturized format significantly reduces the required amount of both the target analyte and each unique immobilized ligand. In addition, glycan microarrays represent ideal platforms for the screening of glycan-dependent molecular interactions by allowing the multivalent display of immobilized ligands on a solid surface, which mimics the weak and reversible nature of cell–cell interactions. Additionally, glycan microarrays are highly reproducible, cost-effective, and allow the fast screening of multiple samples.

Depending on the biological question, different types of macromolecules can be immobilized on the array solid substrate (e.g. glycans, glycoproteins, glycopeptides, mAbs, or lectins), giving rise to a diversified set of platforms. A standard-size microscope glass slide represents the original solid surface for ligand immobilization and remains the most widely used [61]. In the case of glycan microarrays, the immobilized library of pure carbohydrate structures can be either chemically synthesized or isolated from natural sources [53, 55, 58, 62, 63]. Chemoenzymatic strategies combining automated glycan assembly with selected enzymatic steps allow the controlled synthesis of glycosaminoglycans, branched *N*-glycans, and sialylated structures in a linkage-specific manner [64–68]. The selection of the most appropriate strategy for the covalent or noncovalent immobilization of glycans onto the solid phase depends on the synthetic method used in the preparation of a given glycan library or the source from which naturally occurring ligands are isolated. Glycans produced by fully chemical or chemoenzymatic synthesis are derivatized with orthogonal bi-functional linkers to allow their covalent attachment to a functionalized solid surface [69, 70]. On the other hand, naturally occurring carbohydrates, such as milk oligosaccharides and free reducing glycans released from glycoproteins by enzymatic digestion or chemical hydrolysis, also require proper derivatization prior to immobilization [55, 62, 71, 72]. Although, in principle, polysaccharides can be directly attached to the array by simple adsorption, enzymatically released glycans require the introduction of a functional group at their reducing end, which can be accomplished through different conjugation strategies.

Immobilized ligand libraries are spotted with micrometric resolution onto the selected surface by automated arraying robots in a prespecified, spatially resolved manner. For a single carbohydrate ligand, several replicates of serial concentrations are printed. Since glycans naturally establish low-affinity interactions with other macromolecules, the density at which individual glycan structures are spotted onto the microarray surface is a determinant in the generation of detectable signals and subsequent data interpretation [73]. After spotting is concluded, a washing step is performed for the removal of unbound carbohydrate molecules. A single-array slide may contain up to 20 000 spots [74]. Despite the significant advances in the chemical synthesis of structurally defined glycan chains made over the last two decades, both the number and structural diversity of the glycans that can be immobilized onto a single microarray remain far from representative of the human glycome's estimated dimension [72, 73].

The glycan microarray technology has been most extensively applied in the dissection of protein–carbohydrate interactions, namely in the characterization of the binding specificity of mAbs and GBPs [54, 75–77]. There are several possible strategies for the detection and quantification of bound molecular partners at the microarray surface. In the case of glycan microarrays, the analytes of interest, either in a pure isolated form (e.g. single GBP) or as a complex mixture (e.g. serum), are labeled and incubated on the solid substrate to allow ligand binding. Following the washing of unbound macromolecules, the signal emitted by the labeled bound analytes can be detected and quantified in individual spatially resolved spots, each one corresponding to a unique glycan structure. Fluorescence-based quantification of ligand binding remains the most widely used method for signal detection due to its high sensitivity and wide availability of fluorescence scanning equipment [62, 69, 70, 78]. Of note, the labeling of proteins with fluorescent tags may lead to protein denaturation or alterations in their carbohydrate-binding domain. Alternatively, a fluorescently tagged antibody for the recognition of bound analytes can be used. An adaptation of this detection strategy, known as sandwich array, relies on the use of fluorescently labeled secondary antibodies or lectins for the detection of ligands bound to the microarray solid substrate via a primary set of immobilized antibodies [79]. Due to the limited availability of specific carbohydrate-recognizing antibodies and lectins, label-free detection methods, such as quantitative on-chip MS and surface plasmon resonance (SPR), have become increasingly popular for signal quantification [80]. Interestingly, SPR-based analysis of glycan microarray signals provides valuable quantitative parameters on the reaction kinetics, including association and dissociation constants. As the volume and complexity of information generated by glycan microarrays grow exponentially, so does the need for software tools and algorithms capable of comprehensive and integrative data analysis.

The glycan microarray technology has been successfully applied in the cancer research field. The aberrant glycan signatures expressed at the cell surface of neoplastic cells actively support the establishment of an immune-suppressive microenvironment that facilitates tumor onset and progression [12]. Cancer-specific abnormal glycan antigens act as natural ligands for specific membrane-bound or secreted immune receptors. These receptors, which include Siglecs, galectins, and C-type lectins, constitute GBPs bearing unique glycan specificities that tightly control the activation threshold of several immune cell populations. Importantly, glycan microarrays have been used to determine the degree of *N*-glycan branching as well as the Neu5Ac linkage specificity and sulfation status regulating the binding preferences of several receptors of the Siglec family [58, 81–83]. The same strategy was applied to characterize the binding specificity of distinct galectins, including their branching preferences and linkage-dependent tolerance to Neu5Ac motifs [58, 69, 84]. The comprehensive profiling of these multivalent interactions, in particular the identification of which glycan epitopes are bound by specific immune receptors, is fundamental for the establishment of functional associations between specific glycosylation signatures of cancer cells and well-defined outputs of immune response. Such knowledge can not only be the stepping stone for the design of novel anticancer immunotherapies but also an invaluable source of predictive and stratifying biomarkers of immunotherapeutic response. The glycan microarray technology has also been extensively applied for cancer biomarker validation by allowing the miniaturized, multiplexed, and parallel screening of large numbers of analytes, samples, and replicates in a single experiment. In particular, glycan microarrays constitute useful tools to screen and compare the serum antibody repertoire of healthy individuals and cancer patients, which is indispensable for the development of highly specific anticancer vaccines targeting tumor-associated glycan antigens and the identification of glycan-based cancer biomarkers [75, 85–90]. A microarray of aberrantly glycosylated MUC1 *O*-glycopeptides has been developed to assess the antigenicity of these tumor-associated structures through the screening of MUC1-targeting mAbs and cancer patient serum samples [91, 92]. The glycan microarray technology has been applied to the rapid screening of hybridomas for the selection of the most specific mAb candidates against a target glycan antigen [75]. Of relevance to the cancer research field, glycan microarrays have been successfully used to infer the substrate specificity of individual glycosyltransferases (GTs) [93, 94]. Furthermore, this technology may be used to assess the efficacy of therapeutic inhibitors targeting glycan-dependent molecular interactions [95].

6.6 Glycoengineered *In Vitro*, *In Vivo*, and *Ex Vivo* Models

The nontemplate-driven and highly dynamic nature of protein glycosylation poses several challenges to the study of the functional role of complex carbohydrates in biological systems. Additionally, the partially overlapping substrate specificities and compensatory activity of related GTs hinder the systematic dissection of specific glycosylation pathways and the establishment of robust associations between one enzyme's activity and the unequivocal expression of specific cancer-associated glycan epitopes. Such limitations have motivated the development of glycoengineered *in vitro* models of disease, in which the glycome's complexity and heterogeneity are simplified so that the exact phenotypical contributions of specific glycosylation patterns can be pinpointed [96]. The precise editing of genomic *loci* codifying for various elements of the cellular glycosylation machinery (e.g. GTs, glycosidases, molecular chaperones, and monosaccharide transporters) has generated an unprecedented catalog of isogenic cell lines depicting the loss or gain of selected glycosylation capacities [97–99]. These *in vitro* platforms can be exploited for the systematic functional dissection of the human glycome at the single-cell level [100]. Moreover, glycoengineering strategies have found wide applicability in the design and recombinant expression of carbohydrate-based therapeutics, such as enzymes and immunoglobulins, which require highly controlled, well-defined, and homogeneous glycosylation profiles [101–104]. Furthermore, glycoengineered mammalian cell lines in which the expression of a target gene of interest has been either silenced or activated represent appealing platforms for the screening of selective inhibitors targeting specific glycosylation pathways and enzymes.

Initial studies based on random mutagenesis, plasmid-based overexpression, and targeted homologous recombination knock-out (KO) of glycosylation-related genes in both mammalian cell lines and animal models have led to the identification of GTs playing essential roles in early embryonic development, homeostasis, and disease [96, 105, 106]. Furthermore, these studies have provided invaluable knowledge on which glycosylation steps are catalyzed by individual GTs and which ones can be performed by partially redundant isoenzymes. Additionally, site-directed mutagenesis has been extensively used for the selective and site-specific abrogation of glycosylation in target peptidic sequences to interrogate the functional contribution of glycans to the oncogenic or tumor-suppressive features of several relevant cancer-associated glycoproteins, including the E-cadherin adhesion molecule and the epidermal growth factor receptor (EGFR) RTK [47, 107]. However, such approaches are extremely laborious and generate highly unpredictable and inconsistent genomic and phenotypical outcomes. Additionally, noncontrolled plasmid-based GT overexpression often produces unforeseen compensatory perturbations in multiple glycosylation pathways, which prevents the unequivocal assignment of observed glycophenotypes to the function of specific GT-coding genes. On the other hand, RNA interference-based strategies frequently lead to incomplete downregulation of the target protein, which translates into highly variable glycophenotypes.

The stable genetic engineering of glycan-related genes has been significantly simplified with the advent of precise genome editing technologies, characterized by high precision, consistency, and speed at a reduced cost. These include zinc-finger nucleases (ZFNs) [108], transcription activator-like effector nucleases (TALENs) [109], and the clustered regularly interspaced short palindromic repeats with the CRISPR-associated protein 9 (CRISPR/Cas9) [110, 111]. All of the aforementioned genome editing tools can be used to achieve efficient silencing of a selected gene through the introduction of site-specific double-strand breaks in the DNA sequence of a target genomic *locus*. In the CRISPR/Cas9 system, a complementary short guide RNA (gRNA) molecule is used to direct the activity of the Cas9 nuclease to a pre-specified site of the target nucleotide sequence [111]. The produced double-strand breaks are then repaired by the error-prone nonhomologous end joining (NHRJ) cellular mechanism, which results in mutations or small insertions/deletions (indels) at the cleavage site that disrupt the expression of the target gene by shifting its open reading frame (ORF). Currently, multiple engineering events, including both KOs and knock-ins (KIs), can be easily and rapidly performed in the same cell line, which allows the generation of catalog *in vitro* models depicting virtually any desired glycophenotype [99]. Importantly, the successful application of genetic engineering strategies requires comprehensive knowledge of the target glycosylation pathways in the selected cell host, namely the cell line-dependent expression profiles of the target GTs, but also of possible isoenzymes catalyzing redundant biosynthetic steps. This data not only aids in the prediction of structural glycomic outcomes but also facilitates the rational selection of the most directed and appropriate genome editing pipeline. Recently, comprehensive lentiviral and plasmid-based libraries of validated high-efficiency gRNAs for the specific targeting of every known GT-coding gene have been constructed and made globally available to the scientific community [98, 112]. Moreover, the CRISPR/Cas9 system has been adapted to allow the transcription activation of target dormant genes through the fusion of an inactive Cas9 nuclease with specific transcription factors. This strategy has been successfully used to activate the expression of *ST6GAL1* and *MGAT3* GT-coding genes in the CHO cell line [113].

Nuclease-based genome editing strategies also support the stable integration, or KI, of exogenous glycosylation-related genes provided to the cell system under the form of a homologous DNA template [99, 101]. This technology has been successfully used in the unbiased dissection of the substrate specificity and functional role of partially redundant GT isoenzymes, as shown for multiple polypeptide *N*-acetylgalactosaminyltransferases (ppGalNAcTs) [114]. More recently, a ZFN-based KI system was adapted to allow the inducible expression of individual ppGalNAcTs over a KO background in HEK293 cells and demonstrated that isoform-specific substrates are glycosylated in a dose-dependent manner [115]. A similar KO/rescue system was used for the stable integration of the ppGalNAcT6-coding gene in colon cancer cells, which led to the disclosure of this enzyme's active role in regulating cellular growth and differentiation [116].

Genetic glycoengineering has proven a useful tool when combined within glycoproteomic pipelines for glycosite and glycopeptide discovery and validation by reducing glycosylation diversity and heterogeneity and, thus, facilitating glycan-directed enrichment strategies. In the SimpleCell system, the KO of either the *COSMC* or *C1GALT1* transferase-coding genes through precise genome editing prevents the extension of mucin-type *O*-glycan chains beyond the single *N*-acetylgalactosamine (GalNAc) Tn antigen [43, 117, 118]. These isogenic cell lines depicting simplified *O*-glycosylation were used towards the comprehensive MS-based mapping and characterization of the GalNAc-type *O*-glycoproteome by allowing the efficient capture and enrichment of *O*-glycopeptides by the Tn-binding *Vicia villosa* agglutinin

(VVA) and have led to the unprecedented identification of thousands of *O*-glycoproteins and *O*-glycosites across multiple *in vitro* cancer models. The same strategy has been applied to the identification of *O*-mannosylated substrates [119].

One of the major drawbacks of glycan microarrays is related to the inability of such platforms to present the printed and immobilized glycan ligands in their natural cell surface context. The wide application of precise genome editing technologies targeting virtually all known mammalian glycosylation-related genes has allowed the establishment of comprehensive cell-based libraries composed of isogenic cell lines systematically modified to depict any intended glycosylation features by the use of combinatorial KO/KI strategies [21, 82, 120]. The replacement of isolated glycan-based ligands by glycoengineered cell lines allows for the screening of glycan-dependent molecular interactions while preserving their natural cell-dependent context and is conveniently compatible with high-throughput cell-based assays. Cell-based arrays composed of glycoengineered cell lines have been successfully applied to gain insights on the functional role and binding specificities of endothelial selectins, galectins, and Siglec receptors [82, 121]. Precise genome editing technology has been successfully used to completely deplete the HEK293 cell line of its sialylation capacities [82]. The generated Neu5Ac-null HEK293 cells then served as a "white canvas" cellular platform in which the combinatorial loss/gain of individual sialyl- and sulfotransferases allowed the comprehensive characterization and structural elucidation of the selective glycan-binding preferences and protein-specific context underlying the Siglec–glycan human interactome. Information on the fine glycan-binding specificities of human Siglecs is indispensable for the functional dissection of molecular resistance mechanisms hampering the clinical development and implementation of antitumor immunotherapeutic strategies.

The combination of chemical and genetic engineering methodologies toward precise metabolic glycan labeling has provided invaluable insights on the dynamics of the cell surface glycoproteome and allowed the unequivocal assignment of isoform-specific substrate preferences depicted by partially redundant isoenzymes, in particular the ppGalNAcT family [122–127]. Glycosylation metabolic labeling and tagging are based on the use of highly selective bioorthogonal reactions between two complementary, chemically modified functional groups that do not occur naturally in living systems. These chemically engineered probes must be biologically inert, i.e. they must not perturb the homeostatic function of target macromolecules in which they are incorporated while remaining highly selective toward each other, a strategy otherwise known as "click chemistry" [128]. Such bioorthogonal metabolic labeling probes are equipped with either visualization or enrichment tags that support a diverse range of downstream applications, including live visualization through imaging techniques and MS-based structural analysis. In the "bump-and-hole" strategy, the catalytic pocket of an individual ppGalNAcT enzyme (hole) is enlarged by targeted amino acid substitutions through genetic engineering to accommodate a complementary, chemically modified UDP-GalNAc analog (bump) that is incorporated by the selected cell host [123]. The fact that this modified substrate is uniquely recognized by the mutant enzyme makes the interacting pair orthogonal to all other competing GTs within the same cell host, which allows the live detection of isoform-specific functions within a complex living experimental system. Indeed, the compatibility of oligosaccharide metabolic engineering strategies with animal models allows for *in vivo* labeling, noninvasive imaging, and tracking of glycan-mediated interactions in their native environment [129, 130]. A click chemistry-based strategy using alkyne-tagged Neu5Ac monosaccharides has been successfully employed in the live tracking of sialylation to probe the specificity of Siglec-based interactions in HEK293 cells [82]. A chemoenzymatic glycan editing strategy was successfully employed for the generation of natural killer (NK) cells capable of the multivalent presentation of high-affinity ligands for the CD22 cell surface receptor overexpressed in B-cell lymphomas [131].

Glycoengineered animal models, carrying genetic defects on central glycosylation-related genes through homozygous targeted gene mutation, have unveiled the essential function of several GTs and their respective glycan products in mammalian embryonic development [105, 132–135]. Moreover, the organ- and tissue-specific conditional deletion of glycogenes has disclosed their function in central physiological and pathological processes, such as neurogenesis [136], fertility [137, 138], immunity [139, 140], and neoplastic transformation [141–148]. The CFG has comprehensively characterized the phenotype of 36 mutant mouse strains carrying genetic deficiencies in different GTs and GBPs [106]. Despite glycoengineered cell lines providing important mechanistic insights on glycan regulation and function, the comprehensive study of the functional role played by cancer-associated glycan epitopes and specific GTs in human malignant transformation requires more complex *in vivo* models in which organ-specific carcinogenesis can be recapitulated. Importantly, the crossbreeding of glycogene-deficient strains with mice models of spontaneous carcinogenesis has significantly expanded the *in vivo* toolbox for conditional, organ-specific glycoengineering. Such animal systems have highlighted TACAs and the enzymes responsible for their biosynthesis as major regulators of the intricate network of cellular and molecular interactions occurring within the tumor microenvironment, including the interaction of cancer cells with immune cell populations and ECM components. Moreover, they offer an integrative and systemic view of the biological activity of glycosylation-targeting therapeutic agents and represent, therefore, appealing platforms for their preclinical testing.

Targeted gene mutation of *Mgat5*, which codifies for a GT catalyzing the β1,6-branching of *N*-glycan antennae, has disclosed branched complex *N*-glycans as active drivers of tumor growth and metastatic dissemination in a viral-induced mice model of breast carcinogenesis [146]. The development of *Mgat5*-null mice models of spontaneous intestinal carcinogenesis has further established the pivotal role played by β1,6-branched *N*-glycan structures in the immune-based recognition and elimination of colorectal cancer cells [144]. The genetic silencing of *ST6Gal1* in mice bearing spontaneous, viral-induced mammary tumors has functionally implicated this enzyme, responsible for the terminal capping of *N*-glycan chains with α2,6Neu5Ac, in the maintenance of tumor stemness and differentiation [147]. These observations were further confirmed through the conditional restoration of ST6Gal1 activity in cell lines derived from *ST6Gal1*-null tumors. Mice with the inducible expression of the human *FUT3* and *β3GALT5* transgenes restricted to the gastrointestinal tract exhibit a significant increase in the serum levels of the CA 19.9 (sialyl Lewis a [sLe^a]) glycoconjugate and spontaneously develop severe pancreatitis, which progresses to aggressive pancreatic adenocarcinoma in the presence of the *Kras* oncogene [141]. The homozygous KO of *Fut2* in mice, through genetic recombination of embryonic stem cells, abrogates the expression of the α1,2 di-fucosylated Lewis b (Le^b) antigen, which successfully prevents the adhesion of BabA-dependent *H. pylori* strains to the stomach's epithelial lining [49, 148]. In addition, the conditional silencing of *C1Galt1* in the mouse gastric epithelium leads to the spontaneous development of gastritis with progression to Tn-overexpressing adenocarcinomas [145]. Such observations demonstrate that genetic defects in single GTs are sufficient to trigger tumor onset and progression and portray glycoengineered mouse strains as robust *in vivo* models of organ-specific spontaneous neoplastic transformation.

Patient-derived xenografts (PDXs) and PDX-derived organoids reliably recapitulate the genomic and phenotypical features of the individual tumors from which they originated, including therapeutic drug response (reviewed in [149, 150]). By retaining inter- and intratumor molecular heterogeneity, PDX models constitute, therefore, appealing platforms for translational and preclinical research. *N*-glycomic analysis of pancreatic cancer-derived PDX models demonstrated their ability to accurately recapitulate the *in vivo* glycosylation landscape [151]. Furthermore, PDX models provide a nonhuman background, which constitutes an advantage in proteomics-based biomarker discovery. Several unique *N*-glycopeptides have been identified in the sera of mice bearing PDX established from high-grade serous ovarian carcinoma patients [152].

6.7 Structural Elucidation of Glycoconjugates: Glycomic and Glycoproteomic Strategies

The mammalian cell glycome is characterized by formidable complexity and diversity. Although complex glycan biosynthesis relies on the enzymatic assembly of a limited number of monosaccharide units, comprehensive glycan identification remains technically challenging. The structural characterization of carbohydrates by MS-based analytical methods is of major importance since the majority of glycan-mediated interactions occurring within biological systems are strictly dependent on the glycan's tridimensional structure. However, several unique features of glycans hamper their exact structural elucidation. These include the large number of possible branching sites; the presence of closely related isomers with identical monosaccharide composition within a biological sample; different anomeric configurations of glycosidic linkages; and further glycan modifications with chemically diverse functional groups (e.g. sulfate and phosphate). Furthermore, the functional role of a given glycan species is often determined by its nonglycan molecular carrier and the specific site to which the glycan is attached. Moreover, posttranslational glycosylation includes several extremely labile groups, such as terminal Neu5Ac motifs, which are often lost during analyte fragmentation [153]. Therefore, the unequivocal identification and structural characterization of glycan species in a complex biological sample (e.g. cell line, tumor specimen, and body fluid) requires the coupling of methods for efficient glycan separation with high-resolution and sensitive MS-based analytical workflows. However, most analytical methods still require the laborious and time-consuming manual interpretation of complex fragmentation data.

Depending on the biological question, glycans can be analyzed either separately from their nonglycan counterparts (MS-based glycomics) or while still attached to their protein carriers (MS-based glycoproteomics). In the typical *N*-glycomic workflow, the biological sample to be analyzed is usually homogenized and lysed, proteins are irreversibly denaturated in the presence of reducing agents such as β-mercaptoethanol or dithiothreitol, and *N*-glycans are released through enzymatic digestion with PNGase F [154, 155]. PNGase F digestion is fast, robust, and highly efficient in the liberation of all classes of *N*-glycans. Due to the lack of known enzymes capable of cleaving *O*-glycan chains from glycoproteins, chemical-based strategies such as reductive β-elimination and hydrazinolysis remain the gold standard for total *O*-glycan release [156].

Released glycans can then undergo chemical derivatization at their reducing end, or be directly analyzed in their native form [157]. Released glycan species contain a reactive carbonyl group at their reducing end that can be readily derivatized. Derivatization methods equalize the chemical properties of glycans and are especially suitable for: glycan labeling with fluorescent tags to allow sensitive analyte detection; stabilization of the most labile groups of glycan chains, in particular Neu5Ac moieties, which can be easily lost during sample ionization; and providing every released glycan species with a uniform charge, which improves their ionization efficiency. This is of particular importance since carbohydrates are not as easily transferred to the gas phase as proteins or peptides, which are ubiquitously protonated, and do not ionize as efficiently, being particularly susceptible to ion suppression when using positive ion mode analysis [155]. Moreover, during the ionization process of an analyte mixture, more hydrophobic biomolecules will cause the ion suppression of less hydrophobic ones. It is, thus, advisable to separate glycans according to their acidity prior to MS analysis. The most widely used derivatization strategies include reductive amination and glycan permethylation [157]. The latter method precludes the formation of intermolecular hydrogen bonds, which increase glycan hydrophobicity, volatility, and, consequently, ion signal intensity. Besides, it favors the formation of diagnostic fragment ions, i.e. monosaccharide oxonium ions, which provide invaluable structural information. Additionally, glycan permethylation prevents intramolecular rearrangements, such as fucose transfer between the *N*-glycan core and antennae, which preclude unequivocal glycan structural assignment [158]. Finally, by eliminating the negative charge from Neu5Ac moieties, glycan derivatization stabilizes sialylated groups for positive ion mode analysis [159]. Recently, a novel method for linkage-specific Neu5Ac derivatization was developed based on the ethyl esterification of α2,6Neu5Ac groups, followed by the lactonization and sequential stable amidation of α2,3Neu5Ac motifs [160]. This Neu5Ac derivatization method has been applied to confirm the abrogation of α2,6-sialylated *N*-glycan species from the ErbB2 glycome in *ST6GAL1* KO gastric cancer cells [48].

Especially when dealing with complex samples characterized by a diverse repertoire of isomeric or closely related structures, glycan separation is required prior to MS-based analysis. Indeed, the selected separation method can also facilitate isomeric glycan resolution. The combination of high-performance liquid chromatography (HPLC)-based separation with the speed and sensitivity of tandem MS represents one of the most widely used glycan-directed analytical strategies [161]. Online LC-MS/MS provides distinct types of information that aid in the qualitative and semiquantitative analysis of glycan structures based on analyte retention time, accurate mass, and tandem fragmentation spectra. MS-generated data provides insights into glycan class and composition, glycan structure, and relative abundance. Porous graphitized carbon (PGC) [162, 163], solid-phase hydrophilic interaction liquid chromatography (SPE-HILIC) [160, 164], and capillary electrophoresis (CE) [160, 164–166] represent robust methods to achieve well-resolved glycan isomeric separation. Analyte retention and separation power provided by each method are determined by features such as hydrophobicity, polarity, and the establishment of weak electrostatic and ionic interactions between the analyte and distinct stationary phases. A PGC-LC-MS/MS workflow has been used for the structural elucidation of *N*-glycans released from the heavily glycosylated CEA adhesion molecule, which led to the identification of CEA as a molecular carrier of sLe^x in highly glycoengineered cell models of metastatic gastric cancer [46]. Isomeric ion separation can also be achieved in the gas phase through the modern ion-mobility spectrometry method, which determines the size-, shape-, and charge-dependent mobility of gas-phased ions depending on their collisional cross-section size [167, 168]. Furthermore, the performance of tandem MS allows the accurate identification of isomeric species based on distinctive fragmentation patterns.

Over the last decades, a wide array of MS instrumentation and methods have been employed for the structural elucidation of glycoconjugates. However, MALDI and electrospray ionization (ESI) are amongst the most commonly used. MALDI constitutes a soft ionization technique in which analytes are embedded within a crystallized energy-absorbent acidic matrix that, upon excitation by a short UV laser pulse, promotes analyte evaporation and the generation of protonated ions with minimal or no fragmentation [169]. When using MALDI to analyze nonderivatized native glycans, labile acidic groups such as Neu5Ac, fucose, sulfate, and phosphate are often lost during the analyte ionization process due to the significant vibrational excitation of generated ions [170]. Loss of glycan labile groups can be circumvented through glycan stabilization by derivatization protocols, such as permethylation. On the other hand, in ESI-based ionization, a solution containing the target analytes is infused through an electroconductive needle to which an electric field is applied, causing the formation of extremely fine analyte-containing droplets, which readily evaporate to form MS-detectable charged ions [171]. In this method, acidic glycans and other labile ions undergo significantly less fragmentation and can, thus, be subjected to direct MS analysis [172]. In ESI, however, hydrophilic native glycans preferably occupy the center of formed droplets and have their ionization efficiency suppressed by more hydrophobic species occupying the droplet periphery [173]. Ion suppression can, however, be minimized by significantly reducing droplet size and, consequently, the concentration of suppressing molecules through the use of nanoscale ESI setups [160]. Moreover, when compared with MALDI, ESI produces better resolved peaks for glycoconjugates due to the absence of matrix-derived adduct peaks.

In the case of glycoproteins, site-specific mapping and structural characterization of attached glycan species are usually warranted. This type of structural analysis can be performed on intact glycoproteins (top-down glycoproteomics) or glycopeptides resulting from proteolytic digestion (bottom-up glycoproteomics). Although trypsin remains the gold standard enzyme for the digestion of glycoproteins into glycosylated and nonglycosylated peptides, tryptic glycopeptides may harbor multiple glycosylation sites or be too long to be readily analyzed by MS [174]. Indeed, post-translational glycosylation often protects the peptidic backbone from enzymatic digestion (miscleavage), yielding longer glycopeptides that are not as easily ionizable [175]. Thus, the use of other proteases bearing distinct cleavage specificities or the combination of multiple digesting enzymes may allow for wider and more comprehensive glycoproteome coverage [176, 177]. However, the use of less specific proteases may generate glycopeptides that are not unique and that produce poorly understood fragmentation patterns, thus hampering precise glycopeptide assignment.

Ideally, glycoproteomic methods should allow the comprehensive elucidation of tridimensional glycan structures and their unequivocal assignment to specific amino acid residues within a given glycopeptide. The format in which glycoproteins are analyzed and the selection of the most appropriate analytical method depend on the complexity of the biological sample and the depth of characterization required. Such considerations are of particular importance when analyzing glycoproteins harboring multiple glycosylation sites, depicting both macro- and microheterogeneity [172]. Importantly, the post-translational glycosylation of proteins increases their hydrophilicity and surface activity and severely decreases their ionization efficiency [178]. In fact, in a complex mixture of enzymatically digested peptides, the high ionization efficiency of less acidic nonglycosylated peptides causes the ion suppression of low-abundancy, more acidic, and hydrophilic glycopeptides, especially when using positive ion mode analysis. For this reason, glycopeptide enrichment strategies are often required to achieve confident MS-based glycopeptide identification. Lectin affinity chromatography has been widely used for glycopeptide enrichment purposes [179]. The wide range of glycan structures recognized by less specific lectins, such as wheat germ agglutinin (WGA), concanavalin A (ConA), and jacalin (JAC), makes these GBPs ideal for the less strict enrichment of large portions of the target glycoproteome. Lectins bearing narrower carbohydrate specificity, such as the ones binding to linkage-specific sialylated glycans (e.g. *Sambucus nigra* agglutinin [SNA] and *Maackia amurensis* lectin [MAL]), can be used for the selective enrichment of defined subsets of oligosaccharides. Interestingly, *Wisteria floribunda* agglutinin (WFA) was used to guide the direct dissection of lectin-stained neoplastic tissue sections of cholangiocarcinoma for subsequent agarose-bound lectin chromatography toward cancer biomarker discovery [180]. Furthermore, multiple lectins can be serially combined to achieve improved glycopeptide purification (multi-lectin affinity chromatography) [181]. This strategy has been successfully used in the identification of cancer-specific serum biomarkers in multiple solid tumors, including prostate [182], breast carcinoma [183, 184], and cholangiocarcinoma [185]. The Tn-binding VVA lectin was used for the affinity capture of simplified mucin-type *O*-glycosylated peptides generated by the SimpleCell glycoengineering strategy and allowed the identification of over 600 distinct glycoproteins and close to 3000 *O*-glycosites [117]. However, regardless of their glycan specificity, no single lectin or combination of lectins can encompass the entire glycoproteome. Other affinity-based chemical strategies, including titanium dioxide (TiO_2) enrichment and hydrazide chemistry, have been used for the enrichment of sialylated glycopeptides. TiO_2 interacts with Neu5Ac through multipoint binding while voiding neutral glycopeptides or nonglycosylated peptides [186]. TiO_2 enrichment followed by MS-based identification of sialylated glycopeptides was successfully performed in the serum of bladder cancer patients to identify cancer-specific circulating biomarkers and to illustrate the overall increase in sialylation triggered by malignant transformation [187]. When using this enrichment strategy, however, elution of polysialylated glycans is difficult [188]. Hydrazide chemistry is based on aldehyde formation from oxidized glycopeptides [189]. Glycans are then covalently immobilized on a solid hydrazide substrate, followed by the release of the peptide moiety by PNGase F digestion, which allows for peptide sequencing and identification of glycosylated residues. However, this enrichment strategy does not provide structural information on glycan chains since it fails to produce intact glycoconjugates for subsequent MS analysis. Also, by being PNGase F-dependent, hydrazide chemistry is not suitable for the MS-based analysis of *O*-glycoproteins [189, 190]. Hydrazide chemistry has been employed to dissect the tumor-specific glycoproteome of hepatocellular [191], lung [192], prostate [193], and breast carcinomas [194]. SPE-HILIC and SPE-PGC, which can be directly coupled to different mass spectrometers, represent widely used strategies for glycopeptide separation, clean-up, and enrichment and are based on weak molecular interactions (e.g. hydrogen bonding, ionic, and dipole–dipole interactions) established between carbohydrate-containing analytes and a polar stationary phase. HILIC-based glycopeptide enrichment has been used to illustrate differences in the sialylation levels of selected protein biomarker candidates, including haptoglobin and transferrin, in the serum of gastric cancer patients [195], as well as to quantify differences in the fucosylation levels of the α1-acid-glycoprotein in the serum of individuals bearing pancreatic cancer and precancerous pancreatic lesions [196]. HILIC-based enrichment of fluorescently labeled *N*-glycans led to the discovery of the RON oncogenic RTK as a molecular carrier of the sLe^x tumor-associated glycan antigen in ST3Gal4-overexpressing gastric cancer cell lines [45]. HILIC has been used for glycoprotein purification from colorectal cancer clinical tissue specimens, followed by MALDI-(time-of-flight)^2-MS (MALDI-TOF/TOF-MS) analysis of total released *N*-glycans [197]. Analyzed tumor tissues harbored a decrease in glycan species carrying bisecting GlcNAc and an increase in sulfated, paucimannosidic, and sialofucosylated glycan determinants when compared to nontransformed healthy control tissues.

Proper glycopeptide assignment requires intact glycopeptide analysis and thorough glycan structural characterization. Detailed fragmentation information from both the glycan and peptide portions can be obtained through distinct yet complementary tandem fragmentation strategies. Indeed, hybrid fragmentation methods combining parallel (alternating) or sequential (MS^3) fragmentation approaches provide more comprehensive structural information on both the peptidic and glycan moieties of glycopeptides [198, 199]. Collision-induced dissociation (CID) represents one of the most prevalent approaches for the tandem fragmentation of glycans and glycoconjugates and for the generation of diagnostic ions. Lower energy CID-based fragmentation preferentially cleaves the glycosidic bonds between monosaccharides, which are more labile than peptidic amide bonds, generating B- and Y-ions that inform glycan composition and sequence [200]. Higher collision energies, on the other hand, generate cross-ring A- and X-ions and internal double cleavage fragments for the identification of glycan branching sites and the resolution of isobaric structures [199, 201]. This kind of fragmentation does not produce information on the glycopeptide peptidic sequence and, thus, is rarely used individually for glycopeptide assignment [201]. Indeed, CID-derived parent B- and Y-ions with sufficient signal intensity can be selected to undergo sequential fragmentation cycles (MS^3), where complementary *b*- and *y*-ions from the peptidic backbone can be generated by higher collision energies. Since distinct collision energies give rise to differential fragmentation, current glycan- and glycopeptide-directed MS workflows make use of collision energy-stepping, i.e. the use of incremental CID fragmentation energies on the same set of precursor ions within a single mass spectrum scan, which yields richer and more informative fragmentation spectra [198, 202]. Higher energy collisional dissociation (HCD), a higher energy variant of CID, is unique to more modern Orbitrap-based instruments. HCD offers improved accuracy in the detection of small diagnostic oxonium ions, particularly in the low m/z range [201]. HCD produces both glycan B- and Y-oxonium ions and *b*- and *y*-type ions derived from peptide backbone fragmentation, which tend to lose all or part of their glycan modifications during fragmentation. HCD-based stepping workflows not only produce both glycan and peptide backbone fragments from a single fragmentation event but also generate more detectable cross-ring fragment ions, which are fundamental to pinpoint glycosidic linkages and identify isomeric species [198, 199]. Electron transfer dissociation (ETD) constitutes a widely used fragmentation method for intact glycopeptide analysis. In ETD-based fragmentation, protonated glycopeptides receive electrons generated from radical anions, which cause fragmentation along the peptidic backbone in the form of *c*- and *z*-type ions [203]. Distinctively from CID-based fragmentation, ETD retains labile post-translational modifications such as glycosylation on the protein backbone and, therefore, does not generate valuable information on glycan composition or sequence [199, 204]. However, generated *c*- and *z*-type ions are extremely useful to pinpoint glycosylation sites and retrieve peptidic sequences.

Aiming to gather more comprehensive structural information on both peptide and glycan moieties of glycopeptides, many modern-day Orbitrap-based mass spectrometers are able to combine complementary dissociation workflows. For instance, in a product-triggered targeted method, the detection of HCD-derived glycan oxonium ions generated by glycopeptide fragmentation triggers subsequent ETD fragmentation events of the respective parent ions, while nonglycosylated peptides are not selected for tandem MS analysis [205]. This strategy has been employed to comprehensively characterize the site-specific glycosylation profiles of the ErbB2 and EGFR RTKs in gastric and colorectal carcinoma cells, respectively, using the HexNAc oxonium ion (m/z 204.087) as a product ion trigger to select HexNAc-containing parent ions for additional fragmentation rounds [48, 206]. Product-triggered fragmentation approaches have been successfully used to unravel the target-specific *O*-glycoproteome of colorectal and leukemia cancer cell lines [207, 208].

The manual interpretation of data generated from the MS-based analysis of free glycans or glycoproteins is challenging, laborious, time-consuming, and requires significant expertise. The inclusion of fragmentation spectral data further increases the complexity of this analytical exercise. Contrarily to genomic databases, the establishment of both glycomic and glycoproteomic software-based tools represents a more challenging task since glycans constitute secondary gene products whose structure cannot be fully predicted by a given genetic sequence. The establishment of comprehensive glycan structural databases rather relies on the global deposition of glycan analytical data and the development of bioinformatic tools capable of querying this data in a high-throughput manner for the quick and targeted prediction of a given glycan 3D structure, binding partner, biosynthetic pathway, associated pathologies, and host organism [209]. Several open-source and commercial software tools have been developed to aid in the interpretation of glycopeptide mass spectra. These include: SimGlycan [210], Glyco-DIA [211], GlypID 2.0/GlycoFragWork [212, 213], ByOnic [214], MassyTools [215], GlycoPepGrader [216], SweetSEQer [217], ArMone [218], GlycoPepDetector [219], GRIP [220], MAGIC [221], GP Finder [222], Sweet-Heart [223], GPQuest [224], GlycoFinder [201], GlycopeptideSearch (GPS) [225], pGlyco 2.0 [226], O-Pair Search [227], GPSeeker [228], pMatch-Glyco [229], glyXtoolMS [230], SugarQB [231], GlycoPAT [232], I-GPA [233], and GlycoMasterDB [234].

6.8 Concluding Remarks

Glycans undoubtedly play an essential role in maintaining mammalian cell homeostasis. Strategically localized at the interface of a cell and its surrounding microenvironment, glycans act as crucial molecular gatekeepers, actively governing virtually every cellular process. Moreover, since the posttranslational glycosylation cellular process is highly sensitive to a cell's physiological state, glycans have been widely established as robust molecular reporters of pathological conditions and, in particular, of the neoplastic transformation of human tissues. The aberrant expression of cancer-specific glycan antigens by neoplastic cells, which stems from a variety of dysregulated mechanisms affecting the cellular glycosylation machinery, has been linked to the majority of malignant features supporting tumor cell survival, proliferation, differentiation, immune evasion, and local or distant organ dissemination. At the cellular level, TACAs drive malignant transformation by either disrupting the homeostatic functions of tumor-suppressive elements or by enhancing the oncogenic properties of proteins controlling cell adhesion, programmed cell death, mitogenic signaling, and extracellular communication. However, the mechanistic basis underlying such alterations, as well as the functional impact that these may have on tumor onset, development, and metastatic spread, warrant further investigation. Indeed, several biosynthetic and structural features of glycans pose unique challenges to their compositional characterization, structural elucidation, carrier assignment, and functional study. The vast array of monosaccharide substrate donors and transporters and the partial redundant and overlapping specificities depicted by GTs and glycosidases exponentially increase the structural complexity of the human glycosylation landscape. Moreover, the nontemplate-driven nature of glycosylation greatly hinders the unequivocal analytical-based structural elucidation of complex glycan chains. In recent years, we have witnessed the exponential technological development of cellular-, imaging-, and analytical-based tools and platforms that have been instrumental in the unprecedented association between a glycan structure and its function within complex biological systems. The rapidly expanding tool box of gene editing technologies has allowed the generation of cellular and animal models depicting cleaner and well-defined glycosylation backgrounds in which the study of glycan biosynthesis and function can be performed in the absence of confounding factors such as GT redundant and compensatory activity. Increasingly complex imaging systems have prompted the comprehensive characterization of glycan expression patterns and distribution across histologically defined regions, as well as the identification and mapping of tumor-relevant glycan–protein interactions at the single complex level, in tissue specimens from oncological patients. The emerging glycan microarray technology has fueled the high-throughput interrogation and dissection of the glycan-binding specificities of disease-relevant carbohydrate-based molecular interactions actively underpinning numerous features of malignant cells, including evasion to immune recognition and surveillance. The MS-based elucidation of the composition, 3D structure, and site of attachment of cancer-associated glycan antigens has the potential to revolutionize the structure-guided design and optimization of cancer therapeutics. In conclusion, glycans undeniably represent a novel set of tools that will become indispensable in the field of modern-era precision oncology, either as robust molecular biomarkers bearing diagnostic, prognostic, predictive, and stratifying value, or as promising targets for the delivery of personalized anticancer therapeutic agents.

List of Abbreviations

capillary electrophoresis	CE
carcinoembryonic antigen	CEA
Consortium for Functional Glycomics	CFG
collision-induced dissociation	CID
concanavalin A	ConA

CRISPR-associated protein 9	CRISPR/Cas9
extracellular matrix	ECM
epidermal growth factor receptor	EGFR
electrospray ionization	ESI
electron transfer dissociation	ETD
formalin-fixed paraffin-embedded	FFPE
N-acetylgalactosamine	GalNAc
glycan-binding protein	GBP
N-acetylglucosamine	GlcNAc
guide RNA	gRNA
glycosyltransferase	GT
higher energy collisional dissociation	HCD
high-performance liquid chromatography	HPLC
hematoxylin/eosin	H&E
Helicobacter pylori	*H. pylori*
imaging mass spectrometry	IMS
jacalin	JAC
knock-in	KI
knock-out	KO
Lewis b	Le^b
monoclonal antibody	MAb
Maackia amurensis lectin	MAL
matrix-assisted laser desorption/ionization	MALDI
MALDI-(time-of-flight)^2-MS	MALDI-TOF/TOF-MS
mass spectrometry	MS
N-acetylgalactosaminyltransferases	ppGalNAcT
porous graphitized carbon	PGC
proximity ligation assay	PLA
peptide-*N*-glycosidase F	PNGase F
nonhomologous end joining	NHRJ
sialic acid	Neu5Ac
open reading frame	ORF
patient-derived xenograft	PDX
receptor tyrosine kinase	RTK
sialyl Lewis a	sLe^a
sialyl Lewis x	sLe^x
Sambucus nigra agglutinin	SNA
solid-phase hydrophilic interaction liquid chromatography	SPE-HILIC
surface plasmon resonance	SPR
sialyl Tn	STn
tumor-associated carbohydrate antigen	TACA
transcription activator-like effector nucleases	TALEN

titanium dioxide	TiO_2
tissue microarray	TMA
Vicia villosa agglutinin	VVA
wheat germ agglutinin	WGA
Wisteria floribunda agglutinin	WFA
zinc-finger nuclease	ZFN

References

1 Varki, A., Cummings, R.D., Esko, J.D. et al. (2015). *Essentials of Glycobiology*. Cold Spring Harbor, NY: Cold Spring Harbor Laboratory Press.

2 Schjoldager, K.T., Narimatsu, Y., Joshi, H.J., and Clausen, H. (2020). Global view of human protein glycosylation pathways and functions. *Nature Reviews Molecular Cell Biology* 21 (12): 729–749.

3 Varki, A. (2017). Biological roles of glycans. *Glycobiology* 27 (1): 3–49.

4 Taniguchi, N. and Kizuka, Y. (2015). Glycans and cancer: role of *N*-glycans in cancer biomarker, progression and metastasis, and therapeutics. *Advances in Cancer Research* 126: 11–51.

5 Julien, S., Ivetic, A., Grigoriadis, A. et al. (2011). Selectin ligand sialyl-Lewis x antigen drives metastasis of hormone-dependent breast cancers. *Cancer Research* 71 (24): 7683–7693.

6 Blanas, A., Sahasrabudhe, N.M., Rodríguez, E. et al. (2018). Fucosylated antigens in cancer: an alliance toward tumor progression, metastasis, and resistance to chemotherapy. *Frontiers in Oncology* 8: 39.

7 Stowell, S.R., Ju, T., and Cummings, R.D. (2015). Protein glycosylation in cancer. *Annual Review of Pathology: Mechanisms of Disease* 10: 473–510.

8 Radhakrishnan, P., Dabelsteen, S., Madsen, F.B. et al. (2014). Immature truncated O-glycophenotype of cancer directly induces oncogenic features. *Proceedings of the National Academy of Sciences of the United States of America* 111 (39): E4066–E4075.

9 Pinho, S.S. and Reis, C.A. (2015). Glycosylation in cancer: mechanisms and clinical implications. *Nature Reviews Cancer* 15 (9): 540–555.

10 Mereiter, S., Balmaña, M., Campos, D. et al. (2019). Glycosylation in the era of cancer-targeted therapy: where are we heading? *Cancer Cell* 36 (1): 6–16.

11 Magalhães, A., Duarte, H.O., and Reis, C.A. (2021). The role of O-glycosylation in human disease. *Molecular Aspects of Medicine* 79: 100964.

12 Rodríguez, E., Schetters, S.T., and van Kooyk, Y. (2018). The tumour glyco-code as a novel immune checkpoint for immunotherapy. *Nature Reviews Immunology* 18 (3): 204–211.

13 Rodrigues, J.G., Duarte, H.O., Reis, C.A., and Gomes, J. (2021). Aberrant protein glycosylation in cancer: implications in targeted therapy. *Biochemical Society Transactions* 49 (2): 843–854.

14 Flynn, R.A., Pedram, K., Malaker, S.A. et al. (2021). Small RNAs are modified with *N*-glycans and displayed on the surface of living cells. *Cell* 184 (12): 3109–3124.

15 Reis, C.A., Osorio, H., Silva, L. et al. (2010). Alterations in glycosylation as biomarkers for cancer detection. *Journal of Clinical Pathology* 63 (4): 322–329.

16 Marcos, N.T., Pinho, S., Grandela, C. et al. (2004). Role of the human ST6GalNAc-I and ST6GalNAc-II in the synthesis of the cancer-associated sialyl-Tn antigen. *Cancer Research* 64 (19): 7050–7057.

17 Warnock, M., Stoloff, A., and Thor, A. (1988). Differentiation of adenocarcinoma of the lung from mesothelioma. Periodic acid-Schiff, monoclonal antibodies B72.3, and Leu M1. *The American Journal of Pathology* 133 (1): 30.

18 O'Boyle, K.P., Markowitz, A.L., Khorshidi, M. et al. (1996). Specificity analysis of murine monoclonal antibodies reactive with Tn, sialylated Tn, T, and monosialylated (2 → 6) T antigens. *Hybridoma* 15 (6): 401–408.

19 Köhler, G. and Milstein, C. (1975). Continuous cultures of fused cells secreting antibody of predefined specificity. *Nature* 256 (5517): 495–497.

20 Cummings, R.D. (2009). The repertoire of glycan determinants in the human glycome. *Molecular BioSystems* 5 (10): 1087–1104.

21 Briard, J.G., Jiang, H., Moremen, K.W. et al. (2018). Cell-based glycan arrays for probing glycan–glycan binding protein interactions. *Nature Communications* 9 (1): 1–11.

22 Blaschke, C.R.K., McDowell, C.T., Black, A.P. et al. (2021). Glycan imaging mass spectrometry: progress in developing clinical diagnostic assays for tissues, biofluids, and cells. *Clinics in Laboratory Medicine* 41 (2): 247–266.

23 Powers, T.W., Neely, B.A., Shao, Y. et al. (2014). MALDI imaging mass spectrometry profiling of *N*-glycans in formalin-fixed paraffin embedded clinical tissue blocks and tissue microarrays. *PLoS One* 9 (9): e106255.

24 Everest-Dass, A.V., Briggs, M.T., Kaur, G. et al. (2016). *N*-glycan MALDI imaging mass spectrometry on formalin-fixed paraffin-embedded tissue enables the delineation of ovarian cancer tissues. *Molecular and Cellular Proteomics* 15 (9): 3003–3016.

25 Kunzke, T., Balluff, B., Feuchtinger, A. et al. (2017). Native glycan fragments detected by MALDI-FT-ICR mass spectrometry imaging impact gastric cancer biology and patient outcome. *Oncotarget* 8 (40): 68012.

26 Briggs, M.T., Condina, M.R., Ho, Y.Y. et al. (2019). MALDI mass spectrometry imaging of early-and late-stage serous ovarian cancer tissue reveals stage-specific *N*-glycans. *Proteomics* 19 (21, 22): 1800482.

27 Heijs, B., Holst-Bernal, S., de Graaff, M.A. et al. (2020). Molecular signatures of tumor progression in myxoid liposarcoma identified by *N*-glycan mass spectrometry imaging. *Laboratory Investigation* 100 (9): 1252–1261.

28 Boyaval, F., Van Zeijl, R., Dalebout, H. et al. (2021). *N*-glycomic signature of stage II colorectal cancer and its association with the tumor microenvironment. *Molecular and Cellular Proteomics* 20: 100057.

29 Holst, S., Heijs, B., De Haan, N. et al. (2016). Linkage-specific in situ sialic acid derivatization for *N*-glycan mass spectrometry imaging of formalin-fixed paraffin-embedded tissues. *Analytical Chemistry* 88 (11): 5904–5913.

30 Norris, J.L., Porter, N.A., and Caprioli, R.M. (2005). Combination detergent/MALDI matrix: functional cleavable detergents for mass spectrometry. *Analytical Chemistry* 77 (15): 5036–5040.

31 Lemaire, R., Tabet, J., Ducoroy, P. et al. (2006). Solid ionic matrixes for direct tissue analysis and MALDI imaging. *Analytical Chemistry* 78 (3): 809–819.

32 Zhang, H., Shi, X., Vu, N.Q. et al. (2020). On-tissue derivatization with Girard's reagent P enhances *N*-glycan signals for formalin-fixed paraffin-embedded tissue sections in MALDI mass spectrometry imaging. *Analytical Chemistry* 92 (19): 13361–13368.

33 Arentz, G., Mittal, P., Zhang, C. et al. (2017). Applications of mass spectrometry imaging to cancer. *Advances in Cancer Research* 134: 27–66.

34 Powers, T.W., Jones, E.E., Betesh, L.R. et al. (2013). Matrix assisted laser desorption ionization imaging mass spectrometry workflow for spatial profiling analysis of N-linked glycan expression in tissues. *Analytical Chemistry* 85 (20): 9799–9806.

35 Powers, T.W., Holst, S., Wuhrer, M. et al. (2015). Two-dimensional *N*-glycan distribution mapping of hepatocellular carcinoma tissues by MALDI-imaging mass spectrometry. *Biomolecules* 5 (4): 2554–2572.

36 Drake, R.R., Powers, T.W., Norris-Caneda, K. et al. (2018). In situ imaging of *N*-glycans by MALDI imaging mass spectrometry of fresh or formalin-fixed paraffin-embedded tissue. *Current Protocols in Protein Science* 94 (1): e68.

37 Drake, R.R., Powers, T.W., Jones, E.E. et al. (2017). MALDI mass spectrometry imaging of N-linked glycans in cancer tissues. *Advances in Cancer Research* 134: 85–116.

38 Heijs, B., Holst, S., Briaire-de Bruijn, I.H. et al. (2016). Multimodal mass spectrometry imaging of *N*-glycans and proteins from the same tissue section. *Analytical Chemistry* 88 (15): 7745–7753.

39 Fredriksson, S., Gullberg, M., Jarvius, J. et al. (2002). Protein detection using proximity-dependent DNA ligation assays. *Nature Biotechnology* 20 (5): 473–477.

40 Oliveira, F.M.S., Mereiter, S., Lönn, P. et al. (2018). Detection of post-translational modifications using solid-phase proximity ligation assay. *New Biotechnology* 45: 51–59.

41 Söderberg, O., Leuchowius, K.-J., Gullberg, M. et al. (2008). Characterizing proteins and their interactions in cells and tissues using the in situ proximity ligation assay. *Methods* 45 (3): 227–232.

42 Pinto, R., Carvalho, A.S., Conze, T. et al. (2012). Identification of new cancer biomarkers based on aberrant mucin glycoforms by in situ proximity ligation. *Journal of Cellular and Molecular Medicine* 16 (7): 1474–1484.

43 Campos, D., Freitas, D., Gomes, J. et al. (2015). Probing the *O*-glycoproteome of gastric cancer cell lines for biomarker discovery. *Molecular and Cellular Proteomics* 14 (6): 1616–1629.

44 Mereiter, S., Martins, Á.M., Gomes, C. et al. (2019). *O*-glycan truncation enhances cancer-related functions of CD44 in gastric cancer. *FEBS Letters* 593 (13): 1675–1689.

45 Mereiter, S., Magalhães, A., Adamczyk, B. et al. (2016). Glycomic analysis of gastric carcinoma cells discloses glycans as modulators of RON receptor tyrosine kinase activation in cancer. *Biochimica et Biophysica Acta (BBA), General Subjects* 1860 (8): 1795–1808.

46 Gomes, C., Almeida, A., Barreira, A. et al. (2019). Carcinoembryonic antigen carrying SLe(X) as a new biomarker of more aggressive gastric carcinomas. *Theranostics* 9 (24): 7431–7446.

47 Carvalho, S., Catarino, T.A., Dias, A.M. et al. (2016). Preventing E-cadherin aberrant N-glycosylation at Asn-554 improves its critical function in gastric cancer. *Oncogene* 35 (13): 1619–1631.

48 Duarte, H.O., Rodrigues, J.G., Gomes, C. et al. (2021). ST6Gal1 targets the ectodomain of ErbB2 in a site-specific manner and regulates gastric cancer cell sensitivity to trastuzumab. *Oncogene* 40 (21): 3719–3733.

49 Magalhães, A., Rossez, Y., Robbe-Masselot, C. et al. (2016). Muc5ac gastric mucin glycosylation is shaped by FUT2 activity and functionally impacts Helicobacter pylori binding. *Scientific Reports* 6 (1): 25575.

50 Li, X., Jiang, X., Xu, X. et al. (2017). Imaging of protein-specific glycosylation by glycan metabolic tagging and in situ proximity ligation. *Carbohydrate Research* 448: 148–154.

51 Freeze, H.H. (2006). Genetic defects in the human glycome. *Nature Reviews Genetics* 7 (7): 537–551.

52 Wang, D., Liu, S., Trummer, B.J. et al. (2002). Carbohydrate microarrays for the recognition of cross-reactive molecular markers of microbes and host cells. *Nature Biotechnology* 20 (3): 275–281.

53 Fukui, S., Feizi, T., Galustian, C. et al. (2002). Oligosaccharide microarrays for high-throughput detection and specificity assignments of carbohydrate-protein interactions. *Nature Biotechnology* 20 (10): 1011–1017.

54 Gao, C., Wei, M., McKitrick, T.R. et al. (2019). Glycan microarrays as chemical tools for identifying glycan recognition by immune proteins. *Frontiers in Chemistry* 7 (833): 833.

55 Song, X., Heimburg-Molinaro, J., Cummings, R.D., and Smith, D.F. (2014). Chemistry of natural glycan microarrays. *Current Opinion in Chemical Biology* 18: 70–77.

56 Hirabayashi, J., Yamada, M., Kuno, A., and Tateno, H. (2013). Lectin microarrays: concept, principle and applications. *Chemical Society Reviews* 42 (10): 4443–4458.

57 Li, Z. and Feizi, T. (2018). The neoglycolipid (NGL) technology-based microarrays and future prospects. *FEBS Letters* 592 (23): 3976–3991.

58 Smith, D.F., Song, X., and Cummings, R.D. (2010). Chapter Nineteen – Use of glycan microarrays to explore specificity of glycan-binding proteins. In: *Methods in Enzymology*, vol. 480 (ed. M. Fukuda), 417–444. Academic Press.

59 Drickamer, K. and Taylor, M.E. (2002). Glycan arrays for functional glycomics. *Genome Biology* 3 (12): reviews1034.1.

60 Patwa, T., Li, C., Simeone, D.M., and Lubman, D.M. (2010). Glycoprotein analysis using protein microarrays and mass spectrometry. *Mass Spectrometry Reviews* 29 (5): 830–844.

61 Tse-Wen, C. (1983). Binding of cells to matrixes of distinct antibodies coated on solid surface. *Journal of Immunological Methods* 65 (1): 217–223.

62 Song, X., Heimburg-Molinaro, J., Smith, D.F., and Cummings, R.D. (2015). Glycan microarrays of fluorescently-tagged natural glycans. *Glycoconjugate Journal* 32 (7): 465–473.

63 Palma, A.S., Feizi, T., Zhang, Y. et al. (2006). Ligands for the β-glucan receptor, dectin-1, assigned using "designer" microarrays of oligosaccharide probes (neoglycolipids) generated from glucan polysaccharides. *Journal of Biological Chemistry* 281 (9): 5771–5779.

64 Liu, L., Prudden, A.R., Capicciotti, C.J. et al. (2019). Streamlining the chemoenzymatic synthesis of complex *N*-glycans by a stop and go strategy. *Nature Chemistry* 11 (2): 161–169.

65 Fair, R.J., Hahm, H.S., and Seeberger, P.H. (2015). Combination of automated solid-phase and enzymatic oligosaccharide synthesis provides access to α(2,3)-sialylated glycans. *Chemical Communications* 51 (28): 6183–6185.

66 Esposito, D., Hurevich, M., Castagner, B. et al. (2012). Automated synthesis of sialylated oligosaccharides. *Beilstein Journal of Organic Chemistry* 8: 1601–1609.

67 Li, T., Liu, L., Wei, N. et al. (2019). An automated platform for the enzyme-mediated assembly of complex oligosaccharides. *Nature Chemistry* 11 (3): 229–236.

68 Liang, C.F., Hahm, H.S., and Seeberger, P.H. (2015). Automated synthesis of chondroitin sulfate oligosaccharides. *Methods in Molecular Biology* 1229: 3–10.

69 Song, X., Xia, B., Stowell, S.R. et al. (2009). Novel fluorescent glycan microarray strategy reveals ligands for galectins. *Chemistry and Biology* 16 (1): 36–47.

70 Xia, B., Kawar, Z.S., Ju, T. et al. (2005). Versatile fluorescent derivatization of glycans for glycomic analysis. *Nature Methods* 2 (11): 845–850.

71 de Boer, A.R., Hokke, C.H., Deelder, A.M., and Wuhrer, M. (2007). General microarray technique for immobilization and screening of natural glycans. *Analytical Chemistry* 79 (21): 8107–8113.

72 McQuillan, A.M., Byrd-Leotis, L., Heimburg-Molinaro, J., and Cummings, R.D. (2019). Natural and synthetic sialylated glycan microarrays and their applications. *Frontiers in Molecular Biosciences* 6: 88.

73 Temme, J.S., Campbell, C.T., and Gildersleeve, J.C. (2019). Factors contributing to variability of glycan microarray binding profiles. *Faraday Discussions* 219: 90–111.

74 Wang, R., Liu, S., Shah, D., and Wang, D. (2005). A practical protocol for carbohydrate microarrays. In: *Chemical Genomics: Reviews and Protocols* (ed. E.D. Zanders), 241–252. Totowa, NJ: Humana Press.

75 Wang, C.-C., Huang, Y.-L., Ren, C.-T. et al. (2008). Glycan microarray of Globo H and related structures for quantitative analysis of breast cancer. *Proceedings of the National Academy of Sciences of the United States of America* 105 (33): 11661–11666.

76 Gao, C., Liu, Y., Zhang, H. et al. (2014). Carbohydrate sequence of the prostate cancer-associated antigen F77 assigned by a Mucin *O*-glycome designer array. *Journal of Biological Chemistry* 289 (23): 16462–16477.

77 Blixt, O., Head, S., Mondala, T. et al. (2004). Printed covalent glycan array for ligand profiling of diverse glycan binding proteins. *Proceedings of the National Academy of Sciences of the United States of America* 101 (49): 17033–17038.

78 Song, X., Xia, B., Lasanajak, Y. et al. (2008). Quantifiable fluorescent glycan microarrays. *Glycoconjugate Journal* 25 (1): 15–25.

79 Chen, S., LaRoche, T., Hamelinck, D. et al. (2007). Multiplexed analysis of glycan variation on native proteins captured by antibody microarrays. *Nature Methods* 4 (5): 437–444.

80 Park, H., Jung, J., Rodrigues, E. et al. (2020). Mass spectrometry-based shotgun glycomics for discovery of natural ligands of glycan-binding proteins. *Analytical Chemistry* 92 (20): 14012–14020.

81 Padler-Karavani, V., Song, X., Yu, H. et al. (2012). Cross-comparison of protein recognition of sialic acid diversity on two novel sialoglycan microarrays. *The Journal of Biological Chemistry* 287 (27): 22593–22608.

82 Büll, C., Nason, R., Sun, L. et al. (2021). Probing the binding specificities of human Siglecs by cell-based glycan arrays. *Proceedings of the National Academy of Sciences of the United States of America* 118 (17): e2026102118.

83 Wang, S., Chen, C., Guan, M. et al. (2021). Terminal epitope-dependent branch preference of siglecs toward *N*-glycans. *Frontiers in Molecular Biosciences* 8 (142): 645999.

84 Arthur, C.M., Rodrigues, L.C., Baruffi, M.D. et al. (2015). Examining galectin binding specificity using glycan microarrays. *Methods in Molecular Biology* (Clifton, N.J.) 1207: 115–131.

85 Padler-Karavani, V., Hurtado-Ziola, N., Pu, M. et al. (2011). Human xeno-autoantibodies against a non-human sialic acid serve as novel serum biomarkers and immunotherapeutics in cancer. *Cancer Research* 71 (9): 3352–3363.

86 Lawrie, C.H., Marafioti, T., Hatton, C.S.R. et al. (2006). Cancer-associated carbohydrate identification in Hodgkin's lymphoma by carbohydrate array profiling. *International Journal of Cancer* 118 (12): 3161–3166.

87 Huang, W.-L., Li, Y.-G., Lv, Y.-C. et al. (2014). Use of lectin microarray to differentiate gastric cancer from gastric ulcer. *World Journal of Gastroenterology* 20 (18): 5474–5482.

88 Purohit, S., Li, T., Guan, W. et al. (2018). Multiplex glycan bead array for high throughput and high content analyses of glycan binding proteins. *Nature Communications* 9 (1): 258.

89 Pedersen, J.W., Blixt, O., Bennett, E.P. et al. (2011). Seromic profiling of colorectal cancer patients with novel glycopeptide microarray. *International Journal of Cancer* 128 (8): 1860–1871.

90 Zhao, J., Patwa, T.H., Qiu, W. et al. (2007). Glycoprotein microarrays with multi-lectin detection: unique lectin binding patterns as a tool for classifying normal, chronic pancreatitis and pancreatic cancer sera. *Journal of Proteome Research* 6 (5): 1864–1874.

91 Burford, B., Gentry-Maharaj, A., Graham, R. et al. (2013). Autoantibodies to MUC1 glycopeptides cannot be used as a screening assay for early detection of breast, ovarian, lung or pancreatic cancer. *British Journal of Cancer* 108 (10): 2045–2055.

92 Matsushita, T., Takada, W., Igarashi, K. et al. (2014). A straightforward protocol for the preparation of high performance microarray displaying synthetic MUC1 glycopeptides. *Biochimica et Biophysica Acta (BBA), General Subjects* 1840 (3): 1105–1116.

93 Peng, W., Nycholat, C.M., and Razi, N. (2013). Glycan microarray screening assay for glycosyltransferase specificities. *Methods in Molecular Biology* 1022: 1–14.

94 Ban, L., Pettit, N., Li, L. et al. (2012). Discovery of glycosyltransferases using carbohydrate arrays and mass spectrometry. *Nature Chemical Biology* 8 (9): 769–773.

95 Muthana, S.M. (2020). Glycan microarray: toward drug discovery and development. In: *Carbohydrates in Drug Discovery and Development* (ed. V.K. Tiwari), 267–282. Elsevier.

96 Narimatsu, Y., Büll, C., Chen, Y.-H. et al. (2021). Genetic glycoengineering in mammalian cells. *Journal of Biological Chemistry* 296: 100448.

97 Chen, Y.-H., Narimatsu, Y., Clausen, T.M. et al. (2018). The GAGOme: a cell-based library of displayed glycosaminoglycans. *Nature Methods* 15 (11): 881–888.

98 Narimatsu, Y., Joshi, H.J., Yang, Z. et al. (2018). A validated gRNA library for CRISPR/Cas9 targeting of the human glycosyltransferase genome. *Glycobiology* 28 (5): 295–305.

99 Narimatsu, Y., Joshi, H.J., Nason, R. et al. (2019). An atlas of human glycosylation pathways enables display of the human glycome by gene engineered cells. *Molecular Cell* 75 (2): 394–407.e5.

100 Lonowski, L.A., Narimatsu, Y., Riaz, A. et al. (2017). Genome editing using FACS enrichment of nuclease-expressing cells and indel detection by amplicon analysis. *Nature Protocols* 12 (3): 581–603.

101 Yang, Z., Wang, S., Halim, A. et al. (2015). Engineered CHO cells for production of diverse, homogeneous glycoproteins. *Nature Biotechnology* 33 (8): 842–844.

102 Schulz, M.A., Tian, W., Mao, Y. et al. (2018). Glycoengineering design options for IgG1 in CHO cells using precise gene editing. *Glycobiology* 28 (7): 542–549.

103 Tian, W., Ye, Z., Wang, S. et al. (2019). The glycosylation design space for recombinant lysosomal replacement enzymes produced in CHO cells. *Nature Communications* 10 (1): 1785.

104 Čaval, T., Tian, W., Yang, Z. et al. (2018). Direct quality control of glycoengineered erythropoietin variants. *Nature Communications* 9 (1): 3342.

105 Stanley, P. (2016). What have we learned from glycosyltransferase knockouts in mice? *Journal of Molecular Biology* 428 (16): 3166–3182.

106 Orr, S.L., Le, D., Long, J.M. et al. (2012). A phenotype survey of 36 mutant mouse strains with gene-targeted defects in glycosyltransferases or glycan-binding proteins. *Glycobiology* 23 (3): 363–380.

107 Takahashi, M., Yokoe, S., Asahi, M. et al. (2008). *N*-glycan of ErbB family plays a crucial role in dimer formation and tumor promotion. *Biochimica et Biophysica Acta (BBA), General Subjects* 1780 (3): 520–524.

108 Miller, J.C., Holmes, M.C., Wang, J. et al. (2007). An improved zinc-finger nuclease architecture for highly specific genome editing. *Nature Biotechnology* 25 (7): 778–785.

109 Miller, J.C., Tan, S., Qiao, G. et al. (2011). A TALE nuclease architecture for efficient genome editing. *Nature Biotechnology* 29 (2): 143–148.

110 Cong, L., Ran, F.A., Cox, D. et al. (2013). Multiplex genome engineering using CRISPR/Cas systems. *Science* 339 (6121): 819–823.

111 Adli, M. (2018). The CRISPR tool kit for genome editing and beyond. *Nature Communications* 9 (1): 1911.

112 Zhu, Y., Groth, T., Kelkar, A. et al. (2020). A GlycoGene CRISPR-Cas9 lentiviral library to study lectin binding and human glycan biosynthesis pathways. *Glycobiology* 31 (3): 173–180.

113 KJlC, K., Hefzi, H., Xiong, K. et al. (2020). Awakening dormant glycosyltransferases in CHO cells with CRISPRa. *Biotechnology and Bioengineering* 117 (2): 593–598.

114 Schjoldager, K.T.-B.G., Vakhrushev, S.Y., Kong, Y. et al. (2012). Probing isoform-specific functions of polypeptide GalNAc-transferases using zinc finger nuclease glycoengineered SimpleCells. *Proceedings of the National Academy of Sciences of the United States of America* 109 (25): 9893–9898.

115 Hintze, J., Ye, Z., Narimatsu, Y. et al. (2018). Probing the contribution of individual polypeptide GalNAc-transferase isoforms to the *O*-glycoproteome by inducible expression in isogenic cell lines. *Journal of Biological Chemistry* 293 (49): 19064–19077.

116 Lavrsen, K., Dabelsteen, S., Vakhrushev, S.Y. et al. (2018). De novo expression of human polypeptide *N*-acetylgalactosaminyltransferase 6 (GalNAc-T6) in colon adenocarcinoma inhibits the differentiation of colonic epithelium. *Journal of Biological Chemistry* 293 (4): 1298–1314.

117 Steentoft, C., Vakhrushev, S.Y., Joshi, H.J. et al. (2013). Precision mapping of the human *O*-GalNAc glycoproteome through SimpleCell technology. *The EMBO Journal* 32 (10): 1478–1488.

118 Steentoft, C., Vakhrushev, S.Y., Vester-Christensen, M.B. et al. (2011). Mining the *O*-glycoproteome using zinc-finger nuclease–glycoengineered SimpleCell lines. *Nature Methods* 8 (11): 977–982.

119 Vester-Christensen, M.B., Halim, A., Joshi, H.J. et al. (2013). Mining the *O*-mannose glycoproteome reveals cadherins as major O-mannosylated glycoproteins. *Proceedings of the National Academy of Sciences of the United States of America* 110 (52): 21018–21023.

120 Büll, C., Joshi, H.J., Clausen, H., and Narimatsu, Y. (2020). Cell-based glycan arrays – a practical guide to dissect the human glycome. *STAR Protocols* 1 (1): 100017.

121 Joeh, E., O'Leary, T., Li, W. et al. (2020). Mapping glycan-mediated galectin-3 interactions by live cell proximity labeling. *Proceedings of the National Academy of Sciences of the United States of America* 117 (44): 27329–27338.

122 Wang, H. and Mooney, D.J. (2020). Metabolic glycan labelling for cancer-targeted therapy. *Nature Chemistry* 12 (12): 1102–1114.

123 Schumann, B., Malaker, S.A., Wisnovsky, S.P. et al. (2020). Bump-and-hole engineering identifies specific substrates of glycosyltransferases in living cells. *Molecular Cell* 78 (5): 824–34.e15.

124 Agatemor, C., Buettner, M.J., Ariss, R. et al. (2019). Exploiting metabolic glycoengineering to advance healthcare. *Nature Reviews Chemistry* 3 (10): 605–620.

125 Debets, M.F., Tastan, O.Y., Wisnovsky, S.P. et al. (2020). Metabolic precision labeling enables selective probing of O-linked *N*-acetylgalactosamine glycosylation. *Proceedings of the National Academy of Sciences of the United States of America* 117 (41): 25293–25301.

126 Cioce, A., Malaker, S.A., and Schumann, B. (2021). Generating orthogonal glycosyltransferase and nucleotide sugar pairs as next-generation glycobiology tools. *Current Opinion in Chemical Biology* 60: 66–78.

127 Prescher, J.A. and Bertozzi, C.R. (2006). Chemical technologies for probing glycans. *Cell* 126 (5): 851–854.

128 Thirumurugan, P., Matosiuk, D., and Jozwiak, K. (2013). Click chemistry for drug development and diverse chemical–biology applications. *Chemical Reviews* 113 (7): 4905–4979.

129 Laughlin, S.T., Baskin, J.M., Amacher, S.L., and Bertozzi, C.R. (2008). In vivo imaging of membrane-associated glycans in developing zebrafish. *Science* 320 (5876): 664–667.

130 Xie, R., Dong, L., Huang, R. et al. (2014). Targeted imaging and proteomic analysis of tumor-associated glycans in living animals. *Angewandte Chemie International Edition* 53 (51): 14082–14086.

131 Hong, S., Yu, C., Wang, P. et al. (2021). Glycoengineering of NK Cells with glycan ligands of CD22 and selectins for B-cell lymphoma therapy. *Angewandte Chemie International Edition* 60 (7): 3603–3610.

132 Ge, C. and Stanley, P. (2008). The *O*-fucose glycan in the ligand-binding domain of Notch1 regulates embryogenesis and T cell development. *Proceedings of the National Academy of Sciences of the United States of America* 105 (5): 1539–1544.

133 Sung, Y.H., Baek, I.-J., Seong, J.K. et al. (2012). Mouse genetics: catalogue and scissors. *BMB Reports* 45 (12): 686–692.

134 Ioffe, E. and Stanley, P. (1994). Mice lacking *N*-acetylglucosaminyltransferase I activity die at mid-gestation, revealing an essential role for complex or hybrid N-linked carbohydrates. *Proceedings of the National Academy of Sciences of the United States of America* 91 (2): 728–732.

135 Metzler, M., Gertz, A., Sarkar, M. et al. (1994). Complex asparagine-linked oligosaccharides are required for morphogenic events during post-implantation development. *The EMBO Journal* 13 (9): 2056–2065.

136 Ye, Z. and Marth, J.D. (2004). *N*-glycan branching requirement in neuronal and postnatal viability. *Glycobiology* 14 (6): 547–558.

137 Shi, S., Williams, S.A., Seppo, A. et al. (2004). Inactivation of the Mgat1 gene in oocytes impairs oogenesis, but embryos lacking complex and hybrid *N*-glycans develop and implant. *Molecular and Cellular Biology* 24 (22): 9920–9929.

138 Batista, F., Lu, L., Williams, S.A., and Stanley, P. (2012). Complex *N*-glycans are essential, but core 1 and 2 mucin *O*-glycans, *O*-fucose glycans, and NOTCH1 are dispensable, for mammalian spermatogenesis. *Biology of Reproduction* 86 (6): 179.

139 Zhou, R.W., Mkhikian, H., Grigorian, A. et al. (2014). N-glycosylation bidirectionally extends the boundaries of thymocyte positive selection by decoupling Lck from Ca^{2+} signaling. *Nature Immunology* 15 (11): 1038–1045.

140 Fonseca, K.L., Maceiras, A.R., Matos, R. et al. (2020). Deficiency in the glycosyltransferase Gcnt1 increases susceptibility to tuberculosis through a mechanism involving neutrophils. *Mucosal Immunology* 13 (5): 836–848.

141 Engle, D.D., Tiriac, H., Rivera, K.D. et al. (2019). The glycan CA19-9 promotes pancreatitis and pancreatic cancer in mice. *Science* 364 (6446): 1156–1162.

142 Zavareh, R.B., Sukhai, M.A., Hurren, R. et al. (2012). Suppression of cancer progression by MGAT1 shRNA knockdown. *PLoS One* 7 (9): e43721.

143 Song, Y., Aglipay, J.A., Bernstein, J.D. et al. (2010). The bisecting GlcNAc on *N*-glycans inhibits growth factor signaling and retards mammary tumor progression. *Cancer Research* 70 (8): 3361–3371.

144 Silva, M.C., Fernandes, Â., Oliveira, M. et al. (2020). Glycans as immune checkpoints: removal of branched *N*-glycans enhances immune recognition preventing cancer progression. *Cancer Immunology Research* 8 (11): 1407–1425.

145 Liu, F., Fu, J., Bergstrom, K. et al. (2020). Core 1–derived mucin-type O-glycosylation protects against spontaneous gastritis and gastric cancer. *The Journal of Experimental Medicine* 217 (1): e20182325.

146 Granovsky, M., Fata, J., Pawling, J. et al. (2000). Suppression of tumor growth and metastasis in Mgat5-deficient mice. *Nature Medicine* 6 (3): 306–312.

147 Hedlund, M., Ng, E., Varki, A., and Varki, N.M. (2008). α2-6–linked sialic acids on *N*-glycans modulate carcinoma differentiation in vivo. *Cancer Research* 68 (2): 388–394.

148 Magalhães, A., Gomes, J., Ismail, M.N. et al. (2009). Fut2-null mice display an altered glycosylation profile and impaired BabA-mediated Helicobacter pylori adhesion to gastric mucosa. *Glycobiology* 19 (12): 1525–1536.

149 Yoshida, G.J. (2020). Applications of patient-derived tumor xenograft models and tumor organoids. *Journal of Hematology & Oncology* 13 (1): 4.

150 Byrne, A.T., Alférez, D.G., Amant, F. et al. (2017). Interrogating open issues in cancer precision medicine with patient-derived xenografts. *Nature Reviews Cancer* 17 (4): 254.

151 Huang, L., Bockorny, B., Paul, I. et al. (2020). PDX-derived organoids model in vivo drug response and secrete biomarkers. *JCI Insight* 5 (21): e135544.

152 Sinha, A., Hussain, A., Ignatchenko, V. et al. (2019). *N*-Glycoproteomics of patient-derived xenografts: a strategy to discover tumor-associated proteins in high-grade serous ovarian cancer. *Cell Systems* 8 (4): 345–51.e4.

153 Selman, M.H., Hoffmann, M., Zauner, G. et al. (2012). MALDI-TOF-MS analysis of sialylated glycans and glycopeptides using 4-chloro-α-cyanocinnamic acid matrix. *Proteomics* 12 (9): 1337–1348.

154 Pralow, A., Cajic, S., Alagesan, K. et al. (2020). *State-of-the-Art Glycomics Technologies in Glycobiotechnology*, 1–33. Berlin, Heidelberg: Springer Berlin Heidelberg.

155 Wuhrer, M. (2013). Glycomics using mass spectrometry. *Glycoconjugate Journal* 30 (1): 11–22.

156 Kotsias, M., Kozak, R.P., Gardner, R.A. et al. (2019). Improved and semi-automated reductive β-elimination workflow for higher throughput protein O-glycosylation analysis. *PLoS One* 14 (1): e0210759.

157 Ruhaak, L., Zauner, G., Huhn, C. et al. (2010). Glycan labeling strategies and their use in identification and quantification. *Analytical and Bioanalytical Chemistry* 397 (8): 3457–3481.

158 Wuhrer, M., Koeleman, C.A., Hokke, C.H., and Deelder, A.M. (2006). Mass spectrometry of proton adducts of fucosylated *N*-glycans: fucose transfer between antennae gives rise to misleading fragments. *Rapid Communications in Mass Spectrometry: An International Journal Devoted to the Rapid Dissemination of Up-to-the-Minute Research in Mass Spectrometry* 20 (11): 1747–1754.

159 de Haan, N., Reiding, K.R., Haberger, M. et al. (2015). Linkage-specific sialic acid derivatization for MALDI-TOF-MS profiling of IgG glycopeptides. *Analytical Chemistry* 87 (16): 8284–8291.

160 Lageveen-Kammeijer, G.S., de Haan, N., Mohaupt, P. et al. (2019). Highly sensitive CE-ESI-MS analysis of *N*-glycans from complex biological samples. *Nature Communications* 10 (1): 1–8.

161 Kalay, H., Ambrosini, M., van Berkel, P.H. et al. (2012). Online nanoliquid chromatography–mass spectrometry and nanofluorescence detection for high-resolution quantitative *N*-glycan analysis. *Analytical Biochemistry* 423 (1): 153–162.

162 Hua, S., Nwosu, C.C., Strum, J.S. et al. (2012). Site-specific protein glycosylation analysis with glycan isomer differentiation. *Analytical and Bioanalytical Chemistry* 403 (5): 1291–1302.

163 Jensen, P.H., Karlsson, N.G., Kolarich, D., and Packer, N.H. (2012). Structural analysis of *N*- and *O*-glycans released from glycoproteins. *Nature Protocols* 7 (7): 1299–1310.

164 Lu, G., Crihfield, C.L., Gattu, S. et al. (2018). Capillary electrophoresis separations of glycans. *Chemical Reviews* 118 (17): 7867–7885.

165 Selman, M.H.J., Hemayatkar, M., Deelder, A.M., and Wuhrer, M. (2011). Cotton HILIC SPE microtips for microscale purification and enrichment of glycans and glycopeptides. *Analytical Chemistry* 83 (7): 2492–2499.

166 Jensen, P.H., Mysling, S., Højrup, P., and Jensen, O.N. (2013). *Glycopeptide Enrichment for MALDI-TOF Mass Spectrometry Analysis by Hydrophilic Interaction Liquid Chromatography Solid Phase Extraction (HILIC SPE). Mass Spectrometry of Glycoproteins*, 131–144. Springer.

167 Jin, C., Harvey, D.J., Struwe, W.B., and Karlsson, N.G. (2019). Separation of isomeric *O*-glycans by ion mobility and liquid chromatography–mass spectrometry. *Analytical Chemistry* 91 (16): 10604–10613.

168 Pallister, E.G., Choo, M.S.F., Walsh, I. et al. (2020). Utility of ion-mobility spectrometry for deducing branching of multiply charged glycans and glycopeptides in a high-throughput positive ion LC-FLR-IMS-MS workflow. *Analytical Chemistry* 92 (23): 15323–15335.

169 Singhal, N., Kumar, M., Kanaujia, P.K., and Virdi, J.S. (2015). MALDI-TOF mass spectrometry: an emerging technology for microbial identification and diagnosis. *Frontiers in Microbiology* 6: 791.

170 Nie, H., Li, Y., and Sun, X.-L. (2012). Recent advances in sialic acid-focused glycomics. *Journal of Proteomics* 75 (11): 3098–3112.

171 Konermann, L., Ahadi, E., Rodriguez, A.D., and Vahidi, S. (2013). Unraveling the mechanism of electrospray ionization. *Analytical Chemistry* 85 (1): 2–9.

172 Kolarich, D., Jensen, P.H., Altmann, F., and Packer, N.H. (2012). Determination of site-specific glycan heterogeneity on glycoproteins. *Nature Protocols* 7 (7): 1285–1298.

173 Zaia, J. (2010). Mass spectrometry and glycomics. *Omics: A Journal of Integrative Biology* 14 (4): 401–418.

174 Kim, M.S., Zhong, J., and Pandey, A. (2016). Common errors in mass spectrometry-based analysis of post-translational modifications. *Proteomics* 16 (5): 700–714.

175 An, H.J., Peavy, T.R., Hedrick, J.L., and Lebrilla, C.B. (2003). Determination of N-glycosylation sites and site heterogeneity in glycoproteins. *Analytical Chemistry* 75 (20): 5628–5637.

176 Swaney, D.L., Wenger, C.D., and Coon, J.J. (2010). Value of using multiple proteases for large-scale mass spectrometry-based proteomics. *Journal of Proteome Research* 9 (3): 1323–1329.

177 Giansanti, P., Tsiatsiani, L., Low, T.Y., and Heck, A.J. (2016). Six alternative proteases for mass spectrometry–based proteomics beyond trypsin. *Nature Protocols* 11 (5): 993–1006.

178 Sutton, C.W., Oneill, J.A., and Cottrell, J.S. (1994). Site-specific characterization of glycoprotein carbohydrates by exoglycosidase digestion and laser-desorption mass spectrometry. *Analytical Biochemistry* 218 (1): 34–46.

179 Zhu, R., Zacharias, L., Wooding, K.M. et al. (2017). Glycoprotein enrichment analytical techniques: advantages and disadvantages. *Methods in Enzymology* 585: 397–429.

180 Matsuda, A., Kuno, A., Matsuzaki, H. et al. (2013). Glycoproteomics-based cancer marker discovery adopting dual enrichment with Wisteria floribunda agglutinin for high specific glyco-diagnosis of cholangiocarcinoma. *Journal of Proteomics* 85: 1–11.

181 Totten, S.M., Kullolli, M., and Pitteri, S.J. (2017). Multi-lectin affinity chromatography for separation, identification, and quantitation of intact protein glycoforms in complex biological mixtures. In: *Proteomics* (ed. L. Comai, J.E. Katz, and P. Mallick), 99–113. Springer.

182 Totten, S.M., Adusumilli, R., Kullolli, M. et al. (2018). Multi-lectin affinity chromatography and quantitative proteomic analysis reveal differential glycoform levels between prostate cancer and benign prostatic hyperplasia sera. *Scientific Reports* 8 (1): 6509.

183 Zeng, Z., Hincapie, M., Pitteri, S.J. et al. (2011). A proteomics platform combining depletion, multi-lectin affinity chromatography (M-LAC), and isoelectric focusing to study the breast cancer proteome. *Analytical Chemistry* 83 (12): 4845–4854.

184 Yang, Z., Harris, L.E., Palmer-Toy, D.E., and Hancock, W.S. (2006). Multilectin affinity chromatography for characterization of multiple glycoprotein biomarker candidates in serum from breast cancer patients. *Clinical Chemistry* 52 (10): 1897–1905.

185 Na, K., Lee, E.Y., Lee, H.J. et al. (2009). Human plasma carboxylesterase 1, a novel serologic biomarker candidate for hepatocellular carcinoma. *Proteomics* 9 (16): 3989–3999.

186 Palmisano, G., Lendal, S.E., and Larsen, M.R. (2011). Titanium dioxide enrichment of sialic acid-containing glycopeptides. *Methods in Molecular Biology* 753: 309–322.

187 Larsen, M.R., Jensen, S.S., Jakobsen, L.A., and Heegaard, N.H. (2007). Exploring the sialiome using titanium dioxide chromatography and mass spectrometry. *Molecular and Cellular Proteomics* 6 (10): 1778–1787.

188 Thaysen-Andersen, M., Larsen, M.R., Packer, N.H., and Palmisano, G. (2013). Structural analysis of glycoprotein sialylation–part I: pre-LC-MS analytical strategies. *RSC Advances* 3 (45): 22683–22705.

189 Zhang, H., Li, X.-j., Martin, D.B., and Aebersold, R. (2003). Identification and quantification of N-linked glycoproteins using hydrazide chemistry, stable isotope labeling and mass spectrometry. *Nature Biotechnology* 21 (6): 660–666.

190 Wang, L., Aryal, U.K., Dai, Z. et al. (2012). Mapping N-linked glycosylation sites in the secretome and whole cells of *Aspergillus niger* using hydrazide chemistry and mass spectrometry. *Journal of Proteome Research* 11 (1): 143–156.

191 Chen, R., Jiang, X., Sun, D. et al. (2009). Glycoproteomics analysis of human liver tissue by combination of multiple enzyme digestion and hydrazide chemistry. *Journal of Proteome Research* 8 (2): 651–661.

192 Yang, S., Chen, L., Chan, D.W. et al. (2017). Protein signatures of molecular pathways in non-small cell lung carcinoma (NSCLC): comparison of glycoproteomics and global proteomics. *Clinical Proteomics* 14 (1): 1–15.

193 Chen, J., Xi, J., Tian, Y. et al. (2013). Identification, prioritization, and evaluation of glycoproteins for aggressive prostate cancer using quantitative glycoproteomics and antibody-based assays on tissue specimens. *Proteomics* 13 (15): 2268–2277.

194 Tian, Y., Esteva, F.J., Song, J., and Zhang, H. (2012). Altered expression of sialylated glycoproteins in breast cancer using hydrazide chemistry and mass spectrometry. *Molecular & Cellular Proteomics* 11 (6): M111.011403.

195 Bones, J., Mittermayr, S., O'Donoghue, N. et al. (2010). Ultra performance liquid chromatographic profiling of serum *N*-glycans for fast and efficient identification of cancer associated alterations in glycosylation. *Analytical Chemistry* 82 (24): 10208–10215.

196 Balmaña, M., Giménez, E., Puerta, A. et al. (2016). Increased α1-3 fucosylation of α-1-acid glycoprotein (AGP) in pancreatic cancer. *Journal of Proteomics* 132: 144–154.

197 Balog, C.I., Stavenhagen, K., Fung, W.L. et al. (2012). N-glycosylation of colorectal cancer tissues: a liquid chromatography and mass spectrometry-based investigation. *Molecular and Cellular Proteomics* 11 (9): 571–585.

198 Reiding, K.R., Bondt, A., Franc, V., and Heck, A.J.R. (2018). The benefits of hybrid fragmentation methods for glycoproteomics. *TrAC, Trends in Analytical Chemistry* 108: 260–268.

199 Riley, N.M., Malaker, S.A., Driessen, M.D., and Bertozzi, C.R. (2020). Optimal dissociation methods differ for *N*- and *O*-glycopeptides. *Journal of Proteome Research* 19 (8): 3286–3301.

200 Wuhrer, M., Catalina, M.I., Deelder, A.M., and Hokke, C.H. (2007). Glycoproteomics based on tandem mass spectrometry of glycopeptides. *Journal of Chromatography B* 849 (1, 2): 115–128.

201 Cao, L., Tolić, N., Qu, Y. et al. (2014). Characterization of intact N-and O-linked glycopeptides using higher energy collisional dissociation. *Analytical Biochemistry* 452: 96–102.

202 Stavenhagen, K., Hinneburg, H., Kolarich, D., and Wuhrer, M. (2017). Site-specific *N*- and *O*-glycopeptide analysis using an integrated C18-PGC-LC-ESI-QTOF-MS/MS approach. In: *High-Throughput Glycomics and Glycoproteomics: Methods and Protocols* (ed. G. Lauc and M. Wuhrer), 109–119. New York, NY, Springer New York.

203 Syka, J.E., Coon, J.J., Schroeder, M.J. et al. (2004). Peptide and protein sequence analysis by electron transfer dissociation mass spectrometry. *Proceedings of the National Academy of Sciences of the United States of America* 101 (26): 9528–9533.

204 Riley, N.M. and Coon, J.J. (2018). The role of electron transfer dissociation in modern proteomics. *Analytical Chemistry* 90 (1): 40–64.

205 Wu, S.-W., Pu, T.-H., Viner, R., and Khoo, K.-H. (2014). Novel LC-MS2 product dependent parallel data acquisition function and data analysis workflow for sequencing and identification of intact glycopeptides. *Analytical Chemistry* 86 (11): 5478–5486.

206 Rodrigues, J.G., Duarte, H.O., Gomes, C. et al. (2021). Terminal α2, 6-sialylation of epidermal growth factor receptor modulates antibody therapy response of colorectal cancer cells. *Cellular Oncology* 1–16.

207 Pirro, M., Rombouts, Y., Stella, A. et al. (2020). Characterization of macrophage galactose-type lectin (MGL) ligands in colorectal cancer cell lines. *Biochimica et Biophysica Acta (BBA), General Subjects* 44 (4): 835–850.

208 Pirro, M., Schoof, E., van Vliet, S.J. et al. (2018). Glycoproteomic analysis of MGL-binding proteins on acute T-cell leukemia cells. *Journal of Proteome Research* 18 (3): 1125–1132.

209 Rojas-Macias, M.A., Mariethoz, J., Andersson, P. et al. (2019). Towards a standardized bioinformatics infrastructure for *N*- and *O*-glycomics. *Nature Communications* 10 (1): 3275.

210 Apte, A. and Meitei, N.S. (2010). Bioinformatics in glycomics: glycan characterization with mass spectrometric data using SimGlycan™. In: *Functional Glycomics* (ed. J. Li), 269–281. Springer.

211 Ye, Z., Mao, Y., Clausen, H., and Vakhrushev, S.Y. (2019). Glyco-DIA: a method for quantitative *O*-glycoproteomics with in silico-boosted glycopeptide libraries. *Nature Methods* 16 (9): 902–910.

212 Mayampurath, A.M., Wu, Y., Segu, Z.M. et al. (2011). Improving confidence in detection and characterization of protein N-glycosylation sites and microheterogeneity. *Rapid Communications in Mass Spectrometry* 25 (14): 2007–2019.

213 Mayampurath, A., Yu, C.-Y., Song, E. et al. (2014). Computational framework for identification of intact glycopeptides in complex samples. *Analytical Chemistry* 86 (1): 453–463.

214 Bern, M., Kil, Y.J., and Becker, C. (2012). Byonic: advanced peptide and protein identification software. *Current Protocols in Bioinformatics* 40 (1): 13.20.1–13.20.14.

215 Jansen, B.C., Reiding, K.R., Bondt, A. et al. (2015). MassyTools: a high-throughput targeted data processing tool for relative quantitation and quality control developed for glycomic and glycoproteomic MALDI-MS. *Journal of Proteome Research* 14 (12): 5088–5098.

216 Woodin, C.L., Hua, D., Maxon, M. et al. (2012). GlycoPep grader: a web-based utility for assigning the composition of *N*-linked glycopeptides. *Analytical Chemistry* 84 (11): 4821–4829.

217 Serang, O., Froehlich, J.W., Muntel, J. et al. (2013). SweetSEQer, simple de Novo filtering and annotation of glycoconjugate mass spectra. *Molecular and Cellular Proteomics* 12 (6): 1735–1740.

218 Cheng, K., Chen, R., Seebun, D. et al. (2014). Large-scale characterization of intact *N*-glycopeptides using an automated glycoproteomic method. *Journal of Proteomics* 110: 145–154.

219 Zhu, Z., Hua, D., Clark, D.F. et al. (2013). GlycoPep detector: a tool for assigning mass spectrometry data of N-linked glycopeptides on the basis of their electron transfer dissociation spectra. *Analytical Chemistry* 85 (10): 5023–5032.

220 Liu, M., Zhang, Y., Chen, Y. et al. (2014). Efficient and accurate glycopeptide identification pipeline for high-throughput site-specific N-glycosylation analysis. *Journal of Proteome Research* 13 (6): 3121–3129.

221 Lynn, K.-S., Chen, C.-C., Lih, T.M. et al. (2015). MAGIC: an automated N-linked glycoprotein identification tool using a Y1-ion pattern matching algorithm and in silico MS2 approach. *Analytical Chemistry* 87 (4): 2466–2473.

222 Strum, J.S., Nwosu, C.C., Hua, S. et al. (2013). Automated assignments of N- and O-site specific glycosylation with extensive glycan heterogeneity of glycoprotein mixtures. *Analytical Chemistry* 85 (12): 5666–5675.

223 Wu, S.-W., Liang, S.-Y., Pu, T.-H. et al. (2013). Sweet-heart – an integrated suite of enabling computational tools for automated MS2/MS3 sequencing and identification of glycopeptides. *Journal of Proteomics* 84: 1–16.

224 Toghi Eshghi, S., Shah, P., Yang, W. et al. (2015). GPQuest: a spectral library matching algorithm for site-specific assignment of tandem mass spectra to intact *N*-glycopeptides. *Analytical Chemistry* 87 (10): 5181–5188.

225 Chandler, K.B., Pompach, P., Goldman, R., and Edwards, N. (2013). Exploring site-specific N-glycosylation microheterogeneity of haptoglobin using glycopeptide CID tandem mass spectra and glycan database search. *Journal of Proteome Research* 12 (8): 3652–3666.

226 Liu, M.-Q., Zeng, W.-F., Fang, P. et al. (2017). pGlyco 2.0 enables precision *N*-glycoproteomics with comprehensive quality control and one-step mass spectrometry for intact glycopeptide identification. *Nature Communications* 8 (1): 438.

227 Lu, L., Riley, N.M., Shortreed, M.R. et al. (2020). O-pair search with metamorpheus for *O*-glycopeptide characterization. *Nature Methods* 17 (11): 1133–1138.

228 Xiao, K. and Tian, Z. (2019). GPSeeker enables quantitative structural *N*-glycoproteomics for site- and structure-specific characterization of differentially expressed N-glycosylation in hepatocellular carcinoma. *Journal of Proteome Research* 18 (7): 2885–2895.

229 An, Z., Shu, Q., Lv, H. et al. (2018). *N*-linked glycopeptide identification based on open mass spectral library search. *BioMed Research International* 2018: 1564136.

230 Pioch, M., Hoffmann, M., Pralow, A. et al. (2018). glyXtoolMS: an open-source pipeline for semiautomated analysis of glycopeptide mass spectrometry data. *Analytical Chemistry* 90 (20): 11908–11916.

231 Stadlmann, J., Taubenschmid, J., Wenzel, D. et al. (2017). Comparative glycoproteomics of stem cells identifies new players in ricin toxicity. *Nature* 549 (7673): 538–542.

232 Liu, G., Cheng, K., Lo, C.Y. et al. (2017). A comprehensive, open-source platform for mass spectrometry-based glycoproteomics data analysis. *Molecular and Cellular Proteomics* 16 (11): 2032–2047.

233 Park, G.W., Kim, J.Y., Hwang, H. et al. (2016). Integrated glycoproteome analyzer (I-GPA) for automated identification and quantitation of site-specific N-glycosylation. *Scientific Reports* 6 (1): 21175.

234 He, L., Xin, L., Shan, B. et al. (2014). GlycoMaster DB: software to assist the automated identification of N-linked glycopeptides by tandem mass spectrometry. *Journal of Proteome Research* 13 (9): 3881–3895.

7

Carbohydrate-Specific Monoclonal Antibody Therapeutics

Matthew Lohman, Hannah Rowe, and Peter R. Andreana

University of Toledo, School of Green Chemistry and Engineering, Department of Chemistry and Biochemistry, 2801 W. Bancroft St., Toledo, Ohio, 43606, USA

7.1 Introduction

Carbohydrates are a major class of biomolecules in living organisms that are necessary for several important biological processes, such as cell–cell adhesion, cell signaling, protein folding, and trafficking. In a clinical setting, carbohydrates are valuable targets as therapeutics and diagnostic tools because abnormal glycosylation is a common marker of malignant cellular transformation [1, 2]. These cancerous cells and tissues display tumor-associated carbohydrate antigens (TACAs) that have proven to be involved in tumor proliferation, invasion, angiogenesis, metastasis, and immunity [3]. In this chapter, we will discuss monoclonal antibodies (mAbs) and their applicability, focusing on carbohydrate-specific mAbs. It will discuss varying types of mAbs along with their production and what goes into the process of humanizing antibodies to be used in a clinical setting. Furthermore, details about carbohydrate-based applications with mAbs, focusing on Unituxin®, which was the first drug approved by the Food and Drug Administration (FDA) for the market that targets a specific cancer carbohydrate, namely glycolipid disialoganglioside (GD2), will be examined. This landmark therapeutic drug represented an important advancement in postconsolidation treatment of high-risk neuroblastoma patients. Although this drug comes with high-risk boxed warnings, it was approved by the FDA because of limited therapeutic alternatives and because the treatment potential outweighed the risks for patients with poor prognoses [4].

Carbohydrate-Based Therapeutics, First Edition. Edited by Roberto Adamo and Luigi Lay.

7.2 Types of Monoclonal Antibodies

To understand carbohydrate-based mAbs in better detail, we will go over some of the basics of the variuos types of antibodies that are produced. There are five different antibodies that our body generates, namely IgM, IgG, IgD, IgA, and IgE. In the following sections, we will discuss the characteristics of IgG and subclasses along with IgM, scFv, and Fab fragments.

7.2.1 IgG Antibodies

IgG antibodies are the most commonly used proteins in research and are the preferred antibody for development of therapeutics. They are the most abundant of all the classes and have the highest affinity toward antigens. The differences in each of the antibodies affect multiple aspects of what makes a good therapeutic. The structural differences in each antibody are what affect its ability to be more stable and to have varying half-lives [4]. IgG antibodies are composed of four subclasses: IgG1, IgG2, IgG3, and IgG4, as shown in Figure 7.1. The main differences in each of the subclasses are at the constant region, their hinges, and the upper CH2 domain, making IgG1, IgG2, and IgG4 150 kDa and IgG3 170 kDa [8, 9]. These differences alter the way each IgG reacts and binds. For example, IgG3 has roughly 62 amino acids in its hinge region, allowing it to bind and interact differently than the rest. This in turn makes it more susceptible to degradation and causes its half-life to be shorter than the rest, depending on the allotype. The average half-life of IgG1, IgG2, and IgG4 is roughly 21 days, while IgG3s can be anywhere from 7 to 21 days [5, 10].

The best IgG to be used for therapeutic development has been an extensively studied subject, with a preference for their affinity toward the Fc receptor. That affinity is as follows: IgG1>IgG3>IgG4>IgG2. IgG1 being the best suited toward activating effector mechanisms, along with IgG3. Antibodies IgG4 and IgG2 show decreased function in comparison [10]. The main feature of each therapeutic drug going to market is its ability to elicit an antibody-dependent cell-mediated cytotoxicity (ADCC) or complement-dependent cytotoxicity (CDC) effect. This can be accomplished by all IgG subclasses; however, IgG1 is still the most highly sought-after in development [7, 11, 12].

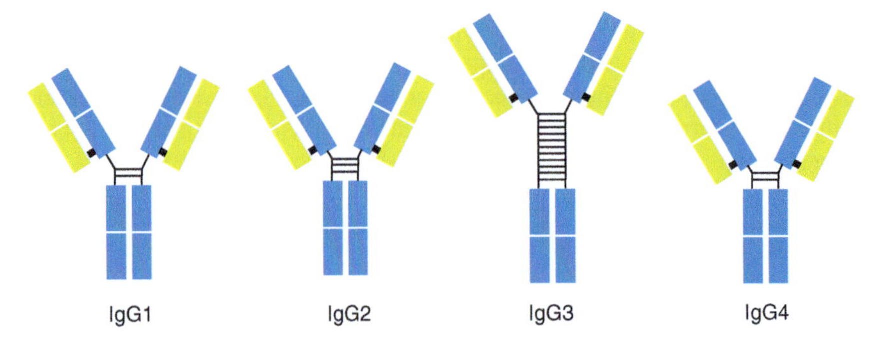

Figure 7.1 Difference in structure between subclasses of IgGs [5–7].

7.2.2 IgM Antibodies

Although IgGs are the preferrable target as a therapeutic, IgMs are still widely studied and show great promise as therapeutic drug candidates. One ascertainable difference between the two is their relative size. IgMs are around 900 kDa on average, compared to IgG at 150 kDa [4]. IgMs form multivalent complexes that can allow for up to 10 or 12 binding sites compared to IgG with only 2. The outcome is that the affinity of IgMs is somewhat lower in comparison, but they are still able to neutralize foreign antigens by binding to multiple antigens at once, which is the concept of avidity[13, 14]. IgMs do not elicit engagement of effector mechanisms as IgGs do because IgMs do not bind to the Fc gamma receptor and do not exhibit ADCC activation. However, IgMs have strong CDC activity and have a better binding affinity toward C1q than IgG [15]. IgMs binding to C1q have been shown by Sharp et al., and once IgM binds to its target antigen, it causes a conformational change in IgM. The conformational change allows for C1q to bind and trigger complement fixation [16]. Therefore, IgMs are also intensively studied as possible therapeutics alongside IgGs. IgM's potential as a therapeutic drug has been shown in many patients' case studies. Currently, there are roughly 20 IgM antibodies in clinical trials that target a variety of diseases, autoimmune disorders, and oncology [17]. The main issue that arises with the clinical use of IgMs is their lower efficacy rates in patient trials. This causes a problem with getting regulatory approval. After a closer look at the clinical trials of IgMs, it was concluded that the cause of lowered efficacy arose from the natural origins of the IgMs being tested; the IgMs tested had not gone through enough somatic mutations [17, 18]. Common antigens that are used consist of proteins, proteolipids, glycolipids, and glycoproteins. However, IgM antibodies are very advantageous when paired with glycans as an antigen due to IgM's avidity and the fact that glycans are found to be exceptionally repetitive, especially on cancer cells. This allows for IgM antibodies to bind to multiple antigens on the outside of a tumor cell [19–21].

7.2.3 ScFv and Fab Fragments

The last group of antibodies to be discussed are the so-called fragmented antibodies, or single-chain fragment variable (scFv), which were first discovered in 1988 and are comprised of a heavy chain (VH) and a light chain (VL). These antibodies are held together by a peptide chain commonly made out of serine and glycine with hydrophilic residues. The antibody fragments discussed in this section have multiple advantages over traditional mAbs, such as their weight being substantially less at around 30 kDa [22, 23]. Furthermore, the production of scFVs is much easier and has lower production costs compared to traditional mAbs. The small size of the antibody fragments allows for facile penetration into tumor cells compared to bulkier immunoglobulins [24]. Another advantage of antibody fragments is their low risk of activating the patient's immune system due to the lack of the Fc region; therefore, they do not cause activation of ADCC and CDC from the patient. The absence of an Fc domain also harbors negative attributes, which include being less stable, causing a shorter half-life, and needing more frequent dosing with larger doses [25].

7.3 Humanization of Monoclonal Antibodies

The mAbs discussed thus far that have been used in a clinical setting have been humanized to modulate immunogenicity [26]. Therapeutic antibodies, when used clinically, are known to need a high concentration per dosage and multiple doses for a complete treatment. This makes the immunogenicity of the monoclonal antibody a critical concern, especially when developing antibody-based drugs [27]. Antibody immunogenicity means the ability of an individual's immune system to recognize and react to therapeutic agents. Several mAbs are generated from mice and, if used, will cause a human anti-mouse antibody (HAMA) response generated by the immune system to combat mouse antibody-based therapeutics administered to patients. HAMA causes several adverse effects, the first being their ability to neutralize therapeutic antibodies, which can decrease the efficacy of the drugs [28]. The other adverse effect patients can experience is an increase in cytokines, causing skin rashes and systemic inflammatory responses. The main causes that can lead to adverse effects are drug dosage, contamination, impurities, administration, and structural features such as sequence and glycosylation patterns [29, 30]. Sections 7.3.1 and 7.3.2 will discuss some of the main techniques that are utilized in developing a humanized monoclonal antibody.

7.3.1 CDR Grafting

Decreasing immunogenicity is done by making the murine monoclonal antibody more human-like, which causes the drug to be better tolerated by the immune system along with lowering the immunogenicity [31]. This involves changing the constant region and the Fv region of the antibody from a mouse sequence to human sequences, with the exception of the complementarity-determining region (CDR), which is still murine to keep affinity for the desired target [32, 33]. CDR grafting is one of the most used techniques to humanize antibodies. CDR grafting uses a donor and acceptor antibody. The donor will have the antigen-specific CDR, and the acceptor will pose as the human backbone for the new antibody [34]. A critical part of CDR grafting is selecting the most appropriate acceptor; a poor choice in an acceptor will lead to loss of affinity and specificity of the newly formed antibody. The best choice for an acceptor is to find a human framework with the highest homology to the chosen donor [34, 35].

7.3.2 Transgenic Animals

A vast amount of new therapeutics that have hit the market are coming from transgenic animals that can produce fully human antibodies. There are three main categories of transgenic animals: (i) fully human Ig transloci, (ii) chimeric human Ig transloci, and (iii) Ig knock-out strains. Fully human transloci contain only human V, D, J, and C genes [36]. This technique for fully human Ig loci took off with the implementation of bacterial artificial chromosomes (BACs) and yeast artificial chromosomes (YACs). This allows for much larger (~1 Mb) human IgH and IgL regions to be inserted into an animal genome [37]. Even with this improvement to YAC, it remains difficult to obtain efficient hybridoma production. Due to this challenge, cross-breeding with mice that had a knock-out mutation was invented. The antibodies produced did not perform as well as expected and had little-to-no isotype switching occur, producing mainly IgM antibodies [38]. To overcome these issues, chimeric constructs were suggested as a solution. They are comprised of human V, D, and J linked to a rodent C_H gene. This strategy proved to work much better, allowing for human V genes to efficiently rearrange in transgenic mammals, and while linked to endogenous C_H genes, it gave rise to a more efficient system that showed its ability to interact and signal facilitating isotype switching [39]. This process allows for the ability to integrate human V_H, D_H, and J_H segments using BAC to insert V_H upstream from rodent C_H. Doing so allows for complete replacement of the rodents V_H with human equivalent from BAC. This approach was done to try and engineer a V_H, D_H, and J_H region that was as close to identical to humans as possible [40].

7.4 Breakthrough Research

The only two FDA-approved therapeutics to date are Danyelza® (naxitamab) and Unituxin (dinutuximab). Dinutuximab is currently the golden standard for carbohydrate research and has set the bar for being able to target tumor-associated carbohydrate pentasaccharide antigen GD2. Naxitamab has been shown to be just as effective at targeting GD2 in neuroblastomas using a combination of granulocyte–macrophage colony-stimulating factor (GM-CSF) for its use on patients, which showed 68% complete response [41]. Prior to Unituxin use, there were three therapeutic mAbs that were approved by the FDA to target protein-based tumor antigens. These included (i) Rituxan® (rituximab), (ii) Herceptin® (trastuzumab), and (iii) Avastin® (bevacizumab). Those mAbs lack affinity toward monovalent glycans, as cancer carbohydrates are often expressed in the form of clusters on the target peptide. This was shown by Oyelaran and Gildersleeve at the National Cancer Institute (NCI) in 2007, who utilized an array with 27 glycan-specific mAbs with varying glycan specificity that showed little-to-no binding to monovalent glycan antigens. When glycan antigen clusters were used in the array with the same mAbs, binding was observed [42]. Among the tested mAbs, the commercially available anti-Tn B1.1 (IgM) failed to bind to monovalent Tn. This issue has revealed a need for mAbs with increased specificity toward their intended glycan or glycopeptide target. In 2018, the Andreana group produced KT-IgM-8 (IgM), a unique mAb, made from a fully carbohydrate-based immunogen [43]. Tn-PS A1 was used to create the required immune response in an attempt to overcome the pitfalls of previously developed mAbs. PS A1 is a zwitterionic polysaccharide (ZPS) capable of triggering a T-cell immune response in the absence of peptides, proteins, or other nonsugar-based material. When compared to commercially available anti-Tn IgM monoclonal Tn-218, using a glycan array, KT-IgM-8 showed a dramatic difference in being able to bind to monovalent Tn at varying concentrations, while commercially available Tn-218 showed virtually no binding to monovalent Tn *in vitro* and induced protection through complement-mediated cytotoxicity in *in vivo* tumor mouse models [43]. MCF-7, a human breast cancer cell line, was xenographed into SCID mice, and KT-IgM-8 showed 40% tumor reduction compared to the control [43].

7.5 mAbs from Preclinical to Clinical Studies

Over the recent years, several mAbs targeting a variety of different types of TACAs have given promising results in preclinical studies and are progressing to clinical trials [44]. Among the TACAs that are being studied here, we will discuss the globo series, blood group, and mucin-attached glycans in more detail. As Unituxin is an anti-GD2 mAb and is part of the gangliosides, we will discuss this section later with Unituxin [45].

7.6 Globo Series

This group is made up of stage-specific embryonic antigen 3 (SSEA), SSEA-4, and Globo-H and is found in several types of cancers, such as breast, gastric, lung, ovarian, endometrial, pancreatic, and prostate [46, 47]. Globo-H is the most common antigen in the series [48]. Anti-Globo-H mAb OBI-888 is currently in phase II clinical trials and has shown promising results toward more than five types of cancer. OBI-888 has shown excellent tumor growth inhibition, which tested for 85% toward breast cancer being the highest and the lowest being 43% for lung cancer [49, 50]. This study showed inhibition of tumor growth in over 150 patients tested [51].

7.6.1 Blood Group

Type I and Type II Lewis antigens are terminal fucosylated carbohydrate structures belonging to the human histoblood group system. LewisA (LeA), LewisB (LeB), LewisX (LeX), and LewisY (LeY) are synthetized in exocrine epithelial cells by fucosyltransferase (FUT) enzymes [52]. The antigens only differ in their glycosidic bonds (Galβ1-3GlcNAc and Galβ1-4GlcNAc) [53]. Teng and coworkers showed that Mab216 is highly selective for blood group antigen CDIM that is located on human B cells. This antibody was then tested in a Phase 1 clinical trial, showing good efficacy against lymphoblastic leukemia [20]. When the Mab216 was progressed to phase 2, the scale-up and production proved to be challenging and prevented further development. The only FDA-approved IgM antibody was called NeutroSpec™, otherwise known as fanolesomab-Tc99M, which was approved for use in 2004. Fanolesomab was a murine-based monoclonal IgM antibody labeled with technetium-99m and directed toward 3-fucosyl-*N*-acetyl lactosamine found on CD15 [19]. When clinical trials were first conducted, the results showed sufficient public safety, so fanolesomab was approved for further clinical trials that eventually led to FDA approval. The main use was for scintigraphic imaging of patients who showed symptoms of appendicitis. Ultimately, fanolesomab was removed from the market by the FDA in 2005, when reports arose that patients taking the drug were suffering from serious cardiopulmonary events, leading to fanolesomab being discontinued by Palatin Technologies in 2008 [18, 19, 21].

7.6.2 Mucin-Attached Glycans

One of the more studied groups are the mucin-attached glycans, which consist of Tn (GalNAc), Tf (Galβ1-3GalNAc), and STn (Neu5Acα2-6GalNAc). These antigens are attractive targets because of their absence from healthy tissues [52–55]. STn is the most targeted of all the mucin glycans, as it has a lot of notoriety with Theratope, which was tested in Phase 3 for breast cancer [56]. The first anti-STn mAb was initially used for cancer detection rather than vaccine [57]. After being humanized, the mAb CC49 showed a modest immune response and a 40% increase in survival vs. controls [58]. An anti-Tn mAb, Gatipotuzumab, was tested in a phase 1 study and showed that it was safe and well tolerated. Gatipotuzumab was moved to phase 2 and failed to demonstrate an improvement vs. the placebo in progression-free survival [59, 60]. Gatipotuzumab in combination with anti-epidermal growth factor receptor (EGFR) antibody was tested in a Phase 1 study and showed promising results for colorectal cancer patients. This study went onto Phase 2 and is waiting for the results [61].

7.7 New Treatment Options for Neuroblastoma

In the United States, neuroblastomas affect 1 in 100 000 children, and approximately 700 children under the age of 15 suffer from the disease annually [62, 63]. Approximately 50% of the diagnosed patients are in advanced stage IV of the disease, and the five-year survival rate of these patients is 20–25% [64]. Neuroblastoma is the most common extracranial solid tumor in children [65]. It accounts for about 7% of children with cancer who are under the age of 15. About 90% of patients diagnosed with neuroblastoma are under the age of 5, and about 40% of diagnosed patients will have high-risk neuroblastoma [66]. Relapse in neuroblastoma tumor cells is not only probable but also highly incurable [63, 67]. Therefore, novel therapeutic strategies are highly needed. Frequent relapses are often caused by the loss of heterozygosity of 1p and 11p chromosomes and partial deletions of other chromosomes, which can lead to a loss of critical genes, leaving patients more susceptible to further gene mutations, cancer, and other diseases. Other major problems can be due to amplification of tumor cell growth because of mutations in the MYCN oncogene, which controls the growth, proliferation, and apoptosis of cell, and metastasis of cancerous cells in the liver, bone, and bone marrow [67]. Treatments using dinutuximab or naxitamab are necessary because the complex and aggressive nature of neuroblastoma tumor cells allows them to avoid regular immune responses by evading T cells through down-regulation or losing human leukocyte antigen (HLA) expression. This is a challenge because it interferes with the afferent arm, with the T cells' ability to find neuroblastoma tumor cells, and with the cytotoxic T lymphocyte (CTL) effector phase of adaptive immunity. Cellular immunity is further impaired by soluble inhibitors such as FAS ligand (FASL) and gangliosides, pro-tumor macrophages, myeloid suppressor cells, and regulatory T cells [68]. Neuroblastoma patients tend to be young with underdeveloped or compromised immune systems because of the burden of living with an aggressive, high-risk disease and receiving intensive chemotherapy, radiation, or other rigorous treatment that can strain the patient [69]. Because neuroblastoma is so poorly immunogenic for T cells, the patient's natural immune responses and common cancer treatments, such as chemotherapy, surgery, or radiation, are not as effective in high-risk neuroblastoma patients [66]. Before immunotherapy options like Unituxin were available, about 40% of high-risk neuroblastoma patients were able to achieve long-term remission, and the 5-year event-free survival (EFS) rates were less than 50% [62, 65, 70]. Because complete eradication of neuroblastoma is rarely possible, dinutuximab is used as a second-line treatment to target any surviving cancer cells and prevent relapse, especially in patients with an advanced version of the disease. In therapeutic trials, patients who received Unituxin had significantly higher EFS rates and overall survival (OS) rates when compared to standard therapy options without Unituxin. EFS rates increased from about 46% to 66%, and OS rates increased from about 75% to 86% [62, 66, 71]. Therefore, including Unituxin treatment cycles as part of the postconsolidation therapy for a patient with high-risk neuroblastoma would help improve patient survival [70].

7.7.1 History of Unituxin

Milstein and Koehler invented monoclonal antibody technology in 1975, but Unituxin was discovered in the late 1980s by Alice Yu, M.D., at the University of California, San Diego Medical Center [72–75]. Dr. Yu's research used laboratory-made mAbs as a possible treatment for neuroblastoma. Many versions of the mAbs were tested before a final version, ch14.18, was settled on [74]. In 2001, Dr. Yu and colleagues in the Children's Oncology Group tested ch14.18 monoclonal antibody on neuroblastoma patients [71]. For this immunotherapy treatment, children were initially treated with chemotherapy, radiation, and stem cell transplantation before ch14.18 was administered [70, 71]. Immune-stimulating agents known as GM-CSF and interleukin 2 (IL-2) were administered with the mAb to increase the drug's effectiveness at killing tumor cells. The NCI's National Clinical Trails Network was a major contributor to these trials [70]. Also, the NCI's Biopharmaceutical Development Program (BDP) produced the ch14.18 for clinical trials and worked on its process development in order to make the drug available for eligible children suffering from high-risk neuroblastoma. In 2010, NCI partnered with a branch of the United Therapeutics Corporation (UTC) to manufacture dinutuximab and gain regulatory approval from the FDA. UTC conducted safety, efficacy, and optimal use studies and gained licensure to manufacture and market dinutuximab under the brand name Unituxin. In 2015, the FDA and European Commission approved Unituxin as a treatment option for pediatric cancer [1, 71, 76].

7.7.2 What is Unituxin?

Dinutuximab, also named ch14.18 or Unituxin, is a chimeric human–mouse IgG mAb [67, 72, 73, 77]. It is composed of murine variable heavy- and light-chain regions and human constant regions [78–81]. All approved therapeutic mAbs are N-glycosylated in the Fc region [80]. For Unituxin, glycosylation of the conserved asparagine position 297 (Asn-297) occurs in the heavy-chain CH2 domain of the Fc region. Because the Fc region is the receptor-binding region, posttranslational modifications in this region create the molecule's high heterogeneity and affect the functions of the key antibody effectors [72].

United Therapeutics manufactures dinutuximab by industrial fermentation using SP2/O, which is a murine myeloma cell line [71, 82]. However, mAbs produced in nonhuman cells tend to express oligosaccharides that are not found in normal IgG serum, and their presence can cause immunogenicity issues in patients [72]. Even with the potential issues, Unituxin is used as a second-line treatment for children with high-risk neuroblastoma because of its ability to target neuroblastoma cells by binding to the glycolipid antigen disialoganglioside, known as GD2, which is highly expressed on the neuroblastoma cell's surface [71, 76]. The chemical structure of GD2 is shown in Figure 7.2. The binding of dinutuximab and GD2 to neuroblastoma induces tumor cell lysis, apoptosis, and proliferation inhibition. During *in vitro* testing, dinutuximab was shown to bind to tumor cells and was more effective at inducing lysis of tumor cells than the murine version 14.G2a [71]. This cell death occurs through two major

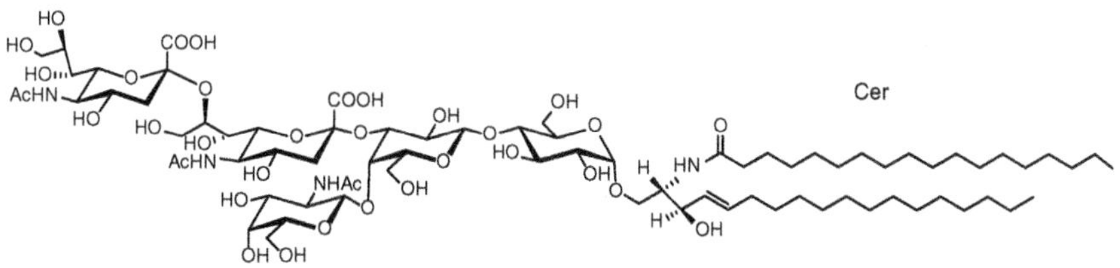

Figure 7.2 Chemical structure of glycolipid disialoganglioside (GD2) [83, 84].

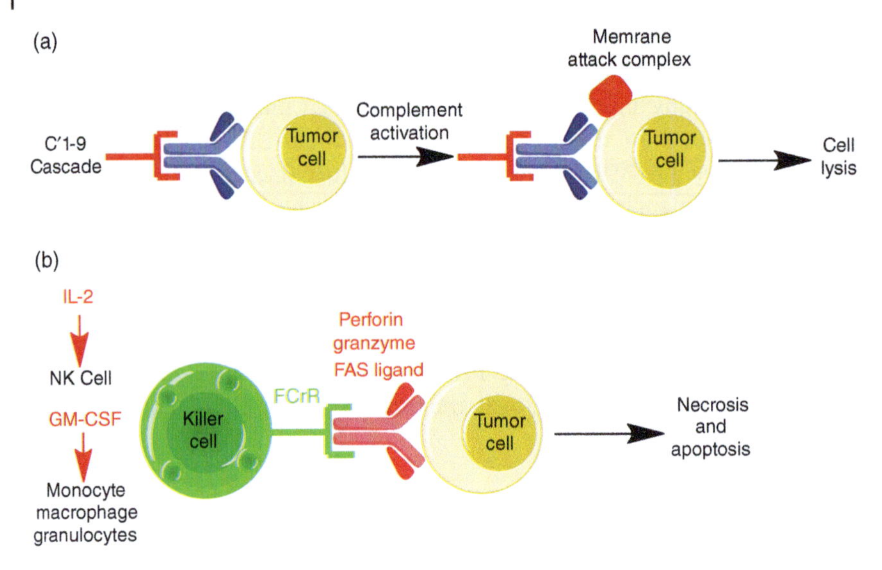

Figure 7.3 Illustration of mechanisms for GD2 antibody-targeted destruction of neuroblastoma. (a) is the CDC pathway, and (b) is the ADCC pathway. (a) CDC: Complement dependent cytotoxicity. (b) ADCC: Antibody-dependent cell-mediated cytotoxicity. Source: Adapted from Matthay et al. [85].

pathways: ADCC and CDC [67, 69]. Figure 7.3 illustrates these two pathways. Following the mechanisms of the ADCC pathway, the mAb's Fc fragment binds the Fc receptors on monocytes, macrophages, granulocytes, or natural killer (NK) cells, allowing these cells to engulf and destroy bound tumor cells. NK cells can also secrete cytokines that lead to cell death [69, 85, 86]. The Fc-receptors for granulocytes are FcgRIIA/CD32, and the Fc-receptors for NK cells are FcRIIIA/CD16A. In CDC, C1q binds to the Fc section of mAbs bound to the tumor cells. This binding activates a complement cascade and forms membrane attack complexes to create pores in the cell membrane and cause lysis of the tumor cells [87]. Following the mechanisms of the CDC pathway, the mAb binds to the receptor and initiates the complement cascade to clear damaged cells and microbes from the system and attack the cell membrane of the identified pathogen. The complement cascade causes the formation of a membrane attack complex to make a hole within the tumor cell membrane, causing cell lysis and death [85, 86]. Unituxin is an intravenous drug that is used as a postconsolidation therapy when combined with GM-CSF, IL-2, and isotretinoin (RA) [78, 82]. The recommended dosage for this drug is 17.5 mg a day administered over a 10-to 20-hour period for four consecutive days in a four-week cycle [41]. When dinutuximab was tested using the Scatchard analysis, it showed acceptable binding to a series of neuroblastoma cell lines and the melanoma M-21 cell line with a K_D value of 11.2 nM and nonspecific binding between 5% and 10% total bound [88].

Despite problems and limitations, several anti-GD2 antibodies have been developed. Four, in particular, have been extensively studied in clinical studies: 3F8, hu3F8, ch14.18, and hu14.18 [83]. Ch14.18 (dinutuximab) was the first antibody to be approved by the FDA for pediatric solid tumor and high-risk neuroblastoma [87]. Anti-GD2 mAbs work by binding to the end-terminal penta-oligosaccharide of GD2 and following the ADCC or CDC pathways to kill neuroblastoma tumor cells. Besides ADCC and CMC pathways, mAbs can enhance nonimmune-mediated effects, such as survival signal blockade and anoikis. Anoikis is the induction of apoptosis in cells. It occurs when abnormal cells detach from the extracellular matrix (ECM) and neighboring cells. In a healthy system, anoikis would remove the unhealthy or abnormal cells, but tumor cells have the ability to escape anoikis by constitutive activation of focal adhesion kinase (FAK). However, anti-GD2 mAbs will dephosphorylate FAK and inhibit activation of PI3K/Akt pathways, which allows the drug to induce apoptosis and cause cancer cell death [87].

7.7.3 Challenges with Unituxin

Unituxin can cause serious adverse side reactions because it binds to GD2 expressed in both benign and malignant tissues. This means neural tissues in the central nervous system, peripheral nerves, and neuroblastoma can be affected by the drug [79]. Dinutuximab is a toxic regimen that must be infused into patients over a 20-hour period in a hospital because there is serious risk of intense pain, serious infusion reactions, capillary leak syndrome, and hypotension [41]. This drug can give patients an increased risk of infection, neurological eye disorders, suppression of bone marrow, electrolyte abnormalities, and atypical hemolytic uremic syndrome. It is limited for use in pediatric patients because adults with melanoma who participated in dose-finding, safety, and tolerability studies experienced severe and possibly irreversible motor neuropathy [41, 79]. However, Unituxin was still approved because of the seriousness of the disease and the lack of alternative treatment options for high-risk neuroblastoma [79]. Murine anti-GD2 mAbs are tolerated in patients and do show antineuroblastoma activity, but the development of HAMA response and hypersensitive reactions to injections limit how often the drug can be administered [74, 77]. Some patients had to wait months for the next round of antibody injections to be administered, which decreased the efficiency and effectiveness of the treatment. HAMA's increase the clearance of murine mAbs and often cause unwanted allergic reactions and tumor penetration [75]. Murine mAbs have a shorter half-life than human mAbs, and the Fc region of murine antibodies is less effective at eliciting ADCC and CDC than human antibodies [73, 77]. These limitations have led to advances in genetic engineering that have allowed for the development of chimeric and humanized anti-GD2 mAbs [87].

7.7.4 mAbs Binding to Neuroblastoma

The anti-GD2 murine antibodies include murine IgG3 (m3F8) and murine IgG2a (14G2a). Murine 3F8 was the first anti-GD2 monoclonal antibody to be tested in patients with neuroblastoma, and it is the murine IgG3 with the highest reported affinity for GD2 with a K_D value of 5 nM. Murine 3F8 has been shown to kill neuroblastoma cells by CDC and by lymphocytes, cultured monocytes, and granulocytes. Murine 3F8 binds to Fc-receptors FcγRII and FcγRIII for neutrophil- and

NK-mediated ADCC. The CR3 receptor also plays an important role in cytotoxicity [83, 87]. CDC can be enhanced by naturally occurring complex polysaccharide β-glucan (BG) [36]. This enhancement increases the adhesion of complement receptors on myeloid cells to natural ligands like iC3b. When m3F8 is combined with the cytokine GM-CSF, there is a greater than 60% long-term survival rate among pediatric patients with stage-4 high-risk neuroblastoma.

Other murine anti-GD2 antibodies include ME36.1 and 14.G2a, which have lower affinities to GD2 with K_D values of 19 and 77 nM, respectively [83, 89]. While ME36.1 was originally obtained as a mouse IgG3, it can be class-switched to IgG2a and IgG1 variants. Also, ME36.1 mostly binds to GD2, but it does have some cross-reactivity to GD3, which means that ME36.1 can be a useful antibody for targeting other tumors, such as melanoma [83]. Preclinical studies showed ME36.1-inhibiting tumor growth at the inoculation site and in the lymph nodes and lungs.

7.7.5 Chimeric and Humanized Anti-GD2 Antibodies

When creating chimeric versions of the anti-GD2 mAbs, the VH and VL domains of the murine antibody are grafted onto human IgG constant domains. When creating humanized versions of the anti-GD2 mAbs, a fully human monoclonal antibody is grafted with murine CDR loops and a few structurally significant residues, or a fully human monoclonal antibody with no murine residues is used [77, 83]. L72, which is a fully human IgM, was the first nonmurine anti-GD2 antibody. It was produced using the Epstein–Barr virus (EBV) to transform B lymphocytes from the peripheral blood lymphocytes of melanoma patients into lymphoblastoid cell lines [83]. L72's clinical studies showed injections caused regression in melanoma tumor cells, except for patients who had tumors with low antigenicity. After that, no further studies were reported.

ch14.18 is the chimeric form of m14.G2a, and hu14.18 is the human form. The names were derived from the original mouse isotype 14.18 IgG3. Phase 1 studies concluded the safety of both forms, with a warning about severe pain during administration and other possible side effects. Phase 3 studies showed that combining ch14.18 with GM-CSF and interleukin-2 can greatly increase the two-year survival rate of patients with high-risk neuroblastoma in comparison to the standard therapy options.

Murine 3F8 was also humanized (hu3F8) through grafting on the CDR, and it is currently in Phase 1 trials. Initial test results show a reduction in the production of HAHA and complement activation in comparison to murine 3F8 [83].

7.7.6 Naxitamab as a Potential Alternative for High-Risk Patients

In 2020, the ongoing research efforts to reduce the toxicity and development of human anti-mouse antibodies (HAMA) in anti-GD2 antibodies like dinutuximab were a success, and naxitamab, marketed as Danyelza®, was approved by the

FDA [90]. Naxitamab is a humanized anti-GD2 monoclonal antibody developed by Memorial Sloan Kettering Cancer Center and Y-mAbs Therapeutics Inc. [41, 91]. Structurally speaking, naxitamab is very similar to dinutuximab, except some mouse components have been substituted with human ones [64]. Naxitamab is a treatment for high-risk neuroblastoma, osteosarcoma, and many other GD2-positive cancers. This drug is injected intravenously in combination with GM-CSF to treat patients with relapsed or refractory high-risk neuroblastoma in the bone or bone marrow [41, 92]. This drug regimen is an improvement over Unituxin because it has a shorter infusion time (naxitamab infuses in 30–60 minutes compared to the 10–20 hours needed for Unituxin), and it can be used in a greater age range of patients, from 1-year-olds to adults [41, 91]. As mentioned above, Unituxin is too risky for most adults to use because of the high probability of side effects. Also, because of the toxicity profiles, dinutuximab is for inpatient administration only, while naxitamab allows for outpatient administration and a better quality of life during treatment [41]. Naxitamab's recommended dosage is $3\,\mathrm{mg\,kg^{-1}\,day^{-1}}$ [41]. This treatment cycle is administered on days 1, 3, and 5 of a four-week-long treatment. The treatment cycle is repeated every four weeks until a complete or partial response is noted [41, 91, 92]. Once a response is noted, the studies recommend at least five more treatment cycles. However, be warned that this treatment still has risks because it may cause severe infusion-related reactions and neurotoxicity [41]. During testing, the drug showed promising binding via surface plasmon resonance to GD2 coated onto CM5 chips. Naxitamab had a k_{on} value of 9.19, a k_{off} value of 1.03E, a K_D value of 11 nM, and low reactivity with gangliosides other than GD2 [91]. In fact, naxitamab's affinity for GD2 is 10 times higher than dinutuximab's affinity [41]. These results, along with the drug's cytotoxicity against the LAN-1 neuroblastoma cell line with an EC_{50} of $5.1\,\mu\mathrm{g\,ml^{-1}}$, allow naxitamab to be a treatment option for patients with high-risk neuroblastoma [91]. When naxitamab is used as a treatment in conjunction with GM-CSF, the three-year EFS is 74.3% and the OS is 91.6%, which is an increase in survival rates compared to dinutuximab [41].

7.7.7 Chimeric Antigen Receptors (CARs) Targeting GD2

Therefore, alternative cancer cell targets are necessary for treatment options. This idea led to the discovery of T-cell-independent carbohydrate differentiation antigens, such as GD2, GD3, O-acetylated GD2 and GD3, and polysialic acid (PSA) [68]. mAbs can be used as chimeric antigen receptors (CARs) or bispecific antibodies, such as anti-GD2 and anti-CD3, to allow polyclonal T cells to target tumors. CARs are single-chain Fv fragments that can fuse through the transmembrane domain of T-cell-activating motifs, such as CD3ζ and CD28 or 41BB, to allow the T cells to target a specific protein [68]. All of these possible neuroblastoma-killing pathways are illustrated in Figure 7.4. There are several CARs that have gone through clinical trials; here is an example of an anti-GD2 CAR that utilized an scFv based on

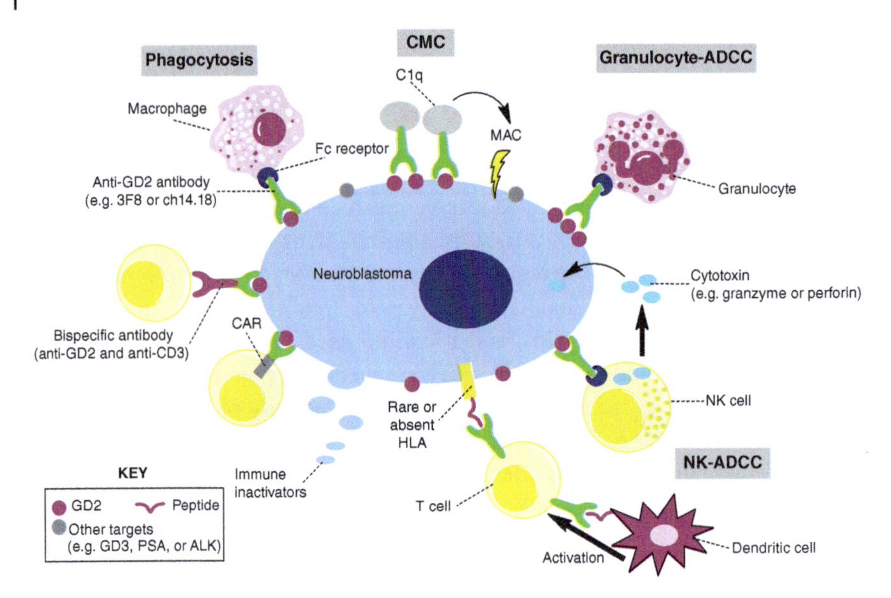

Figure 7.4 Illustration of potential immune responses to neuroblastoma tumor cells. Source: Adapted from Cheyung and Dyer [68].

humanized murine antibody KM8138 that is fused to CD28 [93]. This trial took children with relapsed or refractory neuroblastoma; out of 34 patients, only 15% reported a partial response. In this study, two patients showed significant tumor regression; one had two large tumors and the other patient had one retroperitoneal tumor. Both patients had greater than 90% regression after two months of CAR T-cell therapy [94]. There are drawbacks to CARs; in a different phase 1 clinical trial, anti-GD2 CAR was administered to 11 patients with relapsed or refractory neuroblastoma. The patients were treated with either anti-GD2 CAR T cells alone or in conjunction with lymphodepleting chemotherapy. This trial saw minimal activity with no measurable responses in all patients. CAR T cells aggregated and caused tonic signaling of 14g2a anti-GD2 scFv; this led to T-cell exhaustion and limited antitumor efficacy [95].

7.8 Summary

Overall, carbohydrate-specific mAbs that target carbohydrate antigens have proven to be effective as therapies against cancer, and what waits in the wings is further development against bacterial infections and viruses. Unituxin and naxitamab have paved the path for using immunotherapeutics against carbohydrates a reality. The research community will need to take advantage of carbohydrate antigens further to develop other important therapies in the quest to combat disease more effectively.

List of Abbreviations

tumor-associated carbohydrate antigens	TACAs
antibody-dependent cell-mediated cytotoxicity	ADCC
complement-dependent cytotoxicity	CDC
monoclonal antibodies	mAbs
natural killer	NK
granulocyte–macrophage colony-stimulating factor	GM-CSF
interleukin 2	IL-2
isotretinoin	13-cis-retinoic acid - RA
Biopharmaceutical Development Program	BDP
United Therapeutics Corporation	UTC
Food and Drug Administration	FDA
European Commission	EC
event-free survival	EFS
overall survival	OS
human anti-mouse antibodies	HAMA
human leukocyte antigen	HLA
cytotoxic T lymphocyte	CTL
FAS ligand	FASL
polysialic acid	PSA
NK cell-mediated antibody-dependent cell-mediated cytotoxicity	NK-ADCC
granulocyte-mediated ADCC	granulocyte ADCC
complement-mediated cytotoxicity	CMC
membrane attack complex	MAC
chimeric antigen receptors	CARs
single-chain Fv fragments	scFvs
anaplastic lymphoma receptor tyrosine kinase	ALK
central nervous system	CNS
lactosylceramide	LacCer
asialo-GM2	GA2
cerebrospinal fluid	CSF
extracellular matrix	ECM
focal adhesion kinase	FAK
m3F8	murine IgG3
14G2a	murine IgG2a
β-glucan	BG
Epstein–Barr virus	EBV
humanized m3F8	hu3F8
complementarity-determining region	CDR

References

1 Harris, R.J. (2005). Heterogeneity of recombinant antibodies: linking structure to function. *Developmental Biology* 122: 117–127.

2 Xia, L., Schrump, D.S., and Gildersleeve, J.C. (2016). Whole-cell cancer vaccines induce large antibody responses to carbohydrates and glycoproteins. *Cell Chemical Biology* 23 (12): 1515.

3 Sterner, E., Flanagan, N., and Gildersleeve, J.C. (2016). Perspectives on anti-glycan antibodies gleaned from development of a community resource database. *ACS Chemical Biology* 11 (7): 1773.

4 Pier, G.B., Lyczak, J.B., Wetzler, L.M., and Ruebush, M.J. (2004). Immunology, infection, and immunity. *ASM* 196–197.

5 Milholland, B., Dong, X., Zhang, L. et al. (2017). Differences between germline and somatic mutation rates in humans and mice. *Nature Communications* 8: 15183.

6 Correia, I.R. (2010). Stability of IgG isotypes in serum. *MAbs* 2: 221–232.

7 Muhammed, Y. (2020). The best IgG subclass for the development of therapeutics monoclonal antibodies drugs and their commercial production: a review. *Immunome Research* 16: 173.

8 Ochoa, M.C., Minute, L., Rodriguez, I. et al. (2017). Antibody-dependent cell cytotoxicity: immunotherapy strategies enhancing effector NK cells. *Immunology and Cell Biology* 95: 347–355.

9 Irani, V., Guy, A.J., Andrew, D. et al. (2015). Molecular properties of human IgG subclasses and their implications for designing therapeutic monoclonal antibodies against infectious diseases. *Molecular Immunology* 67: 171–182.

10 Nimmerjahn, F. and Ravetch, J.V. (2005). Divergent immunoglobulin g subclass activity through selective Fc receptor binding. *Science* 310: 1510–1512.

11 Carter, P.J. (2006). Potent antibody therapeutics by design. *Nature Reviews Immunology* 65: 343–357.

12 Vidarsson, G., Dekkers, G., and Rispens, T. (2014). IgG subclasses and allotypes: from structure to effector functions. *Frontiers in Immunology* 5: 520.

13 Strohl, W.R. and Strohl, L.M. (2012). *Therapeutic Antibody Engineering*, 197–223. Woodhead Publishing.

14 Wibroe, P.P., Helvig, S.Y., and Moein Moghimi, S. (2014). The role of complement in antibody therapy for infectious diseases. *Microbiology Spectrum* 2: 63–74.

15 Klimovich, V.B. (2011). IgM and its receptors: structural and functional aspects. *Biochemistry (Moscow)* 76: 534–549.

16 Sharp, T.H., Boyle, A.L., Diebolder, C.A. et al. (2019). Insights into IgM-mediated complement activation based on in situ structures of IgM-C1-C4b. *Proceedings of the National Academy of Sciences* 24: 11900–11905.

17 Shibuya, A., Sakamoto, N., Shimizu, Y. et al. (2000). Fc alpha/mu receptor mediates endocytosis of IgM-coated microbes. *Nature Immunology* 1: 441–446.

18 Weinstein, J.R., Quan, Y., Hanson, J.F. et al. (2015). IgM-dependent phagocytosis in microglia is mediated by complement receptor 3, Not Fc alpha/mu receptor. *Journal of Immunology* 195: 5309–5317.

19 Line, B.R., Breyer, R.J., McElvany, K.D. et al. (2004). Evaluation of human anti-mouse antibody response in normal volunteers following repeated injections of fanolesomab (NeutroSpec), a murine anti-CD15 IgM monoclonal antibody for imaging infection. *Nuclear Medicine Communications* 25: 807–811.

20 Liedtke, M., Twist, C.J., Medeiros, B.C. et al. (2012). Phase I trial of a novel human monoclonal antibody mAb216 in patients with relapsed or refractory B-cell acute lymphoblastic leukemia. *Haematologica* 97: 30–37.

21 Bhat, N.M., Bieber, M.M., Chapman, C.J. et al. (1993). Human anti-lipid A monoclonal antibodies bind to human B cells and the i antigen on cord red blood cells. *Journal of Immunology* 151: 5011–5021.

22 Bird, R.E., Hardman, K.D., Jacobson, J.W. et al. (1988). Single-chain antigen-binding proteins. *Science* 242: 423–426.

23 Huston, J.S., Levinson, D., Mudgett-Hunter, M. et al. (1988). Protein engineering of antibody binding sites: recovery of specific activity in an anti-digoxin single-chain Fv analogue produced in *Escherichia coli*. *Proceedings of the National Academy of Sciences of the United States of America* 85: 5879–5883.

24 Spadiut, O., Capone, S., Krainer, F. et al. (2014). Microbials for the production of monoclonal antibodies and antibody fragments. *Trends in Biotechnology* 32: 54–60.

25 Yokota, T., Milenic, D.E., Whitlow, M., and Schlom, J. (1992). Rapid tumor penetration of a single-chain Fv and comparison with other immunoglobulin forms. *Cancer Research* 12: 3402–3408.

26 Ahmadzadeh, V. (2014). Antibody humanization methods for development of therapeutic applications. *Monoclonal Antibodies in Immunodiagnosis Immunotherapy* 33: 67–73.

27 Dondelinger, M. (2018). Understanding the significance and implications of antibody numbering and antigen-binding surface/residue definition. *Frontiers in Immunology* 9: 2278.

28 Lu, R.M., Hwang, Y.C., Liu, I.J. et al. (2020). Development of therapeutic antibodies for the treatment of diseases. *Journal of Biomedical Science* 27: 1.

29 Almagro, J.C. and Fransson, J. (2008). Humanization of antibodies. *Frontiers in Bioscience: a Journal and Virtual Library* 13: 1619–1633.

30 Harding, F.A., Stickler, M.M., Razo, J., and DuBridge, R.B. (2010). The immunogenicity of humanized and fully human antibodies: residual immunogenicity resides in the CDR regions. *MAbs* 3: 256–265.

31 Ducancel, F. and Muller, B.H. (2012). Molecular engineering of antibodies for therapeutic and diagnostic purposes. *MAbs* 4: 445–457.

32 Hansel, T.T., Kropshofer, H., Singer, T. et al. (2010). The safety and side effects of monoclonal antibodies. *Nature Reviews Drug Discovery* 9: 325.

33 Waldmann, H. (2019). Human monoclonal antibodies: the benefits of humanization. *Methods in Molecular Biology* 1904: 1–10.

34 Safdari, Y. (2013). Antibody humanization methods – a review and update. *Biotechnology & Genetic Engineering Reviews* 29: 175–186.

35 Kashmiri, S. (2005). SDR grafting – a new approach to antibody humanization. *Methods* 36: 25–34.

36 Bruggemann, M., Caskey, H.M., Teale, C. et al. (1989). A repertoire of monoclonal antibodies with human heavy chains from transgenic mice. *Proceedings of the National Academy of Sciences* 86: 6709–6713.
37 Choi, T.K., Hollenbach, P.W., Pearson, B.E. et al. (1993). Transgenic mice containing a human heavy chain immunoglobulin gene fragment cloned in a yeast artificial chromosome. *Nature Genetics* 4: 117–123.
38 Davies, N.P., Rosewell, I.R., Richardson, J.C. et al. (1993). Creation of mice expressing human antibody light chains by introduction of a yeast artificial chromosome containing the core region of the human immunoglobulin kappa locus. *Biotechnology* 11: 911–914.
39 Osborn, M.J., Ma, B., Avis, S. et al. (2013). High-affinity IgG antibodies develop naturally in Ig-knockout rats carrying germline human IgH/Igkappa/Iglambda loci bearing the rat CH region. *Journal of Immunology* 190: 1481–1490.
40 Lee, E.C., Liang, Q., Ali, H. et al. (2014). Complete humanization of the mouse immunoglobulin loci enables efficient therapeutic antibody discovery. *Nature Biotechnology* 32: 356–363.
41 Mora, J., Castañeda, A., Gorostegui, M. et al. (2021). Naxitamab combined with granulocyte-macrophage colony-stimulating factor as consolidation for high-risk neuroblastoma patients in complete remission. *Pediatric Blood & Cancer* 68: e29121.
42 Oyelaran, O. and Gildersleeve, J.C. (2007). Application of carbohydrate array technology to antigen discovery and vaccine development. *Expert Review of Vaccines* (6): 957–969.
43 Trabbic, K.R., Kleski, K.A., Shi, M., and Andreana, P. (2018). Production of a mouse monoclonal IgM antibody that targets the carbohydrate Thomsen-nouveau cancer antigen resulting in in vivo and in vitro tumor killing. *Cancer Immunology, Immunotherapy* 67: 1437–1447.
44 DiMasi, J. et al. (2016). Innovation in the pharmaceutical industry: new estimates of R&D costs. *Journal of Health Economics* 47: 20–33.
45 Suvarna, V. (2010). Phase IV of drug development. *Perspectives in Clinical Research* 2: 57–60.
46 Yu, A.L., Hung, J.T., Ho, M.Y., and Yu, J. (2016). Alterations of glycosphingolipids in embryonic stem cell differentiation and development of glycan-targeting cancer immunotherapy. *Stem Cells and Development* 25: 1532–1548.
47 Yu, J. (2020). Targeting glycosphingolipids for cancer immunotherapy. *FEBS Letters* 594: 3602–3618.
48 Zhang, S. (1997). Selection of tumor antigens as targets for immune attack using immunohistochemistry: I. Focus on gangliosides. *International Journal of Cancer* 73: 42–49.
49 Zhang, S. (1998). Expression of potential target antigens for immunotherapy on primary and metastatic prostate cancers. *Clinical Cancer Research* 4: 295–302.
50 Ruggiero, F.M., Rodríguez-Walker, M., and Daniotti, J.L. (2020). Exploiting the internalization feature of an antibody against the glycosphingolipid SSEA-4 to deliver immunotoxins in breast cancer cells. *Immunology and Cell Biology* 98: 187–202.

51 Yang, M.C., Shia, C.S., Li, W.F. et al. (2021). Preclinical Studies of OBI-999: a novel globo H-targeting antibody-drug conjugate. *Molecular Cancer Therapeutics* 20: 1121–1132.

52 Bennett, E.P., Mandel, U., Clausen, H. et al. (2012). Control of mucin-type O-glycosylation: a classificationof the polypeptide GalNAc-transferase gene family. *Glycobiology* 22: 736–756.

53 Lavrsen, K., Madsen, C.B., Rasch, M.G. et al. (2013). Aberrantly glycosylated MUC1 is expressed on the surface of breast cancer cells and a target for antibody-dependent cell-mediatedcytotoxicity. *Glycoconjugate Journal* 30: 227–236.

54 Colcher, D., Hand, P.H., Nuti, M., and Schlom, J. (1981). A spectrum of monoclonal antibodies reactive with human mammary tumor cells. *Proceedings of the National Academy of Sciences of the United States of America* 78: 3199–3203.

55 Cervoni, G.E., Cheng, J.J., Stackhouse, K.A. et al. (2020). *O*-glycan recognition and function in mice and human cancers. *The Biochemical Journal* 477: 1541–1564.

56 Munkley, J. (2016). The role of sialyl-Tn in cancer. *International Journal of Molecular Sciences* 17: 275.

57 Muraro, R. (1988). Generation and characterization of B72.3 second generation monoclonal antibodies reactive with the tumor-associatedglycoprotein 72 antigen. *Cancer Research* 48: 4588–4596.

58 Gong, Y., Klein Wolterink, R.G.J., Gulaia, V. et al. (2021). Defucosylation of tumor-specific humanized anti-MUC1 monoclonal antibody enhances NK cell-mediated anti-tumor cell cytotoxicity. *Cancers* 13: 2579.

59 Fiedler, W. (2016). A phase I study of PankoMab-GEX, a humanised glyco-optimised monoclonal antibody to a novel tumour-specific MUC1glycopeptide epitope in patients with advanced carcinomas. *European Journal of Cancer* 63: 55–63.

60 Póka, R. (2017). LBA41A double-blind, placebo-controlled, randomized, phase 2 study to evaluate the efficacy and safety of switchmaintenance therapy with the anti-TA-MUC1 antibody PankoMab-GEX after chemotherapy in patients with recurrent epithelial ovarian carcinoma. *Annals of Oncology* 28 (Suppl. 5): v626. Discovery and vaccine development. Expert Review of Vaccines 2007, (6) 957–969.

61 Garralda, E. (2021). Activity results of the GATTO study, a phase Ib study combining the anti-TA-MUC1 antibody gatipotuzumab with the anti-EGFR tomuzotuximab or panitumumab in patients with refractory solid tumors. *Journal of Clinical Oncology* 39: 2522.

62 McGinty, L. and Kolesar, J. (2017). Dinutuximab for maintenance therapy in pediatric neuroblastoma. *American Journal of Health-System Pharmacy* 74 (8): 563.

63 Nallasamy, P., Chava, S., Verma, S.S. et al. (2018). PD-L1, inflammation, non-coding RNAs, and neuroblastoma: Immuno-oncology perspective. *Seminars in Cancer Biology* (52): 53.

64 Kowalczyk, A., Gil, M., Horwacik, I. et al. (2009). The GD2-specific 14G2a monoclonal antibody induces apoptosis and enhances cytotoxicity of chemotherapeutic drugs in IMR-32 human neuroblastoma cells. *Cancer Letters* 281 (2): 171.

65 Greenwood, K.L. and Foster, J.H. (2018). The safety of dinutuximab for the treatment of pediatric patients with high-risk neuroblastoma. *Expert Opinion on Drug Safety* 17 (12): 1257.

66 Ploessl, C., Pan, A., Maples, K.T., and Lowe, D.K. (2016). Dinutuximab: an anti-GD2 monoclonal antibody for high-risk neuroblastoma. *The Annals of Pharmacotherapy* 50 (5): 416.

67 Gur, H., Ozen, F., Can Saylan, C., and Atasever-Arslan, B. (2017). Dinutuximab in the treatment of high-risk neuroblastoma in children. *Clinical Medicine Insights: Therapeutics* 9.

68 Cheung, N.K. and Dyer, M.A. (2013). Neuroblastoma: developmental biology, cancer genomics and immunotherapy. *Nature Reviews. Cancer* 13 (6): 397.

69 Yang, R.K. and Sondel, P.M. (2010). Anti-GD2 strategy in the treatment of neuroblastoma. *Drugs of the Future* 35 (8): 665.

70 Marachelian, A., Desai, A., Balis, F. et al. (2016). Comparative pharmacokinetics, safety, and tolerability of two sources of ch14.18 in pediatric patients with high-risk neuroblastoma following myeloablative therapy. *Cancer Chemotherapy and Pharmacology* 77 (2): 405.

71 Yu, A.L., Eskenazi, A., Strother, D., and Castleberry, R. (2001). A pilot study of anti-idiotype monoclonal antibody as tumor vaccine in patients with high-risk neuroblastoma. Proc Am Soc. *Clinical Oncology* 20: 1470.

72 Boune, S., Hu, P., Epstein, A.L., and Khawli, L.A. (2020). Principles of N-linked glycosylation variations of IgG-based therapeutics: pharmacokinetic and functional considerations. *Antibodies (Basel, Switzerland)* 9 (2): 22.

73 Almagro, J.C., Daniels-Wells, T.R., Perez-Tapia, S.M., and Penichet, M.L. (2017). Progress and challenges in the design and clinical development of antibodies for cancer therapy. *Frontiers in Immunology* 8: 1751.

74 Dinutuximab (Unituxin™). National Cancer Institute Technology Transfer Center (2016). https://techtransfer.cancer.gov/abouttc/successstories/dinutuximabunituxin

75 Lu, R.M., Hwang, Y.C., Liu, I.J. et al. (2020). Development of therapeutic antibodies for the treatment of diseases. *Journal of Biomedical Science* 27 (1): 1.

76 Reichert, J.M. (2016). Antibodies to watch in 2016. *MAbs* 8 (2): 197.

77 Barker, E., Mueller, B.M., Handgretinger, R. et al. (1991). Effect of a chimeric anti-ganglioside GD2 antibody on cell-mediated lysis of human neuroblastoma cells. *Cancer Research* 51 (1): 144.

78 Hoy, S.M. (2016). Dinutuximab: a review in high-risk neuroblastoma. *Targeted Oncology* 11 (2): 247.

79 Keegan, P. (2015). Unituxin injection/dinutuximab. In Division Directory Summary Review, Corporation, U. T., Ed. STN BL 125516: Center for Drug Evaluation and Research, 3713106.

80 Zhang, P., Woen, S., Wang, T. et al. (2016). Challenges of glycosylation analysis and control: an integrated approach to producing optimal and consistent therapeutic drugs. *Drug Discovery Today* 21 (5): 740.

81 Hristodorov, D., Fischer, R., and Linden, L. (2013). With or without sugar? (A)glycosylation of therapeutic antibodies. *Molecular Biotechnology* 54 (3): 1056.

82 Ladenstein, R., Pötschger, U., Valteau-Couanet, D. et al. (2018). Interleukin 2 with anti-GD2 antibody ch14.18/CHO (dinutuximab beta) in patients with high-risk neuroblastoma (HR-NBL1/SIOPEN): a multicentre, randomised, phase 3 trial. *The Lancet Oncology* 19 (12): 1617.

83 Ahmed, M. and Cheung, N.-K.V. (2014). Engineering anti-GD2 monoclonal antibodies for cancer immunotherapy. *FEBS Letters* 588 (2): 288.

84 Tong, W., Sprules, T., Gehring, K., and Saragovi, H. (2012). Rational design of peptide ligands against a glycolipid by NMR studies. *Methods in Molecular Biology (Clifton, N.J.)* (928): 39.

85 Matthay, K.K., George, R.E., and Yu, A.L. (2012). Promising therapeutic targets in neuroblastoma. *Clinical Cancer Research* 18 (10): 2740.

86 Perez Horta, Z., Goldberg, J.L., and Sondel, P.M. (2016). Anti-GD2 mAbs and next-generation mAb-based agents for cancer therapy. *Immunotherapy* 8 (9): 1097.

87 Sait, S. and Modak, S. (2017). Anti-GD2 immunotherapy for neuroblastoma. *Expert Review of Anticancer Therapy* 17 (10): 889.

88 Unituxin. In Assessment report, (CHMP), Committee for Medicinal Products for Human Use, Ed. EMA/CHMP/408316/2015: European Medicines Agency, 2015.

89 Sterner, E., Peach, M.L., Nicklaus, M.C., and Gildersleeve, J.C. (2017). Therapeutic antibodies to ganglioside GD2 Evolved from highly selective germline antibodies. *Cell Reports* 20 (7): 1681.

90 Castel, V., Segura, V., and Canete, A. (2010). Treatment of high-risk neuroblastoma with anti-GD2 antibodies. *Clinical & Translational Oncology* 12 (12): 788.

91 Markham, A. (2021). Naxitamab: first approval. *Drugs* 81 (2): 291.

92 DANYELZA®. 4707781, U.S. Food and Drug Administration (2020).

93 Nakamura, K., Tanaka, Y., Shitara, K., and Hanai, N. (2001). Construction of humanized anti-ganglioside monoclonal antibodies with potent immune effector functions. *Cancer Immunology, Immunotherapy* 50: 275–284.

94 Yang, L., Ma, X., Liu, Y. et al. (2017). Chimeric antigen receptor 4SCAR-GD2-modified T cells targeting high-risk and recurrent neuroblastoma: a phase II multi-center trial in China. *Blood* 130: 335.

95 Quintarelli, C., Orlando, D., Boffa, I. et al. (2018). Choice of costimulatory domains and of cytokines determines CAR T-cell activity in neuroblastoma. *Oncoimmunology* 7: 3518.

8

Carbohydrates in Tissue Engineering

Laura Russo[1,2] and Francesco Nicotra[1]

[1]Università degli Studi di Milano-Bicocca, Department of Biotechnology and Biosciences, Piazza della Scienza 2, Milan, 20126, Italy
[2]National University of Ireland Galway, CÚRAM SFI Research Centre for Medical Devices, Galway, H92 W2TY, Ireland

8.1 Introduction

Carbohydrates cover both the cellular and acellular components of organs and tissues. Their role as signaling molecules is well established today [1]. The diversity and dynamism of the so-defined "glycosignature" on cell surfaces and the extracellular matrix (ECM) partner proteins are still intriguing and not yet fully understood.

In the fields of tissue engineering and medical devices, carbohydrates play a fundamental role in both structural and signaling properties. The first applications in fact took advantage of the structural properties of numerous polysaccharides, prone to generate hydrogels [2, 3]. Hyaluronic acid (HA), for example, has been widely employed for regenerative purposes because of its capacity to generate a wide range of hydrogel formulations. Besides this structural role, carbohydrates can be exploited in the development of medical devices for their unique property to participate in recognition events of physiological and pathological relevance. In this context, it is also important to take into account that glycans are differently expressed in humans in various pathophysiological states and across different species [4], with a consequent problem in terms of antigenicity when a medical device is made of animal-derived materials [5, 6]. Here in this chapter, we will review the glycoengineered solution for both medical devices and tissue engineering applications, taking into consideration not only the advantages but also the dark side that limits their translation in clinics and the challenges for their further development.

Carbohydrate-Based Therapeutics, First Edition. Edited by Roberto Adamo and Luigi Lay.

8.2 Biomaterials and Medical Devices: Natural and Synthetic Strategies

In the last decades, the advancement in medical device and biomaterials research gave rise to different strategies to substitute or replace tissues and organs damaged by pathologies or trauma [7]. Several different biomaterials are available on the market for this purpose, and great efforts are still in place to overcome the current limitations to safely apply them in clinics.

The traditional classification of biomaterials employed as tissue substitutes includes natural polymers, synthetic polymers, hybrid materials, and naturally derived organs and tissues [8–10]. Polysaccharides are relevant components of an ideal biomaterial for their structural properties [11–14]. However, it is important to pay attention to the fact that smaller glycans can act as xenoantigens [15, 16], in particular those exposed in animal-derived prosthesis and medical devices. As often in biomedical research, the current design of biocompatible materials replacing tissue structure and functions takes inspiration from Nature. Polysaccharides are present in natural ECMs as glycosaminoglycans (GAGs) and proteoglycans (PGs), whereas N- and O-glycosylation cover cell surface and act as a signature to interact with protein-based components of the ECM [3, 17, 18]. In this large plethora of actors, glycans cover a multitude of roles fundamental to finely regulating tissue morphogenesis and homeostasis. Depending on the features of the organ or tissue that must be repaired or substituted, the origin of the damage (pathological or traumatic), its morphological features and functionalities, different classes of biomaterials can be exploited. The current approach to repair tissue damage includes both total substitution [19] or the induction of tissue regeneration-exploiting bioresponsive biomaterials able to stimulate the repair of the damaged tissue [20]. The traditional substitution approach requires the use of permanent prosthesis, whereas in regenerative approaches, biodegradable biomaterials able to induce regeneration *in vivo* or *ex vivo* are used. In the following paragraph, we will overview the role of glycans in both approaches, presenting and discussing the devices already used in clinical applications and the strategies studied for future developments.

8.2.1 Carbohydrates as Building Blocks for Medical Device Formulation

Natural polysaccharides are largely employed in the formulation or coating of medical devices [2]. Those of human origin, even if in principle biocompatible, present significant limitations related to availability and scale-up. Production by recombinant methodologies can be a solution, like in the case of HA [21]. A practical alternative consists of their isolation from vegetal or animal sources in which they are abundant, provided that they are biocompatible and functional. The unique way to obtain employable batches of sulfated polysaccharides, for example, is the extraction of them from animal source [22].

8.2.1.1 Human Polysaccharides: Glycosaminoglycans (GAGs) and Proteoglycans (PGs)

Polysaccharides like HA, heparan sulfate (HS), heparin, and chondroitin sulfate (CS) (see Figure 8.1) are naturally expressed in all human tissues and organs and contribute to the physical properties and physiological function of the tissues of reference [23]. They have been used in the formulation of bioactive medical devices to perform their natural functions (structural and functional), properly affected by functionalization and crosslinking that modulate the bioactivity and degradation rate [24].

Since 1934, HA has been the most employed polysaccharide for the development of medical devices, tissue engineering strategies, and cosmetic formulations [25]. Differentially crosslinked HA has been developed to produce dried matrix or injectable hydrogels, depending on the final application of interest [25]. The main fields of application of HA include neurosurgery, orthopedy, and wound-healing treatment of the skin affected by trauma or pathologies [26]. Several examples of commercially available HA-based materials are currently employed in clinical practices The use of HA-based biomaterials for tissue regeneration takes inspiration from its natural functions; the use in wound healing [4–6], for example, is inspired by the massive presence of HA in the normal epidermis, where it plays a fundamental role in the maintenance of tissue mechanical properties and hydration, as far as in cell migration and proliferation. Remarkably, HA-based materials are also effective in extreme disruptive conditions like chronic damage [27]. The fast *in vivo* degradation of HA is still one of the major limitations for its efficacious application in medical devices, therefore requiring stabilization strategies. To stabilize HA and finely tune the structural properties without affecting the biological properties [28], different

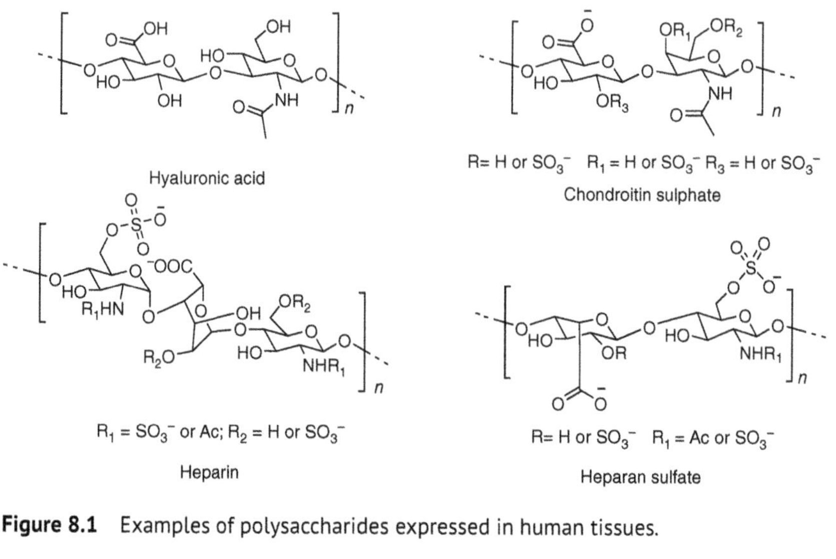

Hyaluronic acid

R= H or SO_3^- R_1 = H or SO_3^- R_3 = H or SO_3^-

Chondroitin sulphate

R_1 = SO_3^- or Ac; R_2 = H or SO_3^-

Heparin

R= H or SO_3^- R_1 = Ac or SO_3^-

Heparan sulfate

Figure 8.1 Examples of polysaccharides expressed in human tissues.

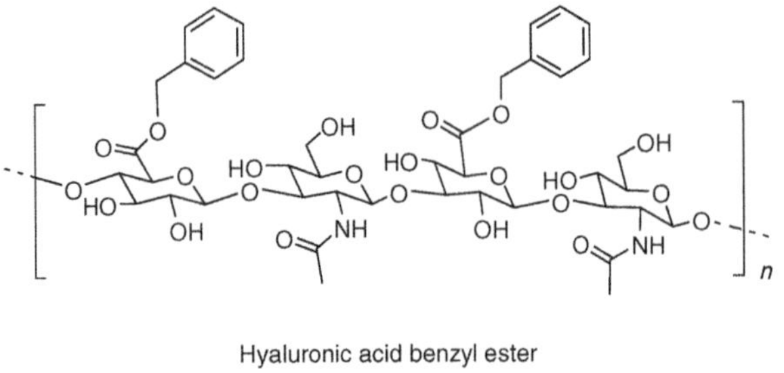

Hyaluronic acid benzyl ester

Figure 8.2 Hyaluronic acid benzyl ester.

degrees of esterification, typically performed with benzyl alcohol, have been experimented with. HYAFF® is a registered HA benzyl ester (see Figure 8.2), employed in particular for wound healing [29] and cartilage tissue repair [30].

Cutaneous lesions generated by burns, ulcers, or trauma are usually cured with skin grafts; however, if the covering process cannot be immediately performed, biomaterial-based patches are employed to induce the correct recovery and regeneration of the damaged area [31]. HYAFF-based medical devices have been successfully employed for the regeneration of cutaneous lesions, and the different esterification degrees and formulation methodologies allow for modulation of the final shape/geometries of the device [32]. HYAFF has also been employed in chondrogenic tissue engineering applications for its ability to induce cartilage formation [33]. *In vitro* studies employing 3D HYAFF-based matrices showed the ability of autologous chondrocytes to restore the physiological signatures of hyaline cartilage [34], whereas *in vivo* studies demonstrated that seeding with autologous chondrocytes induces native cartilage tissue formation with optimal integration in the surrounding tissue [33]. Hyalograft C, a medical device based on cartilage autograft obtained from patient chondrocytes cultured on HYAFF, was introduced for surgical arthroscopic techniques in 1999 [35, 36]. In 2013, following an European Medicines Agency (EMA) report on manufacturing practices, Hyalograft C was withdrawn from the European market [37]. Today, methodologies for treatment of cartilage include new approaches in which autologous chondrocytes are delivered into damaged tissues with HA- or collagen-based injectable biomaterials [38]. HA is also employed, in different formulations, in ophthalmic viscoelastic devices (OVDs) or dermatological therapies [39]. The first use of untreated HA in ophthalmic surgery was reported in cornea transplantation and, today, it is currently employed in ocular surgeries for corneal protection and vitreous replacement [40]. Food and Drug Administration (FDA) approved the first viscosurgical device based on HA in 1983, under the trademark Healon® [41, 42]. Healaflow®, an FDA-approved HA crosslinked with 1,4-butanediol diglycidyl ether (BDDE), is employed in vitreoretinal surgery [43]. Other ophthalmic applications include the modification of contact lenses with HA solutions. Recently, HA has been exploited to improve contact lens performance or obtain a controlled release of HA in the eye [44, 45]. Other FDA-approved products based on crosslinked HA include hydrogels in which the crosslinking is performed with divinyl sulphone (DVS), such as commercial dermal defect fillers (Hylaform®, Captique™, Prevelle®, Lift®, and Varioderm®). In the cosmetic field, the most employed HA-crosslinked hydrogels are based on the mentioned HA-DVS and HA-BDDE (Restylane®, Juvéderm®, Teosyal®, and Hyabell®) as injectable materials [39, 46–48].

Sulfated polysaccharides, normally linked to proteins as PGs, have impressive biochemical and structural roles in the regulation of tissue morphology and cell development and have consequently been extensively employed in medical devices and studied for clinical translation. Among them, CS, a sulfated linear polysaccharide based on repeating units of D-glucuronic acid (GlcA) and *N*-acetylgalactosamine (GalNAc) (Figure 8.1), has been largely employed in the biomaterials field [2, 49–51]. The sulfation pattern of CS, usually at the positions-4 and -6, affects the physiological role. CS is found in many tissues and organs of the human body, from central nervous system (CNS) to cartilage, skin, bones, and blood vessels. In particular, the major efforts in the development of CS-based medical devices are focused on the replacement or treatment of cartilage defects and the regeneration of CNS functions. CS is often employed in combination with other natural polymers to generate hybrid medical devices. Esoxx® is a hybrid medical device employed as mucosal barrier for the treatment of gastro-esophageal reflux symptoms based on HA and CS embedded in Lutrol® F 127, a poloxamer with bioadhesive properties [52]. HA/CS-based medical devices are also employed as injectable hydrogel to treat osteoarthritic (OA) knee, as reported for Structovial CS® [53]. In this formulation, CS has a double role: optimizing the HA rheological properties and regulating the cartilage metabolism and remodeling [53]. CS is also extensively employed in formulations with glucosamine for oral administration as Symptomatic Slow Acting Drugs for Osteoarthritis (SYSADOA), even if this use is still debated and several clinical studies are ongoing [54]. Other sulfated polysaccharides exploited in clinical translation are HS and heparin (Figure 8.1). HS is heavily expressed in the ECM and on the cell surface as HSPGs . The structure and length of the polysaccharide chains are highly variable due to the different sulfation degrees, the different combinations of disaccharide components, and, like for other ECM polysaccharides, the variable molecular weight [55]. The typical disaccharide components consist of 40–60% of β-D-glucuronic acid (GlcA) or α-L-iduronic acid (IdoA) linked by $(1 \rightarrow 4)$ glycosidic bonds to D-glucosamine, which can be N-acetylated or N-sulfated [56, 57]. Mimetics of HS have been developed by derivatizing dextran with carboxymethyl, carboxymethyl benzylamide, and carboxymethyl benzylamide sulfonate groups [58, 59]. An example is a commercially available product named RGTA®, currently employed for the treatment of chronic skin damages (CACIPLIQ20®) and for corneal regeneration (Cacicol20®). Other applications of RGTA are under study and include osteochondral, gastro-enterological, cardiac, and nervous system applications [60]. Heparin contains the same monosaccharides of HS, which can also be, in this case, sulfated as Ido(2S) and GlcA(2S), and the GlcN unit can be N-sulfated (GlcNS) or N-acetylated (GlcNAc) with differential patterns and degrees [61]. The substitution patterns with *O*-sulfate and *N*-acetyl groups result in a wide range of complex structures. Also, heparin has been largely investigated and employed in the pharmaceutical and material device fields for different applications, even if, considering its main role as an anticoagulant, the use of heparin-based medical devices is mainly focused on blood-contacting materials [61, 62]. Temporary or permanent devices like hemodialysis catheters, coronary or vascular stents and grafts, bypass devices, and extracorporeal circulation devices exploit heparin-coated materials [57, 62–64]. The two general methods employed for the development of heparin-coated devices are based on eluting technologies and noneluting technologies. Eluting technologies include all the approaches in which heparin is released in a controlled way, taking advantage of noncovalent coatings based on physical or ionic interactions [62, 65, 66]. These methods are employed when heparin delivery is needed to prevent acute and local device-related blood clots. The other approach is based on the covalent functionalization of the medical device surface with heparin [67, 68]. This approach is desirable when long-term thromboresistance is needed at the material surface. Different conjugation methodologies exploit the direct coupling of complementary functional groups already present in heparin and the material surface counterpart, or alternatively introduce new chemoselective orthogonal functional groups and properly tailored linkers to better control heparin exposition and consequently the antithrombotic activity. The chemical nature of the medical device will influence the choice in light of the functional groups eventually present at the surface or, alternatively, introducible upon proper treatment. The most employed conjugation methodologies are based on carbodiimide chemistry, exploiting the carboxyl group of heparin, or oxidation mediated by periodate, for controlled oxidative degradation to generate some aldehydes [69]. A coating material based on heparin, polyethylene oxide, and sulphonate groups covalently conjugated to hydrophilic modified polyethyleneimine (ASTUTE™ technology) has been developed by BioInteractions Ltd and licensed to Medtronic as Trillium® to cover cardiopulmonary bypass and hemodialysis catheters [62, 64]. Another example is Bioline®, developed by Jostra AG and now a property of Maquet Medical, a material obtained with heparin linked to human albumin employed for coating vascular grafts and cardiopulmonary bypass [63, 70]. The conjugation of heparin to polyamine carriers, performed by introducing etherofunctional crosslinkers, is currently employed as Corline® Heparine Surface Technology, developed by Corline System AB. Many other coating methodologies employed for the modification of medical device surfaces to control the interface between blood components and material surface, and to minimize blood clot formation and thrombolytic events are reported with other commercial examples in Table 8.1.

8.2.1.2 Polysaccharides from Plants, Algae, Animal, and Microbial Fermentation

The large availability of polysaccharides derived from plants, algae, and animals and the possibility to produce them by microbial fermentation make them very attractive and cheaper starting materials [13]. Several studies are focused on the biocompatibility of such polysaccharides, and many of them are already in clinical applications. These polysaccharides present an interesting variety of biochemical and physical properties that can eventually be advantageously exploited in

Table 8.1 Examples of commercial heparin-coating technologies.

Polymer	Commercial names	Application	References
HA benzyl ester	HYAFF®	Osteochondral regeneration, wound healing	[29, 30, 32–34]
Sodium hyaluronate	Healon®	Viscosurgical devices	[41, 42]
HA - BDDE	Healaflow®	Vitreoretinal surgery [43]	[43]
HA - DVS and HA - BDDE	HA-DVS (Hylaform®, Captique™, Prevelle®, Lift®, Varioderm®) HA-BDDE (Restylane®, Juvéderm®, Teosyal®, Hyabell®)	Dermal defect fillers	[39, 46–48]
HA - CS embedded in Lutrol®	Esoxx®	Treatment of gastro-esophageal reflux symptoms	[52]
HA - CS	Structovial CS®	Treatment of osteoarthritic (OA) knee	[53]
Mimetics of HS - derivatized dextran with sulfonate groups	RGTA® (CACIPLIQ20® and Cacicol20®)	Chronic skin damages and corneal regeneration	[58–60]
Heparin, polyethylene oxide, and sulfonate groups covalently conjugated to polyethyleneimine	ASTUTE™/TRILLIUM® coating	Hemodialysis catheters, by-pass devices	[62, 64]
Heparin linked to human albumin	BIOLINE® Coating	Vascular grafts, extracorporeal circulation devices	[63, 70]
Heparin linked by reductive amination to matrices bearing amino groups	CARMEDA® BioActive Surface (CBAS® Heparin Surface)	Extracorporeal circulation devices, stent-grafts, vascular grafts, ventricular assist devices	[62, 71]
Heparin UV photo-crosslinked	PHOTOLINK®	Various medical devices	[62]
Layer by layer coatings. The last layer (4) is coated with polyethyleneimine linked to Heparin by reductive amination	Hepamed™	Coronary stents	[62, 72]
Heparin and benzalkonium chloride blended	DURAFLO II®	Extracorporeal circulation devices	[62]

biomedical applications. Furthermore, they can be properly modified to control the properties and final morphologies of a large plethora of formulations. The main nonhuman polysaccharides exploited for biomedical devices are alginates, carrageenan, and fucoidan (derived from algae), dextran, pullulan, and gellan (derived from microbes), and chitosan (derived from animals) (see Figure 8.3).

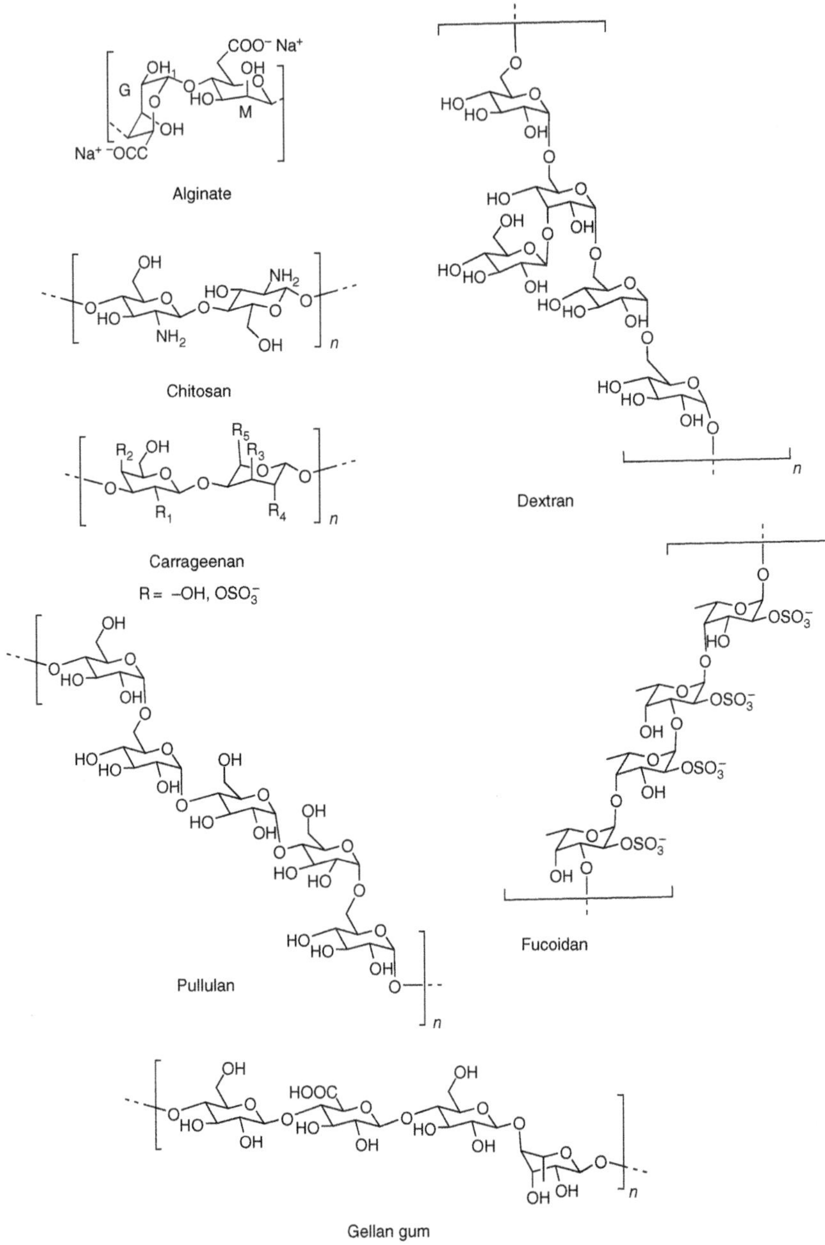

Figure 8.3 Examples of nonhuman polysaccharides employed for biomedical device formulations.

Alginates are linear polysaccharides obtained from seaweed composed of β-D-mannuronate (M) and α-L-guluronate (G) residues linked by (1,4) glycosidic bond [73]. The polysaccharide structures are organized in blocks with consecutive M (MMMMMM) or G (GGGGGG) residues or alternating M and G residues (GMGMGM), depending on the algae from which they are extracted. The hydrogel properties of alginates are related to the M/G ratio and sequence, the G-block length, and finally the molecular weight [74]. Alginates are widely investigated for biomedical applications thanks to their biocompatibility and the ability to form hydrogels by simple addition of divalent cations like Ca^{2+}, and several of them have found applications even in clinics. Alginate dressings have been experimented in wound healing, and are commercially available as Algicell™, AlgiSite M™, Comfeel Plus™, Kaltostat™, Sorbsan™, and Tegagen™; they show an improved ability to support rapid healing through epithelialization [75]. Alginate-based biomaterials are also employed for dental impression applications [76] and for the treatment of gastroesophageal reflux [77, 78]. Other interesting biopolymers derived from algae are carrageenan and fucoidan. Both are sulfated polysaccharides, derived, respectively, from red and brown algae [79]. Carrageenans are sulfate galactans composed of D-galactose (D-Gal) and 3,6-anhydrogalactose with alternating α-(1→3) and β-(1→4) glycosidic bonds; depending on the structure and sulfation patterns, they are classified as γ-, β-, δ-, α-, μ-, κ-, λ-, ν-, ι- and θ-carrageenans. Carrageenans are currently employed as food additives and nutraceuticals; however, several studies highlight their potential use in drug-delivery systems, tissue engineering, and wound healing [80]. Fucoidans are highly sulfated polysaccharides based on repeating units of α-(1→3)-L-fucopyranose, often alternated with α-(1→4)-L-fucopyranose units. They can also contain acetate groups and various glycosyl side branches (i.e. glucuronic acid, mannose, glucose, galactose, or xylose), depending on the type of brown algae. Like carrageenan, fucoidan is currently employed in the nutraceutical and food industry; however, several biomedical applications in the medical devices, drug delivery, and tissue engineering fields are under investigation. [81].

Polysaccharides derived from microbial sources are extensively studied for cosmetic, pharmaceutical, nutraceutical, and biomedical applications. Dextran is one of the most employed; it is α-1,6-poly-α-D-glucoside with α-1,3 branches produced by microbial fermentation employing nonpathogenic *Leuconostoc mesenteroides*, *Saccharomyces cerevisiae*, *Lactobacillus plantarum*, and *Lactobacillus sanfrancisco*. Dextran is used in vascular surgery and antiplatelet applications thanks to its antithrombotic properties; furthermore, it finds application in ophthalmic and diagnostic applications. Dextran 70® (MMW 70.000 kDa) and Dextran 40® (LMW 40.000 kDa) are employed, respectively, as plasma substitutes and to increase blood flow in ischemic limbs [82, 83]. Low-molecular-weight dextran sulfate (LMW-DS) is also under study (Phase II clinical trial) to reduce graft rejection for pancreatic islet transplantation [84]. Dextran is also extensively investigated as coating for diagnostic and therapeutic nanoparticles; in particular, dextran-coated magnetic nanoparticles are employed as diagnostic agents for MRI and as therapeutic nanoparticles for iron deficiency [85]. Other interesting examples of microbial polysaccharides include pullulan, employed for oral care and cosmetic applications [86], bacterial cellulose, employed for topical wound-healing applications [87], and gellan gum, employed in the development of oral, ophthalmic, and spray nasal formulations for biomedical, cosmetic, and nutraceutical applications [88, 89]. All the cited bacterial polysaccharides are also under investigation for other biomedical applications, including wound healing and regenerative medicine [88].

Chitosan is the most commonly used polysaccharide from animal sources; it is obtained from crustacean shell waste. Many chitosan-based biomaterials are commercially available, and some are in clinical trials for wound healing, pharmaceutical, diagnostic, and cosmetic applications. Formulations and medical devices based on chitosan are usually obtained by ionic assembling with anionic polymers or by covalent linkages performed on the amino groups by carbodiimide chemistry. Another approach is based on the partial oxidation of chitosan with periodate-generating aldehyde groups for subsequent chemoselective reactions (Table 8.2).

Table 8.2 Examples of commercial medical devices and formulations produced from vegetal, microbial, or animal-derived polysaccharides.

Polysaccharide	Commercial names	Origin	Application	References
Alginate	Algicell™, AlgiSite M™, Comfeel Plus™, Kaltostat, Sorbsan™, Tegagen™	Algae	Wound dressing	[75]
Sodium alginate	Neocolloi®, Zhermack®; Palgat Plus®, 3M ESPE®	Algae	Dental impression	[76]
Potassium alginate	Blueprint Cremix®, Dentsply DeTrey®		Dental impression	[76]
Sodium alginate	Liquid Gaviscon®	Algae	Gastro-esophageal reflux	[77, 78]
Carrageenan	Carragelose®	Algae	Antiviral eye drops, nasal spray	
Dextran MMW	Dextran 70®	Microbial	Plasma substitute	[82, 83]
Dextran LMW	Dextran 40®	Microbial	Plasma substitute	[82, 83]
Dextran sulfate LMW	Ibsolvmir®	Microbial	Islet transplant	[84]
Dextran-coated iron nanoparticles	Endorem®, Combidex®	Microbial	Imaging liver cancer (endoderm) and lymph node metastases (Combidex)	[85]
Dextran-coated iron nanoparticles	CosmoFer, INFeD, Ferrisat, DexFerrum, and DexIron	Microbial	Iron deficiency therapy	[85]

Table 8.2 (Continued)

Polysaccharide	Commercial names	Origin	Application	References
Pullulan	Listerin®	Microbial	Oral care	[88]
Bacterial cellulose	Bioprocess®, XCell®, and Biofill®	Microbial	Wound healing	[87, 90]
Gellan Gum	Grindsted® Gellan	Microbial	Nutraceutical	[91]
Gellan Gum	Gelrite® and Kelcogel© many others	Microbial	Drug release, ophthalmic, nasal formulations, cosmetic formulations	[89, 92]
Chitosan	HemCon®, Chitoflex®, Celox®	Animal	Hemostatic bleeding treatment and bandages	[93]
Chitosan and cellulose	Chitoseal®	Animal and microbial	Bleeding wounds	[93]
Chitosan – glycerophosphate	BST-CarGel®	Animal	Cartilage repair	[93]
Chitosan-dextran	CD gel	Animal and microbial	Surgical applications, wound healing	[94, 95]

8.2.2 Carbohydrates as Signaling Molecules: Opportunities in Tissue Engineering and Regenerative Medicine

Tissue engineering and regenerative medicine strategies are based on two main pillars: (i) biomaterial scaffolds mimicking the ECM and (ii) stem cells able to regenerate tissues damaged from trauma or pathologies. The biomaterial scaffold can be implanted or just used to culture the human cells that will be implanted/administered subsequently. The role of the biomaterial scaffold in this case consists in homing the cells and driving their fate to the required regeneration. In this context, glycans are deeply investigated not only for their structural role to generate cell-hospitable hydrogels but also for their role as signaling molecules capable of modulating the cell's fate. Recent studies have demonstrated that differential cell fate modulation is strictly dependent on the glycan epitopes present in the ECM [96, 97]. Considering that both stem cells and ECM are heavily glycosylated, it is easily understandable that a "glycocode" plays a key role in cell-signaling induction and transduction [4, 98]. Morphological and functional tissue morphogenesis, related to healthy and pathological states and even aging, is related to N- and O-glycosylation of ECM proteins and cell–ECM communications [99]. Alteration in glycosylation of ECM and extracellular components has been characterized in pathological conditions like cancer [100–103], fibrosis [104, 105], and inflammatory diseases [106–108].

Collagen, laminin, and fibronectin are the most known examples of ECM proteins in which glycosylation varies between physiological and pathological conditions and in tissue morphogenesis [23, 109, 110]. ECM–cell crosstalk events are mediated by carbohydrate–carbohydrate and, more significantly, by carbohydrate–protein (lectins) interactions [3, 111–113]. The crosstalk will result in different structural organizations of the ECM responsible for morphogenesis. Therefore, once the correlation between cell fate and ECM morphological and biomolecular properties (including glycosylation) is clarified, it will be possible to design artificial ECM able to regenerate the required tissue. Even if glycoproteomic and lectin studies are in rapid development, we still need a clear and reliable picture of the main glycosignature changes related to morphogenesis and the occurrence of pathologies.

8.3 Carbohydrates in Animal-Derived Medical Devices: Friends or Foes?

Among implantable medical devices, animal-derived organs or tissues are frequently the best choice for tissue substitution or surgical applications [114]. They not only provide the nanometrically correct architecture for the required functionality but also possess the biochemical composition ideal for cell adhesion and consequent tissue regeneration. As a matter of fact, even in synthetic medical devices, animal-derived polymers are often employed as coatings to optimize the interactions between tissue and material surfaces or to control the medical device performances [115–117]. Bovine pericardium or bovine/porcine patches are employed for the reinforcement of shapes and lines in soft tissue surgery applications. Collagen- and gelatin-derived meshes and coatings, as well as bovine/porcine-based devices, are employed in wound dressing and vascular surgery applications. Animal-derived dural grafts [118, 119] and nerve conducts [120, 121] are also employed in neurosurgery.

Animal-derived materials, however, suffer from a big inconvenience: the presence of xenoantigens like α-Gal, Neu5Gc (N-Glycolylneuraminic acid), and Forssman antigen (GalNAcα1, 3GalNAcβ1, 3Galα1, 4Galβ1, and 4Glc-Cer) [122, 123] (see Figure 8.4) that elicit an immune response in humans.

Strategies like genomic editing and chemical immunosuppression are under investigation to limit xenoantibody production and avoid undesirable early graft rejection [122, 124, 125]. In December 2020, transgenic pigs with depletion of α-Gal – named GalSafe pigs – were approved by FDA for human food consumption and possible biomedical uses [126]. Decellularized animal tissues were also developed with the aim to limit xenoantigens from cellular components; however, the immunological response of these devices is often still in place, probably due to the glycosylation of ECM components [125]. The most efficient solution is probably a coating able to mask the xenoantigens responsible for immunoresponse. Glutaraldehyde fixation [127, 128] is an option. Treatment of animal-derived biological tissues with a low concentration of glutaraldehyde enhances structural performance, maintains sterility, and reduces antigenicity. Although

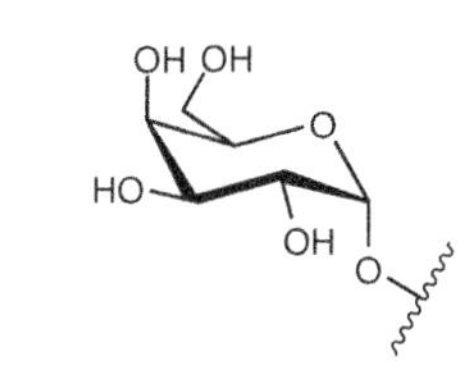

α-Gal

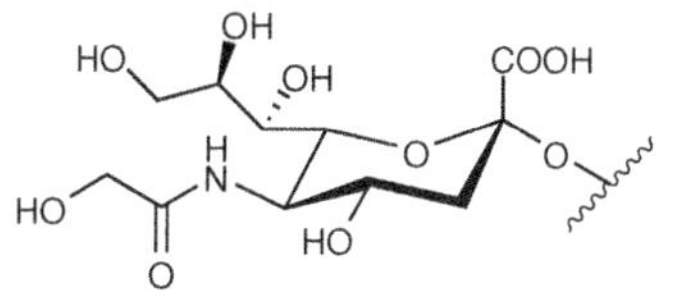

Neu5Gc

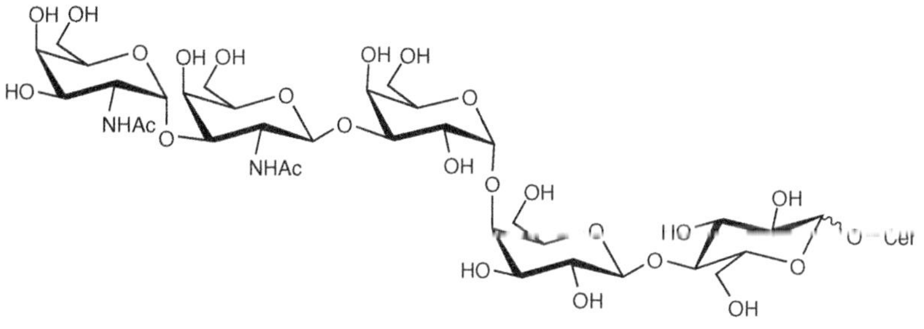

Forssman antigen

Figure 8.4 Xenoantigens responsible for immunoresponse in humans.

the advantages of this kind of coating are well-established, the calcification phenomena induced at the surfaces of the devices are still a remarkable limitation [129]. To reduce the risk of calcification, antimineralization treatments have been developed, such as the use of alpha-amino oleic acid (AOA) employed in Mosaic® bioprosthesis (Medtronic), which also improves durability and hemodynamic properties. The AOA reacts with the free aldehyde groups of glutaraldehyde by Schiff base covalent linkages, avoiding calcium nucleation and precipitation [130]. Glutaraldehyde fixation is able to mask around 50% of α-Gal antigens [131], but at the same time, it increases oxidative and enzymatic degradation of the tissue. Recently, a new treatment available for glutaraldehyde-coated tissues has been patented (FACTA®), able to totally mask α-Gal epitopes in animal-derived medical devices, reducing oxidative degradation and thrombotic risk [132].

8.4 Glycoengineering Application to Regenerative Medicine

Metabolic glycoengineering (MGE) approaches were developed in the late 1980s and are revealed today as valuable tools to study the expression of selected glycans in pathological phenomena and to modulate glycosignature expression [133–135].

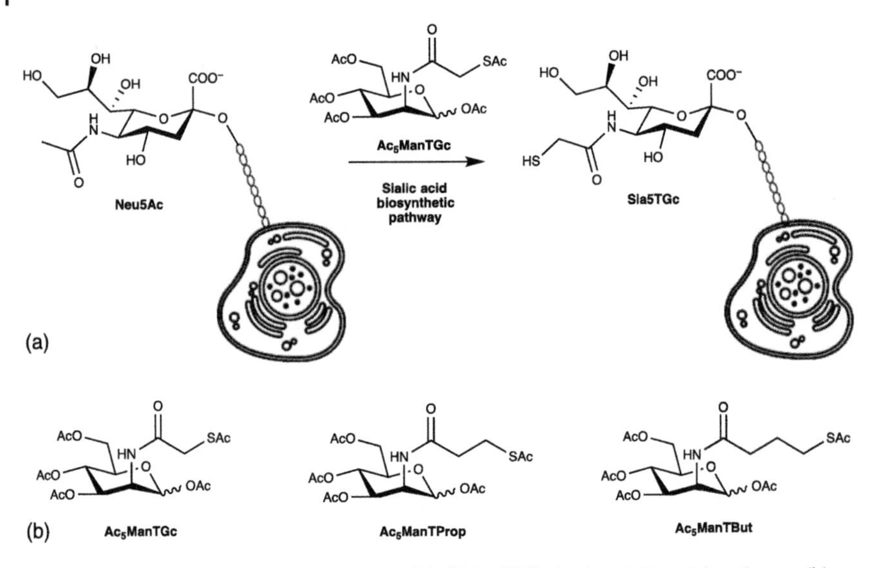

Figure 8.5 (a) Metabolic incorporation of Ac5ManNTGc in the sialic acid pathway; (b) thiol-modified analogs. Source: Adapted from Du et al. [136].

Current approaches to MGE for regenerative medicine applications are mostly focused on the replacement or modulation of cell surface glycosignature. MGE approaches have been performed *in vitro* using different polyacetylated ManNAc analogs in which the *N*-acetyl group was replaced by a thiolated acyl moiety. The thiolated and polyacetylated ManNAc analogs were metabolized, generating cell surface glycans in which Neu5Ac was replaced by a thiolated analog (see Figure 8.5). Neu5(2′-thiolAc), for example, was able to induce clustering of nonadhesive Jurkat cells and morphological neuronal differentiation of human embryoid body-derived (hEBD) stem cells [137]. Other Neu5Ac-thiolated analogs showed interesting bioactivity in human stem-cell modulation, including the induction of neuronal differentiation in human Neuronal Stem Cells (hNSCs) and the suppression of adipogenic differentiation in human adipose stem cells (hASCs) [136].

Gutmann et al. proposed a MGE approach employing tetraacylated monosaccharides containing an azido group, such as 1,3,4,6-tetra-*O*-acetyl-*N*-azidoacetylgalactosamine (Ac4GlcNAz), to obtain ECMs suitable for subsequent click chemistry protocols in order to conjugate different functional biomolecules in NIH 3T3 fibroblasts [138]. Also, fibroblast cell-derived ECM was modified with azido groups by MGE introducing Ac4GalNAz in the culture medium [139]. Another MGE approach was developed by Nellinger et al. [140] by the incorporation of alkene-functionalized monosaccharides in the ECM, able to chemoselectively react with dienophiles as reporter groups. The alkene-functionalized monosaccharides showed good cytocompatibility, ensuring the correct exposition of unnatural sugar on the ECM.

8.5 Future Opportunities and Major Challenges

The role of carbohydrates in medical device and tissue engineering applications is undoubtedly impressive and merits being studied and implemented. Mimicking nature, scientists have exploited polysaccharides as scaffolds to generate artificial tissues, often in combination with proteins, and shorter glycans as signaling molecules to induce cell fate. Where the structural role ends and the biological signal begins is not entirely clear. It will have to be thoroughly investigated. The possibility to synthetically generate human tissues and even organs is a big challenge, requiring multiple expertise to control structural, mechanical, and biomolecular properties, in all of which even carbohydrates play their role. Polysaccharides and glycosylated proteins, derived from human and nonhuman sources, are already widely used as scaffolds to generate medical devices, but there is still room for significant improvements. The clinical translation of glycoengineered materials able to restore damaged tissues by exploiting the glycosignature as modulators of tissue morphogenesis is still in its infancy. The advancement in manufacturing technologies, like 3D printing and 3D bioprinting, will have a significant impact on the generation of polysaccharide-based materials with advanced micrometric physical and morphological properties. The high structural variability of glycans in ECM and cell surfaces and the large number of physiological and pathological events that they can specifically induce require the collection of a lot of information to have a complete picture of their role. The research, therefore, requires robotic platforms for the synthesis and artificial intelligence approaches to elaborate on the multitude of data and develop predictive programs. Once the picture is sufficiently clear, the gap between glycoscience and clinical translation will be covered, revealing the complex mosaic in which glycans are related to age, health, or pathological states. A big challenge for glycoscientists.

Conflict of Interest

LR and FN are partners and members of the Advisory Scientific Board of Biocompatibility Innovation srl (BCI).

References

1 Smith, B.A.H. and Bertozzi, C.R. (2021). *Nature Reviews Drug Discovery* 20: 1–27.

2 Sampaolesi, S., Nicotra, F., and Russo, L. (2019). *Future Medicinal Chemistry* 11: 43–60.

3 Nicolas, J., Magli, S., Rabbachin, L. et al. *Biomacromolecules* https://doi.org/10.1021/acs.biomac.0c00045.

4 Reily, C., Stewart, T.J., Renfrow, M.B., and Novak, J. (2019). *Nature Reviews Nephrology* 15: 346–366.

5 Naso, F., Gandaglia, A., Iop, L. et al. (2012). *Xenotransplantation* 19: 215–220.

6 Seddon, I., Venincasa, M., Farber, N., and Sridhar, J. *International Ophthalmology Clinics* 60 (4): 61–75.

7 Kowalski, P.S., Bhattacharya, C., Afewerki, S., and Langer, R. (2018). *ACS Biomaterials Sciences and Engineering* 4: 3809–3817.

8 Al-Maawi, S., Rutkowski, J.L., Sader, R. et al. (2020). *The Journal of Oral Implantology* 46: 190–207.

9 Duncan, E. (2020). Regulatory constraints for medical products using biomaterials. In: *Biomaterials Science: An Introduction to Materials in Medicine* (ed. W.R. Wagner, S.E. Sakiyama-Elbert, G. Zhang, and M.J.B.T.-B.S. Yaszemski), 1463–1473. Academic Press.

10 Bernard, M., Jubeli, E., Pungente, M.D., and Yagoubi, N. (2018). *Biomaterials Science* 6: 2025–2053.

11 Rinaudo, M. (2008). *Polymer International* 57: 397–430.

12 Kirschning, A., Dibbert, N., and Drager, G. (2018). *Chemistry* 24: 1231–1240.

13 Mohammed, A.S.A., Naveed, M., and Jost, N. *Journal of Polymers and the Environment* https://doi.org/10.1007/s10924-021-02052-2.

14 Tchobanian, A., Van Oosterwyck, H., and Fardim, P. (2019). *Carbohydrate Polymers* 205: 601–625.

15 Breimer, M.E. and Holgersson, J. (2019). *Frontiers in Molecular Biosciences* 6: 57.

16 Joziasse, D.H. and Oriol, R. (1999). *Biochimica et Biophysica Acta, Molecular Basis of Disease* 1455: 403–418.

17 Hinderer, S., Layland, S.L., and Schenke-Layland, K. (2016). *Advanced Drug Delivery Reviews* 97: 260–269.

18 Dzamba, B.J. and DeSimone, D.W. (2018). *Current Topics in Developmental Biology* 130: 245–274.

19 Dang, T. T., Nikkhah, M., Memic, A., and Khademhosseini, A. (2014). Polymeric biomaterials for implantable prostheses. In: *Natural and Synthetic Biomedical Polymers*, 1e (ed. S.G. Kumbar, C.T. Laurencin, and M. Deng), 309–331. Oxford: Elsevier.

20 Li, C., Guo, C., Fitzpatrick, V. et al. (2020). *Nature Reviews Materials* 5: 61–81.

21 Liu, L., Liu, Y., Li, J. et al. (2011). *Microbial Cell Factories* 10: 99.

22 Taylor, S.L., Hogwood, J., Guo, W. et al. (2019). *Scientific Reports* 9: 2679.

23 Pignatelli, C., Cadamuro, F., Magli, S. et al. (2021). Glycans and hybrid glycomaterials for artificial cell microenvironment fabrication. In: *Carbohydrate Chemistry: Chemical and Biological Approaches.* vol. 44 (ed. A.P. Rauter, T.K. Lindhorst, and Y. Queneau), 250–276. https://doi.org/10.1039/9781788013864

24 Reddy, N., Reddy, R., and Jiang, Q. (2015). *Trends in Biotechnology* 33: 362–369.

25 Amorim, S., Reis, C.A., Reis, R.L., and Pires, R.A. (2021). *Trends in Biotechnology* 39: 90–104.

26 Price, R.D., Berry, M.G., and Navsaria, H.A. (2007). *Journal of Plastic, Reconstructive & Aesthetic Surgery* 60: 1110–1119.

27 Hussain, Z., Thu, H.E., Katas, H., and Bukhari, S.N.A. (2017). *Polymer Reviews* 57: 594–630.

28 Campoccia, D., Hunt, J.A., Doherty, P.J. et al. (1996). *Biomaterials* 17: 963–975.

29 Simman, R. (2018). *Journal of the American College of Clinical Wound Specialists* 8: 10–11.

30 Nehrer, S., Domayer, S., Dorotka, R. et al. (2006). *European Journal of Radiology* 57: 3–8.

31 Longinotti, C. (2014). *Burns and Trauma* 2: 162–168.

32 Benedetti, L., Cortivo, R., Berti, T. et al. (1993). *Biomaterials* 14: 1154–1160.

33 Iwasa, J., Engebretsen, L., Shima, Y., and Ochi, M. (2009). *Knee Surgery, Sports Traumatology, Arthroscopy* 17: 561–577.

34 Deszcz, I., Lis-Nawara, A., Grelewski, P. et al. (2020). *Regenerative Biomaterials* 7: 543–552.

35 Marcacci, M., Zaffagnini, S., Kon, E. et al. (2002). *Knee Surgery, Sports Traumatology, Arthroscopy* 10: 154–159.

36 Pavesio, A., Abatangelo, G., Borrione, A. et al. (2003). *Novartis Foundation Symposium* 249: 203–241.

37 Wylie, J.D., Hartley, M.K., Kapron, A.L. et al. (2015). *Clinical Orthopaedics and Related Research* 473: 1673–1682.

38 Makris, E.A., Gomoll, A.H., Malizos, K.N. et al. (2015). *Nature Reviews Rheumatology* 11: 21–34.

39 Huynh, A. and Priefer, R. (2020). *Carbohydrate Research* 489: 107950.

40 Rah, M.J. (2011). *Optometry* 82: 38–43.

41 Modi, S.S., Davison, J.A., and Walters, T. (2011). *Clinical Ophthalmology* 5: 1381–1389.

42 Kretz, F.T.A., Limberger, I.-J., and Auffarth, G.U. (2014). *Journal of Cataract and Refractive Surgery* 40: 1879–1884.

43 Barth, H., Crafoord, S., and Ghosh, F. (2021). *Current Eye Research* 46: 373–379.

44 Samsom, M., Korogiannaki, M., Subbaraman, L.N. et al. (2018). *Journal of Biomedial Materials Research Part B Applied Biomaterials* 106: 1818–1826.

45 Maulvi, F.A., Singhania, S.S., Desai, A.R. et al. (2018). *International Journal of Pharmaceutics* 548: 139–150.

46 Kablik, J., Monheit, G.D., Yu, L. et al. (2009). *Dermatologic Surgery* 35 (Suppl 1): 302–312.

47 Stocks, D., Sundaram, H., Michaels, J. et al. (2011). *Journal of Drugs in Dermatology* 10: 974–980.

48 Bogdan Allemann, I. and Baumann, L. (2008). *Clinical Interventions in Aging* 3: 629–634.

49 Nicolas, J., Magli, S., Rabbachin, L. et al. (2020). *Biomacromolecules* 21: 1968–1994.

50 Henrotin, Y., Mathy, M., Sanchez, C., and Lambert, C. (2010). *Therapeutic Advances in Musculoskeletal Disease* 2: 335–348.

51 Oliveira, J.T. and Reis, R.L. (2008). Natural-based polymers for biomedical applications. In: *Woodhead Publishing Series in Biomaterials* (ed. R.L. Reis, N.M. Neves, J.F. Mano, et al.) H. S. B. T.-N.-B. P. for B. A. Azevedo, 485–514. Woodhead Publishing.

52 Iannitti, T., Morales-Medina, J.C., Merighi, A. et al. (2018). *Drug Delivery and Translational Research* 8: 994–999.

53 Henrotin, Y., Hauzeur, J.-P., Bruel, P., and Appelboom, T. (2012). *BMC Research Notes* 5: 407.

54 Henrotin, Y., Marty, M., and Mobasheri, A. (2014). *Maturitas* 78: 184–187.

55 Shriver, Z., Capila, I., Venkataraman, G., and Sasisekharan, R. (2012). Heparin and heparan sulfate: analyzing structure and microheterogeneity. *Handbook of Experimental Pharmacology,* 207: 159–176.

56 Meneghetti, M.C.Z., Hughes, A.J., Rudd, T.R. et al. (2015). *Journal of the Royal Society, Interface* 12: 589.

57 Fu, L., Suflita, M., and Linhardt, R.J. (2016). *Advanced Drug Delivery Reviews* 97: 237–249.

58 Meddahi, A., Lemdjabar, H., Caruelle, J.P. et al. (1996). *International Journal of Biological Macromolecules* 18: 141–145.

59 Barbosa, I., Morin, C., Garcia, S. et al. (2005). *Journal of Cell Science* 118: 253–264.

60 Barritault, D., Gilbert-Sirieix, M., Rice, K.L. et al. (2017). *Glycoconjugate Journal* 34: 325–338.

61 Rabenstein, D.L. (2002). *Natural Product Reports* 19: 312–331.

62 Biran, R. and Pond, D. (2017). *Advanced Drug Delivery Reviews* 112: 12–23.

63 Preston, T.J., Ratliff, T.M., Gomez, D. et al. (2010). *The Journal of Extra-Corporeal Technology* 42: 199–202.

64 Wendel, H.P. and Ziemer, G. (1999). *The European Journal of Cardio-Thoracic Surgery* 16: 342–350.

65 Puranik, A.S., Dawson, E.R., and Peppas, N.A. (2013). *International Journal of Pharmaceutics* 441: 665–679.

66 Rykowska, I., Nowak, I., and Nowak, R. *Molecules* https://doi.org/10.3390/molecules25204624.

67 Liang, Y. and Kiick, K.L. (2014). *Acta Biomaterialia* 10: 1588–1600.

68 Cheng, C., Sun, S., and Zhao, C. (2014). *Journal of Materials Chemistry B* 2: 7649–7672.

69 Bedini, E., Laezza, A., and Iadonisi, A. (2016). *European The Journal of Organic Chemistry* 2016: 3018–3042.

70 Tayama, E., Hayashida, N., Akasu, K. et al. (2000). *Artificial Organs* 24: 618–623.

71 Larm, O., Larsson, R., and Olsson, P. (1983). *Biomaterials, Medical Devices, and Artificial Organs* 11: 161–173.

72 Blezer, R., Cahalan, L., Cahalan, P.T., and Lindhout, T. (1998). *Blood Coagulation and Fibrinolysis* 9 (5): 435–440.

73 Lee, K.Y. and Mooney, D.J. (2012). *Progress in Polymer Science* 37: 106–126.

74 George, M. and Abraham, T.E. (2006). *Journal of Controlled Release* 114: 1–14.

75 Ahmed, S. (ed.) (2019). *Alginates: Applications in the Biomedical and Food Industries*. Wiley.

76 Nallamuthu, N.A., Braden, M., and Patel, M.P. (2012). *Dental Materials* 28: 756–762.

77 Dettmar, P.W., Sykes, J., Little, S.L., and Bryan, J. (2006). *International Journal of Clinical Practice* 60: 275–283.

78 Kwiatek, M.A., Roman, S., Fareeduddin, A. et al. (2011). *Alimentary Pharmacology & Therapeutics* 34: 59–66.
79 Cunha, L. and Grenha, A. (2016). *Marine Drugs* 14: 42.
80 Yegappan, R., Selvaprithiviraj, V., Amirthalingam, S., and Jayakumar, R. (2018). *Carbohydrate Polymers* 198: 385–400.
81 Ale, M.T. and Meyer, A.S. (2013). *RSC Advances* 3: 8131–8141.
82 Aronson, J.K. (ed.) (2016). Dextrans. In: *Meyler's Side Effects of Drugs*, 893–898. Elsevier, Oxford.
83 Sobolewski, K., Radparvar, S., Wong, C., and Johnston, J. (2018). *A Worldwide Yearly Survey of New Data in Adverse Drug Reactions*, S. D. B. T.-S. E. of D. A. Ray, vol. 40, 415–429. Elsevier.
84 von Zur-Mühlen, B., Lundgren, T., Bayman, L. et al. (2019). *Transplantation* 103: 630–637.
85 Tassa, C., Shaw, S.Y., and Weissleder, R. (2011). *Accounts of Chemical Research* 44: 842–852.
86 Coltelli, M.-B., Danti, S., De Clerck, K. et al. (2020). *Journal of Functional Biomaterials* 11: 20.
87 Petersen, N. and Gatenholm, P. (2011). *Applied Microbiology and Biotechnology* 91: 1277–1286.
88 Moscovici, M. (2015). *Frontiers in Microbiology* 6: 1012.
89 Osmałek, T., Froelich, A., and Tasarek, S. (2014). *International Journal of Pharmaceutics* 466: 328–340.
90 Czaja, W.K., Young, D.J., Kawecki, M., and Brown, R.M.J. (2007). *Biomacromolecules* 8: 1–12.
91 Finn Madsen, Frédéric Liot. USE OF HIGH ACYL GELLAN IN WHIPPING CREAM. WO2014184134 (2014).
92 J. Gupta, R. Rathour, K. Medhi, B. Tyagi and I. S. Thakur, eds. R. P. Kumar, E. Gnansounou, J. K. Raman and G. B. T.-R. B. R. for S. E. and B. Baskar, Academic Press, 2020, pp. 51–85.
93 Zhang, J., Xia, W., Liu, P. et al. (2010). *Marine Drugs* 8: 1962–1987.
94 Aziz, M.A., Cabral, J.D., Brooks, H.J.L. et al. (2012). *Antimicrobial Agents and Chemotherapy* 56: 280–287.
95 M. S. Islam, M. S. Rahman, T. Ahmed, S. Biswas, P. Haque and M. M. Rahman, eds. S. Gopi, S. Thomas and A. B. T.-H. of C. and C. Pius, Elsevier, 2020, pp. 721–759.
96 Erginer, M., Akcay, A., Coskunkan, B. et al. (2016). *Carbohydrate Polymers* 149: 289–296.
97 Liu, Q., Lyu, Z., Yu, Y. et al. (2017). *ACS Applied Materials & Interfaces* 9: 11518–11527.
98 Berger, R.P., Dookwah, M., Steet, R., and Dalton, S. (2016). *BioEssays* 38: 1255–1265.
99 Lu, P., Takai, K., Weaver, V.M., and Werb, Z. *Cold Spring Harbor Perspectives in Biology* https://doi.org/10.1101/cshperspect.a005058.
100 Deb, B., Patel, K., Sathe, G., and Kumar, P. (2019). *Journal of Clinical Medicine* 8: 1303.

101 Peixoto, A., Relvas-Santos, M., Azevedo, R. et al. (2019). *Frontiers in Oncology* 9: 380.

102 Mereiter, S., Balmaña, M., Campos, D. et al. (2019). *Cancer Cell* 36: 6–16.

103 Martins, Á.M., Ramos, C.C., Freitas, D., and Reis, C.A. (2021). *Cell* 10: 109.

104 Merl-Pham, J., Basak, T., Knüppel, L. et al. (2019). *Matrix Biology Plus* 1: 100005.

105 Lecca, M.R., Maag, C., Berger, E.G., and Hennet, T. (2011). *Journal of Cellular and Molecular Medicine* 15: 1788–1796.

106 Sorokin, L. (2010). *Nature Reviews Immunology* 10: 712–723.

107 Maverakis, E., Kim, K., Shimoda, M. et al. (2015). *Journal of Autoimmunity* 57: 1–13.

108 Groux-Degroote, S., Cavdarli, S., Uchimura, K. et al. (2020). *Inflammatory Disorders*, Part A, ed. R. B. T.-A. in P. C. and S. B. Donev, vol. 119, 111–156. Academic Press.

109 Marsico, G., Russo, L., Quondamatteo, F., and Pandit, A. *Trends in Cancer* https://doi.org/10.1016/j.trecan.2018.05.009.

110 Rebelo, A.L., Chevalier, M.T., Russo, L., and Pandit, A. *Cell Reports Physical Sciences* https://doi.org/10.1016/j.xcrp.2021.100321.

111 Hughes, R.C. (1992). *Biochemical Society Transactions* 20: 279–284.

112 Tian, E., Hoffman, M.P., and Ten Hagen, K.G. (2012). *Nature Communications* 3: 869.

113 Marsico, G., Russo, L., Quondamatteo, F., and Pandit, A. (2018). *Trends in Cancer* 4: 537–552.

114 Schoelles, K., Snyder, D.L., and Sullivan, N. (2012). *Skin Substitutes for Treating Chronic Wounds*. Rockville (MD): Agency for Healthcare Research and Quality (US) Results https://www.ncbi.nlm.nih.gov/books/NBK248357/.

115 Corduas, F., Lamprou, D.A., and Mancuso, E. (2021). *Bio-Design and Manufacturing* 4: 278–310.

116 Cornwell, K.G., Landsman, A., and James, K.S. (2009). *Clinics in Podiatric Medicine and Surgery* 26: 507–523.

117 Cornwell, K.G., Jessee, C.B., and Adelman, D.M. (2020). *British Journal of Hospital Medicine (London, England)* 81: 1–10.

118 Liu, W., Wang, X., Su, J. et al. (2021). *Frontiers in Bioengineering and Biotechnology* 9: 105.

119 Knopp, U., Christmann, F., Reusche, E., and Sepehrnia, A. (2005). *Acta Neurochirurgica* 147: 877–887.

120 Saeki, M., Tanaka, K., Imatani, J. et al. (2018). *Injury* 49: 766–774.

121 Klein, S., Vykoukal, J., Felthaus, O. et al. (2016). *Materials (Basel, Switzerland)* 9: 219.

122 Ezzelarab, M., Ayares, D., and Cooper, D.K.C. (2005). *Immunology and Cell Biology* 83: 396–404.

123 Du, J. and Yarema, K.J. (2010). *Advanced Drug Delivery Reviews* 62: 671–682.

124 Adams, A.B., Kim, S.C., Martens, G.R. et al. (2018). *Annals of Surgery* 268: 564–573.

125 Wong, M.L. and Griffiths, L.G. (2014). *Acta Biomaterialia* 10: 1806–1816.

126 C&EN Glob. Enterp. (2021). First GM pigs for allergies. Could xenotransplants be next? Elie Dolgin. *Nature Biotechnology*, 39: 397–400.

127 Jayakrishnan, A. and Jameela, S.R. (1996). *Biomaterials* 17: 471–484.

128 Griffiths, L.G., Choe, L.H., Reardon, K.F. et al. (2008). *Biomaterials* 29: 3514–3520.

129 Saporito, W.F., Pires, A.C., Cardoso, S.H. et al. (2011). *BMC Surgery* 11: 37.

130 Riess, F.-C., Cramer, E., Hansen, L. et al. (2010). *The European Journal of Cardio-Thoracic Surgery* 37: 145–153.

131 Naso, F., Gandaglia, A., Bottio, T. et al. (2013). *Xenotransplantation* 20: 252–261.

132 Naso, F., Stefanelli, U., Buratto, E. et al. (2017). *Tissue Engineering. Part A* 23: 1181–1195.

133 Mahal, L.K., Yarema, K.J., and Bertozzi, C.R. (1997). *Science* 276: 1125–1128.

134 Dube, D.H. and Bertozzi, C.R. (2003). *Current Opinion in Chemical Biology* 7: 616–625.

135 Agatemor, C., Buettner, M.J., Ariss, R. et al. (2019). *Nature Reviews Chemistry* 3: 605–620.

136 Du, J., Agatemor, C., Saeui, C.T. et al. *Cell* https://doi.org/10.3390/cells10020377.

137 Sampathkumar, S.-G., Li, A.V., Jones, M.B. et al. (2006). *Nature Chemical Biology* 2: 149–152.

138 Gutmann, M., Bechold, J., Seibel, J. et al. (2019). *ACS Biomaterials Sciences and Engineering* 5: 215–233.

139 Ruff, S.M., Keller, S., Wieland, D.E. et al. (2017). *Acta Biomaterialia* 52: 159–170.

140 Nellinger, S., Rapp, M.A., Southan, A. et al. *ChemBioChem* https://doi.org/10.1002/cbic.202100266.

9

Carbohydrate-Based Therapeutics for Lysosomal Storage Disorders

Camilla Matassini, Francesca Clemente, and Francesca Cardona

Department of Chemistry "U. Schiff" (DICUS), University of Florence, via della Lastruccia 3-13, 50019 Sesto F.no (FI), Italy

9.1 An Introduction to Lysosomal Storage Disorders (LSDs)

Lysosomal Storage Disorders (LSDs) are a group of 70 inherited metabolic disorders [1] that affect the function of the lysosome, a membrane-bound organelle with a key role in processes involved in degrading and recycling cellular waste, cellular signaling, and energy metabolism [2]. LSDs are monogenic disorders caused by the mutation in a gene encoding for a lysosomal protein (such as lysosomal glycosidases, proteases, integral membrane proteins, transporters, enzyme modifiers, or activators), resulting in lysosomal failure and subsequent accumulation of non-metabolized substrates. This steady accumulation of substrates in the lysosome, the "storage," hence the name of these disorders, ultimately leads to cell dysfunction and cell death (Figure 9.1).

LSDs are considered a class of genetically and clinically heterogeneous disorders, even though some common clinical features, such as a pediatric onset and the enlargement of abdominal organs, mainly liver and spleen (visceromegaly), can be identified. Other frequent symptoms, related to the specific genetic mutation and the consequent stored substrate, are skeletal dysmorphia, developmental delay, or other central nervous system (CNS) deficits. Diagnosis of LSDs is based on clinical symptoms and subsequent confirmation with biochemical and genetic tests, such as enzymatic dosage and gene sequencing [3]. Unfortunately, especially for the less severe forms with longer survival, the overlapping clinical features that are not specific for LSDs lead to considerable diagnostic delay and missed cases, while early diagnosis is a priority for the application of new disease-modifying therapies.

Carbohydrate-Based Therapeutics, First Edition. Edited by Roberto Adamo and Luigi Lay.

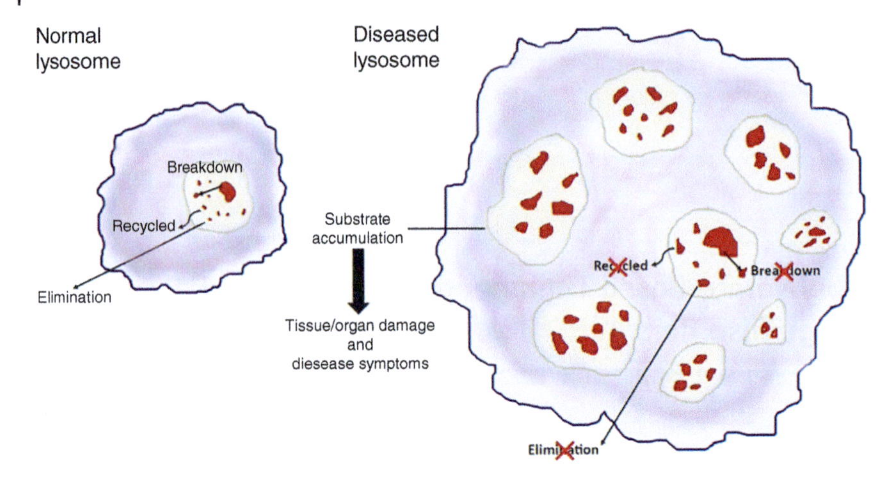

Figure 9.1 Comparison of a healthy lysosome with a defective one of LSD patients: the genetic mutation results in lysosome failure, hampering the breakdown and recycling of macromolecules and leading to tissue and organ dysfunctions with consequent disease symptoms.

Epidemiology and diagnosis of LSDs, as well as challenges for the screening of carriers, high-risk and newborn populations have been carefully reviewed by Wijburg and coworkers [4].

Information about the incidence of LSDs is relatively limited. However, although rare if taken individually (one case per 50 000–250 000 individuals), they are common as a group, with an estimated incidence of 1 in 5000–5500 live births [5]. The most common LSDs are Fabry, Gaucher, and Pompe diseases (PDs), with up to 2–2.5 cases per 100 000 individuals [6]. Moreover, despite a general low occurrence of individual LSDs, the incidence may be higher in specific ethnicities, such as in the case of Gaucher disease (GD), for which a frequency of 1 in 500 to 1 in 1000 is observed among Ashkenazi Jews [7].

LSDs related to enzyme deficiencies, which are the focus of this chapter, can be subclassified according to the stored materials (see Table 9.1):

1- Mucopolysaccharidoses (MPS) (such as MPS I, II, III e IVA)
2- Sphingolipidoses (such as Fabry, Niemann–Pick (NP) and GD, GM1 and GM2 gangliosidosis, and Krabbe disease (KD))
3- Glycogen Storage Disorders (namely, Pompe disease)
4- Glycoproteinoses (such as α-Mannosidosis and Fucosidosis)

Each disease can be in turn classified into different types based on symptoms, affected organs, and age of onset, such as congenital or infantile (which usually have the most severe presentation), late-infantile, juvenile, and adult types.

Table 9.1 Lysosomal storage disorders object of this chapter, classified according to the stored material: for each disorder the name(s), the deficient enzyme, and the approved therapy are listed.

Disease	Deficient lysosomal enzyme	Approved therapies
Mucopolysaccharidoses (MPS)		
MPS I: Hurler syndrome	α-L-iduronidase	HSCT, ERT
MPS II: Hunter syndrome	Iduronate-2-sulfatase (IDS)	ERT
MPS IIIA: Sanfilippo syndrome A	*N*-sulfoglucosamine sulfohydrolase	NONE[a]
MPS IIIB: Sanfilippo syndrome B	*N*-acetyl-α-glucosaminidase (NAGLU)	NONE[a]
MPS IIIC (Sanfilippo syndrome)	Acetyl CoA glucosamine *N*-acetyltransferase	NONE[a]
MPS IIID (Sanfilippo syndrome)	*N*-acetyl-glucosamine-6-sulfatase	NONE[a]
MPS IVA: Morquio syndrome	*N*-acetylgalactosamine-6-sulfatase (GALNS)	ERT
MPS IVB (Morquio B syndrome)[b]	β-galactosidase	NONE[a]
MPS VI (Maroteaux–Lamy syndrome)	Arylsulfatase B	ERT
MPS VII (Sly syndrome)	β-glucuronidase	ERT
MPS IX	Hyaluronidase	NONE[a]
Sphingolipidoses		
Fabry disease	α-galactosidase A	ERT, PCT
Gaucher disease	β-glucocerebrosidase, also known as acid β-glucosidase (GCase)	ERT, SRT
Niemann–Pick	Sphingomyelin-degrading enzyme acid sphingomyelinase and cholesterol-transport proteins	ERT
GM1 gangliosidosis	β-galactosidase	NONE[a]
GM2 gagliosidosis: Tay–Sachs disease and Sandhoff disease	β-hexosaminidase	NONE[a]
Krabbe disease	Galactosylceramidase	HSCT
Glycogen storage disease		
Pompe disease	α-glucosidase	ERT
Glycoproteinoses		
Fucosidosis	α-L-fucosidase	NONE[a]
α-mannosidosis	α-mannosidase	ERT

a) "NONE" means that only symptomatic and supportive therapies are provided to patients.
b) Being β-galactosidase the deficient enzyme in MPS IVB, this disorder will be discussed together with GM1 (Sphingolipidoses section) instead of in the Mucopolysaccharidoses section.
ERT, enzyme replacement therapy; SRT, substrate reduction therapy; PCT, pharmacological chaperone therapy.

9.2 Available Treatments for LSDs: The Role of Carbohydrate-Based Therapeutics

The promulgation of U.S. Orphan Drug Act (ODA) in 1983, followed by analogous "orphan" legislation in other countries, undoubtedly stimulated the investments in novel therapies for rare conditions, including LSDs, that otherwise would not be profitable due to the small target population [8]. The consequently growing understanding of the pathophysiology of LSDs reached in the last decades, allowed to identify various potential clinical intervention points, thus developing innovative therapies, some of which passed all the steps from preclinical stage to regulatory approval, and they were ultimately made available for patients. The most relevant therapeutic approaches for LSDs are briefly described below.

The first treatment of an LSD, which showed any success was the hematopoietic stem cell transplantation (HSCT) in a one-year-old MPS I patient, in 1981 [9]. In particular, the transplanted bone marrow-derived cells with normal enzyme levels were thought to donate enzymes to the patient's deficient cells [10]. Unfortunately, this therapy resulted efficient only in MPS I patients younger than nine months of age, and presents important issues related to donor selection and adverse effects of transplantation [11, 12]. HSCT is currently the standard care for MPS I younger patients, despite the high risks of mortality, and it has been also used for other LSDs without available treatments, such as KD [1].

Recently, gene therapy methods, such as gene replacement, antisense oligonucleotide (ASO) therapies, or gene editing, are also emerging for the treatment of LSDs [13–15].

In particular, some variants of adeno-associated virus (AAV) resulted in excellent vectors to deliver the gene encoding the deficient enzymes to cells [16]. Once manufactured for human use, the vector can be directly infused into the bloodstream or target organs (*direct gene transfer*) or it can be applied *ex-vivo* to stem cells in culture and subsequently re-implanted into the donor (*indirect gene transfer*) [1]. These cutting-edge therapies, all still in clinical trials for very few disorders, seem to be effective in achieving remission of neurological manifestations. However, the major drawbacks are related to establishing safe administration regime, the invasive route of administration, and the difficulty of foreseeing long-term undesired effects [17].

Enzyme replacement therapy (ERT) consists of intravenous infusion of a recombinant wild-type enzyme, which enters the cell through membrane receptors (typically the mannose-6-phosphate receptor (M6PR)) and replaces the catalytic action of the mutated lysosomal enzyme [18, 19]. Available since 1996 [20], ERT is the only treatment approved for many LSDs (Table 9.1). The main limitations associated with ERT are its high cost, possible immunologic response, and frequent hospitalization of patients, which have been now partially overcome with domiciliary treatments. In addition, the recombinant enzymes used are not able to cross the blood–brain barrier (BBB) and reach the CNS, so that ERT lacks efficacy in neuronopathic forms of disorders.

The aim of the substrate reduction therapy (SRT) is to reduce the storage in the lysosomes of undegraded substrates by using small molecules able to inhibit their biosynthesis [21]. SRT could be in principle developed for all LSDs, but, to date, is approved for Gaucher and NP diseases only, because, for these glycosphingolipids, the knowledge of their biosynthetic pathways allowed to identify specific inhibitors of the enzyme that synthesizes the substrates or one of their precursors.

Although trials in other glycosphingolipids and MPS are ongoing, the limitation to very few pathologies represents the main drawback of SRT. In addition, SRT has a slower efficacy than ERT and several side effects or complications, such as diarrhea, flatulence, abdominal pain, or weight loss.

In several LSDs, gene mutations result in lysosomal enzymes prone to a rapid turnover or with a reduced catalytic activity [22]. However, since the substrate accumulation occurs when residual enzyme activity decays below a certain threshold, a small improvement in the enzyme activity can be sufficient to slow down the disease progression and to ameliorate clinical evidence. This is the concept on which the pharmacological chaperone therapy (PCT) is based: it employs small molecules (pharmacological chaperones, PCs) able to stabilize the tertiary structure of the mutant enzyme, thus improving its catalytic activity and reducing the storage in the lysosomes (Figure 9.2).

The main advantages of PCs are their low-cost, the oral administration, and the possibility to cross the BBB exerting their action on the CNS [23].

Several PCs were designed and synthesized based on the structure of the natural substrate or alternatively identified by high-throughput screening (HTS) studies. To date, the only PC that reached the drug market is Migalastat (Galafold®, Amicus

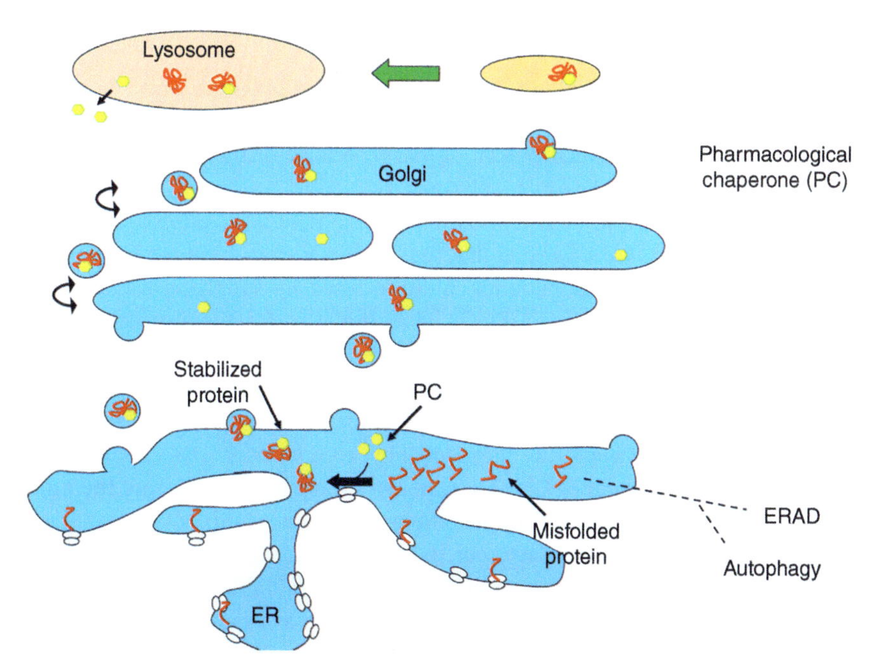

Figure 9.2 Cartoon representing the Pharmacological Chaperone (PC) concept. Source: Reproduced with permission from Boyd et al. [23]. © 2013 American Chemical Society Publications.

Therapeutics) [24, 25] for the treatment of Fabry disease (FD), but many other compounds showed encouraging results in preclinical studies for the treatment of other LSDs, including GD, Pompe disease, or gangliosidosis.

Surprisingly, except for cell-based therapies (both the oldest bone marrow transplantation and the emerging gene therapy), the most currently employed treatments, namely ERT, SRT, and PCT, all involve the use of iminosugars, sugar analogs with a nitrogen replacing the endocyclic oxygen atom [26].

Iminosugars have been isolated from natural sources, including plants and microorganisms, for over 50 years. A great effort for their total synthesis has been pursued by many researchers all around the world. For their synthesis, which is very challenging due to the presence of several contiguous stereocenters with a well-defined configuration, both chiral pool strategies and enantioselective syntheses have been developed [27–33].

Their resemblance to carbohydrates allows them to be recognized by, and interact with carbohydrate receptors. In particular, iminosugars have been extensively investigated in the last 30 years as glycosidase [34–36] and glycosyltransferase inhibitors [37–41].

They are more chemically and metabolically stable than their parent sugars, are highly water soluble, can cross the BBB, and are normally excreted from the body unmodified. Such properties distinguish them from other small polar molecules and provide significant advantages for their use as drugs, especially in the treatment of LSDs.

9.2.1 Enzyme Replacement Therapy (ERT)

In the 1980s the ingestion of plants belonging to the *Leguminosae* family was considered responsible for inducing lysosomal storage diseases in animals (e.g. swainsona toxicosis [42] or locoweed poisoning [43]) because of their high content of iminosugars able to inhibit lysosomal glycosidases.

The observation that the endogenous enzyme activity was restored when the inhibitor was removed [44] suggested that the lysosomal storage caused by the deficient enzyme in human LSD patients could be faced by supplying a natural or recombinant functional enzyme, thus paving the way for the introduction of the ERT.

ERT is based on the replacement of the mutated protein with normal protein to restore its wild-type function. Although initially purified from human tissues, nowadays the deficient enzymes are produced by recombinant DNA technology. Mammal-derived cells, such as Chinese Hamster Ovary (CHO) cells, are the most widely used platform to produce ERT recombinant enzymes [45].

After their efficacy and safety have been confirmed by clinical trials, to date, there are 10 approved therapeutic enzymes as ERT for 7 LSDs (Table 9.1). Nevertheless, there are some drawbacks related to this therapy, such as immune reactions, low efficiency of lysosomal targeting, and difficulty reaching CNS. To overcome these limitations, several strategies have been developed with the aim of increasing the targeting efficiency and the ability to cross the BBB. Many of these approaches

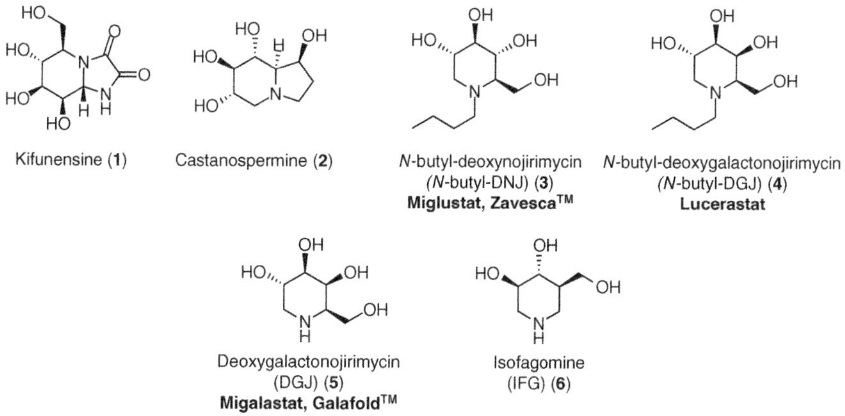

Kifunensine (**1**)

Castanospermine (**2**)

N-butyl-deoxynojirimycin (*N*-butyl-DNJ) (**3**) **Miglustat, Zavesca™**

N-butyl-deoxygalactonojirimycin (*N*-butyl-DGJ) (**4**) **Lucerastat**

Deoxygalactonojirimycin (DGJ) (**5**) **Migalastat, Galafold™**

Isofagomine (IFG) (**6**)

Figure 9.3 Structures of iminosugars involved in ERT, SRT, and PCT for LSDs.

involve glycan remodeling in order to improve the cellular uptake and lysosomal targeting.

Indeed, recombinant human lysosomal enzymes for ERT contain a mixture of N-linked glycans, including complex, hybrid, and at least one high mannose with the mannose-6-phosphate (M6P) motif, which allow the enzyme to enter the cell by exploiting the interaction between M6P and M6PR [46]. The only exception is ERT for GD, which does not utilize the M6PR system but targets the enzyme to the mannose-receptor on macrophages. For this reason, recombinant β-glucocerebrosidase has to be processed chemically and with high costs after production, to expose mannose-residues on its N-linked glycans. Today, this goal is achieved by treating the culture medium of recombinant β-glucocerebrosidase (Velaglucerase alpha, VPRIV®, Shire Human Genetic therapies Inc.) with the iminosugar kifunensine (**1**, Figure 9.3), obtaining high mannose-type glycans [47].

More recently, a further improvement in the ERT was provided by nanomedicine: nano-carriers, such as liposomes or biocompatible polymers, are being developed for sustained delivery and efficient intracellular targeting [48].

9.2.2 Substrate Reduction Therapy (SRT)

In the late 1980s Winchester and coworkers reported that co-incubating wild-type fibroblasts with castanospermine (**2**, Figure 9.3), a powerful inhibitor of α-and β-glucosidases, resulted in the appearance of additional glycosphingolipids in the treated cells, providing the first evidence that an iminosugar could affect the metabolism of glycosphingolipids [49]. Few years later, Butter and coworkers demonstrated that the α-glucosidase (α-glu) inhibitor *N*-butyl-deoxynojirimycin (*N*-butyl-DNJ, **3**, Figure 9.3) also inhibited ceramide glucosyltransferase, the enzyme involved in the first step of the biosynthesis of most glycosphingolipids, thus acting as a mimic of ceramide [50]. This suggested that *N*-butyl-DNJ (**3**) could be used as drug for reducing the synthesis of all glycosphingolipids for which glucosylceramide is the precursor and consequently decrease their accumulation in those disorders where their degradation is impaired; this is the basis of SRT.

N-butyl-DNJ (**3**), known under the commercial name of Zavesca® (Miglustat, Actelion-Janssen-Cilag International) has been licensed in Europe (2002) and USA (2003) for treatment of non-neuronopathic GD type 1. Unfortunately, *N*-butyl-DNJ (**3**) did not have any appreciable effect on the neurological manifestations of neuronopathic GD type [51]. Clinical trials of SRT with *N*-butyl-DNJ (**3**) for other glycosphingolipidoses, such as Fabry, Tay–Sachs, and Sandhoff diseases (SDs), did not show clinical benefit and have been abandoned. Conversely, *N*-butyl-DNJ (**3**) has been approved by EMA in 2009 for the treatment of Niemann–Pick type C (NPC) disease [52, 53], for which it was proven to cross the BBB and slow the progression of neurological symptoms [54, 55]. Several other compounds including some iminosugars are being evaluated in preclinical trials for Fabry (*N*-butyl-deoxygalactonojirimycin, **4**) disease and MPS III. Finally, SRT can be combined with other therapies in view of a more personalized care program for patients [56].

9.2.3 Pharmacological Chaperone Therapy (PCT)

Although among enzymologists it was known that simple sugars protected glycosidases during their extraction from tissues, and it was observed that sometimes iminosugars activated rather than inhibited glycosidases *in vitro* assays, suggesting some stabilization [57]. It was only in 1999 that the concept of "active site-specific chaperone" (ASSC) was introduced [58].

An ASSC, later referred to as PC, is a small molecule able to bind the misfolded enzyme in the endoplasmic reticulum (ER) and, by serving as a folding template, prevent its premature degradation by endoplasmic reticulum-associated degradation (ERAD) pathway. The PC stabilization favors the trafficking of the properly folded mutant enzyme to the Golgi apparatus for maturation, and the final transport to the lysosome. Once the complex enzyme-PC reaches the lysosome, the PC is replaced by the natural substrate (present in high concentration due to the pathology), which starts to be hydrolyzed, thus resulting in a rescue of the enzyme activity [59–62].

Albeit it may seem contradictory, most PCs have also been competitive inhibitors of their target enzyme [63] and were essentially identified among sugar mimetics [64].

Indeed, in their seminal work Fan and collaborators showed that sub-inhibitory concentrations of the potent α-galactosidase inhibitor deoxygalactonojirimycin (DGJ, **5**, Figure 9.3) increased the residual activity in lymphoblasts from FD patients, by seven- to eight-fold over five days. Surprisingly, even infusions of simple D-galactose, a weak inhibitor of the α-galactosidase, led to clinical improvement in a Fabry patient with residual enzyme activity, supporting the use of chaperone-mediated therapy for late-onset LSDs [65]. DGJ (**5**) was considered a better candidate as a practicable therapeutic agent because of its higher affinity than D-galactose toward mutant lysosomal α-galactosidase [66]. Nevertheless, almost 20 years of further studies were necessary for DGJ (**5**) to pass Phase III clinical trial and be approved and commercialized as Galfold™ (Migalastat, Amicus Therapeutics Inc.) [67] by EMA (2016) [68] and subsequently by FDA (2018) [69].

To date, Galafold is the only approved PC drug for the treatment of a LSD that is FD. For GD, caused by mutated enzyme acid-β-glucosidase (glucocerebrosidase or GCase), many efforts have been made to develop a PC, especially with the natural iminosugar isofagomine (IFG (**6**), Figure 9.3). Unfortunately, **6**, although appearing very promising in stabilizing mutant GCase [70], did not succeed in significantly reducing the accumulation of glucosylceramide [71, 72].

In general, major difficulties in the translation to clinic are due to the identification of optimal dosing and regimen of PC-based drug administration, which aim to maximize the enhancing effect (leading to substrate turnover) while avoiding the inhibitory effect [22, 73, 74].

This goal can be achieved through different options, the first being an administration of the PC at alternative intervals (also called "on-off administration"), which results in greater substrate clearance with respect to daily administration.

A more original and less investigated alternative is the development of pH-responsive PCs able to change their affinity for the mutated enzyme from the ER (pH = 7) to the lysosome (pH = 5). For example, Ortiz Mellet and coworkers recently reported that the insertion of an acid-labile moiety onto the iminosugar scaffold allows the PC to switch from amphiphilic (higher affinity for the mutated enzyme) to hydrophilic (low affinity for the mutated enzyme) in this biologically useful pH window. Therefore, the PC stabilizes the enzyme in the ER, while it experiences self-inactivation once it reaches the lysosome, not interfering with substrate processing [75].

Alternatively, it is possible to use PCs directed to a lysosomal enzyme domain, which does not include the active site but has a crucial role in the stabilization of the protein (allosteric sites). In particular, in the last few years, the combination of experimental multiple solvent crystal structure methods and computational fragment mapping allowed to identify previously unknown binding pockets for several lysosomal enzymes. These studies led to the development of noncompetitive inhibitory PCs (which do not bind to the active site, offering an interesting alternative to ASSC) or, even better, of non-inhibitory PCs able to guarantee the re-folding function, without affecting the enzyme hydrolytic action. While the former usually belong to the family of L-iminosugars [76], the latter are not sugar mimetic compounds and are therefore out of the scope of this chapter [77]. However, if on one hand, the non-inhibitory PCs limit the risk of adverse enzymatic inhibition, their identification is extremely challenging and can be pursued only through a massive screening-based approach.

Due to the several available options for the design and the use of PCs, as well as the early stage understanding of their mechanism of action, it is not surprising that most of the studies on iminosugars for the treatment of LSDs reported in this chapter focused on the development of PCs. In particular, the structure of the most promising compounds, together with their development phases (*in vivo* studies, *in vitro* studies, clinical trials, approval) [78, 79] for selected MPS, sphingolipidoses, glycogen storage disorders, and glycoproteinoses have been discussed.

9.2.4 Combined ERT/PC Therapy

For those LSDs for which the ERT is available, low half-life and uneven distribution of the recombinant enzyme remain inherent disadvantages of this symptomatic treatment and are responsible for frequent patients' hospitalization and high governmental costs. A promising solution that is being explored is the coadministration of ERT with PCs, which is able to increase the lifetime of the infused enzyme, resulting in improved rate of substrate degradation and consequently ameliorating the treatment efficacy. For example, significant increase in α-galactosidase activity in the blood and the skin has been observed in Fabry patients orally treated with Galafold (DGJ, **5**) before the ERT infusion [21, 23].

9.3 Mucopolysaccharidoses

The MPS are genetic disorders caused by the deficiency of 1 of 11 lysosomal enzymes involved in the metabolism of glycosaminoglycans (GAGs) (Table 9.1). Enzyme deficiencies may lead to the accumulation of the GAGs heparan sulfate (HS), dermatan sulfate (DS), keratan sulfate (KS), chondroitin sulfate (CS), or hyaluronan within the lysosomes. The clinical phenotype of MPS varies and is characterized by progressive multisystemic involvement affecting the brain, eye, ear, upper and lower airways, liver, spleen, heart, bone, cartilage, and joints [80]. Currently, intravenous ERT is the standard of care for non-neurological manifestations of MPS I, MPS II, MPS IVA, MPS VI, and MPS VII patients (Table 9.2). ERTs have been shown to be

Table 9.2 Classification of the mucopolysaccharidoses (MPS) and available ERT.

Disorder	Enzyme deficiency	ERT
MPS I (Hurler, Hurler–Scheie, and Scheie syndrome; X-linked disease)	α-l-iduronidase	**Laronidase** (Aldurazyme®; Genzyme Europe B.V., Gooimeer 10, NL-1411 DD Naarden, The Netherlands), available since 2003
MPS II (Hunter syndrome)	Iduronate-2-sulfatase (IDS)	**Recombinant human idursulphase** (Elaprase®; Shire Human Genetic Therapies, Inc., Cambridge, MA, USA), available since 2006
MPS IVA (Morquio A syndrome)	*N*-acetyl-glucosamine-6-sulfatase sulfatase (GALNS)	**Elosulphase alpha** (Vimizim™; BioMarin Pharmaceutical, Inc., Novato, CA, USA), available since 2014
MPS VI (Maroteaux–Lamy syndrome)	Arylsulfatase B	**Galsulphase** (Naglazyme®; BioMarin Pharmaceutical, Inc., Novato, CA, USA), available since 2005
MPS VII (Sly syndrome)	β-glucuronidase	**Vestronidase alfa** (Mepsevii™; Ultragenyx Pharmaceutical Inc., Novato, CA, USA), approved in 2018

effective in reducing urinary GAGs and liver and spleen volume for attenuate and severe phenotypes [81]. Although intravenous ERT is an effective treatment, the amelioration of CNS with ERT is limited by the BBB. HSCT is primarily used to treat CNS manifestations in MPS I Hurler [82], while this effect is less clear for the other MPS disorders [83].

Promising new therapeutic strategies, including intrathecal/intracerebroventricular injection of recombinant enzyme, the gene therapy, and nanotechnology as enzyme and nucleic acid delivery systems have been developed and are being tested. *In vivo* tests and clinical trials are in progress to determine the safety and efficacy of these strategies in MPS patients [84–86].

Recently, a PC was described as a potential treatment alternative for MPS. In the case of MPS II, a sulfated disaccharide derived from heparin **7** (**D2S0,** Δ-unsaturated 2-sulfouronic acid-*N*-sulfoglucosamine), which mimics the structure of the natural substrate of IDS, was described as a potential PC for this enzyme (Table 9.3). The compound was a competitive inhibitor (IC_{50} of 30.1 μM) and increased the thermal stability of human IDS *in vitro* and improved the function of intracellular IDS between 1.6- and 39.6-fold in a mutation-dependent manner in patient fibroblasts and HEK293T cells expressing mutated IDS without a significant reduction in GAG levels [87]. MPS III is characterized by four types: MPS IIIA, IIIB, IIIC, and IIID. For MPS IIIB, α- (**8**), and β- (**9**) *N*-acetyl-D-glucosaminidases (NAGLU), a weak competitive inhibitor of NAGLU, demonstrated a significant PC activity on mutant NAGLU (Table 9.3) [88]. For MPS IIIC, the use of glucosamine **10**, a competitive inhibitor of the enzyme, allowed for an increase of between 1.2- and 2.5-fold in the activity of glucosaminide *N*-acetyltransferase in eight out of nine tested mutations (Table 9.3) [89]. For MPS IVA, three GALNS PCs, bromocriptine, ezetimibe, and pranlukast (noncarbohydrate-based compounds), were identified by molecular docking-based virtual screening [90, 91].

The identification of inhibitors of MPS recombinant enzymes represents a starting point for the development of enzyme stabilizers not only for PCT but also for combined PC/ER therapy. In this context, Cardona and coworkers investigated the multivalent effect. This phenomenon is defined as the increase in the biological response obtained with compounds that possess more than one bioactive unit linked to a common scaffold, compared to the sum of the contributions given by the individual bioactive molecules [92]. The multivalent effect can be quantified by determining the affinity enhancement per bioactive unit (rp/*n*), namely by dividing the relative potency (rp) by valency (*n*). Dimers and trimers of DAB-1 (1,4- dideoxy-1,4-imino-D-arabinitol) (compounds **11** and **12**) [93] and other multimeric iminosugars such as the nonavalent compounds **13** and **14** [94], and gold glyconanoparticles (AuGNPs) based on the same bioactive unit (**15**) were identified as excellent selective inhibitors of GALNS [95], together with simple AuGNPs decorated with monosaccharides and/or sulfate-ended ligands (**16**, unpublished data). The best results are reported in the Table 9.4.

Table 9.3 Pharmacological chaperones for MPSs.

Disorder	PC	IC_{50} or K*i*	Mutation	Max. activity enhancement	References
MPS II	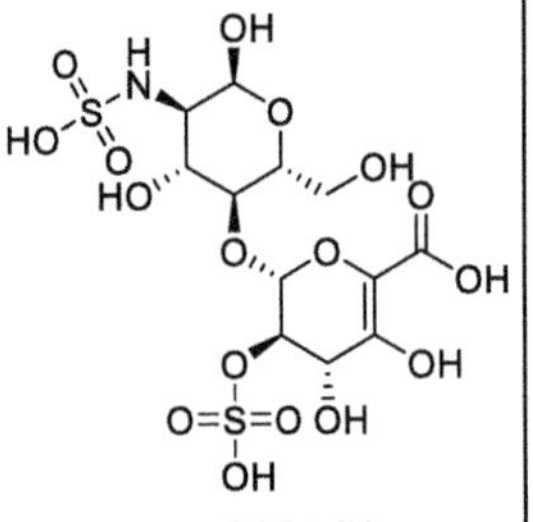 D2S0 (7)	$IC_{50} = 30.1\,\mu M$	*Patient fibroblasts*		[87]
			P231L/ P231L	1.7-fold at 0.1 μM	
			HEK293T cells		
			N63D	1.6-fold at 10 μM	
			L67P	3-fold at 10 μM	
			R88H	3-fold at 10 μM	
			Y108S	1.8-fold at 10 μM	
			P231L	39.6-fold at 10 μM	
MPS IIIB	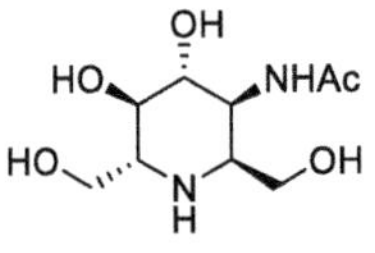 8	$IC_{50} = 67\,\mu M$ K*i* = 11.4 μM	*Patient fibroblasts*		[88]
			P358L/ P358L	1.8-fold at 100 μM	
			E153K/ E153K	1.1-fold at 10 and 100 μM	
			Y140C/ Y140C	1.1-fold at 10 μM	
	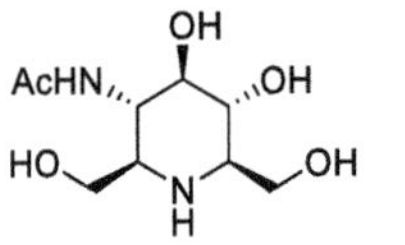 9	$IC_{50} = 374\,\mu M$ K*i* = 130 μM	*Patient fibroblasts*		
			P358L/ P358L	1.6-fold at 0.01 μM	
			E153K/ E153K	2.4-fold at 1 μM	
			Y140C/ Y140C	1.5-fold at 0.1 μM	
MPS IIIC	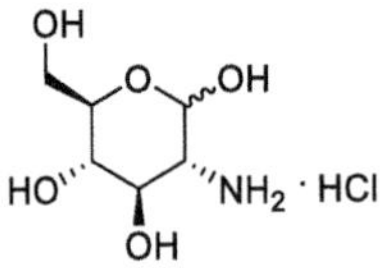 D-(+)-glucosamine hydrochloride (10)	K*i* = 0.28 mM	*Patient fibroblasts*		[89]
			L137P/ S518F	2.3-fold at 14 mM	
			S541L/ c.234+1G.A	2.1-fold at 14 mM	
			P283L/ R344C	2.5-fold at 7 mM	
			S518F/ S518F	2.5-fold at 7 mM	
			R344H/ R384X	1.2-fold at 7 mM	
			N273K/ N273K	2.1-fold at 7 mM	
			R344C/ R344C	1.7-fold at 7 mM	
			S518F/ S518F	1.5-fold at 7 mM	

Table 9.4 Inhibitors of IDS and GALNS (human enzymes).

Compounds	IDS, IC_{50} (rp/*n*)	GALNS, IC_{50} (rp; rp/*n*)	References
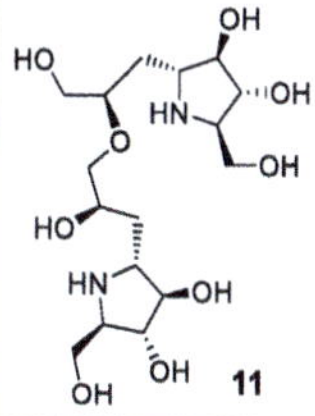 11	n.d.[a]	0.3 μM (rp; rp/*n* not calculated)	[93]
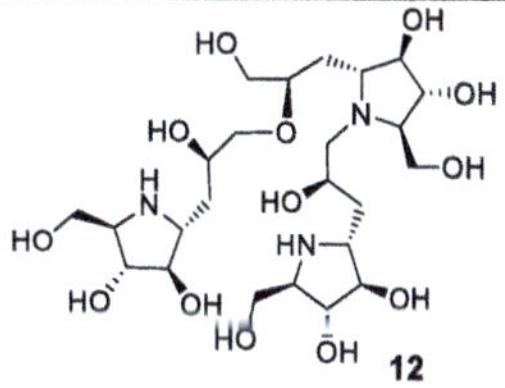 12	n.d.[a]	0.2 μM (rp; rp/*n* not calculated)	
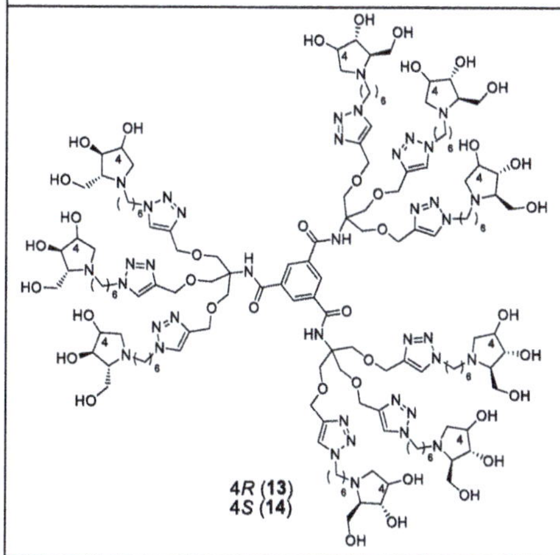 4*R* (13) 4*S* (14)	**13** = 140 μM (rp = 23; rp/*n* = 3) **14** = 31 μM (rp = 177; rp/*n* = 20)	**13** = 47 μM (rp = 83; rp/*n* = 9) **14** = 85 μM (rp = 59, rp/*n* = 7)	[94]
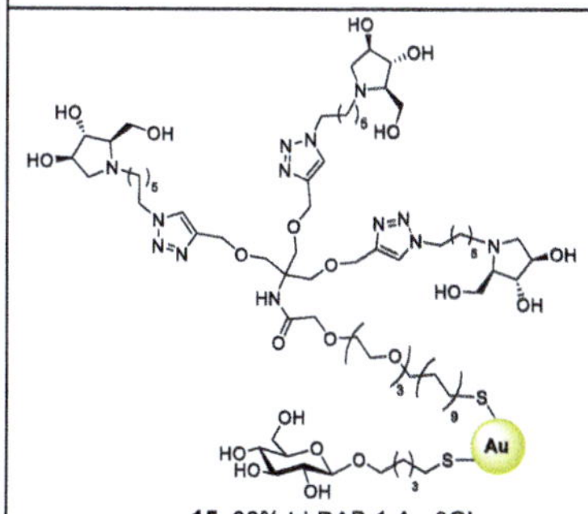 **15**: 30% tri-DAB-1-Au-βGlc	n.i.[b]	0.004 mg ml^{-1} of Au GNPs [corresponding to 0.52 μM of iminosugar] (rp = 7500)	[95]
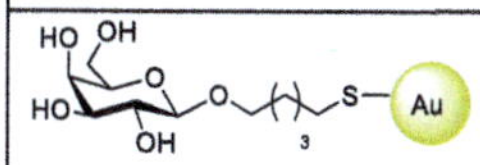 **16:** βGal-Au	n.d.[a]	9 μg ml^{-1} of Au GNPs [corresponding to 13 μM of sugar] (rp not calculated)	Data not published

a) n.d. = non determined. n.i. = no inhibition at 1 mM.

b) rp = IC_{50} of the monovalent reference/IC_{50} of the multivalent compound.

9.4 Sphingolipidoses

9.4.1 Fabry Disease

FD is a X-linked recessive LSD with generally more severe manifestations in males. This progressive and multisystemic disease is caused by mutations in *GLA* gene resulting in deficiency of the lysosomal enzyme α-galactosidase-A (α-Gal A), which leads to the accumulation of globotriaosylceramide (Gb3) and related glycosphingolipids in lysosomes in cells throughout the body. The prevalence of FD is estimated at 1 : 40 000–170 000 live births, with increased incidence in Nova Scotia (9.3 : 100 000 males) [1]. Phenotypically, FD can be distinguished into the more severe classical form, predominantly affecting juvenile males or adult females, and a nonclassical form, more prominent in males with residual enzyme activity [96]. In classically affected patients, early symptoms include angiokeratoma, anhidrosis, neuropathic pain, gastrointestinal symptoms, and microalbuminuria, while progressive renal failure, heart failure, and stroke generally occur later in life. In nonclassically affected patients, the disease presents a more variable and smoother course. In general, the shortened life expectancy of FD patients is related to the organ damage degree [97–99].

Since 2011, Fabry patients have benefited from two ERT products: Agalsidase alfa (Replagal®, Shire HGT) and Agalsidase beta (Fabrazyme®, Genzyme Inc.), which have the same amino acid sequence, but a different glycosylation. The latter is the only one approved by the FDA, while both are approved by the EMA. Although the initial clinical trials showed that ERT produced beneficial effects on neuropathic pain, cardiac mass, and kidney function, disease complications may still occur in patients treated with this therapy [100, 101].

In addition, due to the very short plasma half-life of the enzyme, its therapeutic activity is characterized by very short "on" peaks and long "off" activity valleys, which recently prompted the development of two second-generation ERTs for FD. These are recombinant α-Gal A derived from plants, namely moss-α-Gal A (Phase I clinical trials) and tobacco-α-Gal A modified with polyethylene glycol (PEG) chains (pegunigalsidase, Phase II clinical trials), essentially aimed at reducing the costs of the treatment. More interestingly, in the last 10 years, nanotechnology offered a series of platforms able to improve ERT administration, by addressing important issues such as the protection of the naked enzyme, the increase of enzymes' half-life in plasma, and a targeted delivery of the protein for improving cell internalization. Trimethyl-chitosan nanoparticles, cholesterol-dipalmitoylphosphatidylcholine nanoliposomes functionalized with RGD tripeptide moieties, and extracellular vesicles are only some of the exiting tools developed to improve the delivery of α-Gal A enzyme and recently reviewed by Schwartz and coworkers [102].

Regarding SRT for FD, glucosylceramide synthase (GCS) inhibitors have been developed to reduce the accumulation of precursors of Gb3 at the biosynthetic cascade, but not specifically the FD toxic substrate Gb3. To date, there are two products under clinical investigation: the ceramide mimetic Venglustat (Ibiglustat, Sanofi Genzyme) and the iminosugar Lucerastat (*N*-butyl-deoxygalactonojirimycin,

4, Figure 9.3, Idorsia Pharmaceuticals, Switzerland). The latter is the D-galactose configured analog of Miglustat (**3**, Figure 9.3), the first drug marketed as SRT for GD. Lucerastat has been proven to avoid the accumulation of Gb3 in tissue and is currently under Phase 3 of clinical trials, mostly centered on patients with neuropathic pain and gastrointestinal symptomatology [103, 104].

PC studies for FD mostly used DGJ iminosugar **5** (Table 9.5), now known as Galafold™ (Migalastat, Amicus Therapeutics), which is a potent inhibitor of α-Gal A, but when administered at sub-inhibitory doses increases enzymatic activity for some *GLA* gene mutations [58]. Being an analog of the terminal galactose of Gb3, **5** binds and stabilizes wild-type and mutant forms of α-Gal A [113]. It was also proven to alleviate the Gb3 storage in mouse kidneys [114] and in recent years, the several clinical trials conducted to validate **5** as PC for FD confirmed its beneficial effects on organ function, Gb3 clearance, and α-Gal A activity [25, 115, 116]. Once proven to be safe and well tolerated, Galafold was approved in USA, EU, Israel, Australia, and Canada as an oral treatment for Fabry patients with amenable mutations. Patients' eligibility for treatment with Galafold is determined using an *in vitro* enzyme activity assay and within the eligible group (*GLA* mutations leading to the loss of gross structural protein domains and loss of α-Gal A expression are not amenable), an increase ranging from 1.2 up to 30.4-fold was observed in the α-Gal A activity [105].

Some structural modifications of DGJ (**5**) resulted in an improved PC activity as reported by Ortiz Mellet and collaborators. Indeed, DGJ thioureas **17** and **18** resulted in less potent inhibitors than **5** (Table 9.5) but were able to equalize its chaperoning activity (threefold increase of α-Gal A activity) at lower concentration (3 μM vs. 20 μM) in skin fibroblasts derived from Fabry patients homozygous for the R301G mutation. In addition, a maximal enhancement of fivefold was obtained by increasing to 30 μM the concentration of **17** and **18**, which did not show cytotoxicity up to 500 μM [106]. Later, among a new family of 1-deoxygalactonojirimycin-aryl thioureas (DGJ-ArTs), compounds **19** and **20** exhibited a significantly higher chaperoning efficiency than **5** at 30 μM in SV40-mediated transformed cell lines from normal and Q279E FD fibroblasts, as well as the ability to reduce the accumulation of Gb3 in FD cells. In addition, PCs **19** and **20** act in a synergistic manner with the proteostasis regulator 4-phenylbutyric acid (Table 9.5) [108].

Among different pyrrolidine and piperidine iminosugars isolated from the roots of *Adenophora triphylla* and fully characterized, the pyrrolidine iminosugar 2,5-dideoxy-2,5-imino-D-altritol (DIA, **21**) showed strong competitive inhibition toward α-Gal A and was proven to stabilize the enzyme *in vitro* assay of thermal denaturation (α-Gal A activity was lost within 60 minutes heating at 48 °C, while it remained over 70% in the presence of 100 μM **21**). In addition, the treatment with **21** for three days dose-dependently increased intracellular α-Gal A activity, with a maximal increase of 9.6-fold at 500 μM (Table 9.5) [109].

The only example of a multivalent PC for FD was reported in 2020 by Carmona and coworkers. In particular, the nonavalent pyrrolidine iminosugar **22** (Table 9.5), a potent competitive inhibitor of α-Gal A, increased by 5.2-fold the activity of the misfolded enzyme in R301G patient cells at 2.5 μM, demonstrating the potential use of multivalent ligands in treatment of FD [110].

Table 9.5 PCs and combined ERT/PC for Fabry disease.

PC	IC_{50} or K*i*	Mutation	Max. activity enhancement	References
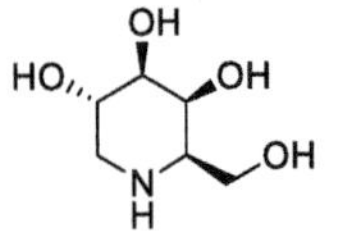 Deoxygalactonojirimycin (DGJ) (**5**) **Migalastat, Galafold™**	$IC_{50} = 60\,nM$ (pH = 5) $IC_{50} = 10\,nM$ (pH = 7) $Ki = 15.1\,nM$	*FD patients (Ph3 clinical trials)*		[105]
		1.2 up to 30.4-fold		
		Patient fibroblasts		[106, 107]
		R301G	3-fold at 20 μM	
		R301Q	10.8-fold at 10 μM 20-fold at 100 μM	
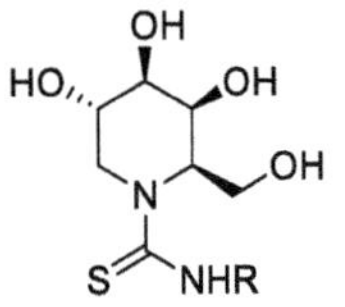 R = Ph, **17** R = *n*-octyl, **18**	IC_{50} (**17**) = 4.5 μM (pH = 5) IC_{50} (**17**) = 0.2 μM (pH=7) IC_{50} (**18**) = 37 μM (pH = 5) IC_{50} (**18**) = 5 μM (pH = 7)	R301G	3-fold at 3 μM 5-fold at 30 μM	[106]
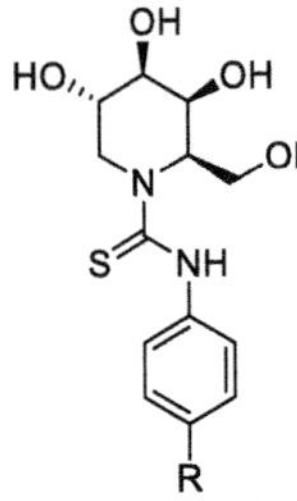 R = F, DGJ-*p*F PhT, **19** R = OMe, DGJ-*p*MeO PhT, **20**	IC_{50} (**19**) = 0.34 μM (pH=5) IC_{50} (**19**) = 0.043 μM (pH = 7) IC_{50} (**20**) = 0.074 μM (pH = 5) IC_{50} (**20**) = 0.016 μM (pH = 7)	Q279E	3-fold at 3 μM 4-fold at 30 μM	[108]
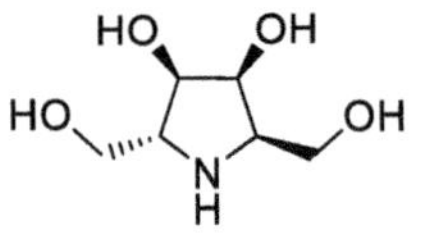 2,5-dideoxy-2,5-imino-D-altritol (DIA) (**21**)	$IC_{50} = 0.69\,μM$	R301Q lymphoblasts	9.6-fold at 500 μM	[109]
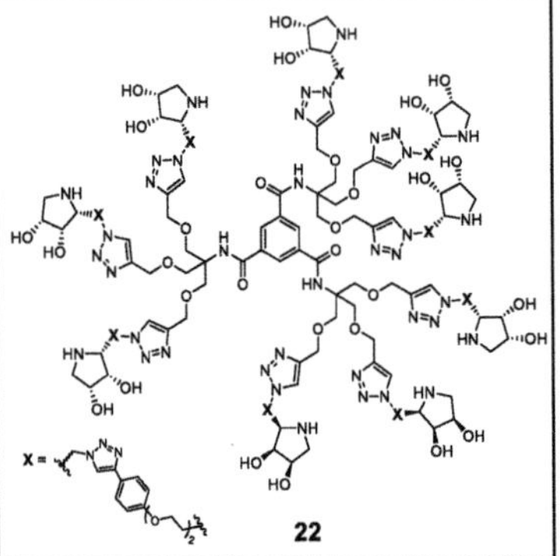 **22**	$IC_{50} = 1.2\,μM$ $Ki = 0.20\,μM$	R301G	5.2-fold at 2.5 μM	[110]

Table 9.5 (Continued)

PC	IC_{50} or K_i	Mutation	Max. activity enhancement	References
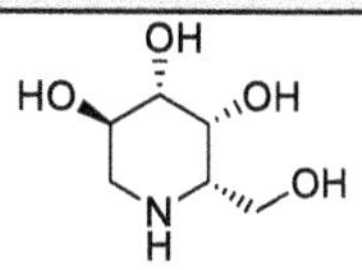 L-deoxygalactonojirimycin (L-DGJ) (**23**)	$Ki = 38.5\,\mu M$	R301Q	10.8-fold at10 mM	[107]
PC/ERT	**IC_{50} or K_i**	**Mutation**	**Max. activity enhancement**	**References**
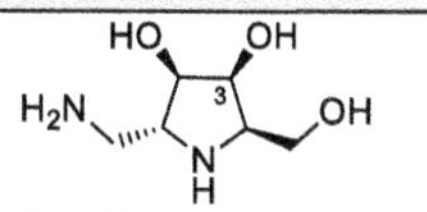 3-*epi*-1-aminodeoxy-2,5-dihydroxymethyl 3,4 dihydroxypyrrolidine (3-*epi*-ADMDP) (**24**)	**IC_{50} = 0.67 μM (pH = 5)** $IC_{50} = 0.053\,\mu M$ (pH = 7)	N215S	9-fold at 50 μM	[111]
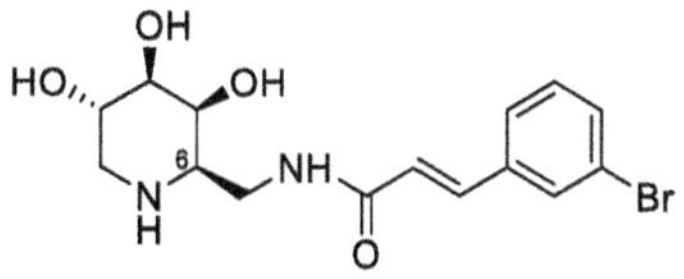 **25**	$Ki = 3.5\,\mu M$	W162X	12-fold at 100 μM	[112]

Regarding PCs not specific for the enzyme active site, Fleet and coworkers reported in 2011 that the enantiomer of **5**, L-DGJ (**23**, Table 9.5), was a noncompetitive inhibitor of α-Gal A, about 1000-fold weaker than **5** (competitive). Compound **23** still behaved as chaperone with a 10.8-fold activity enhancement in Fabry R301Q fibroblasts at 10 mM, which was like that observed with a 1000-fold lower concentration of **5** (10 μM). When administered simultaneously, the mixture of enantiomers clearly showed dose-response synergistic effects, enhancing α-Gal A up to 14-fold, thus suggesting that the concomitant binding to two different sites might further stabilize the enzyme conformation [107].

More recently, thanks to *in silico* docking, an allosteric hot spot for ligand binding was identified, and 2,6-dithiopurine, which preferentially bonds this site, was demonstrated to stabilize recombinant human α-Gal A (rh-α-Gal A) *in vitro* and to rescue the A230T mutant α-Gal A that is not responsive to **5** in a cell-based assay [117].

Regarding the ERT/PC therapy, co-formulation of α-Gal A and DGJ (**5**) for treatment of FD was patented in 2014 by Khanna et al. [118].

As an example of the utility of natural product-inspired combinatorial chemistry in the search for stabilizers of rh-α-Gal A, Cheng and coworkers identified two lead compounds belonging to pyrrolidine and piperidine iminosugar families, respectively. Indeed, coadministration of 50 μM concentration of 3-*epi*-ADMDP (**24**, Table 9.5) with rh-α-Gal A (1 nM) in the Fabry N215S cell line was found to enhance overall α-Gal A activity of approximately ninefold, while α-Gal A alone (ERT) or **24** alone (PC) are able to only enhance overall α-Gal A activity twofold [111]. More recently, structural modifications and bioevaluations performed on a series of C-2 and C-6 derived (3*S*,4*S*,5*S*)-trihydroxylated piperidines allowed to identify derivative **25** (Table 9.5), which showed the best improvement of rh-α-Gal A (12-fold increase at 100 μM) of this co-treatment study in W162X patient cell line, without any detectable cytotoxicity toward normal lymphocytes, or inhibition of other human glycosidases [112].

9.4.2 Gaucher Disease

GD is the most common LSD with an incidence of two cases per 100 000 individuals, which dramatically increases in Ashkenazi Jews (100 per 100 000 individuals), owing to the so-called founder effect [6]. GD is caused by mutations in the *GBA* gene (chromosome: 1q21–22), which encodes for the lysosomal enzyme acid-β-glucosidase (glucocerebrosidase or GCase). GCase catalyzes the hydrolysis of glucosylceramide (GlcCer) to glucose and ceramide in the lysosomes [119].

More than 350 mutations of *GBA* have been reported for GD patients [120], the N370S and L144P missense mutations being the most frequent ones. Three clinical types of GD are distinguished on the basis of the age onset and the severity of the associated symptoms. Type 1, the most common form, causes liver and spleen enlargement, bone pain and fractures (broken bones), and, sometimes, lung and kidney problems. It does not affect the brain and can occur at any age. Type 2, which causes severe brain damage, appears in infants. Most children who have it die by age 2: this is the rarest and most severe form. In type 3, there may be liver and spleen enlargement, the brain is gradually affected, and it usually starts in childhood or adolescence. Recently, a pathological loop between GD patients and carriers and Parkinson's disease emerged. Although the connection between *GBA* mutations and Parkinson's development is far to be fully understood, therapeutic interventions aimed at enhancing GCase activity to treat Parkinson's disease are already under investigation [121].

ERT is effective only for type I GD (the non-neuronopathic phenotype) and there are three drugs available to date: Cerezyme® (imiglucerase, Sanofi Genzyme, from 1994), VPRIV (velaglucerase alfa, Shire Human Genetic Therapies, from 2010), and Elelyso® (aliglucerase alfa, Pfizer, from 2012). Imiglucerase is a modified form of human GCase, produced by recombinant DNA technology using a mammalian CHO cell culture. Velaglucerase alfa has the native-enzyme sequence produced in a human cell line, while taliglucerase alfa is plant-cell-derived and produced in an inexpensive platform [122].

Regarding SRT, the first drug developed was the iminosugar-based drug Zavesca™ (Miglustat, *N*-butyl DNJ, **3**), which is able to reversibly inhibit GCS and consequently reduce the production of GlcCer, representing an appropriate choice for type 1 GD patients.

Since **3** mechanism was first demonstrated in 1994, its safety and efficacy have been extensively investigated and non-negligible adverse effects have been unfortunately identified, especially gastrointestinal disturbances and tremors. In addition, Zavesca™ is contraindicated in pregnancy, in anticipation of pregnancy and breastfeeding, because maternal death and infertility were observed in mouse models [97].

Later, the more selective GCS Eliglustat™ (Cerdelga, Sanofi Genzyme) was introduced and approved both by FDA (2014) and EMA (2015) as a first-line treatment for adults with Type 1 GD [123]. Unfortunately, neither Zavesca™ nor Eliglustat™ can cross the BBB and cannot be applied in the treatment of neuronopathic GD.

The compound that reached the most advanced clinical trial as PC for GD is isofagomine (IFG, **6**, Figure 9.3), which was unfortunately stopped at Phase II trials because it was not effective in reducing the accumulation of GlcCer in GD patients. Although being a strong competitive inhibitor of human lysosomal GCase, ($Ki = 0.016\,\mu M$; $IC_{50} = 0.06\,\mu M$) [124] **6** was found to increase mutant GCase activity up to threefold at 30 μM in fibroblasts with the N370S missense mutation, associated to Type 1 GD (Table 9.6) [125]. IFG failure in clinical trials was attributed to its high hydrophilicity, which might hamper an efficient transport to the cells. For this reason, a series of alkylated iminosugars were later developed, among which the 6-nonyl IFG (**26**) [126, 127], the nonyl-deoxynojirimycin (NN-DNJ, **27**) [128, 129] and the α-1-C-nonyl-DIX (**28**) [130] resulted in the most promising PCs, being able to enhance GCase activity in N370S GD fibroblasts, ranging from 1.5-fold at 3 nM (**26**) to 2-fold at 10 μM (**27**) (Table 9.6).

Moreover, bicyclic nojirimycin (NJ) analogs with structure of sp^2 iminosugars were found to behave as very selective, competitive inhibitors of GCase, and compounds **29–31** also displayed a better chaperoning activity than the parent NN-DNJ (**27**) toward some mutations involved in neuronopathic GD forms. In particular, they resulted in increases in GCase activity of 60–75% (0.3–1 μM) and 125–175% (3–30 μM) in fibroblasts bearing the G202R/L444P mutation and of 30–40% (0.3–1 μM) and 40–120% (3–30 μM) in fibroblasts bearing the F213I/L444P mutation, while **27** showed no effect in these two cell lines (Table 9.6) [131]. More recently, the same group reported several DNJ-based sp^2-iminosugars incorporating an orthoester fragment, which are able to switch from hydrophobic to hydrophilic in the pH 7 to pH 5 window, having a dramatic effect on the enzyme binding affinity, and thus maximizing the chaperone over the inhibitory behavior [75]. pH-sensitive compounds **32, 33,** and **34** showed to be better GCase ligands (1.3 to 200-fold) than Ambroxol ($IC_{50} = 41.5\,\mu M$), a non-glycomimetic PC under clinical trial for GD [138], at the neutral pH (ER), while at acidic pH (lysosome), the product **35**, obtained from the hydrolysis of **32–34**, was a threefold weaker ligand than Ambroxol. More interestingly, compound **32** was able to increase GCase activity by sixfold in N188S/G193W GD fibroblasts, while a modest enhancement was obtained for the N370S mutation (1.5-fold).

Among pyrrolidine iminosugars, the *C*-tridecyl derivative of DAB-1 (1,4-dideoxy-1,4-imino-D-arabinitol) (**36**) showed the same GCase activity enhancement as IFG (**6**) in GD fibroblasts bearing the N370S mutation, but at a 10 times lower concentration (0.5 μM) [132].

Table 9.6 PCs for Gaucher disease.

PC	IC_{50} or K*i*	Mutation	Max. activity enhancement	References
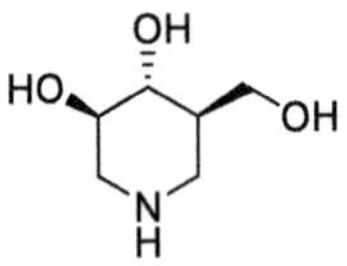 D-IFG (**6**)	$IC_{50} = 0.06\,\mu M$ $Ki = 0.016\,\mu M$ ($Ki = 8.4\,nM$)	N370S	3-fold at 30 μM 1.6-fold at 10 μM	[124, 125]
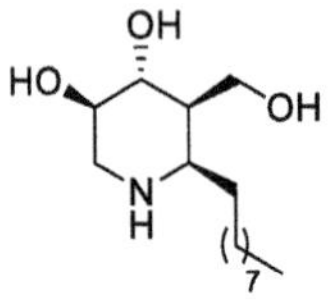 6-nonyl IFG (**26**)	$IC_{50} = 0.6\,nM$	N370S	1.5-fold at 3 nM	[126, 127]
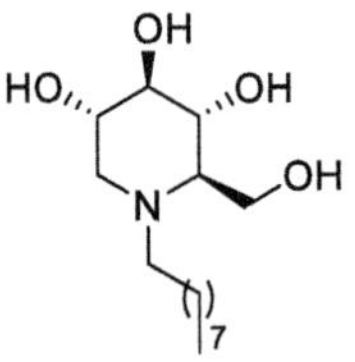 nonyl-deoxynojirimycin (NN-DNJ, **27**)	$IC_{50} = 1\,\mu M$	N370S	2-fold at 10 μM	[128, 129]
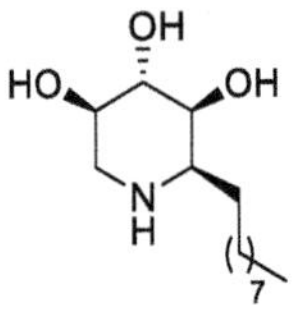 α-1-C-nonyl-DIX (**28**)	$IC_{50} = 6.8\,nM$	N370S	1.8-fold at 10 nM	[130]
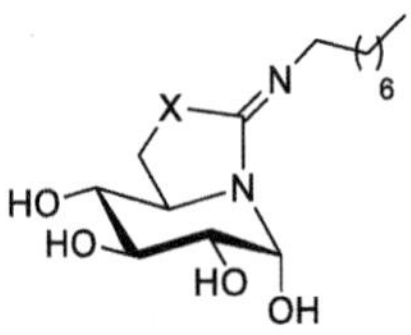 X=O, NOI-NJ, **29** X= S, 6S-NOI-NJ, **30** X=NH, 6N-NOI-NJ, **31**	Ki (**29**) = 5.6 μM Ki (**30**) = 3.5 μM Ki (**31**) = 4.0 μM	N370S	60% at 0.3–1 μM 40–165% at 3–30 μM	[131]
		G202R/L444P	60–75% at 0.3–1 μM 125–155% at 3–30 μM	
		F213I/L444P	30-40% at 0.3–1 μM 40–120% at 3–30 μM	

Table 9.6 (Continued)

PC	IC_{50} or K_i	Mutation	Max. activity enhancement	References
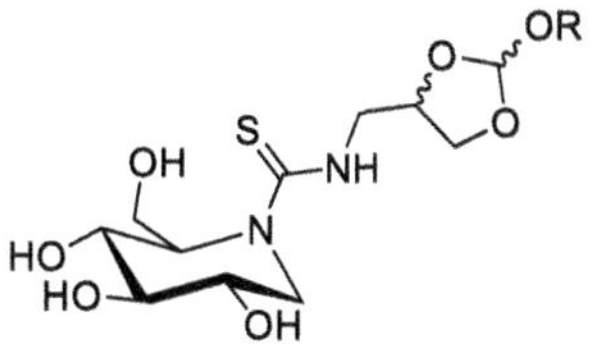 R = *n*-butyl, **32** R = *n*-octyl, **33** R = *n*-dodecyl, **34**	IC_{50} (**32**) = 0.20 μM IC_{50} (**33**) = 0.15 μM IC_{50} (**34**) = 32.6 μM	N188S/ G193 (**32**)	6-fold at 50 μM (**32**)	[75]
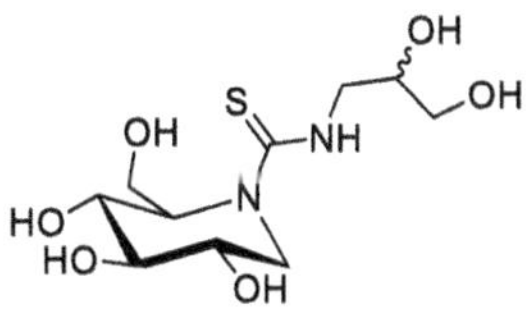 **35** (hydrolysis product of **31–33**)	IC_{50} > 1000 μM	-	-	
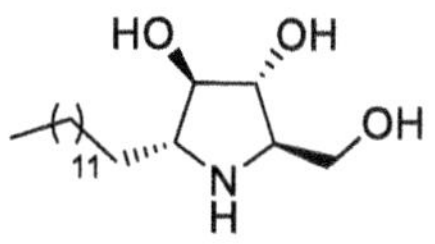 **36**	IC_{50} = 0.77 μM	N370S	1.5-fold at 0.5 μM	[132]
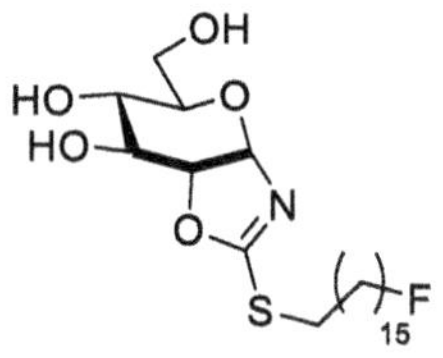 **37**	IC_{50} = 3.9 μM	N370S	62% at 30 μM	[133]
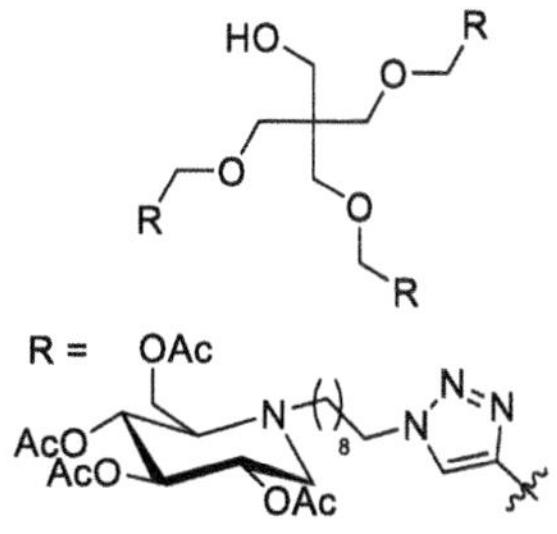 **38**	>5 mM	N370S	3-fold at 1 μM	[134]

(Continued)

Table 9.6 (Continued)

PC	IC_{50} or Ki	Mutation	Max. activity enhancement	References
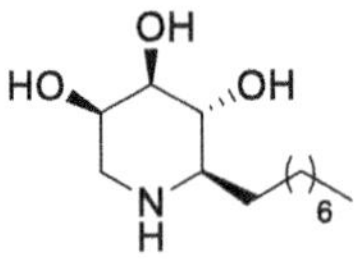 **39**	$IC_{50} = 29.3\,\mu M$	L444P/ L444P	1.8-fold at 100 μM	[135, 136]
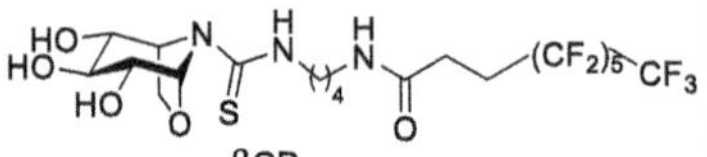 βCD **40**	$Ki = 1.4\,\mu M$	L444P/ L444P	2.8-fold at 20 μM	[137]
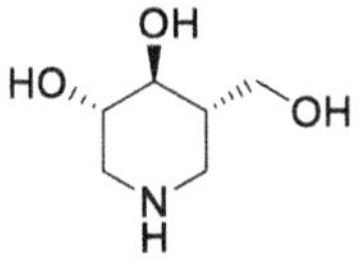 L-IFG (**41**)	$Ki = 6.9\,\mu M^{a}$	N370S	1.6-fold at 500 μM	[124]
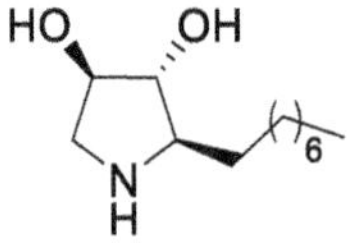 **42**	$IC_{50} = 0.78\,\mu M$ $Ki = 0.40\,\mu M^{a}$	N370S	2-fold at 3 μM	[76]
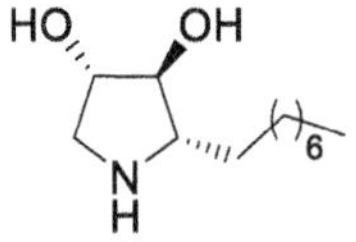 **43**	$IC_{50} = 59.6\,\mu M$ $Ki = 6.87\,\mu M$	N370S	2-fold at 300 μM	

a) Non-competitive inhibitor.

Apart from iminosugars, other carbohydrate-derived analogs have been studied as PCs for the treatment of GD. As a representative example, Díaz and coworkers reported on a series of pyranoid-type glycomimetics with a *cis*-1,2-fused glucopyranose-2-alkylsulfanyl-1,3-oxazoline structure [133].

The best results of the series were obtained with compound **37**, showing a GCase improvement of 62% at 30 μM in homozygous N370S mutated fibroblasts, which is superior to that observed for Ambroxol at the same concentration.

Finally, it should be noticed that Compain and coworkers contributed to this field with the only examples of multivalent PCs for GD reported to date, to the best of our knowledge, using the iminosugar 1-deoxynorijimicin (DNJ) as the bioactive unit. The trivalent acetyl-DNJ-derivative (**38**) provided a threefold increase of GCase residual activity at 1 μM in N370S GD fibroblasts, being more active than the corresponding deprotected analog, thus suggesting an improved permeability and cellular uptake [134].

Although L444P mutation is resistant to most PCs, the recently reported 2-octyl trihydroxypiperidine **39** showed a remarkable 80% activity rescue (1.8-fold GCase enhancement) in fibroblasts bearing this homozygous mutation [135, 136]. A higher enhancement toward this mutation was obtained only with a much more sophisticated system that involved a nortropane iminosugar functionalized with a terminal polyfluorinated fragment in form of β-cyclodextrin (βCD) complex. In particular, the fluorinated iminosugar βCD complex **40** showed 2.8-fold GCase enhancement at 20 μM [137].

Due to the lack of an approved PC, the incidence of GD, and the high number of *GBA* mutations involved in this pathology, there are a multitude of publications reporting on potential PCs for different GD mutations. Their comprehensive review is beyond the scope of this chapter; therefore, only selected examples are cited in this chapter.

Analogously to what already observed with the enantiomeric couple D-DGJ (**5**) and L-DGJ (**23**) for FD (Section 9.4.1), D-IFG (**6**) and L-IFG (**41**) are competitive and noncompetitive inhibitors of GCase, respectively, with **41** being a less potent inhibitor than **6** (Ki = 6.9 μM), but still able to increase GCase activity in the Gaucher N370S cell line by 1.6-fold at 500 μM [124]. Curiously, also hybrid analogs of α-1-C-nonyl-DIX (**28**), obtained by combining the iminosugar scaffold with triazolyl alkyl side chains by means of CuAAC click chemistry reactions, behaved as noncompetitive inhibitors of GCase and could enhance GCase activity up to twofold at 10 nM in GD fibroblasts bearing the homozygous G202R mutation [139].

More recently, further evidence of the impact of inhibitors chirality on their affinity with a target protein and of the efficacy of noncompetitive inhibitors was provided by the study on the C-octyl pyrrolidines **42** and their enantiomers **43.** While **43** is a modest competitive inhibitor of GCase (IC_{50} = 59.6 μM), **42** is a much more potent noncompetitive inhibitor (IC_{50} = 0.78 μM). In addition, both pyrrolidines were also able to enhance GCase residual activity in N370S homozygous Gaucher fibroblasts, with the noncompetitive inhibitor **42** having a chaperoning activity comparable to D-IFG (**6**) and NN-DNJ (**27**) [76].

Regarding the ERT/PC therapy, Murray and coworkers demonstrated that preincubation of Cerezyme (imiglucerase, Sanofi Genzyme) with D-IFG (**6**) significantly increased stability of the human recombinant enzyme to heat, neutral pH, and denaturing agents *in vitro*. Moreover, preincubation of Cerezyme with **6** prior to uptake by cultured cells resulted in increased intracellular GCase activity accompanied by an increase in enzyme protein, thus suggesting that this co-incubation before infusion might improve the effectiveness of ERT for Gaucher patients [140, 141].

9.4.3 Niemann–Pick

NP disorders are different disorders with distinct genetic origins. Types A and B NP disorders are caused by mutations in the gene encoding the lysosomal sphingomyelin (SM)-degrading enzyme acid sphingomyelinase (ASM). Common manifestations of both disease types are hepatosplenomegaly and appearance of cherry-red spots in the retina whereas neurodegeneration is only manifest in patients with

NPA. Type C NP disorder (NPC) is caused by mutations in the genes that encode lysosomal cholesterol-transport proteins NPC1 (95% of the cases) or NPC2. The most common symptoms of NPC include hepatosplenomegaly and neurologic deterioration with ataxia, motor pathologies, and horizontal saccadic eye movements (HSEMs) [142]. The treatment for NP disease was based on different drugs such as antiepileptics, anticholinergic, or antidepressants to alleviate symptoms, i.e. tremor, dystonia, or seizures. Miglustat (**3**, Zavesca), a small iminosugar molecule that reversibly inhibits glycosphingolipid synthesis, is currently available for NPC [143]. In the NPA and NPB types, current research focuses on hematopoietic cell transplantation and enzyme replacement [144].

9.4.4 GM1 Gangliosidosis and Morquio B (β-Gal)

Two lysosomal storage diseases, GM1 gangliosidosis (GM1) and Morquio B disease (MBD), are caused by sequence alterations in a single gene, *GLB1*. They result in functional deficits of acid β-galactosidase (β-Gal), an enzyme that cleaves terminal β-linked galactose residues from complex carbohydrates in the lysosomal compartment. Both diseases are inherited in an autosomal recessive manner. Depending on the mutations, degradation of one or the other of the β-galactosidase substrates is more or less impaired. If degradation of sphingolipidosis GM1-gangliosidosis is predominantly defective, the patients develop the symptomatology of GM1-gangliosidosis, while accumulation of KS is an indication for Morquio disease type B. GM1-gangliosidosis is considered a neurodegenerative disorder and MBD is characterized by marked skeletal abnormalities, corneal clouding, cardiac involvement, and increased urinary excretion of KS but no clinical signs of storage in neural tissues [145].

At present, only symptomatic and supportive therapies are available for patients with GM1 gangliosidosis and Morquio B. For GM1, only symptomatic treatment for some of the neurologic symptoms is available, which does not significantly alter the progression of the condition. For example, anticonvulsants may initially control seizures. Supportive treatments may include proper nutrition and hydration and keeping the affected individual's airway open [146]. For Morquio B, only physical therapy and surgical procedures, such as spinal fusion, may help with scoliosis and other bone and muscle issues [147]. Therapies relying on PCs may constitute a future option for the treatment of these lysosomal diseases. 1-Deoxygalactonojirimycin (DGJ, **5**, Figure 9.3) ($IC_{50} = 25\,\mu M$) was able to rescue the activity of mutant β-Gal in mouse fibroblasts with different mutations (from two- to seven-fold) after culture with 0.5 mM [148]. Several N-alkylated DGJ derivatives were synthesized to enhance the compound specificity and affinity to galactosidases, such as the as *N*-butyl-DGJ (NB-DGJ, **4** Figure 9.3 and Table 9.7) and *N*-nonyl-DGJ (NN-DGJ, **44**, Table 9.7) [148, 149]. A bicyclic DGJ derivative 6S-NBI-DGJ (5N,6S-N′-butyliminomethylidene)-6-thio-1-deoxygalactonojirimycin, **45**, Table 9.7), was evaluated as a novel PC for GLB1. This derivative inhibits human β-Gal with an IC_{50} of 32 μM, and significantly increases the thermostability of the enzyme. Treatment of GM1 patient fibroblasts with 20 and 80 μM **45** showed a significant improvement of GLB1 activity

Table 9.7 PCs for GM1 gangliosidosis and Morquio B disease.

PC	IC_{50} or K_i	Mutation	Max. Activity Enhancement	References
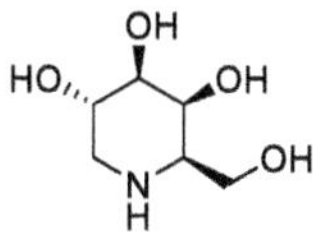 Deoxygalactonojirimycin (DGJ) (**5**)	$IC_{50} = 25\,\mu M$	*Mouse cell lines expressing human β-galactosidase*		[148]
		R201C	5.4-fold at 0.5 mM	
		I51T	2.2-fold at 0.5 mM	
		R201H	2.6-fold at 0.5 mM	
		R457Q	6.0-fold at 0.5 mM	
		W273L	1.8-fold at 0.5 mM	
		Y83H	1.7-fold at 0.5 mM	
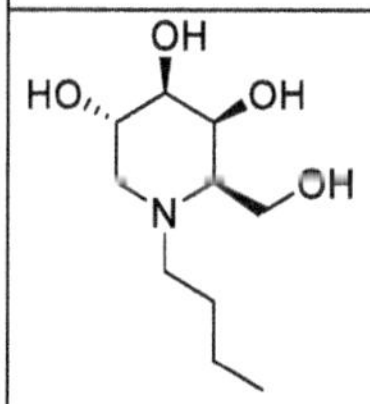 NB-DGJ (**4**)	$IC_{50} = 3.5\,\mu M$	*Mouse cell lines expressing human β-galactosidase*		
		R201C	4.8-fold at 0.5 mM	
		I51T	6.1-fold at 0.5 mM	
		R201H	2.1-fold at 0.5 mM	
		R457Q	5.4-fold at 0.5 mM	
		W273L	1.8-fold at 0.5 mM	
		Y83H	1.1-fold at 0.5 mM	
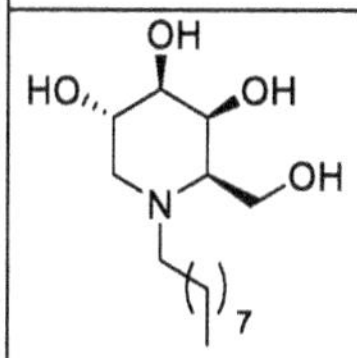 NN-DGJ (**44**)	$IC_{50} = 0.12\,\mu M$	*Patient fibroblasts*		[149]
		R148S/D332N	4.1-fold at 1.2 μM[a]	
			4-fold at 1.2 μM[b]	
		R148S/R482H	4.9-fold at 1.2 μM[a]	
			7.8-fold at 1.2 μM[b]	
		R201H/ IVS14-2A>G	7.3-fold at 1.2 μM[a]	
			13.8-fold at 1.2 μM[b]	
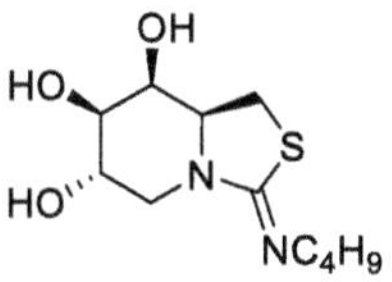 **45**	$IC_{50} = 32\,\mu M$	*Patient fibroblasts*		[150]
		I51T/Y316C	5.5-fold at 80 μM	
		I51T/R457Q	4.9-fold at 80 μM	
		R201C/ R201C	4.9-fold at 80 μM	
		COS7 cells		
		Y444C	2.8-fold at 80 μM	
		R201H	2.5-fold at 80 μM	
		R590H	2-fold at 80 μM	

(Continued)

Table 9.7 (Continued)

PC	IC_{50} or K*i*	Mutation	Max. Activity Enhancement	References
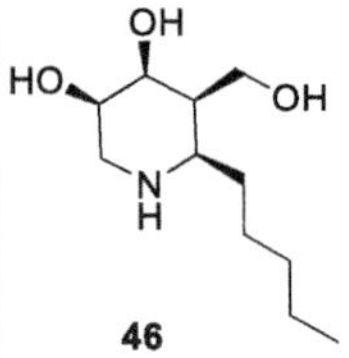 **46**	IC_{50} = 8 nM	*Patient fibroblasts*		[151]
		R201C/R201C	15-fold at 10 μM	
		R201C/H281Y	18-fold at 10 μM	
		Q255H/K578R	20-fold at 10 μM	
		H281Y/splicing	35-fold at 10 μM	
		R457Q/R457Q	7.3-fold at 10 μM	
		S191N/ R351Term	11-fold at 10 μM	
		W273L/ R482H	1.5-fold at 10 μM	
		W273L/ W509C	1.5-fold at 10 μM	
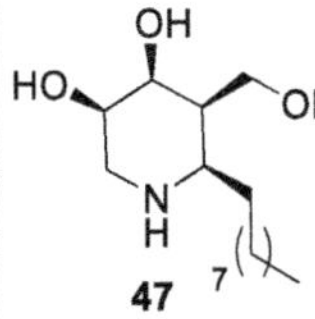 **47**	IC_{50} = 0.4 nM	*Patient fibroblasts*		[152]
		Half-maximal recovery of mutant β-gal activity at 0.01 μM in fibroblast of GM1-gangliosidosis patient		
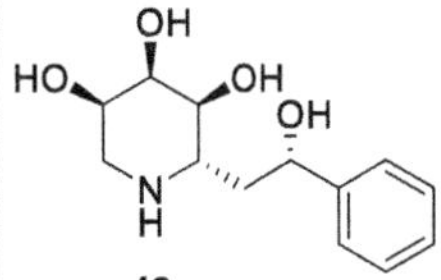 **48**	IC_{50} = 75 μM	*Patient fibroblasts*		[153]
		R201H/ IVS14-2A>G)	6.2-fold at 394 μM	
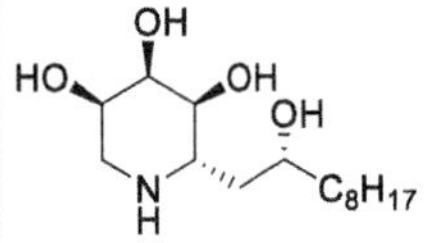 **49**	IC_{50} = 44 μM	*Patient fibroblasts*		
		R201H/ IVS14-2A>G)	2-fold at 100 μM	
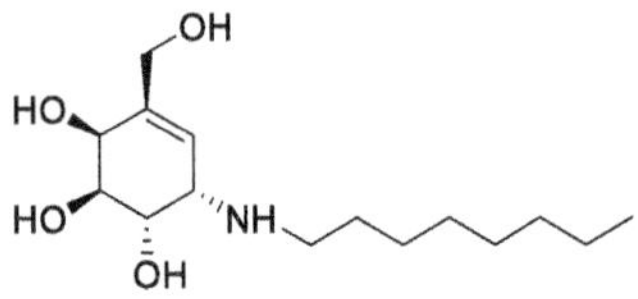 **50**	IC_{50} = 0.2 μM	*Mouse cell lines expressing human β-galactosidase*		[154]
		R201C/ R201C	5.1 at 0.2 μM	
		R201H/ R201H	4.50 at 0.2 μM	
		R457Q/R457Q	2.4 at 0.2 μM	
		W273L/ W273L	2.2 at 0.2 μM	
		Y83H/ Y83H	2.0 at 0.2 μM	

Table 9.7 (Continued)

PC	IC_{50} or K*i*	Mutation	Max. Activity Enhancement	References
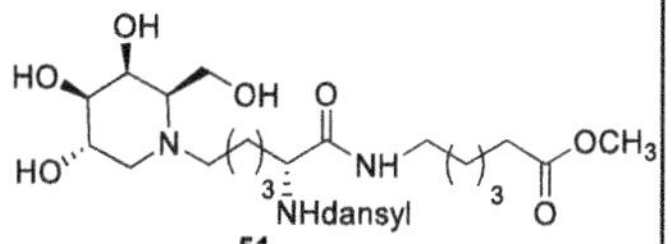 51	$IC_{50} = 6\,\mu M$	*Patient fibroblasts*		[155]
		R201H/H281Y	12.5-fold	
		R201H/S149F	12.3-fold	
		W273L/ W273L	1.3-fold	
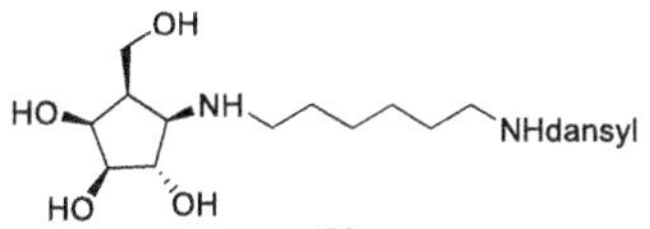 52	$IC_{50} = 0.47\,\mu M$	*Patient fibroblasts*		[156]
		R201C/ R201C	3.5-fold at $2\,\mu M$	

a) NN-DGJ one dose over 5 days.
b) NN-DGJ three doses over 15 days.
Note: the mutations indicated in bold are associated with an MPS IVB phenotype.

(between 2- and 5.9-fold) for human fibroblasts carrying different mutations and on 24 (27%) out of 88 mutated GLB1 enzymes expressed in COS7 cells [150].

In addition, several C-alkylated azasugars displayed a better activity in terms of inhibition and chaperoning activity. Demotz and coworkers identified C-pentyl-4-*epi*-isofagomine (**46**, Table 9.7) as a highly potent and selective inhibitor of human lysosomal β-Gal able to increase the enzyme activity in 56% of the evaluated mutations, which ranged from 1.5- to 35-fold. Specifically, MPS IVB fibroblasts showed a 1.5-fold increase in the GLB1 activity at $10\,\mu M$ [151, 152]. The nonyl analogous **47** (Table 9.7) was a potent inhibitor of lysosomal β-Gal ($IC_{50} = 0.4\,nM$) and was more active than the pentyl derivative as a PC (half-maximal recovery of β-Gal activity was reached at $0.01\,\mu M$ concentration), but it was a very potent inhibitor of lysosomal β-glucosidase (GCase, $IC_{50} = 40\,nM$), an activity, which may cause undesirable side-effects [152]. Moreover, the "all-*cis*" trihydroxypiperidines **48** and **49** (Table 9.7) were good inhibitors of lysosomal β-Gal and were able to increase β-Gal activity in GM1 gangliosidosis patient fibroblasts up to two- to six-fold (at $<100\,\mu M$ concentration) [153]. The valienamine derivative NOEV (*N*-octyl-4-epi-beta-valienamine, **50,** Table 9.7) NOEV has an IC_{50} of $0.2\,\mu M$ against human GLB1 and increased the enzyme activity between 2.0- and 5.1-fold in mouse fibroblasts expressing GLB1 carrying GM1-gangliosidosis (p.R201C, p.R201H, and p.R457Q) or MPS IVB (p.W273L and p.Y83H) mutations. Similar results were observed in human fibroblasts from GM1-gangliosidosis patients. *In vivo* evaluation of NOEV was performed in a model mouse of juvenile GM1-gangliosidosis, expressing a mutant enzyme protein R201C. Oral administration of NOEV led to a significant increase in GLB1 activity, which resulted in significant enhancement of the enzyme activity in the brain and other tissues [154].

A significant contribution also came from the Graz group, starting with the compound coined as DLHEX-DGJ (**51,** Table 9.7), which showed significant activity enhancements (1.3 at 12.5-fold) in GM1-gangliosidosis and MPS IVB patient fibroblasts with 20–500 μM [155]. In parallel to IFG, valienamine, and deoxygalactonojirimycin derivatives, highly functionalized cyclopentane derivatives ("carbasugars") were also investigated as potential PCs. The dansyl aminohexyl derivative **52** was found to be one of the more promising PCs of this series (Table 9.7) [156].

9.4.5 GM2 Gangliosidosis (β-Hexosaminidase)

GM2 gangliosides are catabolized by the lysosomal hydrolases β-hexosaminidases (HEX) through the hydrolysis of the *N*-acetylgalactosamine residues. HEX are a subset of isozymes formed by the dimerization of α and β subunits: HEXA (αβ), HEXB (ββ), and HEXS (αα). In addition, GM2 gangliosides degradation involves the GM2 activator protein (GM2-AP), which presents the gangliosides to α subunit of HEXA. Mutations in the genes encoding for α (HEXA), β (HEXB), or GM2-AP (GM2A) proteins affect the lysosomal degradation of GM2 ganglioside and other glycolipids, causing their accumulation into the lysosome and the GM2 gangliosidoses Tay–Sachs disease (TSD), SD, or GM2-activator protein deficiency (AB variant), respectively [157].

CNS dysfunction is the main characteristic of GM2 gangliosidoses patients that includes neurodevelopmental alterations, neuroinflammation, and neuronal apoptosis. Currently, there is no approved therapy for GM2 gangliosidoses, but several clinical trials with different therapeutic strategies including HSCT, ERT, and gene therapy are ongoing. The BBB presents a developmental challenge for therapeutic agents for these disorders. In this sense, alternative routes of administration of recombinant enzymes (e.g. intrathecal or intracerebroventricular) were evaluated, as were the delivery systems that allow the transport of proteins to the CNS. Yet, none of these tricks has materially altered the course of the disease [158].

A potential approach is SRT using inhibitors of GCS to decrease the synthesis of glucosylceramide and related glycosphingolipids that accumulate in the lysosomes. Miglustat (**3**, Figure 9.3) increased lifespan, improved clinical features, and reduced ganglioside storage in murine models of SD [159] and TSD [50]. While in a clinical trial of late-onset TSD patients, the drug did not meet its efficacy endpoint target [160], Miglustat completed the Phase 3 of clinical trial on patients with acute infantile-onset GM2 gangliosidosis (NCT00672022).

PCs have been identified for treatment of GM2 gangliosidoses. Tropak et al. reported that both adult TSD and SD fibroblasts grown in culture medium containing some HEX inhibitors including the *N*-Acetyl-galactosamine (GalNAc, **53**), *N*-Acetyl-glucosamine-thiazoline (NGT, **54**), 6-Acetamido-6-deoxycastanospermine (ACAS, **55**), 2-Acetamido-2-deoxynojirimycin (ADNJ, **56**), 2-Acetamido-1,2-dideoxynojirimycin (AdDNJ, **57**) showed increase of HEX activity, between 2.6 and 5.8-fold, above untreated fibroblasts (Table 9.8) [161].

Table 9.8 PCs for GM2 gangliosides.

PC	IC_{50} or K*i*	Mutation	Max. activity enhancement	References
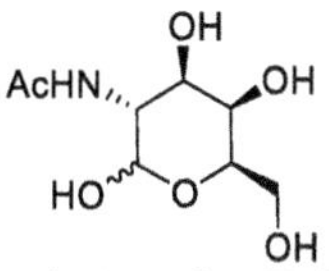 *N*-acetyl-galactosamine (GalNAc, **53**)	K*i* = 1.9 mM[a,c]	*Patient fibroblasts*		[161]
		αG269S/αG269S	2.9-fold (total Hex) at 270 mM	
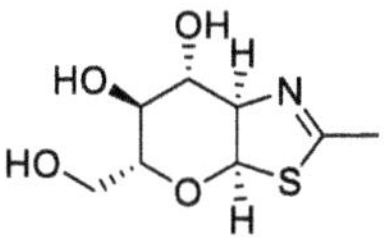 *N*-acetyl-glucosamine-thiazoline (NGT, **54**)	K*i* = 300 nM[a,c]	*Patient fibroblasts*		
		αG269S/αG269S	2.6-fold (total Hex) at 0.9 mM	
		βP504 S/16-kb 5′	5.8-fold (HexA and S) 6.1-fold (total Hex) at 0.9 mM	
		16-kb/16-kb (HexB)	4.3-fold (HexA and S) 2.7-fold (total Hex) at 0.9 mM	
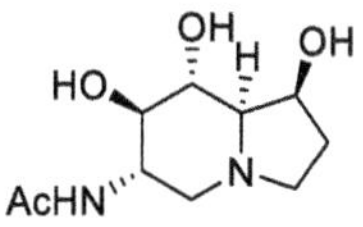 6-acetamido-6-deoxycastanospermine (ACAS, **55**)	IC_{50} = 500 nM[a,c]	*Patient fibroblasts*		
		αG269S/αG269S	3.6-fold (total Hex) at 0.18 mM	
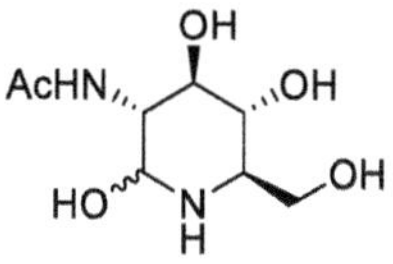 2-acetamido-2-deoxynojirimycin (ADNJ, **56**)	K*i* = 5 nM[b]	*Patient fibroblasts*		
		αG269S/αG269S	2.8-fold (total Hex) at 0.2 mM	
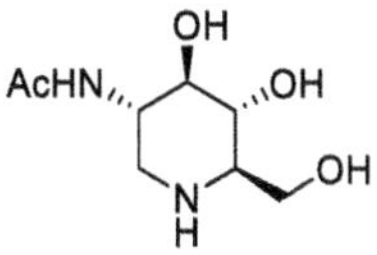 2-acetamido-1,2-dideoxynojirimycin (AdDNJ, **57**)	K*i* = 700 nM[b]	*Patient fibroblasts*		
		αG269S/αG269S	3.3-fold (total Hex) at 0.5 mM	

(Continued)

Table 9.8 (Continued)

PC	IC_{50} or Ki	Mutation	Max. activity enhancement	References
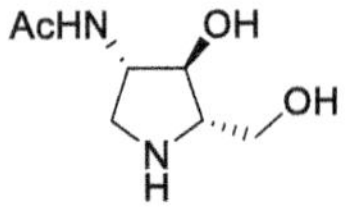 2-acetamido-1,2,4-trideoxy-1,4-imino-L-arabinitol (**58**)	$Ki = 15\,\mu M^a$	n.d.[d]		[162]
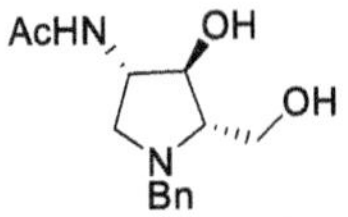 *N*-benzyl-2-acetamido-1,2,4-trideoxy-1,4-imino-L-arabinitol (**59**)	$Ki = 3.7\,\mu M^a$	*Patient fibroblasts*		
		αG269S/ αG269S	1.8-fold (HexA and S) 1.8-fold (total Hex) at 50 μM	
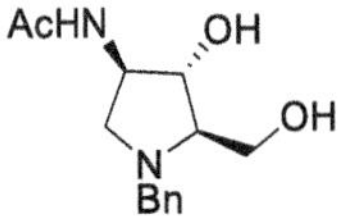 2-acetamido-1,2,4-trideoxy-imino-D-arabinitol (**60**)	$Ki = 180\,\mu M^a$	n.d.[d]		
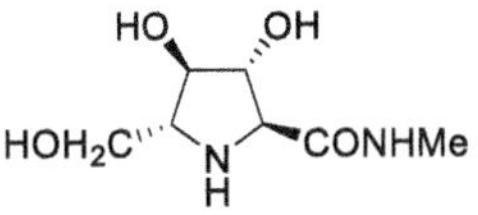 Pyrrolidine 2,5-dideoxy-2,5-imino-D-mannitol (DMDP, **61**)	$Ki = 0.041\,\mu M$	*Patient fibroblasts*		[163]
		αG269S/ αG269S	14.8-fold (HexA) at 100 μM	

a) Value determined using human placental Hex.
b) Value determined using Jack Bean Hex.
c) Compound is able to stabilize HexA under thermal denaturation.
d) n.d. = non determined.

Fleet and coworkers reported 2-acetamido analogs of DAB (1,4-dideoxy-1,4-imino-D-arabinitol) and LAB (1,2,4-trideoxy-1,4-imino-L-arabinitol). In particular, compounds **58**, **59**, and **60** showed modest to good inhibitory activity toward HEX (3.7–180 μM), and only compound **59** (the best inhibitor of the series) was tested as PC providing HEX activity enhancement up to 1.8-fold at 50 μM. The authors did not exclude a potential activity enhancement by compounds **58** and **60** despite their lower inhibition strength, which often results in beneficial for the PC activity (Table 9.8) [162]. In order to identify potential PCs for GM2 gangliosidoses, a molecular docking and dynamics simulation study identified the pyrrolidine 2,5-dideoxy-2,5-imino-D-mannitol (DMDP, **60**) amide as the strongest competitive inhibitor of HEXA. DMDP amide improved the intracellular activity of HEXA up to 14.8-fold at 100 μM in TSD fibroblasts patients (Table 9.8) [163].

The pyrimethamine, a noncarbohydrate derivative, is the most promising PC for GM2 Gangliosidosis, the compound completed the Phase 2 of trial clinic (NCT01102686) [164–168].

9.4.6 Krabbe

KD (also called globoid cell leukodystrophy) is a severe neurological condition caused by defects in the enzyme β-galactocerebrosidase (GALC). It is part of a group of disorders known as leukodystrophies, which result from the loss of myelin (demyelination) in the nervous system. GALC is required for the hydrolysis of galacto-sphingolipids, including the major lipid component of myelin β-galactocerebroside (GalCer) required for lipid turnover and maintenance of the myelin sheath that surrounds and protects neurons. Histological signs of disease include the widespread loss of myelin in the central and peripheral nervous systems, profound neuroinflammation, and axonal degeneration. Patients suffering from KD also display neurological deterioration. The only approved and available treatment option for KD is HSCT. However, combination therapies (HSCT, ERT, gene therapy, and SRT [L-cycloserine]) that target different pathogenic mechanisms/pathways have been more effective at reducing histological signs of disease, delaying disease onset, prolonging lifespan, and improving behavioral/cognitive functions in animal models of KD [169, 170]. Several recent studies have identified GALC inhibitors able to stabilize the enzyme under thermal denaturation, which may have great potential for future PC and ERT/PC therapies for KD (compounds **5** and **62–68**, Table 9.9) [171–173].

9.5 Glycogen Storage Disorders

9.5.1 Pompe Disease

The glycogen storage disease type II (GSD-II), or Pompe disease (PD), is due to the deficit of lysosomal glycogen degradation enzyme acid α-glu characterized by progressive accumulation of lysosomal glycogen in heart and skeletal muscles. Symptoms include muscle weakness, fatigue, dysphagia, respiratory insufficiency, and enlarged liver. ERT is the approved treatment for PD (Table 9.10). A major shortcoming of the current standard of care is the inability of Lumizyme® (alglucosidase alfa; Sanofi) Genzyme to reach skeletal muscle efficiently [174]. This limitation of ERT motivated the scientific community to develop the next generation of therapies for PD based on combined ER/PC therapy, gene therapy, and nanotechnology systems functionalized with receptor-binding molecules, thus promoting long-time circulation and controlled release in muscles of recombinant enzyme. These new approaches have already progressed to the clinic [175].

In recent years, PC treatment with 1-deoxynojirimycin (DNJ, **69**) has become a potential therapeutic treatment for patients with PD. *In vitro* studies have shown

Table 9.9 Inhibitors of GALC.

Compounds	GALC, K_i	References
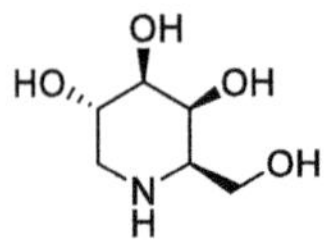 (DGJ) (**5**)	190 μM[a]	[171]
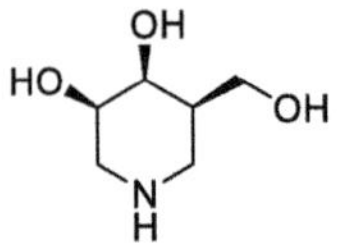 Iso-galacto-fagomine (IGF, **62**)	380 nM[a]	
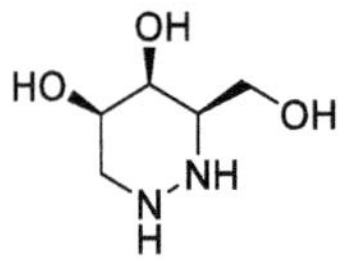 Aza-galacto-fagomine (AGF, **63**)	630 nM[a]	
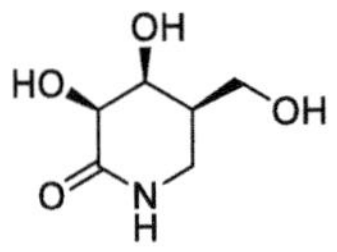 Iso-galacto-fagomine lactam (IGL, **64**)	52 μM[a]	
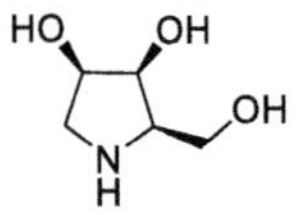 Dideoxyiminolyxitol (DIL, **65**)	130 μM[a]	
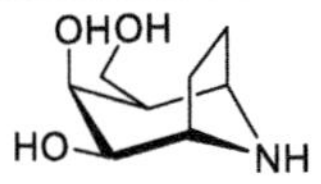 Deoxygalacto-noeurostegine (DGN, **66**)	2.3 mM[a]	
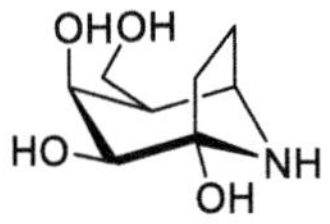 Galacto-noeurostegine (GNS, **67**)	7.0 μM[a]	[172]
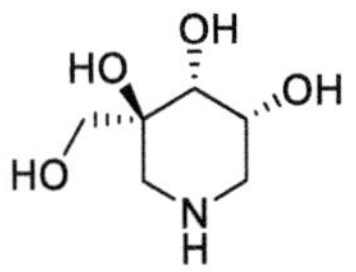 **68**	450 μM[b]	[173]

a) Compound is able to stabilize GALC under thermal denaturation.
b) Good inhibitor also for lysosomal β-galactosidase.

Table 9.10 ERT for Pompe disease.

Disorder	Gene	Enzyme deficiency	ERT	Current indication
Pompe	*GAA*	α-glucosidase	Lumizyme® (alglucosidase alfa; Sanofi Genzyme, Cambridge, MA, USA) available since 2010	Infantile-onset Pompe disease
			Nexviazyme® (avalglucosidase alfa-ngpt; Sanofi Genzyme, Cambridge, MA, USA) available since 2021	Late-onset Pompe disease

that **69** could significantly increase enzyme stability under thermal denaturation, increased enzyme activity and protein levels for different α-glu mutants in patient-derived fibroblasts and in transiently transfected COS-7 cells (the best results are reported in Table 9.11) [176]. Studies on animal models confirmed that DNJ increased the specific activity and lysosomal delivery of mutant α-glu and promoted glycogen reduction in tissues [179]. A clinical trial on DNJ showed that total α-glu activity and protein in plasma were increased 1.2- to 2.8-fold compared to ERT alone in Pompe patients. Moreover, muscle α-glu activity was also increased [180]. Unfortunately, based on the serious adverse events of the latest clinical trial (NCT00688597) the administration of **69** on Pompe patients was terminated. Moreover, the DNJ alkylated derivative (*N*-butyldeoxynojirimycin, NB-DNJ, **3**) was also effective in enhancing α-glu residual activity in fibroblasts from PD patient carrying specific mutations and in HEK293T cells overexpressing mutated *GAA* gene (Table 9.11) [177]. Even more remarkably, the co-incubation of Pompe fibroblasts with recombinant human α-gluc and the chaperone **3** resulted in more efficient stabilization of enzyme activity. Improved enzyme correction was also found *in vivo* in a PD mouse model and PD patients treated with coadministration of infusions of recombinant human α-gluc and oral D-NB-DNJ [181, 182]. L-NBDNJ, the unnatural enantiomer of the iminosugar **3**, showed α-gluc activity rescue, either when administered singularly (1.5-fold at 20 μM) in PD fibroblasts bearing L552P/L552P mutation (Table 9.11) or when co-incubated with the recombinant human α-gluc. In addition, different from its D-NBDNJ, L-NBDNJ (**70**) did not act as a glucosidase inhibitor. The lack of inhibition of the deficient enzyme and of other glycosidases further increases the potential of **70**, especially compared with its enantiomer [178].

9.6 Glycoproteinoses

9.6.1 Fucosidosis

Fucosidosis is caused by mutations of the α-L-fucosidase (*FUCA1*) gene, resulting in deficiency of the α-l-fucosidase enzyme. As a result of the hydrolytic enzyme deficiency, incomplete catabolism of *N*- and *O*-glycosylproteins results in the accumulation of fucose-containing glycolipids and glycoproteins in various tissues and urine.

Table 9.11 PCs for Pompe disease.

PC	IC_{50} or Ki	Mutation	Max. activity enhancement	References
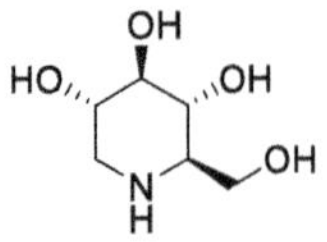 1-deoxynojirimycin (DNJ) (**69**)	$IC_{50} = 1.3\,\mu M$, $Ki = 530\,nM$[a]	*Patient fibroblasts*[b]		[176]
		IVS8+1G>A/ M519V	6.6-fold	
		P545L/P545L	6.4-fold	
		L552P/L552P	17.8-fold	
		L552P/r.spl?	5.1-fold	
		L552P/A445P	4.3-fold	
		G54R/r.0	4.3-fold	
		COS-7 cells[c]		
		P545L	4.2-fold	
		L552P	3.9-fold	
		Y575S	8.5-fold	
		E579K	3.5-fold	
		A610V	5.8-fold	
		H612Q	4.5-fold	
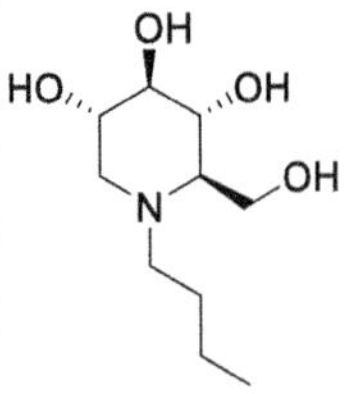 D-NB-DNJ (**3**)	n.d.[d]	*Patient fibroblasts*		[177]
		L552P/L552P	5.6-fold at 20 μM	
		L552P/abn splic	2.7-fold at 20 μM	
		L552P/A445P	1.8-fold at 20 μM	
		G549R/abn splic	3.7-fold at 20 μM	
		HEK293T cells		
		L552P	4-fold at 20 μM	
		G549R	16-fold at 20 μM	
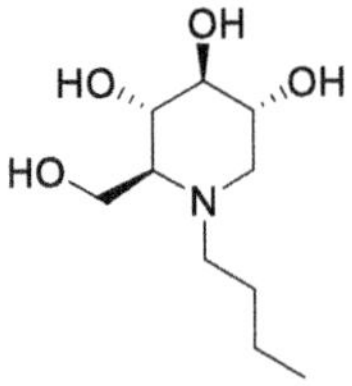 L-NB-DNJ (**70**)	---[e]	*Patient fibroblasts*		[178]
		L552P/L552P	1.5-fold at 20 μM	

a) Compound is able to stabilize α-glu under thermal denaturation;
b) Cell lines were tested at least three times with DNJ concentrations ranging from 50 nM to 1 mM;
c) Cells were treated with 100 μM DNJ;
d) n.d. = not determined;
e) Compound did not act as a glycosidase inhibitor.

The clinical features of fucosidosis are progressive mental retardation and neurological deterioration, coarse facies, growth retardation, recurrent infections, dysostosis multiplex, and angiokeratoma. The treatment of fucosidosis is directed toward the specific symptoms that are apparent in each individual [183]. Correction of the enzymatic deficiency by allogeneic bone marrow transplantation has been first experienced on an animal model [184, 185] and then on patients with an amelioration of the clinical signs [186–188]. PCs for fucosidase have not been proposed to date, but several iminosugar derivatives have proved to be effective inhibitors. The discussion of fucosidase inhibitors is not the aim of this work, but a report of the main compounds is reported in the articles [189, 190].

9.6.2 α-Mannosidosis

α-mannosidosis is an ultra-rare autosomal recessive genetic disorder caused by mutations in the *MAN2B1* gene encoding for α-mannosidase, a lysosomal enzyme involved in glycoprotein catabolism. The result of α-mannosidase deficiency is blockage of the degradation of glycoproteins, leading to an accumulation of mannose-rich oligosaccharides in all tissues [191]. Accumulation of mannose-rich oligosaccharides manifests in a broad variety of symptoms including skeletal abnormalities, motor function impairment, intellectual disability, hearing loss, respiratory dysfunction, recurrent infections, and cellular and humoral immune defects usually presenting in early childhood [192, 193].

Currently, intravenous ERT, Lamzede (Velmanase alfa, Chiesi Italia S.p.A.), is available since 2021 for the treatment of mild–moderate forms of α-mannosidosis in adults, adolescents, and children, but it is not effective treatment for neurological involvement. Preliminary studies demonstrated the ability of HSCT to partially preserve neurocognitive function, stabilize skeletal abnormalities, and prevent early death [194–196].

Recently, a series of PCs combining the 5*N*,6*O*-oxomethylidenemannonojirimycin (OMJ) in either mono- (**71–76**, Table 9.12) or multivalent (β-cyclodextrins as scaffold) fashion (**77–79**, Table 9.12) were reported. Multivalent derivatives exhibited potent enzyme inhibition that prevailed over the chaperone effect. On the contrary, monovalent OMJ derivatives proved effective as activity enhancers for several mutant alfa-man forms in patient fibroblasts and/or transfected MAN2B1-KO cells [197].

9.7 Conclusions

The chapter summarizes the *in vitro* screening, preclinical, and clinical results of carbohydrate-based compounds in the currently available therapeutic approaches (ERT, SRT, and PCT) for LSDs, organized accordingly to LSD classification into MPS, sphingolipidoses, glycogen storage disorders, and glycoproteinoses.

Table 9.12 PCs for α-mannosidosis

PC	IC_{50}	Mutation	Max. activity enhancement[a]
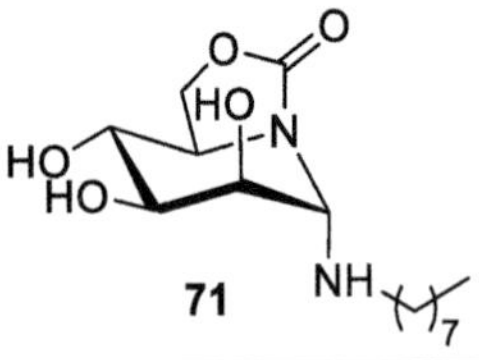 **71**	IC_{50} = 26.8 μM	*Patient fibroblasts*	
		H72L/H72L	8.2-fold at 20 μM
		P356R/P356R	6.4-fold at 2 μM
		R750W/R750W	1.35-fold at 20 μM
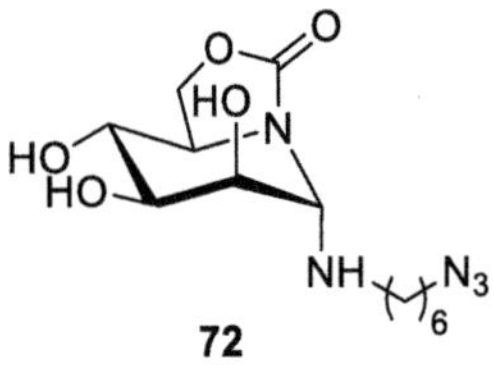 **72**	IC_{50} = 116 μM	*Patient fibroblasts*	
		H72L/H72L	8.0-fold at 20 μM
		P356R/P356R	4.5-fold at 20 μM
		R750W/R750W	1.3-fold at 20 μM
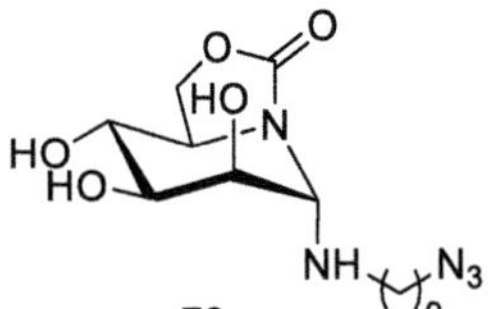 **73**	IC_{50} = 19.3 μM	*Patient fibroblasts*	
		H72L/H72L	7.8-fold at 20 μM
		P356R/P356R	5.7-fold at 20 μM
		R750W/R750W	1.35-fold at 20 μM
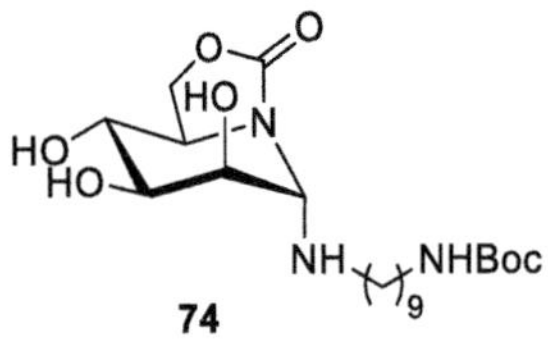 **74**	IC_{50} = 30.7 μM	*Patient fibroblasts*	
		H72L/H72L	11-fold at 20 μM
		P356R/P356R	5-fold at 20 μM
		R750W/R750W	1.6-fold at 2 μM
		MAN21B-KO HAP1 cells	
		C55F	51% increase at 2 μM
		H71L	34% increase at 2 μM
		L352P	8% increase at 2 μM
		L565P	14% increase at 2 μM
		R916C	26% increase at 2 μM
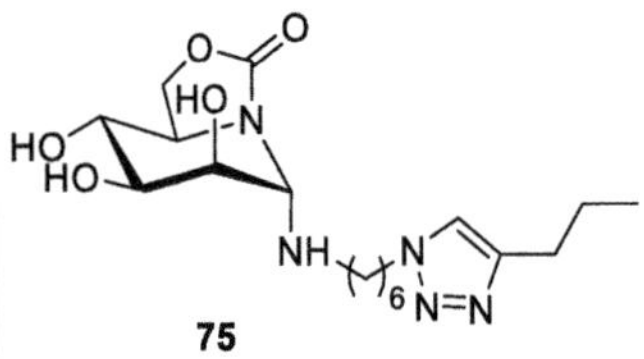 **75**	IC_{50} = 17.3 μM	*Patient fibroblasts*	
		H72L/H72L	4-fold at 0.2 μM
		R750W/R750W	1.7-fold at 20 and 2 nM

Table 9.12 (Continued)

PC	IC_{50}	Mutation	Max. activity enhancement[a]
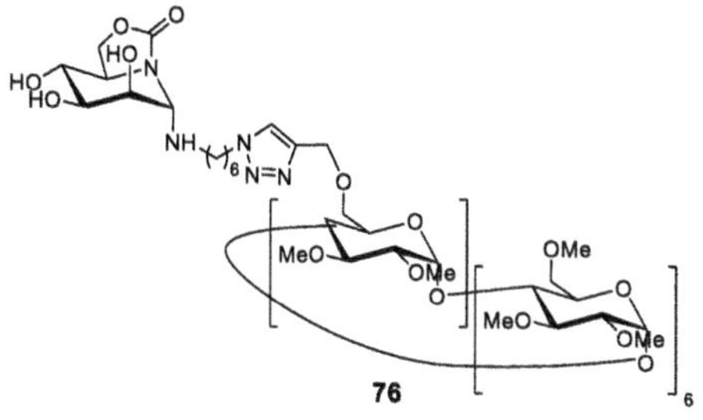 **76**	IC_{50} = 2.2 μM	*Patient fibroblasts*	
		H72L/H72L	7.8-fold at 0.2 μM
		R750W/ R750W	1.7-fold at 2 nM
		MAN21B-KO HAP1 cells	
		C55F	111% increase at 20 μM
		H71L	68% increase at 20 μM
		L352P	10% increase at 0.2 μM
		L565P	8% increase at 2 μM
		R916C	63% increase at 20 μM
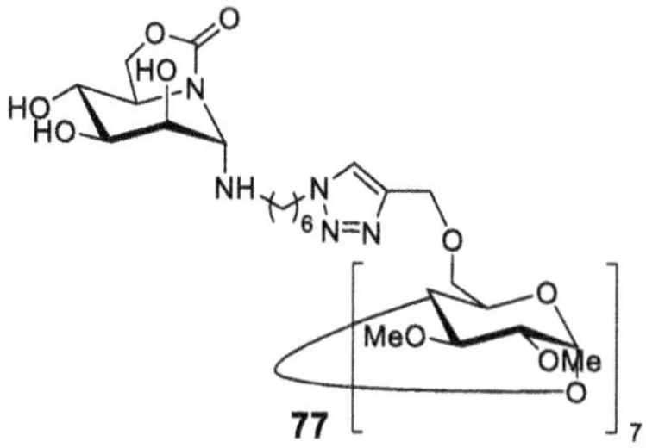 **77**	IC_{50} = 0.44 μM	*Patient fibroblasts*	
		R750W/ R750W	1.5-fold at 2 nM
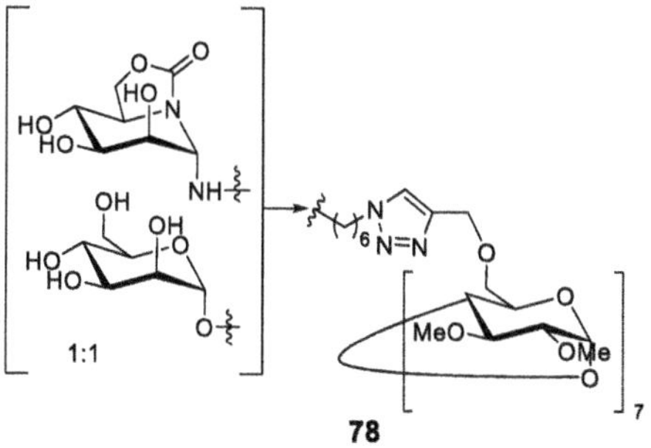 **78**	IC_{50} = 0.45 μM	*Patient fibroblasts*	
		R750W/ R750W	1.3-fold at 2 nM
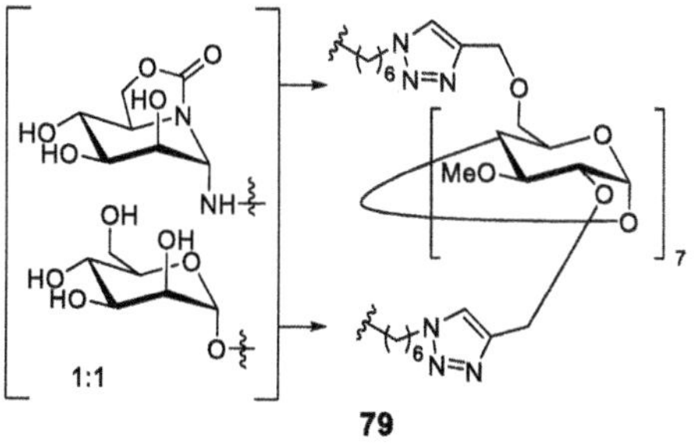 **79**	IC_{50} = 55 μM	*Patient fibroblasts*	
		R750W/ R750W	1.4-fold at 2 nM

a) Ref. [197].

Within the field of MPS, a sulfated disaccharide derived from heparin was reported as a good PC for MPS II (almost 40-fold activity enhancement at 10 μM), but only for a specific mutation, demonstrating that often PCT lacks general applicability. In addition, when searching for MPS enzyme stabilizers to be employed both in PCT and ERT/PC, multivalent sugars and iminosugars emerged as valuable candidates due to their strong affinity toward GALNS and IDS enzymes, paving the way for future investigations in this field. Probably due to the multimeric nature of the deficient enzyme (α-mannosidase), multivalent PCs were developed also for α-mannosidosis, a glycoproteinosis. However, in this case, the strong inhibitory activity of the compounds prevailed over the PC effect. Among sphingolipidoses, FD represents an excellent example of the potential use of iminosugars in all three therapeutic approaches. Indeed, several pyrrolidine and piperidine iminosugars are currently under investigation to stabilize and prolong the ERT enzyme activity (ERT/PC). The iminosugar analog of D-galactose (DGJ) is the only PC, which became a drug to date (Galafold), and further investigations are ongoing to address the few mutations that are not responsive to DGJ. Finally, the *N*-butyl DGJ analog is currently under Phase 3 of clinical trials for SRT. Conversely, for GD, an iminosugar-based SRT (Zavesca™) is available from 2002, while no PC has reached the drug market yet, even though the IFG iminosugar reached Phase 2 of clinical trials. For this reason, many efforts have been devoted to identifying novel PCs for GD, by preparing multivalent compounds, introducing alkyl chains (both on sugars and iminosugars skeleton) to improve cell-permeability and by developing pH-sensitive systems or noncompetitive inhibitors with the aim of favoring the enhancer activity with respect to the inhibitory one. In the last few years, the search for PCs for GD attracted even higher attention due to link between Gaucher and Parkinson disease, suggesting that effective PCs can be potentially applied to all protein misfolded diseases, including neurodegenerative disorders. DGJ analogs, IFG analogs, and aminocyclitols (e.g. valienamine derivatives) were investigated as PCs for GM1 gangliosidosis and MBD, but all studies are at most at a preclinical level (mouse models). Regarding the glycogen storage disorder Pompe disease, the serious adverse effects caused in patients by coadministration of ERT with the natural iminosugar 1-deoxynojirimycin (DNJ) stopped the ER/PC therapy clinical trial, despite the remarkable increase in enzyme activity observed. Lower side effects might be obtained with the *N*-butyl enantiomer of the natural DNJ, which is a modest PC but does not act at all as enzyme inhibitor. The latter case suggests that non-inhibitory chaperones, although more difficult to be identified, might represent a valuable alternative for a faster development of safe and efficient PCs for LSDs, in general.

Acknowledgments

The authors thank Regione Toscana (Bando Salute 2018, project: “Late onset Lysosomal Storage Disorders” [LSDs] in the differential diagnosis of neurodegenerative diseases: development of new diagnostic procedures and focus on potential pharmacological chaperones [PCs], Acronym: Lysolate) and by Università di Firenze and Fondazione CR Firenze (Bando congiunto per il finanziamento di progetti competitivi sulle malattie neurodegenerative 2018, project: A multidisciplinary approach to target Parkinson's disease in Gaucher related population, Acronym: MuTaParGa).

Abbreviations and Acronyms

Lysosomal storage disorders	LSDs
Central nervous system	CNS
Orphan drug act	ODA
Haematopoietic stem cell transplantation	HSCT
Antisense oligonucleotide	ASO
Adeno-associated virus	AAV
Enzyme replacement therapy	ERT
Blood brain barrier	BBB
Substrate reduction therapy	SRT
Pharmacological chaperone therapy	PCT
High-throughput screening	HTS
Chinese hamster ovary	CHO
Active site-specific chaperone	ASSC
Endoplasmic reticulum	ER
Endoplasmic reticulum-associated degradation	ERAD
Deoxynojirimycin	DNJ
Niemann–Pick type C disease	NPC
Deoxygalactonojirimycin	DGJ
Food and drugs administration	FDA
European medicines agency	EMA
Isofagomine	IFG
Mucopolysaccharidoses	MPS
Glycosaminoglycans	GAGs
Heparan sulfate	HS
Dermatan sulfate	DS
Keratan sulfate	KS
Chondroitin sulfate	CS
acid-β-glucosidase (or glucocerebrosidase)	GCase
N-Acetyl-D-glucosaminidase	NAGLU
Iduronate-2-sulfatase	IDS
N-acetyl-glucosamine-6-sulfatase sulfatase	GALNS
gold glyconanoparticles	AuGNPs
Fabry Disease	FD
α-galactosidase A	α-Gal A
globotriaosylceramide	Gb3
polyethylene glycol	PEG

Gaucher disease	GD
1,5-Dideoxy-1,5-imino-D-xylitol	DIX
Nojirimycin	NJ)
1,4-dideoxy-1,4-imino-D-arabinitol	DAB-1
β-cyclodextrin	βCD
copper-catalyzed azide-alkyne cycloaddition	CuAAC
Niemann–Pick	NP
acid sphingomyelinase	ASM
GM1 gangliosidosis	GM1
Morquio B disease	MBD
N-octyl-4-epi-beta-valienamine	NOEV
β-Hexosaminidases	HEX
Tay–Sachs disease	TSD
Sandhoff disease	SD
glucosylceramide synthase	GCS
N-Acetyl-galactosamine	GalNAc
N-Acetyl-glucosamine-thiazoline	NGT
6-Acetamido-6-deoxycastanospermine	ACAS
2-Acetamido-2-deoxynojirimycin	ADNJ
2-Acetamido-1,2-dideoxynojirimycin	AdDNJ
2,5-dideoxy-2,5-imino-D-mannitol	DMDP
rabbe disease	KD
β-galactocerebrosidase	GALC
β-galactocerebroside	GalCer
Pompe disease	PD
5*N*,6*O*-oxomethylidenemannonojirimycin	OMJ

References

1 Platt, F.M., d'Azzo, A., Davidson, B.L. et al. (2018). *Nature Reviews Disease Primers* 4: 27.
2 Bonam, S.R., Wang, F., and Muller, S. (2019). *Nature Reviews. Drug Discovery* 18: 923–948.
3 Parenti, G., Andria, G., and Ballabio, A. (2015). *Annual Review of Medicine* 66: 471–486.
4 Kingma, S.D.K., Bodamer, O.A., and Wijburg, F.A. (2015). *Best Practice & Research. Clinical Endocrinology & Metabolism* 29: 145–157.
5 Meikle, P.J., Hopwood, J.J., Clague, A.E., and Carey, W.F. (1999). *JAMA* 281: 249–254.
6 US National Library of Medicine (2023). Genetics Home Reference. https://medlineplus.gov/genetics/ (May 2023)
7 Charrow, J., Andersson, H.C., Kaplan, P. et al. (2000). *Archives of Internal Medicine* 160: 2835–2843.

8 García Fernández, J.M. and Ortiz Mellet, C. (2019). *ACS Medicinal Chemistry Letters* 10: 1020–1023.

9 Hobbs, J.R., Hugh-Jones, K., Barret, A.J. et al. (1981). *Lancet* 2: 709–712.

10 Marques, A.R.A. and Saftig, P. (2019). *Journal of Cell Science* 2: 132–146.

11 Boelens, J.J., Aldenhoven, M., Purtill, D. et al. (2013). *Blood* 121: 3981–3987.

12 Poe, M.D., Chagnon, S.L., and Escolar, M.L. (2014). *Annals of Neurology* 76: 747–753.

13 Bevan, A.K., Duque, S., Foust, K.D. et al. (2011). *Molecular Therapy* 19: 1971–1980.

14 Goina, E., Peruzzo, P., Bembi, B. et al. (2017). *Molecular Therapy* 25: 2117–2128.

15 Schneller, J.L., Lee, C.M., Bao, G., and Venditti, C.P. (2017). *BMC Medicine* 15: 43–55.

16 Deverman, B.E. (2016). *Nature Biotechnology* 34: 204–209.

17 Ortolano, S., Viéitez, I., Navarro, C., and Spuch, C. (2014). *Recent Patents on Endocrine, Metabolic & Immune Drug Discovery* 8 (1): 9–25.

18 Brady, R.O. (2006). *Annual Review of Medicine* 57: 283–296.

19 Ohashi, T. (2012). *Pediatric Endocrinology Reviews* 10: 26–34.

20 Grabowski, G.A., Barton, N.W., Pastores, G. et al. (1995). *Annals of Internal Medicine* 122: 33–39.

21 Ortolano, S. (2016). *The Journal of Inborn Errors of Metabolism and Screening* 4: 1–11.

22 Parenti, G., Andria, G., and Valenzano, K.J. (2015). *Molecular Therapy* 23: 1138–1148.

23 Boyd, R.E., Lee, G., Rybczynski, P. et al. (2013). *Journal of Medicinal Chemistry* 56: 2705–2725.

24 Markham, A. (2016). *Drugs* 76: 1147–1152.

25 Hughes, D.A., Nicholls, K., Shankar, S.P. et al. (2017). *Journal of Medical Genetics* 54: 288–296.

26 Compain, P. and Martin, O.R. (2007). *Iminosugars: From Synthesis to Therapeutic Applications*. New York: Wiley VCH.

27 Ferjancic, Z. and Saicic, R.N. (2021). *European Journal of Organic Chemistry* 3241–3250.

28 Nicolas, C. and Martin, O.R. (2018). *Molecules* 23: 1612–1642.

29 Harit, V.K. and Ramesh, N.G. (2016). *RSC Advances* 6: 109528–109607.

30 Sousa, C.E., Mendes, R.R., Costa, F.T. et al. (2014). *Current Organic Synthesis* 11: 182–203.

31 Takahata, H. (2012). *Heterocycles* 85: 1351–1376.

32 Stocker, B.L., Dangerfield, E.M., Win-Mason, A.L. et al. (2010). *European Journal of Organic Chemistry* 1615–1637.

33 Benjamin, D.G. (2009). *Tetrahedron: Asymmetry* 20: 652–671.

34 Brás, N.F., Cerqueira, N.M., Ramos, M.J., and Fernandes, P.A. (2014). *Expert Opinion on Therapeutic Patents* 24: 857–874.

35 López, Ó., Merino-Montiel, P., Martos, S., and González-Benjumea, A. (2012). Glycosidase inhibitors: versatile tools in glycobiology. In: *Carbohydrate Chemistry-Chemical and Biochemical Approaches*, vol. 38 (ed. A.P. Rauter and T. Lindhorst), 215–262. RSC.

36 Bols, M., López, Ó., and Ortega-Caballero, F. (2007). Glycosidase inhibitors: structure, activity, synthesis, and medical relevance. In: *Comprehensive Glycoscience-from Chemistry to Systems Biology* (ed. J.P. Kamerling), 815–884. Elsevier.

37 Conforti, I. and Marra, A. (2021). *Organic & Biomolecular Chemistry* 19: 5439–5475.

38 Merino, P., Delso, I., Marca, E. et al. (2009). *Current Chemical Biology* 3: 253–271.

39 Butters, T.D., Dwek, R.A., and Platt, F.M. (2005). *Glycobiology* 15: 43R–52R.

40 Compain, P. and Martin, O.R. (2003). *Current Topics in Medicinal Chemistry* 3: 541–560.

41 Butters, T.D., Dwek, R.A., and Platt, F.M. (2000). *Chemical Reviews* 100: 4683–4696.

42 Dorling, P.R., Huxtable, C.R., and Vogel, P. (1978). *Neuropathology and Applied Neurobiology* 4: 285–295.

43 Molyneux, R.J. and James, L.F. (1982). *Science* 216: 190–191.

44 Cenci di Bello, I., Dorling, P., and Winchester, B. (1983). *Biochemical Journal* 215: 693–696.

45 Solomon, M. and Muro, S. (2017). *Advanced Drug Delivery Reviews* 118: 109–134.

46 Rodríguez, M.C., Ceaglio, N., Antuña, S. et al. (2019). Chapter 2: Production of therapeutic enzymes by lentivirus transgenesis. In: *Therapeutic Enzymes: Function and Clinical Implications, Advances in Experimental Medicine and Biology 1148* (ed. N. Labrou), 25–54. Springer Nature Singapore Pte Ltd.

47 Oh, D.B. (2015). *BMB Reports* 48: 438–444.

48 Sharma, A., Vaghasiya, K., Ray, E., and Verma, R.K. (2018). *Journal of Drug Targeting* 26: 208–221.

49 Cenci di Bello, I., Mann, D., Nash, R., and Winchester, B.G. (1988). Castanospermine-induced deficiency of lysosomal β-D-glucosidase: a model of Gaucher's disease in fibroblasts. In: *Lipid Storage Disorders: Biological and Medical Aspects* (ed. R. Salvayre, L. Douste-Blazy, and S. Gatt), 635–641. New York: Plenum.

50 Platt, F.M., Nieses, G.R., Reinkensmeir, G. et al. (1997). *Science* 276: 428–431.

51 Schiffmann, R., Fitzgibbon, E.J., Harris, C. et al. (2008). *Annals of Neurology* 64: 514–522.

52 Patterson, M.C., Vecchio, D., Prady, H. et al. (2006). *Revista de Neurologia* 43: 8.

53 Patterson, M.C., Vecchio, D., Prady, H. et al. (2007). *Lancet Neurology* 6: 765–772.

54 Walterfang, M., Chien, Y.-H., Imrie, J. et al. (2012). *Orphanet Journal of Rare Diseases* 7: 76–93.

55 Lyseng- Williamson, K.A. (2014). *Drugs* 74: 61–74.

56 Coutinho, M.F., Santos, J.I., and Alves, S. (2016). *International Journal of Molecular Sciences* 17: 1065–1087.

57 Winchester, B.G. (2009). *Tetrahedron: Asymmetry* 20: 645–651.

58 Fan, J.-Q., Ishii, S., Asano, N., and Suzuki, Y. (1999). *Nature Medicine* 5: 112–115.

59 Pereira, D.M., Valentão, P., and Andrade, P.B. (2018). *Chemical Science* 9: 1740–1752.

60 Convertino, M., Das, J., and Dokholyan, N.V. (2016). *ACS Chemical Biology* 11: 1471–1489.

61 Sánchez-Fernández, E.M., García Fernández, J.M., and Ortiz Mellet, C. (2016). *Chemical Communications* 52: 5497–5515.

62 Fan, J.-Q. (2008). *Biological Chemistry* 389: 1–11.

63 Gloster, T.M. and Vocadlo, D.J. (2012). *Nature Chemical Biology* 8: 683–694.

64 Stütz, A.E. and Wrodnigg, T.M. (2016). *Advances in Carbohydrate Chemistry and Biochemistry* 73: 225–302.

65 Frustaci, A., Chimenti, C., Ricci, R. et al. (2001). *The New England Journal of Medicine* 345: 25–32.

66 Tsukimura, T., Chiba, Y., Ohno, K. et al. (2011). *Molecular Genetics and Metabolism* 103: 26–32.

67 McCafferty, E.H. and Scott, L.J. (2019). *Drugs* 79: 543–554.

68 https://www.ema.europa.eu/en/medicines/human/EPAR/galafold (May 2022).

69 Moran, N. (2018). FDA approves Galafold, a triumph for amicus. *Nature Biotechnology* 36: 913.

70 Sun, Y., Liou, B., Xu, Y.H. et al. (2012). *The Journal of Biological Chemistry* 287: 4275–4287.

71 Benito, J.M., García Fernández, J.M., and Mellet, C.O. (2011). *Expert Opinion on Therapeutic Patents* 21: 885–903.

72 Dasgupta, N., Xu, Y.H., Li, R. et al. (2015). *Human Molecular Genetics* 24: 7031–7048.

73 Liguori, L., Monticelli, M., Allocca, M. et al. (2020). *International Journal of Molecular Sciences* 21: 489–509.

74 Boyd, R.E., Benjamin, E.R., Xu, S. et al. (2016). Chapter 15: Pharmacological chaperones as potential therapeutics for lysosomal storage disorders: preclinical research to clinical studies. In: *Lysosome: Biology, Disease and Therapeutics* (ed. F.R. Maxfield, J.M. Willard, and S. Lu), 357–382. John Wiley & Sons, Inc.

75 Mena-Barragán, T., Narita, A., Matias, D. et al. (2015). *Angewandte Chemie, International Edition* 54: 11696–11700.

76 Castellan, T., Garcia, V., Rodriguez, F. et al. (2020). *Organic & Biomolecular Chemistry* 18: 7852–7861.

77 Tran, M.L., Génisson, Y., Ballereau, S., and Dehoux, C. (2020). *Molecules* 25: 3145–3166.

78 Mohamed, F.E., Al-Gazali, L., Al-Jasmi, F., and Ali, B.R. (2017). *Frontiers in Pharmacology* 8: 448.

79 Aymami, J., Barril, X., Rodríguez-Pascau, L., and Martinell, M. (2013). *Pharmaceutical Patent Analyst* 2: 109–124.

80 Neufeld, E. and Muenzer, J. (2001). The mucopolysaccharidoses. In: *The Metabolic and Molecular Bases of Inherited Diseases* (ed. C. Scriver, A. Beaudet, W. Sly, and D. Valle), 3421–3452. New York: McGraw-Hill.

81 Sawamoto, K., Stapleton, M., Alméciga-Díaz, C.J. et al. (2019). *Drugs* 79: 1103–1134.

82 Hobbs, J.R. (1981). *Lancet* 2: 735–739.

83 Scarpa, M., Orchard, P.J., Schulzd, A. et al. (2017). *Molecular Genetics and Metabolism* 122: 25–34.

84 Giugliani, R., Federhen, A., Vairo, F. et al. (2016). *Expert Opinion on Emerging Drugs* 21: 9–26.

85 Giugliani, R., Dalla Corte, A., Poswar, F. et al. (2018). *Expert Opinion on Orphan Drugs* 6: 403–411.

86 Schuh, R.S., Baldo, G., and Teixeira, H.F. (2016). *Expert Opinion on Drug Delivery* 13: 1709–1718.

87 Hoshina, H., Shimada, Y., Higuchi, T. et al. (2018). *Molecular Genetics and Metabolism* 123: 118–122.

88 Zhu, S., Jagadeesh, Y., Tuan Tran, A. et al. (2021). *Chemistry - A European Journal* 44: 11291–11297.

89 Feldhammer, M., Durand, S., and Pshezhetsky, A.V. (2009). *PLoS One* 4: e7434.

90 Almeciga-Diaz, C.J., Hidalgo, O.A., Olarte-Avellaneda, S. et al. (2019). *Journal of Medicinal Chemistry* 62: 6175–6189.

91 Del Olarte-Avellaneda, J.C., Castillo, A.F., Rojas-Rodriguez, O. et al. (2020). *ACS Medicinal Chemistry Letters* 11: 1377–1385.

92 Matassini, C., Parmeggiani, C., Cardona, F., and Goti, A. (2016). *Tetrahedron Letters* 57: 5407–5415.

93 Matassini, C., D'Adamio, G., Vanni, C. et al. (2019). *European Journal of Organic Chemistry* 4897–4905.

94 D'Adamio, G., Matassini, C., Parmeggiani, C. et al. (2016). *RSC Advances* 6: 64847–64851.

95 Matassini, C., Vanni, C., Goti, A. et al. (2018). *Organic & Biomolecular Chemistry* 16: 8604–8612.

96 van der Tol, L., Smid, B.E., Poorthuis, B.J. et al. (2014). *Journal of Medical Genetics* 51: 1–9.

97 Wang, H., Shen, Y., Zhao, L., and Ye, Y. (2021). *Current Medicinal Chemistry* 28: 628–643.

98 van der Veen, S.J., Hollak, C.E.M., van Kuilenburg, A.B.P., and Langeveld, M. (2020). *Journal of Inherited Metabolic Disease* 43: 908–921.

99 Arends, M., Hollak, C.E.M., and Biegstraaten, M. (2015). *Orphanet Journal of Rare Diseases* 10: 77.

100 Weidemann, F., Niemann, M., Stork, S. et al. (2013). *Journal of Internal Medicine* 274: 331–341.

101 Rombach, S.M., Smid, B.E., Linthorst, G.E. et al. (2014). *Journal of Inherited Metabolic Disease* 37: 341–352.

102 Abasolo, I., Seras-Franzoso, J., Moltó-Abad, M. et al. (2021). *WIREs Nanomedicine and Nanobiotechnology* 13: e1684.

103 Guérard, N., Oder, D., Nordbeck, P. et al. (2018). *Clinical Pharmacology and Therapeutics* 103: 703–711.

104 Welford, R., Mühlemann, A., Garzotti, M. et al. (2018). *Human Molecular Genetics* 27: 3392–3403.

105 Benjamin, E.R., Della Valle, M.C., Wu, X. et al. (2017). *Genetics in Medicine* 4: 430–438.

106 Aguilar-Moncayo, M., Takai, T., Higaki, K. et al. (2012). *Chemical Communications* 48: 6514–6516.

107 Jenkinson, S.F., Fleet, G.W.J., Nash, R.J. et al. (2011). *Organic Letters* 13: 4064–4067.

108 Yu, Y., Mena-Barragán, T., Higaki, K. et al. (2014). *ACS Chemical Biology* 9: 1460–1469.

109 Kato, A., Yamashita, Y., Nakagawa, S. et al. (2010). *Bioorganic & Medicinal Chemistry* 18: 3790–3794.
110 Martínez-Bailén, M., Carmona, A.T., Cardona, F. et al. (2020). *European Journal of Medicinal Chemistry* 192: 112173.
111 Cheng, W.-C., Wang, J.-H., Li, H.-Y. et al. (2016). *European Journal of Medicinal Chemistry* 123: 14–20.
112 Li, H.-Y., Lee, J.-D., Chen, C.-W. et al. (2018). *European Journal of Medicinal Chemistry* 144: 626–634.
113 Johnson, F.K., Mudd, P.N.J., DiMino, T. et al. (2015). *Clinical Pharmacology in Drug Development* 4: 256–261.
114 Ishii, S., Chang, H.H., Yoshioka, H. et al. (2009). *The Journal of Pharmacology and Experimental Therapeutics* 328: 723–731.
115 Gaggl, M. and Sunder-Plassmann, G. (2016). *Nature Reviews. Nephrology* 12: 653–654.
116 Germain, D.P., Giugliani, R., Hughes, D.A. et al. (2012). *Orphanet Journal of Rare Diseases* 7: 91.
117 Citro, V., Peña-García, J., den-Haan, H. et al. (2016). *PLoS One* 11: e0165463.
118 Khanna, R., Valenzano, K.J., and Fowles, S.E. (2014). *PCT International Application* WO2014014938A1.
119 Grabowski, G.A. (ed.) (2013). *Advances in Gaucher Disease: Basic and Clinical Perspectives.* Unitec House, 2 Albert Place, London N3 1QB, UK: Future Medicine Ltd.
120 Human Gene Mutation Database (HGMD). http://www.hgmd.cf.ac.uk/ac/index.php (May 2023).
121 Ryan, E., Seehra, G., Sharma, P., and Sidransky, E. (2019). *Current Opinion in Neurology* 32: 589–596.
122 Zimran, A. and Elstein, D. (2014). *Pediatric Endocrinology Reviews* 12: 82–87.
123 Mistry, P.K., Lukina, E., Ben Turkia, H. et al. (2017). *American Journal of Hematology* 92: 1170–1176.
124 Kuriyama, C., Kamiyama, O., Ikeda, K. et al. (2008). *Bioorganic & Medicinal Chemistry* 16: 7330–7336.
125 Chang, H., Asano, N., Ishii, S. et al. (2006). *The FEBS Journal* 273: 4082–4092.
126 Zhu, X., Sheth, K.A., Li, S. et al. (2005). *Angewandte Chemie (International Ed. in English)* 44: 7450–7453.
127 Fan, J.-Q., Zhu, X., and Sheth, K., WO2005046612, 2005 and US20100189708, 2010.
128 Sawker, A.R., Cheng, W.-C., Beutler, E. et al. (2002). *Proceedings of the National Academy of Sciences* 99: 15428–15433.
129 Brumshtein, B., Greenblatt, H.M., Butters, T.D. et al. (2007). *The Journal of Biological Chemistry* 282: 29052–29058.
130 Compain, P., Martin, O.R., Boucheron, C. et al. (2006). *ChemBioChem* 7: 1356–1359.
131 Luan, Z., Higaki, K., Aguilar-Moncayo, M. et al. (2009). *ChemBioChem* 10: 2780–2792.

132 Kato, A., Nakagome, I., Sato, K. et al. (2016). *Organic & Biomolecular Chemistry* 14: 1039–1048.
133 Castilla, J., Rísquez, R., Higaki, K. et al. (2015). *European Journal of Medicinal Chemistry* 90: 258–266.
134 Joosten, A., Decroocq, C., de Sousa, J. et al. (2014). *ChemBioChem* 15: 309–319.
135 Clemente, F., Matassini, C., Goti, A. et al. (2019). *Chemistry Letters* 10: 621–626.
136 Clemente, F., Matassini, C., Faggi, C. et al. (2020). *Bioorganic Chemistry* 98: 103740–103763.
137 Garcia-Moreno, M.I., de la Mata, M., Sánchez-Fernández, E.M. et al. (2017). *Journal of Medicinal Chemistry* 60: 1829–1842.
138 https://clinicaltrials.gov/ct2/show/NCT03950050 (February 2023).
139 Serra-Vinardell, J., Díaz, L., Casas, J. et al. (2014). *ChemMedChem* 9: 1744–1754.
140 Shen, J.-S., Edwards, N.J., Hong, Y.B., and Murray, G.J. (2008). *Biochemical and Biophysical Research Communications* 369: 1071–1075.
141 Kornhaber, G.J., Tropak, M.B., Maegawa, G.H. et al. (2008). *ChemBioChem* 9: 2643–2649.
142 Vanier, M.T. (2013). Chapter 176: Niemann–Pick diseases. In: *Handbook of Clinical Neurology*, vol. 113 (3rd series) Pediatric Neurology Part III, (ed. O. Dulac, M. Lassonde, and H.B. Sarnat), 1717–1721. Elsevier.
143 Pineda, M., Walterfang, M., and Patterson, M.C. (2018). *Orphanet Journal of Rare Diseases* 13: 140–161.
144 Schuchman, E.H. and Desnick, R.J. (2017). *Molecular Genetics and Metabolism* 120: 27–33.
145 Suzuki, Y., Oshima, A., and Namba, E. (2001). β-Galactosidase deficiency (β-galactosidosis) GM1 gangliosidosis and Morquio B disease. In: *The Metabolic and Molecular Bases of Inherited Disease* (ed. C.R. Scriver, A.L. Beaudet, W.S. Sly, and D. Valle), 3775–3809. New York: McGraw-Hill.
146 Brunetti-Pierri, N. and Scaglia, F. (2008). *Molecular Genetics and Metabolism* 94: 391–396.
147 Prat, C., Lemaire, O., Bret, J. et al. (2008). *Joint, Bone, Spine* 75: 495–498.
148 Tominaga, L., Ogawa, Y., Taniguchi, M. et al. (2001). *Brain & Development* 23: 284–287.
149 Rigat, B.A., Tropak, M.B., Buttner, J. et al. (2012). *Molecular Genetics and Metabolism* 107: 203–212.
150 Takai, T., Higaki, K., Aguilar-Moncayo, M. et al. (2013). *Molecular Therapy* 21: 526–532.
151 Front, S., Biela-Banas, A., Burda, P. et al. (2017). *European Journal of Medicinal Chemistry* 126: 160–170.
152 Front, S., Almeida, S., Zoete, V. et al. (2018). *Bioorganic & Medicinal Chemistry* 26: 5462–5469.
153 Siriwardena, A., Sonawane, D.P., Bande, O.P. et al. (2014). *The Journal of Organic Chemistry* 79: 4398–4404.
154 Matsuda, J., Suzuki, O., Oshima, A. et al. (2003). *Proc. Natl. Acad. Sci. USA* 100: 15912–15917.

155 Fantur, K., Hofer, D., Schitter, G. et al. (2010). *Molecular Genetics and Metabolism* 100: 262–268.

156 Schalli, M., Tysoe, C., Fischer, R. et al. (2017). *Bioorganic & Medicinal Chemistry Letters* 27: 3431–3435.

157 Sandho, K. and Harzer, K. (2013). *The Journal of Neuroscience* 33: 10195–10208.

158 Leal, A.F., Benincore-Flórez, E., Solano-Galarza, D. et al. (2020). *International Journal of Molecular Sciences* 21: 6213–6240.

159 Jeyakumar, M., Butters, T.D., Cortina-Borja, M. et al. (1999). *Proceedings of the National Academy of Sciences of the United States of America* 96: 6388–6393.

160 Shapiro, B.E., Pastores, G.M., Gianutsos, J. et al. (2009). *Genetics in Medicine* 11: 425–433.

161 Tropak, M.B., Reid, S.P., Guiral, M. et al. (2004). *Biological Chemistry* 279: 13478–13487.

162 Rountree, J.S.S., Butters, T.D., Wormald, M.R. et al. (2009). *ChemMedChem* 4: 378–392.

163 Kato, A., Nakagome, I., Nakagawa, S. et al. (2017). *Organic & Biomolecular Chemistry* 15: 9297–9304.

164 Maegawa, G.H., Tropak, M., Buttner, J. et al. (2007). *The Journal of Biological Chemistry* 282: 9150–9161.

165 Bateman, K.S., Cherney, M.M., Mahuran, D.J. et al. (2011). *Journal of Medicinal Chemistry* 54: 1421–1429.

166 Chiricozzi, E., Niemir, N., Aureli, M. et al. (2014). *Molecular Neurobiology* 50: 159–167.

167 Clarke, J.T., Mahuran, D.J., Sathe, S. et al. (2011). *Molecular Genetics and Metabolism* 102: 6–12.

168 Osher, E., Fattal-Valevski, A., Sagie, L. et al. (2015). *Orphanet Journal of Rare Diseases* 10: 1–7.

169 Mikulka, C.R. and Sands, M.S. (2016). *Journal of Neuroscience Research* 94: 1126–1137.

170 Bradbury, A.M., Bongarzone, E.R., and Sands, M.S. (2021). *Neuroscience Letters* 752: 135841–135841.

171 Hill, C.H., Viuff, A.H., Spratley, S.J. et al. (2015). Azasugar inhibitors as pharmacological chaperones for Krabbe disease. *Chemical Science* 6: 3075–3086.

172 Viuff, A., Salamone, S., McLoughlin, J. et al. (2021). The bicyclic form of galacto-noeurostegine is a potent inhibitor of β-galactocerebrosidase. *ACS Medicinal Chemistry Letters* 12: 56–59.

173 Biela-Banaś, F., Oulaïdi, S., Front, E. et al. (2014). Martin iminosugar-based galactoside mimics as inhibitors of galactocerebrosidase: SAR studies and comparison with other lysosomal galactosidases. *ChemMedChem* 9: 2647–2652.

174 Do, H.V., Khanna, R., and Gotschall, R. (2019). *Annals of Translational Medicine* 7: 291–306.

175 Meena, N.K. and Raben, N. (2020). *Biomolecules* 10: 1339–1378.

176 Flanagan, J.J., Rossi, B., Tang, K. et al. (2009). *Human Mutation* 30: 1683–1692.

177 Parenti, G., Zuppaldi, A., Pittis, M.G. et al. (2007). *Molecular Therapy* 15: 508–514.

178 D'Alonzo, D., De Fenza, M., Porto, C. et al. (2017). *Journal of Medicinal Chemistry* 60: 9462–9469.

179 Khanna, R., Jr, A.C., Powe, Y. et al. (2014). *PLoS One* 9: e102092.

180 Kishnani, P., Tarnopolsky, M., Roberts, M. et al. (2017). Duvoglustat HCl increases systemic and tissue exposure of active acid α -glucosidase in pompe patients co-administered with alglucosidaseα. *Molecular Therapy* 25: 1199–1208.

181 Porto, C., Cardone, M., Fontana, F. et al. (2009). *Molecular Therapy* 17: 964–971.

182 Parenti, G., Fecarotta, S., la Marca, G. et al. (2014). *Molecular Therapy* 22: 2004–2012.

183 Stepien, K.M., Ciara, E., and Jezela-Stanek, A. (2020). *Genes* 11: 1383–1406.

184 Taylor, R.M., Farrow, B.R., and Stewart, G.J. (1992). *American Journal of Medical Genetics* 42: 628–632.

185 Taylor, R.M., Farrow, B.R., Stewart, G.J. et al. (1988). *Transplantation Proceedings* 20: 89–93.

186 Vellodi, A., Cragg, H., Winchester, B. et al. (1995). *Bone Marrow Transplantation* 15: 153–158.

187 Miano, M., Lanino, E., Gatti, R. et al. (2001). *Bone Marrow Transplantation* 27: 747–751.

188 Gupta, A., Lund, T.C., Anderson, N. et al. (2019). *Molecular Genetics and Metabolism* 126: S66.

189 Moreno-Clavijo, E., Carmona, A.T., Moreno-Vargas, A.J. et al. (2011). *Chimia* 65: 40–44.

190 Moreno-Clavijo, E., Carmona, A.T., Moreno-Vargas, A.J. et al. (2011). *Current Organic Synthesis* 8: 102–133.

191 Malm, D. and Nilssen, O. (2008). *Orphanet Journal of Rare Diseases* 3: 1–21.

192 Desnick, R.J., Sharp, H.L., Grabowski, G.A. et al. (1976). *Pediatric Research* 10: 985–996.

193 Autio, S., Louhimo, T., and Helenius, M. (1982). *Annals of Clinical Research* 14: 93–97.

194 Grewal, S.S., Shapiro, E.G., Krivit, W. et al. (2004). *The Journal of Pediatrics* 144: 569–573.

195 Yesilipek, A.M., Akcan, M., Karuso, G. et al. (2012). *Pediatric Transplantation* 16: 779–782.

196 Mynarek, M., Tolar, J., Albert, M.H. et al. (2012). *Bone Marrow Transplantation* 47: 352–359.

197 Rísquez-Cuadro, R., Matsumoto, R., Ortega-Caballero, F. et al. (2019). *Journal of Medicinal Chemistry* 62: 5832–5843.

10

Carbohydrates and Carbohydrate-Based Therapeutics in Alzheimer's Disease

Ana M. Matos, João Barros and Amélia P. Rauter

Universidade de Lisboa, Centro de Química Estrutural, Institute of Molecular Sciences, Faculdade de Ciências, Department of Chemistry and Biochemistry, Ed. C8, Piso 5, Campo Grande, Lisboa, 1749-016, Portugal

10.1 Introduction

Carbohydrates, the most abundant organic compounds in nature, play unique roles in health and disease. The complex glycans displayed on the cell surface facing the extracellular space are the key molecules for cell recognition and adhesion in signaling pathways and in cell–pathogen interactions [1, 2]. Carbohydrates also trigger immune responses [3–5], and modulate key biological processes in neurodegeneration, namely those related to the *O*-GlcNAc modification (also known as O-GlcNAcylation) of serine (Ser) and threonine (Thr) protein residues in the brain, the dysregulation of which is associated with neurodegenerative diseases [6, 7]. Alzheimer's disease (AD) is the most common one, affecting 60% of the over 50 million people with dementia in 2020, a number expected to double every two years, reaching 82 million in 2030 and 152 million in 2050 [8]. AD is a chronic and progressive disease, resulting from an irreversible degeneration of the brain that causes cognitive impairment, dementia, and ultimately results in death, as no efficient therapeutics are known to control the progression of this pathology. This protein misfolding disease is associated with the formation of soluble Aβ1–42 toxic small oligomers, derived from abnormal cleavage of β-amyloid precursor protein (APP). Their aggregation leads to the formation of fibrils, which give rise to extracellular deposits, the senile plaques [9]. Another AD neuropathological hallmark consists of intracellular neurofibrillary tangles (NFTs) formation due to altered kinase and phosphatase activities, leading to hyperphosphorylation of the Tau protein. This is followed by the aggregation of hyperphosphorylated Tau into paired helical filaments and, finally, into NFTs. The aggregation events and associated aggregation stress leads to cell apoptosis [9, 10] (Figure 10.1).

Carbohydrate-Based Therapeutics, First Edition. Edited by Roberto Adamo and Luigi Lay.

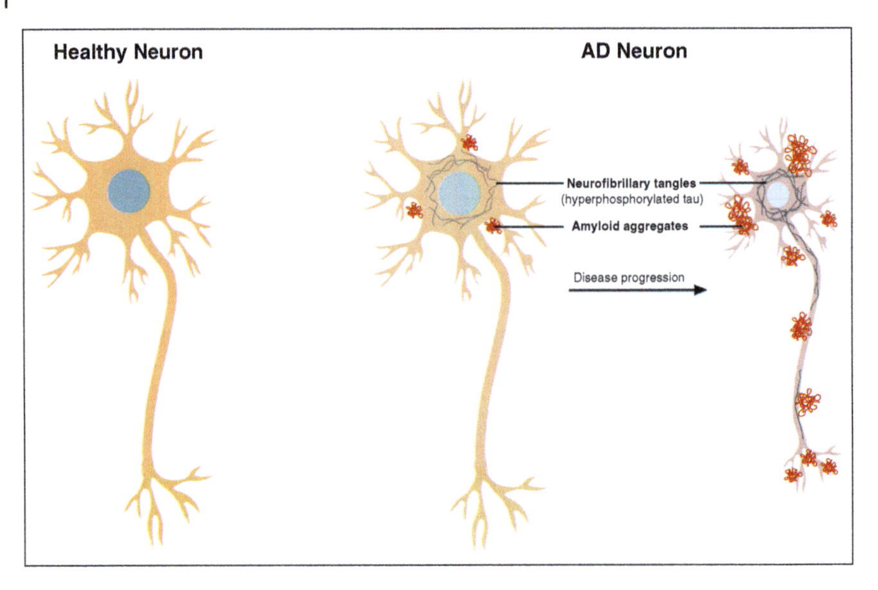

Figure 10.1 Healthy neuron vs. neuron after degeneration. Source: Adapted from Demetrius et al. [9].

Moreover, the cellular prion protein (PrP^C), located in the neuronal cell surface, is a high-affinity binding partner of Aβ oligomers (Aβos) and promotes the activation of Fyn kinase, triggering a cell signaling pathway that culminates in Tau hyperphosphorylation [11]. Accordingly, Fyn activity increases in AD brain when neurons are exposed to Aβo, via PrP^C [12, 13]. Interestingly, prion protein misfolding also promotes prion amyloid formation. Its conformational transition from the α-helix-rich cellular form into the mainly β-sheet containing counterpart initiates an "autocatalytic" reaction that leads to the accumulation of amyloid fibrils (PrP^{Sc}) in the central nervous system (CNS), leading to neurodegeneration. This amyloidogenic process is, among other factors, *O*-glycosyl dependent and can be triggered or prevented by changing the sugar residue attached to Ser/Thr in the PrP core [14].

AD epidemiological data, associated with its devastating nature, unsuccessful treatment options, and high socio-economic impact are challenging for the research of new therapeutics. In 2015, a review on carbohydrates and glycomimetics in AD diagnosis and therapeutics was published [15], covering molecules acting on Aβ amyloid events, namely glycosides containing terpenes, phenolics, peptidomimetics, and metal-ion chelators, free and protected sugars, cyclitols and glycosaminoglycans (GAGs). Carbohydrate-based molecules targeting the cholinergic system were also reviewed [16], namely the acetylcholinesterase (AChE) and butyrrylcholinesterase (BChE) inhibitors, aiming to increase levels of the neurotransmitter acetylcholine (ACh) in the brain of AD patients. More recently, carbohydrate-peptide conjugates were also reviewed as amyloidogenic aggregation inhibitors [17].

In this chapter, we focus on the investigation carried out to uncover the role of carbohydrates in AD and approaches toward carbohydrate-based therapeutics. Dysregulation of O-GlcNAcylation in intracellular proteins is associated with neurodegenerative diseases, including AD, for which *O*-GlcNAc cycling is considered a therapeutic target against the formation of amyloid plaques and NFTs . Patients with AD have *O*-GlcNAc levels 50% lower than normal individuals, and efforts toward the discovery of efficient carbohydrate-based therapeutics to control these levels by inhibition of the *O*-GlcNAc hydrolase enzyme, *O*-GlcNAcase (OGA), are here reviewed. Synthesis and bioactivity of the most promising classes of inhibitors are presented, namely carbohydrate-based thiazolines, 2-acetamido-2-deoxy-D-glucono-1,5-lactone *O*-(phenylcarbamoyl)oxime (PUGNAc), and GlcNAc statins.

The role of GalNAc in neurodegeneration is also revised herein, highlighting reports that show the importance of GalNAc levels in APP to control Aβ production and the significance of GalNAc residues in chondroitin sulfate proteoglycans (PGs) that play a role in regulating the trophic microenvironment of neurons. Moreover, GalNAc containing oligosaccharides and chitosan (CTS) oligosaccharides have neurodegenerative effects and are also covered in this chapter.

AD is a multifactorial disease, with the cholinergic system of AD patients being severely affected and represent a therapeutic target. ACh is a major neurotransmitter in the brain and AD patients have low levels of ACh, emerging the inhibition of cholinesterases (ChEs), enzymes that split ACh into choline and acetate, as a promising option to treat AD. The drugs rivastigmine, galantamine, and donepezil are clinically in use and act as ChE inhibitors. However, they are not efficient to control disease progression and have side effects, such as convulsions, severe nausea, stomach cramps, vomiting, irregular breathing, confusion, muscle cramps, and muscle weakness, among others [18, 19]. New directions in anticholinesterase drug development are encouraging and the latest findings on carbohydrate-based ChE inhibitors are also covered.

Finally, our latest discoveries on carbohydrate-based inhibitors of Fyn kinase, Aβ aggregation, Aβ and prion binding, and of the oxidative stress-induced neurotoxicity are also disclosed. This chapter ends by giving an overview about carbohydrate–protein interactions as potential targets for AD drug discovery.

10.2 *O*-GlcNAc Transferase (OGT) and *O*-GlcNAc Hydrolase (OGA) in Neurodegeneration

The modification of the hydroxy group of Ser and Thr residues to install an *O*-(2-acetamido-β-D-glucopyranosyl, *O*-GlcNAc) functionality (a process termed O-GlcNAcylation) is an essential mechanism that takes place in cell cycle and metabolic processes on nuclear, cytoplasmic, and mitochondrial proteins. It is achieved enzymatically by *O*-GlcNAc transferase (OGT), which transfers *O*-GlcNAc from uridine diphosphate (UDP)-GlcNAc donor to Ser and Thr residues, while the enzymatic hydrolysis to remove the glycan and unveil the free hydroxy group is carried out by OGA (Figure 10.2), a member of the glycoside hydrolase family 84 (GH84) of

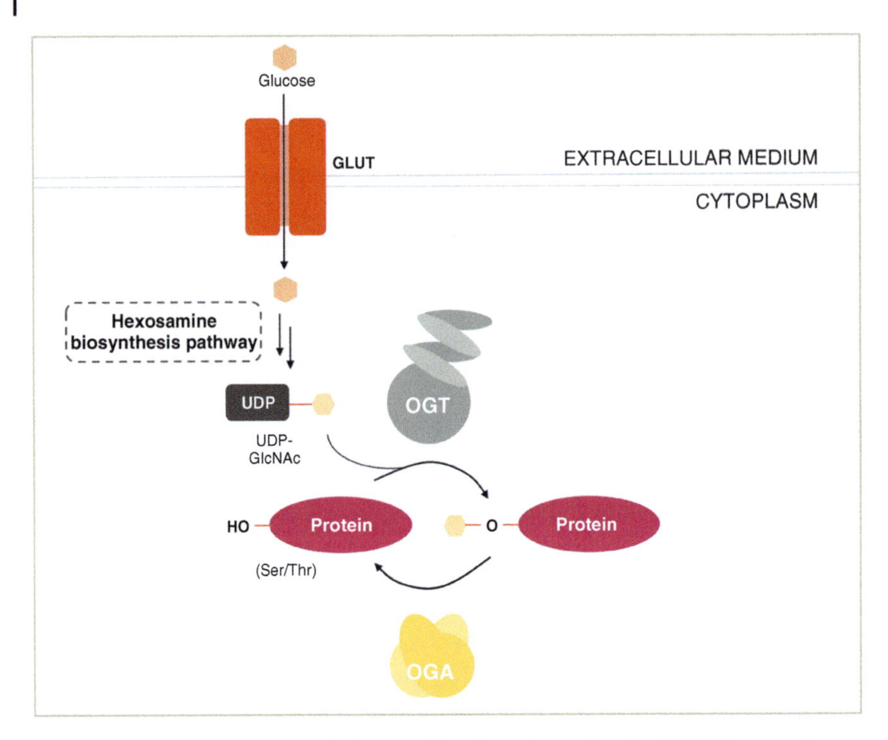

Figure 10.2 Illustration of *O*-GlcNAc modification of proteins via *O*-GlcNAc transferase (OGT) and the hydrolase *O*-GlcNAcase (OGA).

the CAZy classification system [20]. The *O*-GlcNAc modification is particularly abundant in the brain controlling a number of processes, including memory formation [21], and seems to be essential for the normal function of the mammalian nervous system [22–24]. Dysregulation of O-GlcNAcylation is associated with neurodegenerative diseases. As the O-GlcNAcylation of proteins is regulated by the intracellular glucose metabolism, which is reduced in the brain of AD patients, their *O*-GlcNAc levels are nearly 50% lower than in control brains [25, 26].

The role of O-GlcNAcylation in neurodegeneration has been extensively reviewed [6]. The expression and activities of OGT and OGA in the brain are age-dependent and more than four thousand *O*-GlcNAc protein targets have been identified as playing critical roles in many cellular processes [6, 27]. In the mammalian brain, *O*-GlcNAc modification of Tau, which is naturally O-GlcNAcylated [28], decreases its phosphorylation and toxicity, suggesting a neuroprotective role of brain O-GlcNAcylation for AD treatment [6].

10.2.1 *O*-GlcNAc Cycling as a Therapeutic Target Against Alzheimer's Amyloid Plaques and Neurofibrillary Tangles

Approximately 2–5% of all glucose entering the cell is channeled into the hexosamine biosynthetic pathway to generate UDP-GlcNAc [6] (Figure 10.2). Because

O-GlcNAcylation depends on the availability of UDP-GlcNAc, and in turn, intracellular UDP-GlcNAc level determines OGT activity, O-GlcNAcylation is considered a valuable intracellular sensor of glucose metabolism that can be directly regulated in a glucose-responsive manner [29].

Positron Emission Tomography (PET) studies [30], supported by experiments with transgenic mice [31, 32] have shown that brain glucose metabolism declines with age and is more impaired in AD patients. Glucose metabolism impairment may lead to lower UDP-GlcNAc levels and consequently lower O-GlcNAcylation, resulting in impaired GlcNAc cycling. But how does O-GlcNAcylation affect the main pathological hallmarks of AD, i.e. formation of extracellular Aβ amyloid plaques and intracellular NFT that is accompanied by synaptic loss and dementia? The APP is an *O*-GlcNAc-modified protein processed by three different proteases, namely the α-, the β-, and the γ-secretase. APP cleavage by α-secretase in the extracellular domain is non-amyloidogenic, generating a soluble fragment sAPPα and a membrane-bound C-terminal fragment (APP-CTFα) (Figure 10.3). The latter is cleaved by γ-secretase in the transmembrane region to afford the non-amyloidogenic fragment p3 and the β-amyloid precursor protein intracellular domain (AICD). APP β-secretase (BACE1) cleavage generates the soluble sAPPβ fragment and the membrane-bound fragment APP-CTFβ, which is also cleaved by γ-secretase generating amyloid-β peptide monomers (Aβ1–40 and Aβ1–42), together with the AICD [6, 33]. This abnormal cleavage of APP is an important event leading to the overproduction and aggregation of Aβ species. Approximately, 90% of all Aβ fragments generated are Aβ1–40 and a smaller fraction corresponds to the toxic amyloidogenic, Aβ1–42, observed in amyloid plaques in the brain. Therefore, among the

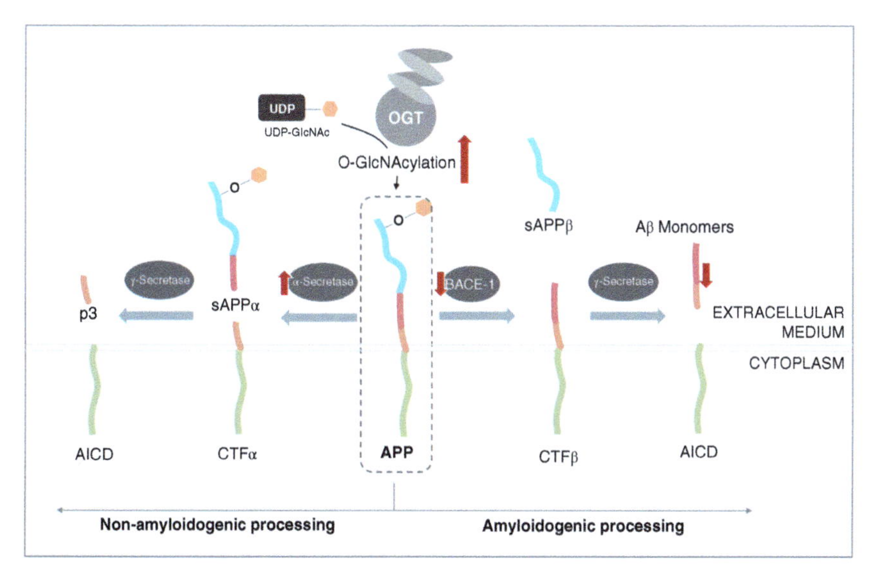

Figure 10.3 Illustration of amyloidogenic and non-amyloidogenic cleavage of APP. Source: Adapted from Wani et al. [6].

products generated by the amyloidogenic (initially cleaved by the BACE1) and non-amyloidogenic (initially cleaved by α-secretase) pathways, only the amyloidogenic products are thought to lead to AD pathology [34].

Interestingly, it was demonstrated that by increasing O-GlcNAcylation, either through pharmacological or genetic interventions, processing by α-secretase is also increased, resulting in enhanced neuroprotective sAPPα levels and decreased Aβ formation. In addition, hyper O-GlcNAcylation, produced in a 5XFAD Aβ mouse model treated with the OGA inhibitor 1,2-dideoxy-2′-propyl-α-D-glucopyranoso-[2,1-*d*]-$\Delta^{2'}$-thiazoline (NButGT Figure 10.4), decreased Aβ1–40 and Aβ1–42 levels, reduced Aβ plaque formation, and improved cognition [26]. These findings suggest that *O*-GlcNAc cycling is indeed a target for the identification of therapeutics for AD.

After the discovery that human brain Tau is O-GlcNAcylated, scientists have focused on finding a link with neurodegeneration, as Tau hyperphosphorylation leads to the formation of NFT [28]. Hence, the relationship between protein O-GlcNAcylation and phosphorylation is key for the investigation of the molecular mechanism leading to brain tauopathies, including AD [35]. Liu et al. [36] found that decreased *O*-GlcNAc levels and increased Tau phosphorylation occur in mice brain obtained from mice sacrificed after injection into the left ventricle of the brain of 6-diazo-5-oxonorleucine (DON), an inhibitor of glutamine:fructose-6-P, the rate-limiting enzyme of hexosamine biosynthesis pathway. The analysis of human brain

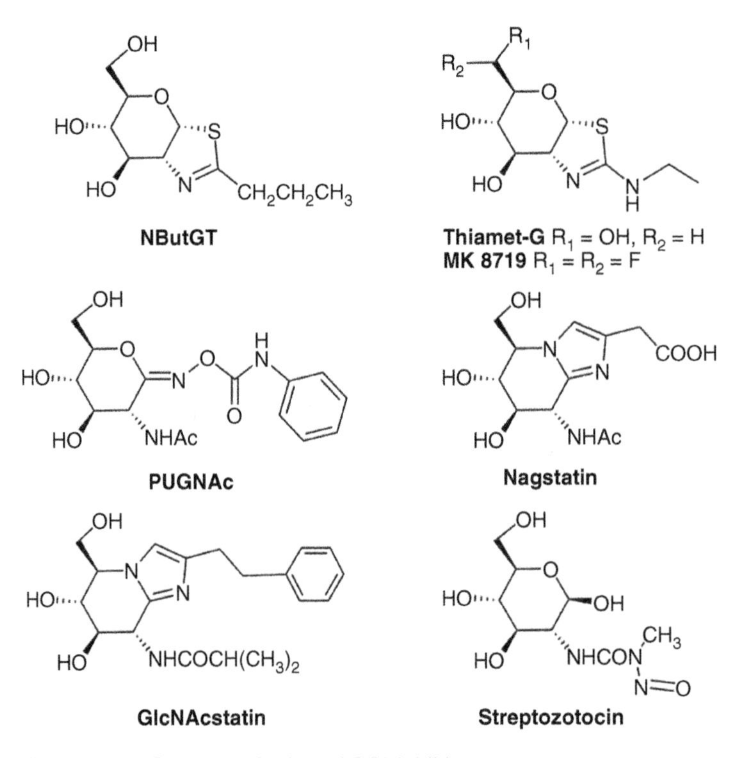

Figure 10.4 Structure of selected OGA inhibitors.

tissue showed that hyperphosphorylated Tau has up to four-fold lower *O*-GlcNAc levels. Interestingly, the reduction of protein O-GlcNAcylation in AD and other Tau-driven pathologies seems to vary in brain regions, as reported by Frenkel–Pinter et al. [37]. They found an increase in O-GlcNAcylation in the hippocampus of AD patients, suggesting that this complex disorder involves lesions in various brain regions, which should be separately investigated for a comprehensive study of AD etiology. Nonetheless, the research carried out so far supports that targeting O-GlcNAcylation represents one of the most promising therapeutic opportunities to control AD neurodegeneration by decreasing Tau hyperphosphorylation and by recovering brain dysmetabolism, leading to reduced brain damage and improved cognition [29].

10.2.2 OGA Inhibitors

The structure of some of the most studied OGA inhibitors is depicted in Figure 10.4. These carbohydrate-based compounds contain either a thiazoline ring, such as NButGT, Thiamet-G, and MK8719, or an imidazole ring fused to the carbohydrate moiety as GlcNAcstatins, which structure was inspired by that of nagstatin. PUGNAc, one of the first OGA inhibitors discovered, shows a different structure as the sugar is linked to a carbamate functionality. Around 2000, PUGNAc and streptozotocin (STZ) were used to increase cellular *O*-GlcNAc levels for the study of related cellular processes, but STZ is a poor OGA inhibitor and is highly toxic to pancreatic β-cells, being used to induce type 1 diabetes in animals [38].

Human *O*-GlcNAcase (hOGA) is expressed in two isoforms. The larger variant OGA, localized in the nucleus and cytosol, is more active than the other isoform, expressed only during embryo formation [27]. The crucial role of OGA for maintaining or decreasing *O*-GlcNAc levels in the brain is, indeed, key for brain neurodegeneration, encouraging the development of OGA inhibitors. In 2005, Vocadlo and coworkers were able to elucidate, for the first time, the catalytic mechanism of hOGA [39]. They described a two-step mechanism involving the nucleophilic participation of the 2-acetamido group, with the formation of an oxazole or oxazolinium intermediate. Later on, in 2016, Vocadlo's group, supported by the experimental work carried out with thiazoline inhibitors (see Section 10.2.2.3), proposed that the intermediate is an oxazoline rather than an oxazolinium ion (Figure 10.5a) [40]. A molecule of water attacks the anomeric center of the intermediate, breaking the oxazole ring and leading to the hemiacetal with retention of configuration. In 2017, the crystal structures of hOGA and its substrates were determined [41], providing important information for further investigation of the mechanism of hOGA.

Recently, Xiong and Xu confirmed the existence of the oxazoline intermediate by studying the whole catalytic process with hOGA using a Quantum Mechanics/Molecular Mechanics approach [42]. Based on the calculated free energy profiles, they showed that the intramolecular cyclization by the 2-acetamido group leading to an intermediate bicyclic oxazolinium ion (I1) is the rate-limiting step, and their simulations further suggest the formation of the oxazoline ring intermediate I2,

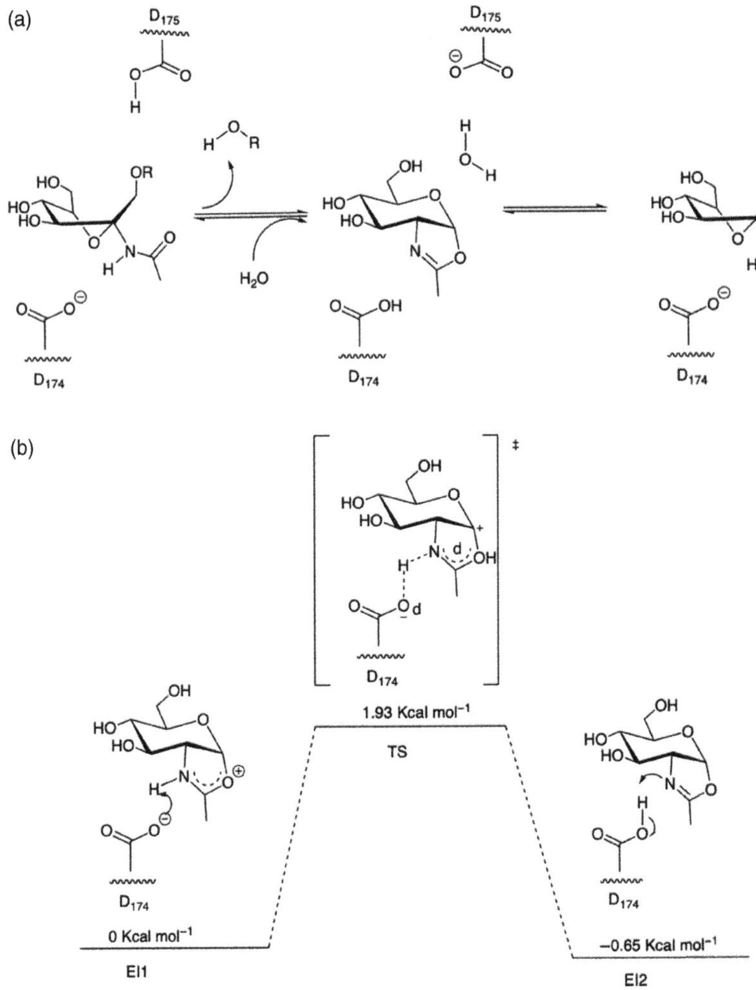

Figure 10.5 Mechanism for hOGA inhibition. (a) Mechanism proposed by Macauley et al. [39] and Cekic et al. [40] (b) Energy profiles (Kcal mol^{-1}) adapted from Xiong and Xu (2020) [41] for the formation of the oxazoline intermediate by proton transfer of the oxazolinium intermediate to the residue D_{174} as deduced by computational simulations.

resulting from proton transfer from the 2-acetamido group to residue D174 (Figure 10.5b). A relatively small barrier was found to connect EI1 and EI2 (5.43 kcal mol^{-1} in SCC-DFTB/MM and 1.93 kcal mol^{-1} in DFT calculations), which might contribute to accelerating the overall reaction rate.

The authors also made mutagenesis simulations of D174A and D175A, which indicate that these catalytic residues mainly affect the observed reaction rate by affecting the stability of the intermediate [42]. These results seem to be in full agreement with the experimental results of Vocadlo's group, who developed aminothiazoline inhibitors (see Section 10.2.2.3). Their potency is strongly pK_a-dependent, leading to picomolar binding. The authors described them as genuine transition state analogs [40]. This substrate-assisted catalytic mechanism is followed by both hOGA and human lysosomal β-hexosaminidases, which are hydrolases of family 20 (GH20) [43], cleaving GlcNAc from terminal glycoproteins/glycosphingolipids. Nonetheless, Vocadlo's group experiments, in 2005 [39], showed that the active site of hOGA tolerates bulkier acetamido substituents than the one of lysosomal β-hexosaminidase, which encouraged them to design new structures aiming at selectivity for hOGA inhibition.

10.2.2.1 PUGNAc

In the search for OGA inhibitors, Vasella, and coworkers, in 1990, were inspired by aldonolactones, known as potent glycosidase inhibitors, and particularly by 2-acetamido-2-deoxyglucono-1,5-lactone, reported to inhibit OGA, although with a μM K_i. As both glucono-1,5-lactone oxime and its phenylcarbamoyl derivative were better inhibitors of β-glucosidase than the corresponding lactone, Vasella synthesized for the first time 2-acetamido-2-deoxy-D-glucono-1,5-lactone *O*-(phenylcarbamoyl)oxime (PUGNAc) (Scheme 10.5) [45] to explore its ability to inhibit OGA. Starting from oxime **1** oxidized with MnO_2, they prepared the cyclized benzylidene protected (*Z*)-**2** in 50% yield, with starting material recovered in 35% yield. Debenzylidenation with Na/NH_3 and acetylation was followed by partial deacetylation with CH_3NH_2 remaining the sugar hydroxy groups acetylated (compound **3**). Reaction with phenyl isocyanate gave carbamate **5**, which was deprotected with ammonia in methanol to afford PUGNAc in 16.8% overall yield starting from oxime **1** (Scheme 10.1).

Vasella and coworkers have then tested PUGNAc for the inhibition of OGA from animal, fungal, and plant origin. They found out that PUGNAc was very effective, particularly for the inhibition of the fungal enzyme, exhibiting $K_i = 40\,nM$ [47]. This potent OGA inhibitor was also the best of any glycosidase inhibitors described at that time. However, the first synthesis has severe problems with the scale-up,

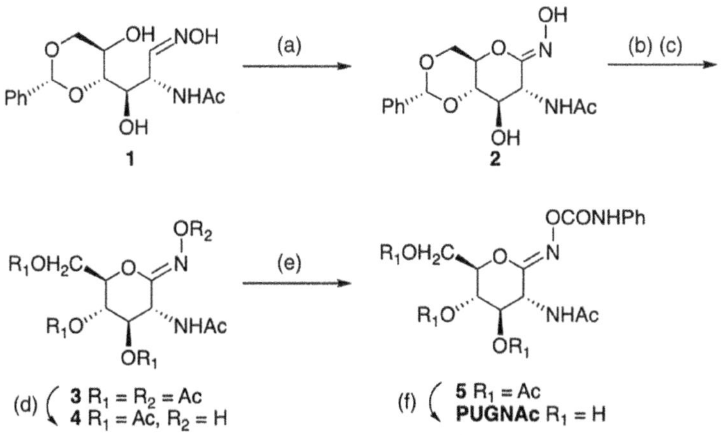

Scheme 10.1 The first synthesis of PUGNAc by Beer et al. [45, 46]. Reagents/solvent and yield: (a) Activated MnO_2, MeOH, 50%; (b) Na, NH_3, MeOH, 86%; (c) Ac_2O, Py, 96%; (d) CH_3NH_2, CH_3OH, 71%, (e) PhNCO, Et_3N, THF, 94%; (f) saturated NH_3, MeOH, 61%.

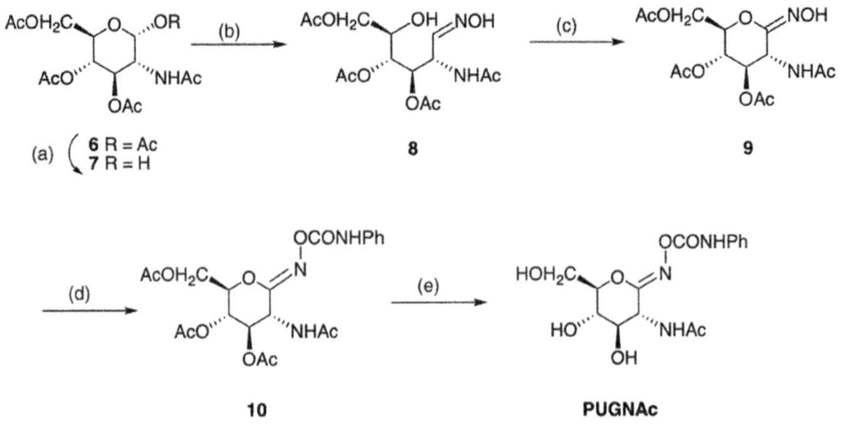

Scheme 10.2 Improved approach to prepare PUGNAc by Vasella and Mohan [48]. Reactions/solvent and yield: (a) $(NH_4)_2CO_3$, THF/MeOH, 74%; (b) $NH_2OH \cdot HCl$, py/MeOH, 75%; (c) DBU, NCS/DCM, 59%; (d) PhNCO, Et_3N, DCM, 70%; (e) Aq. NH_3, 73%.

particularly the oxidation step with MnO_2, which encouraged Vasella group to improve the synthetic approach for PUGNAc (Scheme 10.2) [48]. Starting from 2-acetamido-1,3,4,6-tetra-*O*-acetyl-2-deoxy-α-D-glucopyranose **6**, the anomeric deacetylation was carried out by ammonium carbonate to give **7** in 74% yield from 100 g starting material. The open chain oxime **8** was obtained by treatment of **7** with hydroxylammonium chloride and pyridine in MeOH under reflux. Reaction with *N*-chlorosuccinimide in DBU afforded compound **9**, which reacted with PhNCO in THF to yield phenyl carbamate **10**. De-O-acetylation with saturated aqueous NH_3 afforded PUGNAc in 16.7% overall yield.

As shown by Vocadlo and coworkers [39], PUGNAc is a potent inhibitor of hOGA (K_i = 46 nM) but unfortunately, it is also a potent inhibitor of β-hexosaminidase (K_i = 36 nM). The dysfunction of this enzyme results in the accumulation of gangliosides and other glycoconjugates in the lysosome, causing Tay Sachs and Sandhoff neurodegenerative diseases, as the enzyme is localized in the lysosome. This lack of selectivity encouraged the modification of PUGNAc structure by varying the 2-*N*-acyl chain and Stubbs et al. [49] reported the synthesis of PUGNAc and analogs starting from glucosamine hydrochloride to prepare the 2-*N*-Boc protected precursor **11**, submitted to the reaction conditions previously reported by Mohan and Vasella [48] to prepare **12** (Scheme 10.3) [49]. Treatment with phenyl isocyanate yields the carbamate **13**, whose Boc deprotection is achieved with trifluoroacetic acid (TFA). Further reaction with acyl chloride in pyridine and acetyl deprotection to release the sugar hydroxy groups with NH_3/MeOH afforded PUGNAc in 5.7% overall yield from **11** and its desired analogs. Unfortunately, they showed K_i for hOGA and for human hexosaminidase in the μM range and their selectivity was very poor [49].

The crucial step in both the synthetic approaches followed by Vasella group [48] and Vocadlo group [49] is the *N*-chlorosuccinimide-mediated oxidative ring closure of the oxime precursor in relatively low yields, as a secondary product with a five-membered ring resulting from acetyl migration is formed. Moreover, the separation

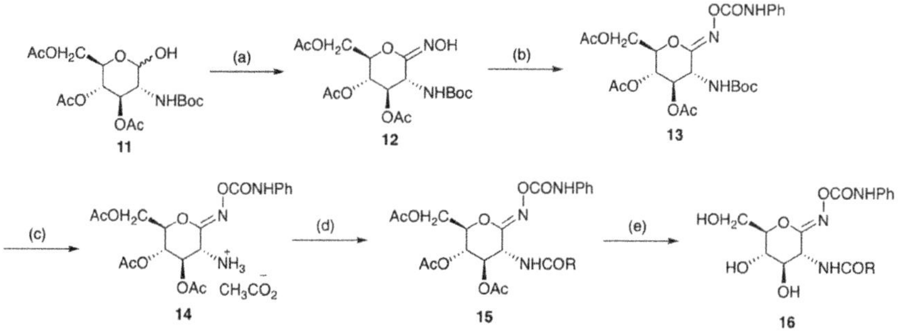

5 R = CH_3
15a R = CH_2CH_3
15b R = $(CH_2)_2CH_3$
15c R = $(CH_2)_3CH_3$
15d R = $(CH_2)_4CH_3$
15e R = $CH(CH_3)_2$
15f R = $CH_2CH(CH_3)_2$

PUGNAc (R=CH_3)
16a R = CH_2CH_3
16b R = $(CH_2)_2CH_3$
16c R = $(CH_2)_3CH_3$
16d R = $(CH_2)_4CH_3$
16e R = $CH(CH_3)_2$
16f R = $CH_2CH(CH_3)_2$

Scheme 10.3 Synthesis of PUGNAc and *N*-acylated derivatives as reported by Stubbs et al. [49] in 2006. Reagents/solvent and yield: (a) i. $NH_2OH·HCl$, py, MeOH; ii. DBU, NCS, DCM, 52% over the two steps; (b) PhNCO, Et_3N, THF, 71%; (c) CF_3COOH, DCM; (d) RCOCl, py, **5** (48%); (e) NH_3, MeOH, PUGNAc (32%), **16a** (32%), **16b** (26%), **16c** (23%), **16d** (26%), **16e** (29%) **16f** (23%) yield over two steps.

of both products was found to be quite difficult [48, 49]. Aiming to overcome this problem, Goddard–Borger and Stubbs [50] developed a new approach, starting from the peracetylated 2-azido precursor **17** (Scheme 10.4), obtained from glucosamine in two steps and good overall yield [48, 50]. This azido sugar reacted with benzylamine to deacetylate the anomeric position. Synthesis of the open chain oxime succeeded in very high yield (97%) and ring closure with DBU and *N*-chlorosuccinimide gave **19** in 93% yield with the desired (*Z*)-stereochemistry. The reductive acylation of the azido

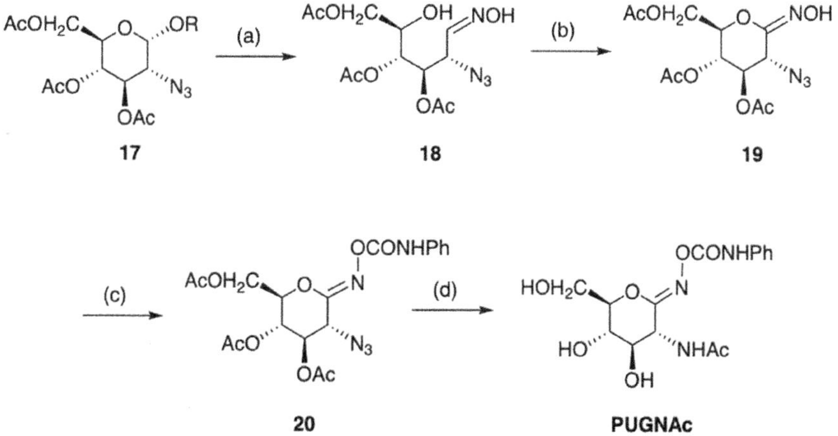

Scheme 10.4 PUGNAc synthesis developed by Goddard-Borger and Stubbs [50]. Reagents/solvent and yield: (a) $NH_2OH·HCl$, py, MeOH, 97%; (b) DBU, NCS, DCM, 93%; (c) PhNCO, Et_3N, THP, 91%; (d) 1. PMe_3, AcOH, 2-$(PhSe)_2$, CH_2Cl_2, toluene, water, 89%; 2. NH_3/MeOH, 61%.

group was carried out by reaction of **19** with acetic acid, trimethylphosphine, and a catalytic amount of 2,2′-diphenyldiselenide to afford PUGNAc in 89% yield. This was, indeed, the most fruitful and high-yielding methodology to synthesize PUGNAc, whose access in reasonable quantities has been required for research, due to its broad spectrum of inhibitory activity acting on enzymes interfering in a wide number of biological processes. Interestingly, it was used for the synthesis of PUGNAc analog with *galacto* configuration, termed Gal-PUGNAc [50], which is a potent and selective inhibitor of lysosomal *exo-N*-acetyl-D-glucosaminidases [51].

Pursuing their research to control PUGNAc selectivity, Stubbs and coworkers [44], in 2016, synthesized new PUGNAc analogs, where the carbamoyl moiety is varied in size, hydrophobicity, and shape. For the purpose, they envisioned three series of compounds. In one of them, the PUGNAc phenyl group is substituted in the *para*-position with methyl or methoxy groups, or with a bromine atom, or replaced by a benzyl group. Another series has a cyclic or acyclic aliphatic structure replacing the phenyl group and the third one has NHPh of the carbamate replaced by an amino acid derivative. To access PUGNAc and the designed 63 compounds, the authors developed a new methodology, in which a colorogenic intermediate carbonate **21** is formed *in situ* by reaction of **9** with 4-nitrophenyl chloroformate, and then transformed into carbamates in good overall yield by reaction with anilines or amines, in a one-pot reaction (Scheme 10.5). The inhibition of OGA and of

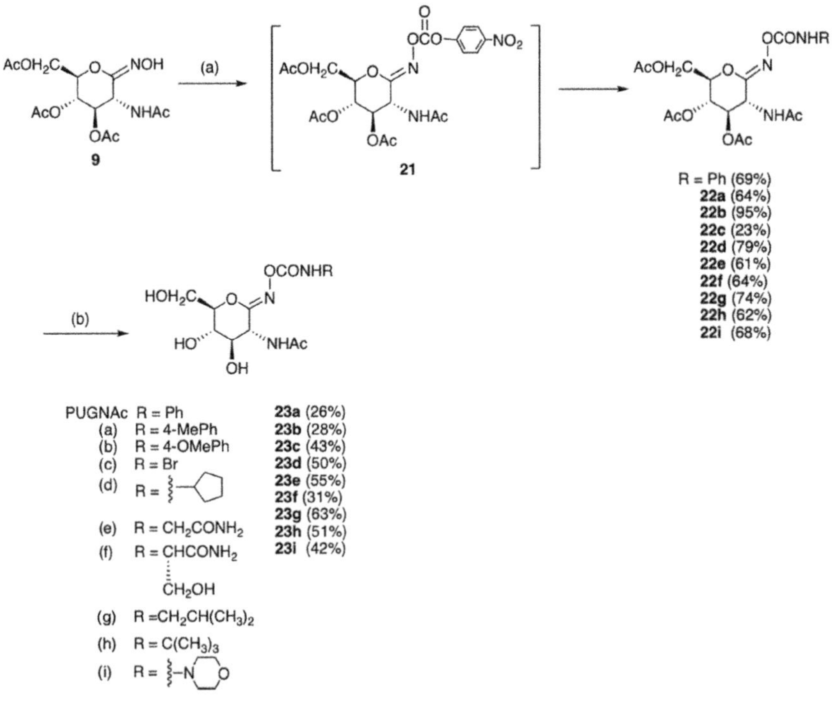

Scheme 10.5 Synthesis of PUGNAc and representative analogs with modified carbamate moiety as developed by Stubbs and coworkers [44]. Reagents/solvent: (a) 1. 4-nitrophenyl chloroformate, DIPEA, THF; 2. Amine or aniline-based compound, DIPEA; (b) NH_3, MeOH.

Table 10.1 Inhibition constant of compounds **23a–23i** over OGA and HEXB.

	K_i (μM)	
Inhibitor	**HEXB**	**OGA**
PUGNAc	0.036 [34]	0.046 [34]
23a	0.021 ± 0.005	0.028 ± 0.007
23b	0.111 ± 0.019	0.178 ± 0.042
23c	0.047 ± 0.002	0.056 ± 0.016
23d	14 ± 3.5	230 ± 78
23e	1.3 ± 0.17	78 ± 26
23f	45 ± 16	420 ± 140
23g	17 ± 4	220 ± 75
23h	161 ± 8	188 ± 72
23i	0.77 ± 0.031	11 ± 4.8

Source: Adapted from Hattie et al. [44].

β-hexosaminidase B found for this small library of compounds and the K_i values of the most representative compounds (Table 10.1) showed that the aryl derivatives have, in general, a similar potency as PUGNAc, although the electron-donating methoxy group decreased potency as compared to PUGNAc. All the aliphatic derivatives were less potent than PUGNAc for both OGA and HEXB, which is the product of the lysosomal hexosaminidase gene HEXB. Nonetheless, the results obtained demonstrate that the envisioned structural changes were efficient in tuning selectivity to HEXB over OGA, leading to the discovery of potent and selective inhibitors of HEXB.

In conclusion, all efforts to achieve selectivity of PUGNAc for OGA failed, either by changing the carbamate moiety, or the sugar *N*-acylation. Also, the Gal-PUGNAc analog, in which sugar position 4 has the opposite configuration of that exhibited by PUGNAc, is a selective inhibitor of the human lysosomal β-hexosaminidases, changing GM2 ganglioside levels in cultured cells, but is not able to affect *O*-GlcNAc levels [51]. Nonetheless, these selective HEXB inhibitors may become powerful small molecules for defining the role played by GM2 in biological processes at the molecular level in diseases with abnormal accumulation of GM2, and for creating models to follow up disease stages.

10.2.2.2 GlcNAcstatins

In 2006, van Aalten and coworkers [52] presented the first rationally designed glucoimidazole GlcNAcstatin, inspired by the structure of nagstatin (Figure 10.6), a natural product and potent inhibitor of hexosaminidase [55]. GlcNAcstatin has a molecular architecture similar to that of nagstatin, bearing an isopropylamido group on position 8 and a phenethyl group at position 2. The authors expected that

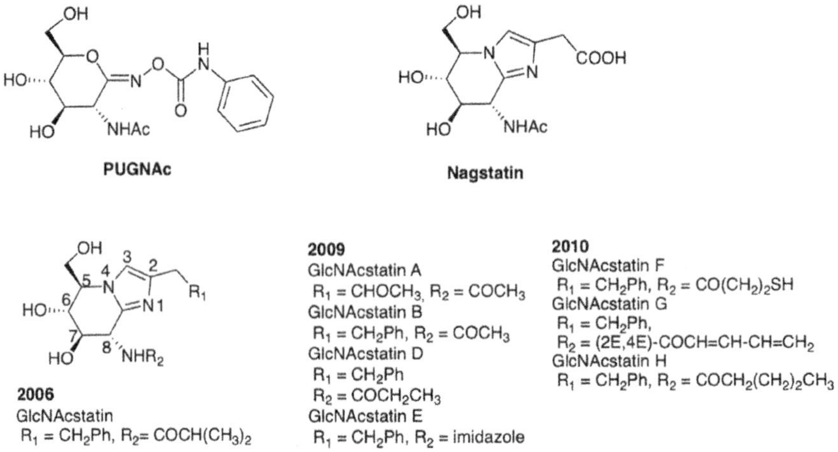

PUGNAc

Nagstatin

2006
GlcNAcstatin
R_1 = CH_2Ph, R_2= $COCH(CH_3)_2$

2009
GlcNAcstatin A
R_1 = $CHOCH_3$, R_2 = $COCH_3$
GlcNAcstatin B
R_1 = CH_2Ph, R_2 = $COCH_3$
GlcNAcstatin D
R_1 = CH_2Ph
R_2 = $COCH_2CH_3$
GlcNAcstatin E
R_1 = CH_2Ph, R_2 = imidazole

2010
GlcNAcstatin F
R_1 = CH_2Ph, R_2 = $CO(CH_2)_2SH$
GlcNAcstatin G
R_1 = CH_2Ph,
R_2 = (2E,4E)-COCH=CH-CH=CH_2
GlcNAcstatin H
R_1 = CH_2Ph, R_2 = $COCH_2(CH_2)_2CH_3$

Figure 10.6 Structure of PUGNAc, nagstatin, and GlcNAcstatins designed and tested by van Aalten and coworkers [52–54].

protonation of the exocyclic nitrogen atom of the imidazole ring would improve OGA inhibitory properties by mimicking the charge distribution in the oxocarbenium ion, as previously reported for the inhibition of β-glycosidases [56]. Indeed, this small molecule GlcNAcstatin is a potent inhibitor of a bacterial OGA isolated from *Clostridium perfringens*, which shares high sequence similarity with that of hOGA, showing a $K_i = 4.6 \pm 0.1$ pM [52]. By structural analysis studies, the authors could confirm a tight interaction between the catalytic site and the presumably protonated imidazole [52].

By combining structural data gathered from the OGA-PUGNAc complex, e.g. the (*Z*)-oxime stereochemistry required for activity [57], and sp^2 hybridization of the C1 carbon, helping PUGNAc pyranose ring to assume a ^{4}E envelope conformation to mimic the transition state [58], van Aalten's group designed and synthesized new derivatives looking for a highly specific inhibition of OGA against hexosaminidases, by varying the *N*-acyl substituents and those of the imidazole ring position 2 [53, 54, 59].

The first synthesis of GlcNAcstatin [52] is illustrated in Scheme 10.6. It was achieved by an elegant but very long synthetic approach with 15 reaction steps starting from dibenzylated L-xylose **24**. Reduction with sodium borohydride and protection of the free hydroxy groups with *tert*-butyldimethylsilyl group afforded the open-chain sugar **25**, which was selectively deprotected by acid hydrolysis to afford the primary alcohol **26**. Swern oxidation gave aldehyde **27**, which reacted with the anion of *N*-tritylimidazole, generated by reaction with butyl lithium, to give a mixture of imidazoles embodying a L-*gulo* chain in **28** and a L-*ido* chain in **29**, isolated in 53% and 18% yield, respectively. Acid detritylation of **28** with triethylsilane was followed by benzoylation of the free hydroxy group leading to the formation of **30**. Reaction with *N*-iodosuccinimide gave the 4,5-diiodoimidazole derivative **31**, submitted to acid hydrolysis to obtain compound **32** with a free hydroxy group. Reaction with trifluoroacetic anhydride to afford

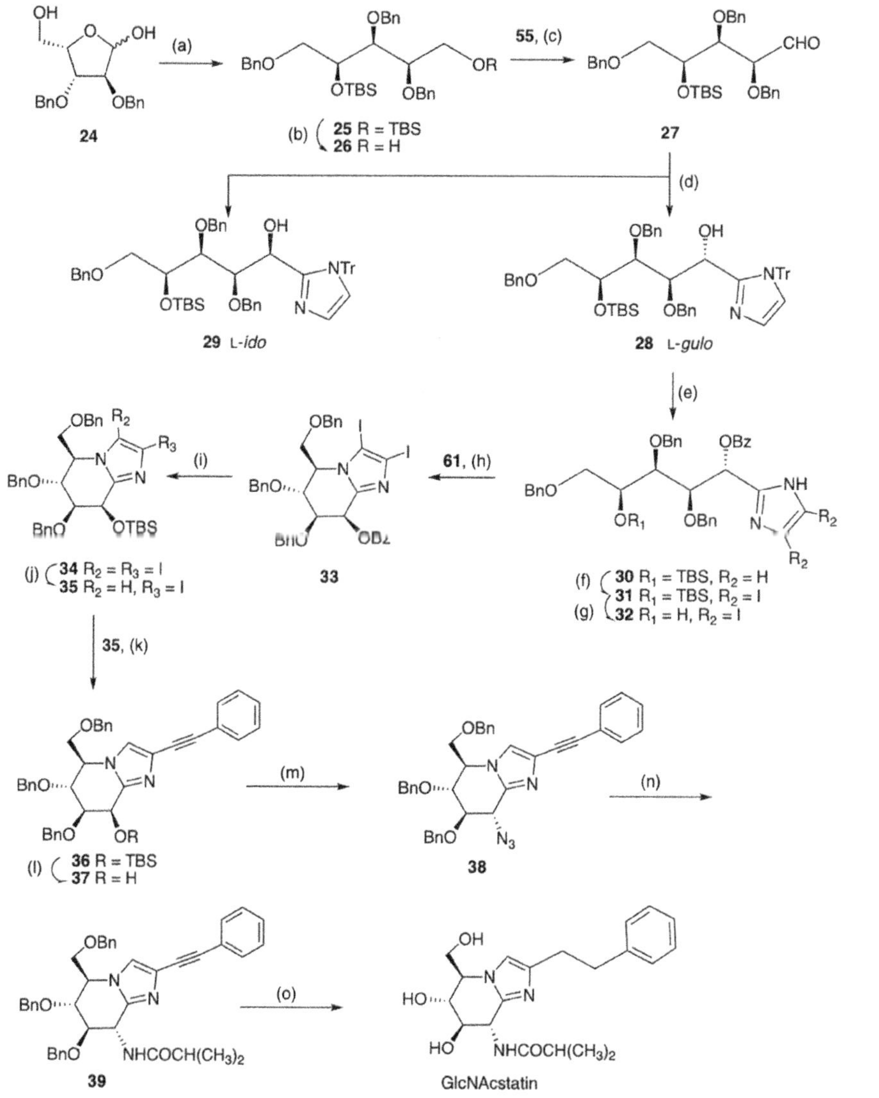

Scheme 10.6 The first synthesis of GlcNAcstatin [52]. Reagents/solvent and yield: (a) 1. NaBH^4, MeOH; 2. TBSCl, ImH, DMF, 95% over the two steps; (b) TFA, H_2O, $CHCl_3$, 65%; (c) $(COCl)_2$, DMSO, Et_3N, DCM, 93%; (d) *N*-tritylimidazole, BuLi, THF, **57**, 53% **58**, 18%; (e) 1. TFA, DCM, then Et_3SiH; 2. BzCl, Py, DMAP cat., 91% over the two steps; (f) *N*-iodosuccinimide, CH_3CN, 90%; (g) HCl,1,4-dioxane, quant.; (h) Tf_2O, Py, DCM, 89%; (i) 1. MeONa, MeOH, DCM; 2. TBSOTf, *i*Pr_2NEt, DCM, 91% over the two steps; (j) EtMgBr, THF, 93%; (k) $C_6H_5C{\equiv}CH$, $Pd(PPh_3)_4$, CuI, Et_3N, DMF, 96%; (l) TBAF, THF, 83%; (m) DPPA, DBU, toluene-THF, 90%; (n) PPh_3, THF-H_2O, then $(iPrCO)_2O$, Et_3N, 95%; (o) $Pd(OH)_2$, H_2 14.5 psi, AcOH, 60%.

the triflate was followed by intramolecular cyclization to give the bicyclic derivative **33**. Benzoate hydrolysis was followed by protection with the *tert*-butyldimethylsilyl group to give derivative **35**, which was then submitted to the Sonogashira coupling with the phenylalkyne to give **36**. Reaction with *tert*-butylammonium fluoride afforded the alcohol **37**, which was converted into the azide **38** with diphenylphosphoryl azide. Staudinger reduction was followed by amine acylation with isobutyric anhydride to afford precursor **39**, the hydrogenation of which gave GlcNAcstatin in 7.7% overall yield from the starting material **24**. With the promising biological results obtained for GlcNAcstatin, van Aalten, and coworkers used this compound as scaffold, and synthesized GlcNAcstatins A-H (Figure 10.6) [53, 54, 59] to further explore the ability of this compound family to inhibit hOGA and to be selective for hOGA against hexosaminidases. As hOGA active site has a cysteine residue (Cys215), which could react irreversibly with the inhibitor, van Aalten and coworkers designed GlcNAcstatins F and G comprising thiol-reactive groups in the acyl side chain (Scheme 10.7) [54]. The synthetic strategy reported in 2010 [59] for GlcNAcstatin and analogs differed from the one developed for the initial GlcNAcstatin [52], although some reactions were also applied, as shown in Scheme 10.7. The starting material used was methyl D-mannopyranoside **40**, which reacted with butane-2,3-dione and trimethyl orthoformate in the presence of catalytic camphorsulfonic acid to afford the diacetal protected mannoside **41**. Selective protection of the secondary alcohol with the *p*-methoxybenzyl group gave **42** in 54% yield, separated from the diprotected compound, obtained in 25% yield. Reaction of OH-6 in **42** with triphenylphosphane and iodine gave the iodo derivative **43**. The reductive opening of the pyranoside ring was carried out with activated zinc in aqueous THF. The intermediate aldehyde formed was treated with a glyoxal solution in methanolic ammonia to afford the imidazole **44**. Its catalytic dihydroxylation with osmium tetraoxide was highly stereoselective but the configuration of the newly formed chiral center was not the required one. After silylation of the primary alcohol, the inversion of this configuration resulted from the introduction of two additional reaction steps, namely Swern oxidation to the ketone and sodium borohydride reduction, giving compound **46** with the desired L-*gulo* stereochemistry. After cyclization to give compound **47**, the strategy followed to introduce the alkyne moiety was similar to that used in the first synthesis of GlcNAcstatin [52]. Oxidative removal of the *p*-methoxybenzyl group with 2,3-dichloro-5,6-dicyano-1,4-benzoquinone, azidation, azide reduction, acylation, hydrogenation, and final deprotection with aqueous TFA for 36 hours gave the target GlcNAcstatin and GlcNacstatins B, D, G, and H in 3.6%, 4.7%, 4.0%, 2.9%, and 4.0% overall yield, respectively, from compound **40**. For GlcNAcstatin F, the sulfanyl group was then deacetylated, providing this GlcNAcstatin derivative with 2.0% overall yield.

The synthesized compounds were tested to evaluate their potency for OGA inhibition and selectivity for hOGA over human hexosaminidases (Table 10.2) [53, 54]. Most of them presented nanomolar K_i values, with the exception of GlcNAcstatin B showing a picomolar inhibition constant, the most potent hOGA GlcNAcstatin inhibitor, and GlcNacstatin E presenting the lowest inhibition (K_i = 8.5 μM). Interestingly, GlcNAcstatin G, incorporating a penta-2,4-dien-1-yl group, is the most selective inhibitor reported so far, with a selectivity over 900 000, eventually resulting from the expected irreversible reaction with the hOGA active site cysteine residue (Cys215) [54]. GlcNAcstatin G also penetrates live cells inducing cellular hyper O-GlcNAcylation with EC_{50} = 20 nM [54].

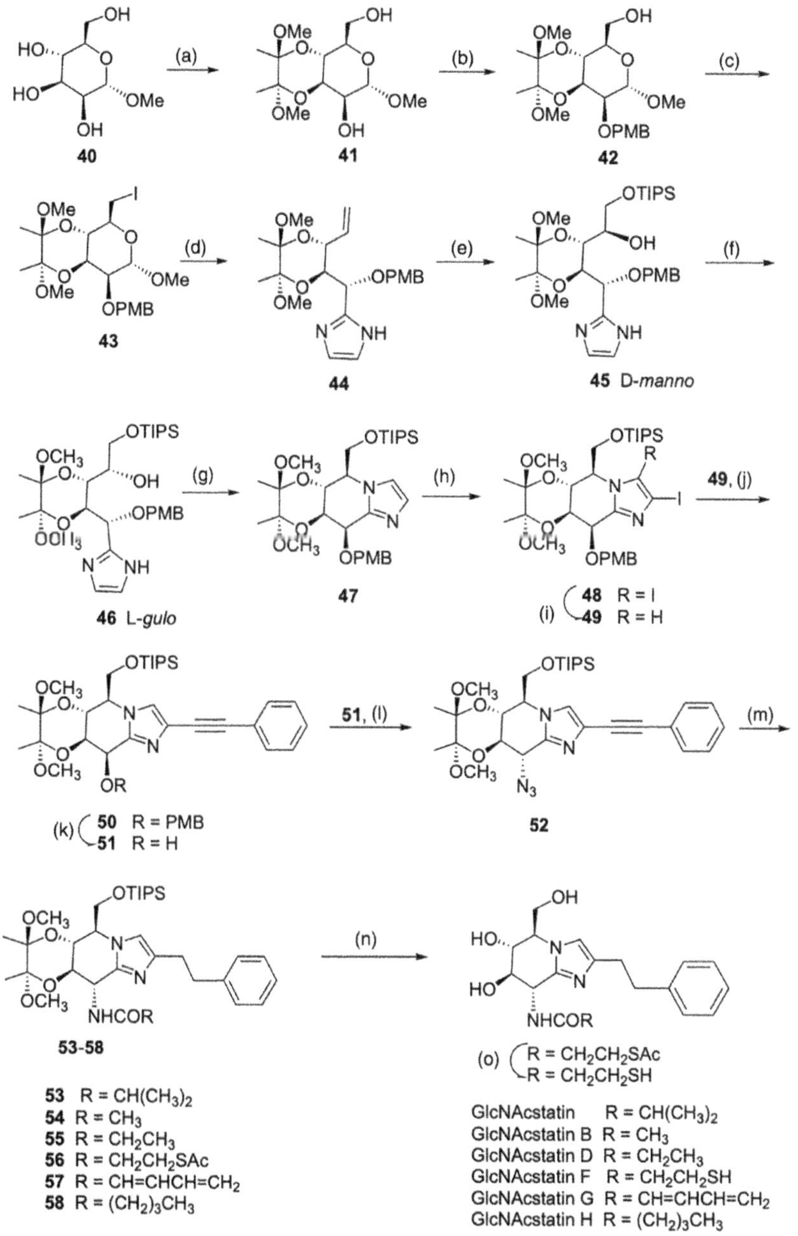

Scheme 10.7 Preparation of GlcNAcstatin and analogs [59]. Reagents/solvent and yield: (a) $CH_3COCOCH_3$, $CH(OCH_3)_3$(3 equiv.), cat. CSA, MeOH, 95%; (b) PMBCl, NaH, *n*-Bu_4NI, DMF, 54%; (c) I_2, ImH, PPh_3, toluene, 78%; (d) 1. Zn, THF/H_2O (10 : 1); 2. 40% aqueous glyoxal, 7 M NH_3/MeOH, 70%; (e) 1. $K_2(OsO_4)$/$K_3[Fe(CN)_6]$, K_2CO_3, $CH_3SO_2NH_2$, *tert*-BuOH/THF/water; 2. TIPSCl, Py, 77%; (f) 1. $(COCl)_2$, DMSO, DCM, then Et_3N; 2. $NaBH_4$, EtOH, 81%; (g) Tf_2O, Py, $C_2H_4Cl_2$, 97%; (h) NIS, DMF, 73% or NIS, MeCN, PPTS, 62%; (i) EtMgBr, THF, 84%; (j) PhC≡CH, CuI, Et_3N, $Pd(PPh_3)_4$, DMF, 93%; (k) DDQ, DCM/H_2O; (l) DPPA, DBU, toluene, 85%; (m) 1. H_2,Pd/C, MeOH or EtOAc, 1 h; 2. $(RCO)_2O$, Et_3N, DCM for R = CH_3, 93%; R = C_2H_5, 83%; i-C_3H_7, 78% or PyBOP, DIPEA, RCO_2H, DCM, for R = C_4H_9, 90%; for R = C_4H_5, 93%, and for R = C_2H_4SAc, 74%; (n) TFA/H_2O (95 : 5), R = CH_3, 77%; R = C_2H_5, 73%; R = *i*C_3H_7, 70%; R = C_4H_9, 68%; R = C_4H_5, 48%; R = C_2H_4SAc, 75%; (o) DMF, MeONa/MeOH, DTT, 56%.

Table 10.2 Pharmacodynamic of brain *O*-GlcNac protein in rats treated with the inhibitor ($3\,mg\,kg^{-1}$, oral dose).

Compound Nr/Structure	hOGA K_i (nM)	*O*-GlcNAc protein[a]		Brain exposure ($nmol\,g^{-1}$)		Brain/Plasma ratio[b]	
		8 h	24 h	8 h	24 h	8 h	24 h
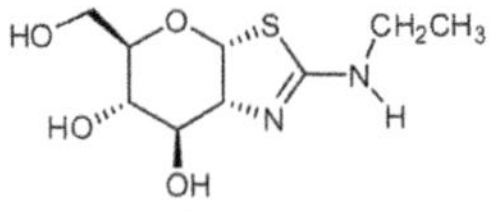 65 (Thiamet-G)	0.41	1.85	1.75	0.185	0.051	0.94	>2.7
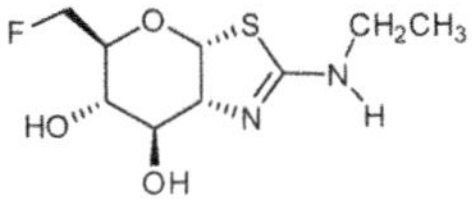 76	20	1.84	1.74	0.228	0.047	1.49	≥18.7
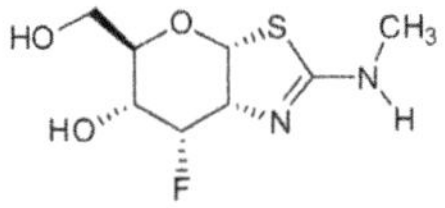 79	0.53	2.13	1.45	0.099	0.074	1.67	≥29.5
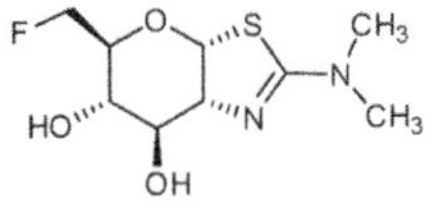 80	9	2.33	1.77	0.222	0.038	0.90	1.5
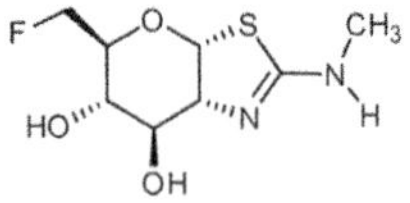 81	0.55	2.44	1.65	0.068	0.025	0.60	≥7.9
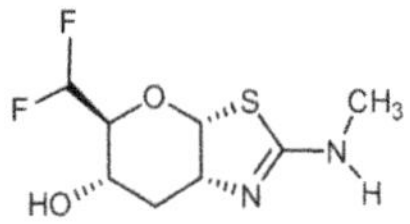 84	28	1.80	—	0.289	—	2.22	—
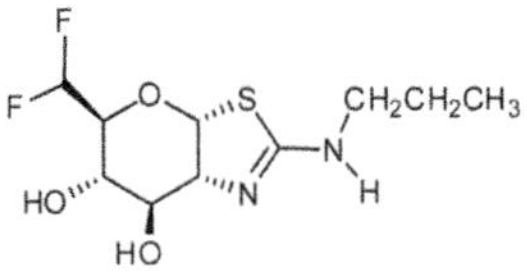 85	5.3	1.81	—	0.193	—	8.81	—

Table 10.2 (Continued)

Compound Nr/Structure	hOGA K_i (nM)	O-GlcNAc protein[a]		Brain exposure ($nmol\,g^{-1}$)		Brain/Plasma ratio[b]	
		8 h	24 h	8 h	24 h	8 h	24 h
F, F, O, S, CH_2CH_3, N, N, H, HO, HO, 86 **86**	7.9	2.06	2.17	0.219	0.088	1.84	≥20.5

a) Fold increase in *O*-GlcNAc protein in treated rats compared to vehicle-dosed rats;
b) Brain/plasma ratio = brain concentration ($nmol\,g^{-1}$)/plasma concentration (μM). The lower limit values 18.7, 7.9, and 20.5 result from plasma exposure below the limit of detection. Source: Adapted from Selnick et al. [67].

In summary, some GlcNAcstatins are, indeed, promising candidates for further studies toward hOGA inhibitor's therapeutics. Nonetheless, the challenge seems to also involve the search for synthetic approaches with less reaction steps to easily access this family of compounds, facilitating an eventual industrial production.

10.2.2.3 Thiazoline Inhibitors

The discovered mechanism of hOGA inhibition inspired Vocadlo's group to develop new transition state mimics by replacing the oxazoline by a thiazoline ring containing a side chain. Linear side chains varying in size and branched chains were investigated, aiming at selectivity for hOGA over lysosomal β-hexosaminidase [39]. For the purpose, a facile synthetic route was envisioned to access compounds **62a–62g** in good overall yields, in only three steps: *N*-acylation with RCOCl of peracetylated 2-amino-2-deoxy-β-D-glucopyranose hydrochloride **59**; treatment with Lawesson's reagent to afford the thiazoline-fused ring; final deprotection of the acetyl groups with sodium methoxide in methanol, and neutralization with glacial acetic acid in methanol (Scheme 10.8). This procedure led to the generation of potent inhibitors, some of them with a remarkable selectivity of hOGA over β-hexosaminidase (Table 10.3). Indeed, the most selective inhibitors are **62c** (NButGT) and **62d** with their thiazoline ring containing a propyl group and a butyl group, respectively (Scheme 10.8, Table 10.3), but NButGT is the most active and selective inhibitor. The activity and selectivity of this series were compared to that of PUGNAc (Figure 10.4), a natural product and one of the first cell-permeable hOGA inhibitors with K_i = 46 nM. Unfortunately, PUGNAc also inhibited the human lysosomal β-hexosaminidase with K_i = 36 nM [39, 60].

The disadvantage of these thiazoline-based inhibitors is their limited chemical stability in solution over periods of days to weeks. Aiming to overcome this issue, Vocadlo's group designed and synthesized Thiamet-G starting by *N*-acylation of

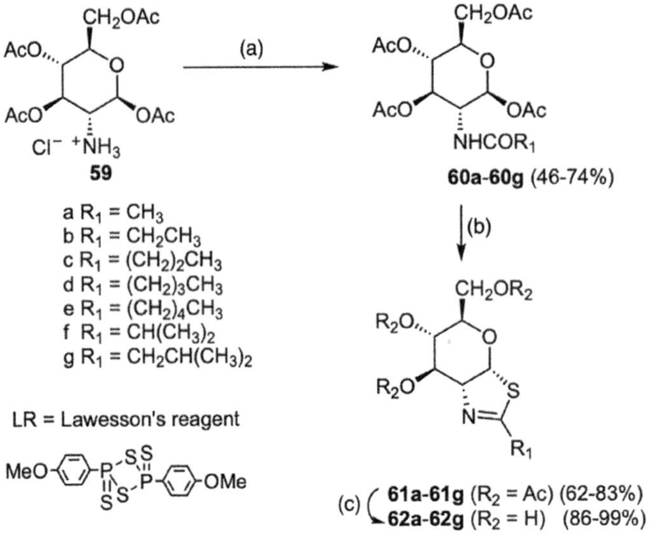

CH2OAc
AcO
O
AcO
OAc
Cl− +NH3
59
(a)
CH2OAc
AcO
O
AcO
OAc
NHCOR1
60a-60g (46-74%)
(b)
a R1 = CH3
b R1 = CH2CH3
c R1 = (CH2)2CH3
d R1 = (CH2)3CH3
e R1 = (CH2)4CH3
f R1 = CH(CH3)2
g R1 = CH2CH(CH3)2
LR = Lawesson's reagent
MeO
OMe
CH2OR2
R2O
O
R2O
S
N
R1
(c)
61a-61g (R2 = Ac) (62-83%)
62a-62g (R2 = H) (86-99%)

Scheme 10.8 Synthesis of thiazoline-based OGA inhibitors **62a–62g** as reported by Macauley et al. [39] Reagents/solvent: (a) RCOCl, NEt_3, DCM (b) Lawesson's reagent, toluene (c) 1. NaOMe, MeOH; 2. AcOH, MeOH. Source: Adapted from Macauley et al. [39].

compound 59 with ethyl isothiocyanate to afford the thiourea derivative 63 in very high yield. After titanium tetrachloride promoted cyclization, the protected bicyclic compound 64 was obtained in 90% isolated yield (Scheme 10.9) [61]. Acetyl group cleavage catalyzed by potassium carbonate gave Thiamet-G 7 in 74% overall yield.

Thiamet-G is highly selective for hOGA being able to remove GlcNAc from *O*-GlcNAc-modified proteins with K_i = 21 nM for hOGA as determined using the Michaelis–Menten method [43]. It crosses the blood-brain barrier and is orally available. In addition, the authors demonstrated that this inhibitor blocks Tau

Table 10.3 Inhibition constants of compounds **62a–62g** for hOGA and lysosomal β-hexosaminidase, and selectivity.

Compound nr.	hOGA K_i (μM)	β-hexosaminidase K_i (μM)	β-hexosaminidase K_i/hOGA K_i
62a (NAG-thiazoline)	0.070	0.070	1
62b	0.12	32	270
62c (NbutGT)	**0.23**	**340**	**1500**
62d	**1.5**	**4600**	**3100**
62e	57	11 000	100
62f	1.6	720	700
62g	5.7	4000	190
PUGNAc	0.046	0.036	0.8

Source: Adapted from Macauley et al. [39].

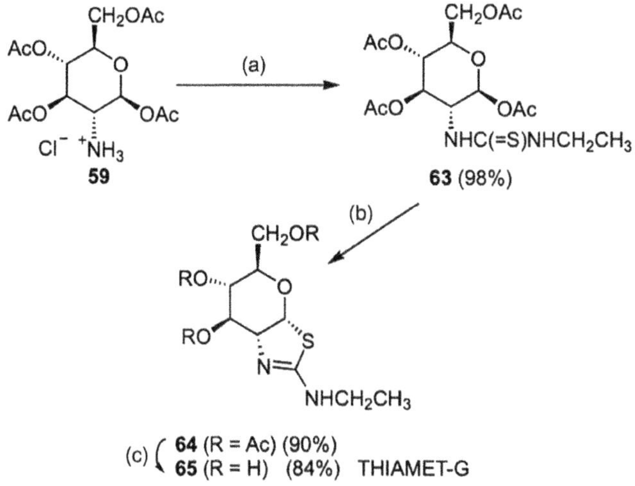

Scheme 10.9 Synthesis of Thiamet-G developed by Yuzwa et al. [61] Reagents/solvent: (a) $CH_3CH_2N{=}C{=}S$, Et_3N, CH_3CN (b) $SnCl_4$, DCM (c) K_2CO_3, MeOH.

phosphorylation in cultured neuron-like cells and decreases phosphorylation of Tau *in vivo*, thus becoming an interesting compound for further investigation of its functional role in AD pathology. In 2014, Yuzwa et al. [62] investigated the role of *O*-GlcNAc on APP and β-amyloid production in mice exhibiting both Tau and β-amyloid pathologies, using Thiamet-G to increase the global levels of *O*-GlcNAc in bigenic Tau/APP mutant mice (TAPP mice) [62], and concluded that pharmacological inhibition of OGA prevents cognitive decline and amyloid plaque formation in the studied mutant mice. They showed that Thiamet-G increases *O*-GlcNAc levels in TAPP mouse brain, leading to reduction of neuritic plaques and amyloidogenic β-amyloid peptides levels, and blocking the onset of cognitive impairment. Intrigued by the role of this inhibitor, in 2016, Cekic et al. [40] explored substitution of the amino group to promote activity and selectivity for hOGA inhibition. They designed and synthesized a series of Thiamet-G derivatives by modifying *N*-substitution, aiming to understand the role of the alkyl side chain size, the influence of altered electronic properties on the binding, and the effect of inhibitor pK_a. The series comprises compounds type **66** (Scheme 10.10), in which the amino group is either free or substituted with methyl, allyl, ethyl (Thiamet-G), propyl, and butyl groups, with 2-fluoroethyl, 2,2-difluoroethyl and 2,2,2-trifluoroethyl groups. The synthetic approach (Scheme 10.10) is inspired by that applied for Thiamet-G (Scheme 10.9). Briefly, reaction of salt **59** either with *N*-fluorenylmethyloxycarbonyl (Fmoc) protected or *N*-allyl protected isothiocyanate afforded intermediates **67a** and **67b**, respectively. Cyclization of **67a** was accomplished by reaction with titanium tetrachloride, while that of the *N*-allyl intermediate was possible with TFA. Deprotection to **66a** was carried out as for Thiamet-G, while deacetylation of **68b** succeeded with potassium carbonate in methanol. The series **66c–66i** and Thiamet-G (**65**) were prepared starting from the isothiocyanate **69**, which reacted

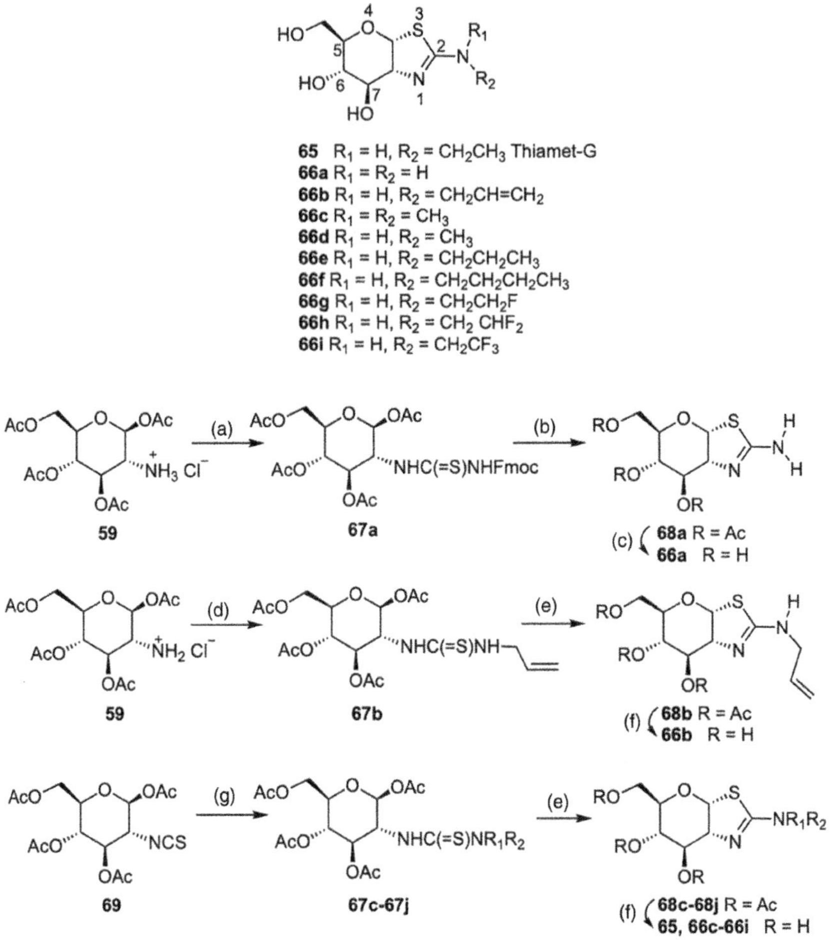
65 R1 = H, R2 = CH2CH3 Thiamet-G
66a R1 = R2 = H
66b R1 = H, R2 = CH2CH=CH2
66c R1 = R2 = CH3
66d R1 = H, R2 = CH3
66e R1 = H, R2 = CH2CH2CH3
66f R1 = H, R2 = CH2CH2CH2CH3
66g R1 = H, R2 = CH2CH2F
66h R1 = H, R2 = CH2 CHF2
66i R1 = H, R2 = CH2CF3
59
67a
(c) 68a R = Ac
66a R = H
59
67b
(f) 68b R = Ac
66b R = H
69
67c-67j
(f) 68c-68j R = Ac
65, 66c-66i R = H

Scheme 10.10 Preparation of Thiamet-G-based thiazoline inhibitors **65** and **66a–66i** as described by Cekic et al. [40] Reagents/solvent: (a) 1. NEt_3, DCM; 2. Fmoc-NCS, py, NEt_3; (b) $SnCl_4$, py, NEt_3; (c) 1. NaOMe, MeOH; 2. AcOH; 3. Piperidine, DMF; (d) allyl-NCS, NEt_3, MeCN; (e) TFA, DCM; (f) K_2CO_3, MeOH; (g) $NHR_1R_2{\cdot}HCl$, NEt_3, CH_3CN.

with the respective ammonium chloride in acetonitrile in the presence of triethylamine to afford the thiourea intermediate. After cyclization promoted by TFA and deacetylation with potassium carbonate in methanol, compounds **65** and **66c–66i** were obtained in good yields.

With compounds in hand, Vocadlo and coworkers determined K_i values for the inhibition of hOGA and for that of human lysosomal hexosaminidases (Table 10.4), which are the products of *HEXA* and *HEXB* genes. They used Michaelis–Menten kinetics for the less potent inhibitors [40], while for the most potent ones, they applied the Copeland modified Morrison method [63–65], which can be used when

Table 10.4 K_i values for hOGA and βHexB, K_i selectivity ratios of inhibitors **65, 66a–66i** for hOGA over hHexB, and pK_a values for Thiamet-G and inhibitors **66g–66i**.

Inhibitor	hOGA K_i^a (nM)	hHexB K_i^b (μM)	hHexB/hOGA[c]	pK_a	Fraction protonated at pH 7.4
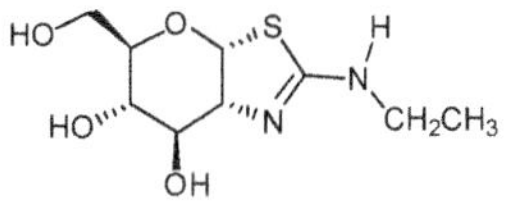 **65 Thiamet-G**	2.1 ± 0.3	740 ± 60 [40]	350 000	7.68	0.66
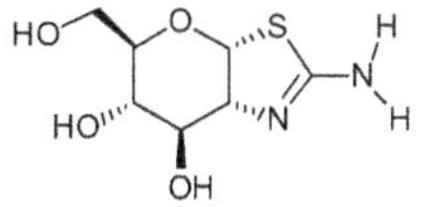 **66a**	4.7 ± 0.3	5.0 ± 0.6^d	1100		
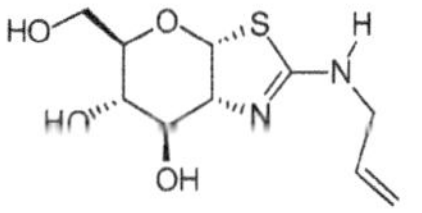 **66b**	3.2 ± 0.4	2850 ± 570	950 000		
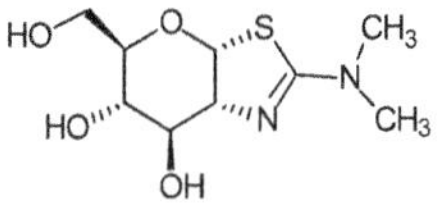 **66c**	2.4 ± 0.2	13.0 ± 3.8	5400		
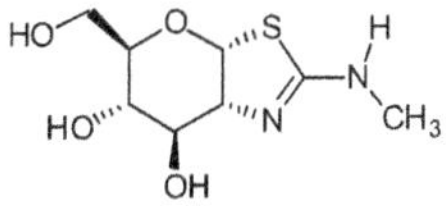 **66d**	0.51 ± 0.05	1.70 ± 0.19^d	3300		
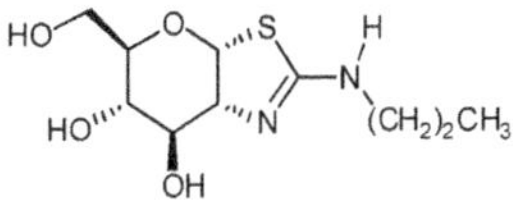 **66e**	2.0 ± 0.2	3700 ± 670	1 850 000		
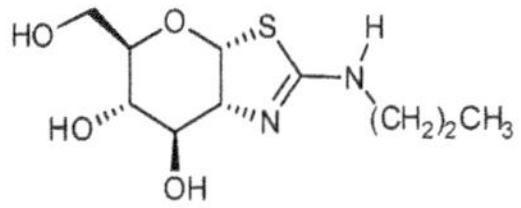 **66f**	350 ± 90^d	4800 ± 763	13 700		
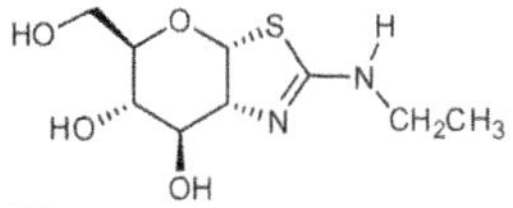 **66g**	15 ± 5^d	180 ± 44	12 000	6.92	0.2

(Continued)

Table 10.4 (Continued)

Inhibitor	hOGA K_i^a (nM)	hHexB K_i^b (μM)	hHexB/ hOGA[c]	pK_a	Fraction protonated at pH 7.4
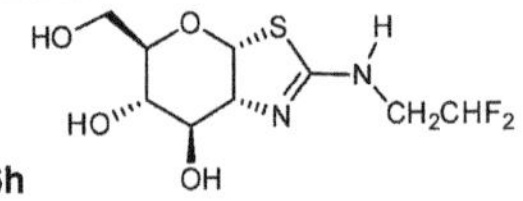 **66h**	60 ± 10^d	150 ± 50	2500	6.18	0.06
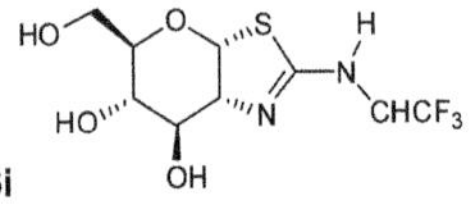 **66i**	1000 ± 200^d	4200 ± 1525	4200	5.33	0.01

a) Determined with the Morrison K_i fit if values are below 5 mM;
b) Determined using Dixon plot analysis;
c) Selectivity ratios indicating favored selectivity for hOGA over hHexB;
d) Determined using Michaelis–Menten inhibition analysis. Source: Adapted from Cekic et al. [40].

K_i values are comparable to the enzyme concentration being studied, and reached some remarkable results. The first one is related to Thiamet-G, in which K_i value is 2.1 nM instead of the 21 nM given previously when evaluated by the Michaelis–Menten method [43]. The first picomolar hOGA inhibitor discovered is LSO, exhibiting $K_i = 510 \pm 50$ pM and a selectivity of 3300 for hOGA over β-hexosaminidase. The size of the chain is also important for binding. Compounds embodying *N*-methyl, *N*-ethyl, *N*-propyl, or *N*-allyl groups have K_i in the range of 2.0–3.2 nM, while the compound with the *N*-butyl group suffered 100-fold decrease of activity ($K_i = 350$ nM). The side chains with two and three carbon atoms are, indeed, the most important to increase selectivity for hOGA inhibition over β-hexosaminidase inhibition, as clearly deduced from selectivity values given in Table 10.5. Interestingly, compounds **8a**

Table 10.5 Inhibition data of hOGA and lysosomal hexosaminidases (HexA/B) by GlcNAcstatins A-G, PUGNAC, Thiamet-G, and selectivity for hOGA [53, 54].

Compound	hOGA (K_i nM)	HexA/B(GH20) (K_i)	Selectivity GH20/hOGA	References
GlcNAcstatin	4.4 ± 0.1	550 ± 10 nM	164	[58]
GlcNAcstatin A	4.3 ± 0.2	0.55 ± 0.05 nM	Not selective	[58]
GlcNAcstatin B	$\mathbf{0.42 \pm 0.06}$	$\mathbf{0.17 \pm 0.05}$ **nM**	**Not selective**	[58]
GlcNAcstatin D	0.74 ± 0.09	2.7 ± 0.4 nM	4	[58]
GlcNAcstatin E	8500 ± 300	1100 ± 100 nM	Not selective	[58]
GlcNAcstatin F	11.2 ± 1.4	11.0 ± 0.6 μM	1000	[59]
GlcNAcstatin G	$\mathbf{4.1 \pm 0.7}$	**>3700**	**>900 000**	[59]
GlcNAcstatin H	2.6 ± 0.3	100 ± 30	35 000	[59]
PUGNAc	35 ± 6	25 ± 2.5	Not selective	[59]
Thiamet-G	21	750	35 000	[43, 59]

and **8d** have some selectivity for hOGA, which is not the case for NAG-thiazoline **4a**, a molecule with a similar size [39, 40]. This result reinforces the key role of the amino group in improving inhibitor selectivity over human β-hexosaminidase.

The contribution of inhibitor pK_a for the potency of the inhibition stood out from the results obtained for the fluorinated inhibitors. pK_a of their conjugate acids was determined by ^{13}C nuclear magnetic resonance (NMR) titration [40, 66]. The correlation obtained by plotting pK_a with the corresponding logK_i value suggested that pK_a dominates the effect of inhibitor binding to the active site, as compared to the steric effect resulting from increased fluor substitution. The key role of pK_a in binding may be due to optimization of hydrogen bond strength or favoring inhibitor protonated form. Quantitative methods carried out by Cekic et al. [40] demonstrated that, through their formal positive charge at physiological pH, these compounds have favorable interactions with the active site, only partly carried out within the transition state for the natural 2-acyl substrates, showing that Thiamet-G and analogs embodying an alkylamino group are tight-binding transition state analogs for hOGA. This work gave, in 2016, a new insight into the catalytic mechanism of hOGA and provided a new picomolar inhibitor representative of this compound series [40].

More recently, in 2019, further studies on this family of compounds were reported, inspired by Thiamet-G, which is well tolerated over extended treatment periods. However, it was found that it has a high polar surface (105 Å), resulting in a low diffusion rate into and out of the CNS from systemic circulation [67]. Aiming to obtain better clinical candidates, the collaboration of the companies Alectos Therapeutics Inc, Merck, and Pharmaron Beijing resulted in the generation of MK8719 (**86**, see Table 10.6, Figure 10.7), the Thiamet G analog, which is a highly potent (K_i = 7.9 nM for hOGA) and selective OGA inhibitor with excellent CNS penetration, and has been advanced to phase I clinical trials [68]. It stood out of a small library of 49 Thiamet G analogs generated by hypothesizing that modification of molecule polar substituents would result in a reduced topological polar surface area (TPSA) and consequently in a greater and faster distribution of the compound into CNS [67].

Focusing on the carbohydrate moiety hydroxy groups and the thiazoline-substituted amino group, a series of analogs with structure type A (Figure 10.7) was synthesized and tested [67]. Regioselective methylation of hydroxy groups or its replacement by hydrogen or by fluor atoms afforded compounds with lower TPSA. However, the potency of hOGA inhibition varied according to inhibitor structure and stereochemistry (Table 10.3). No inhibition occurred when OH-4′ was absent, as shown for compounds **70, 74,** and **77**, indicating that this group is important for activity. Nonetheless, some structural changes are tolerated without complete loss of activity, e.g. the replacement of the primary hydroxyl group by hydrogen (compound **73**) or by fluor (compound **76**), and that of OH-3′ by fluor (compounds **78** and **79**). The absolute configuration was shown to be important for hOGA inhibition, together with the replacement of the *N*-ethyl group by the *N*-methyl group. By combining monofluorination of the primary alcohol with a 7-fluoro substituent in compound **82**, the pharmacokinetic properties were improved when compared to the monofluorinated compound **81**, as well as the selectivity over hexosaminidase inhibition, the TPSA value and the apparent permeability.

Table 10.6 Data obtained in assays to determine the inhibition of hOGA, the concentration required for EC_{50} values for elevation of all protein *O*-GlcNAc levels, the selectivity over hexosaminidases, calculated TPSA, and apparent permeability.

Compound Nr/Structure	hOGA *Ki* (nM)	rOGA cell EC_{50} (nM)	hEX K_i (nM)	TPSA (Å)	Papp (10^{-6} cm s^{-1})
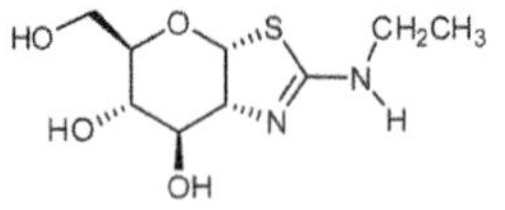 65 (Thiamet-G)	0.41	13.5	>10 000	105	<1.0
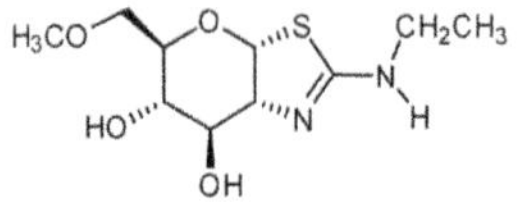 69	190	—	—	94	—
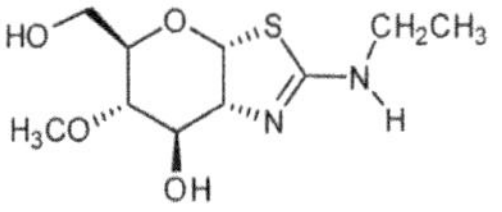 70	>3000	—	—	91	—
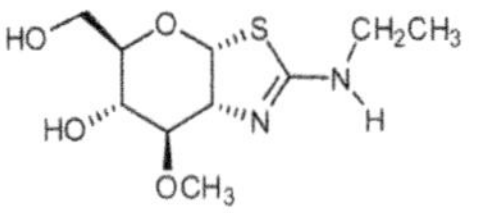 71	270	—	>10 000	91	—
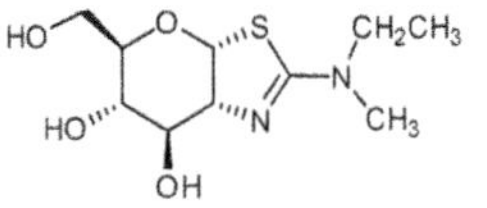 72	5.5	36.7	3600	93	1.1
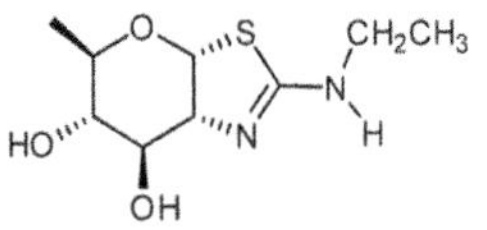 73	69	—	—	84	—
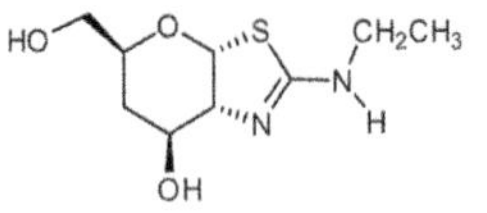 74	>3000	—	—	86	7.0
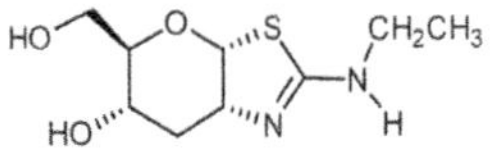 75	44	364	—	84	2.6
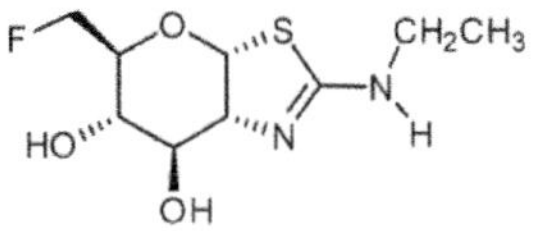 76	20	176	—	84	6.1

(Continued)

Table 10.6 (Continued)

Compound Nr/Structure	hOGA *Ki* (nM)	rOGA cell EC_{50} (nM)	hHEX K_i (nM)	TPSA (Å)	Papp (10^{-6} cm s^{-1})
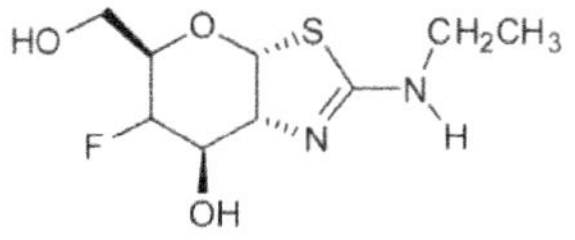 77[a]	>3000	—	>10 000	84	—
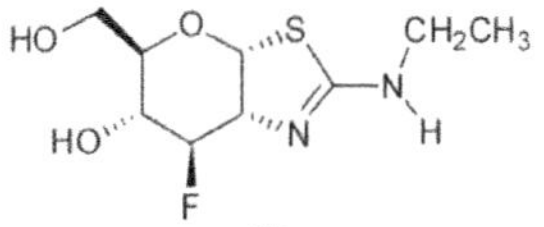 78	29	443	>10 000	84	2.7
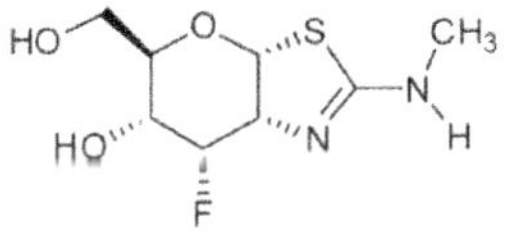 79	0.53	10.6	>10 000	84	—
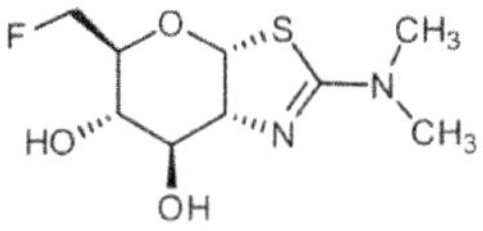 80	9.0	177	1600	69	6.8
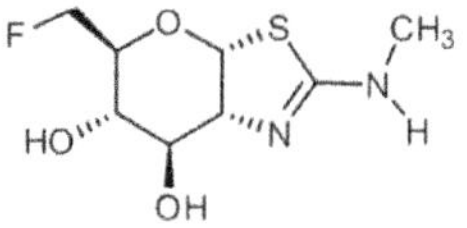 81	0.55	34.4	790	83	5.3
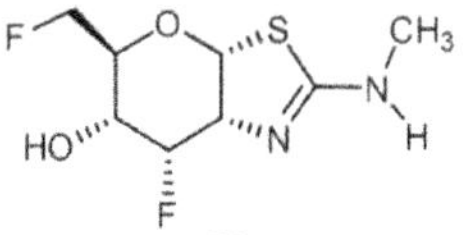 82	35	328	>10 000	61	26
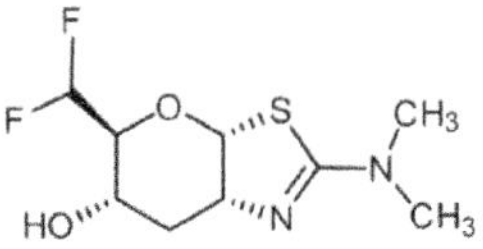 83	>3000	—	>10 000	47	31
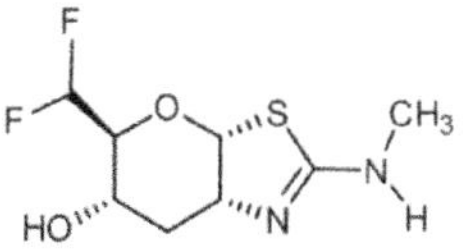 84	28	—	>10 000	47	31

(Continued)

Table 10.6 (Continued)

Compound Nr/Structure	hOGA *Ki* (nM)	rOGA cell EC_{50} (nM)	hHEX K_i (nM)	TPSA (Å)	Papp (10^{-6} $cm\,s^{-1}$)
85	5.3	—	>10 000	61	24
86 (MK8719)	7.9	52.7	>10 000	80	6.4

a) Absolute configuration at position 6 was not assigned. Source: Adapted from Selnick et al. [67].

The high permeability found for the difluorinated compound **82** inspired further research with structures type B (Figure 10.7) and this scaffold was investigated for modifications of the amino group substituent. Compound **83**, bearing a dimethyl-amino group, does not inhibit hOGA, while compounds **84–86**, with a mono-substituted amino group, have restored this activity. Nonetheless, compound **86** (MK8719) is the one showing a good balance of hOGA potency, apparent permeability, and selectivity vs. hHEX.

The collection of *Papp* data for 49 compounds tested was compared to the calculated TPSA values and a reasonable correlation was found, corroborating the hypothesis based on reducing TPSA to obtain compounds with higher permeability

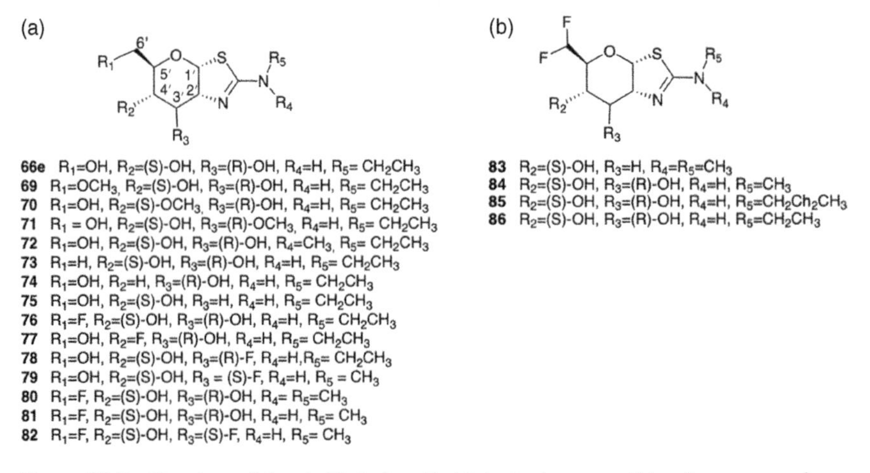

Figure 10.7 Structure of 5-substituted methyl tetrahydropyranothiazole compounds type (a) and of 5-difluoromethyl tetrahydropyranothiazoles type (b) studied by Selnick et al. [67] to illustrate TPSA/permeability and structure/activity relationships. Source: Adapted from Selnick et al. [67].

than that of Thiamet-G. Interestingly, some of the new structures were able to increase protein O-GlcNAc levels (Table 10.6).

As compound bioavailability was moderate to very good, compound differentiation was made through pharmacodynamic effects. For the most promising compounds exhibiting potent hOGA inhibition (K_i <20 nM), EC_{50} <200 nM, and measurable permeability (>2×10^{-6} cm s^{-1}), as well as for the 5-difluoromethyl derivatives **83–86**, the pharmacodynamic effect of OGA inhibition in the CNS was evaluated by measuring the time-dependent accumulation of total O-GlcNAc protein in rat brain after a single oral dose of inhibitor and data was collected after 8 and 24 hours of dose intake [67]. Inhibitor concentration in plasma and in brain homogenate was also evaluated to compare distributional delay in brain exposure (Table 10.2). Compounds **80** and **81** have both a greater effect at eight hours than compound **65** (Thiamet-G) but **81** shows an interesting high level of O-GlcNAc protein for a low brain exposure when compared to compounds **80** and **65**. Compounds **84** and **85** have also demonstrated significant effects and a good brain exposure [67].

In conclusion, the potent and selective hOGA inhibitors, exhibiting biodistribution to the CNS, also afford increased O-GlcNAc protein levels. Among all tested compounds, **86** (MK8719) stood out with a balanced brain and plasma exposure and a greater efficacy over Thiamet-G, possibly resulting from its greater brain exposure, as the brain AUC value (area under the plot of inhibitor concentration in the brain vs. time after dose intake) for a dose of 10 mg kg^{-1} of **86**, at time points 1, 4, 8, 12, and 24 hours was 9.64 nM h g^{-1}, while that of Thiamet-G was only 1.14 nM h g^{-1} [67].

The production of MK8719 (**86**) was carried out starting from compound **65** (Thiamet-G) (Scheme 10.11). Boc protection of the amino group was followed by

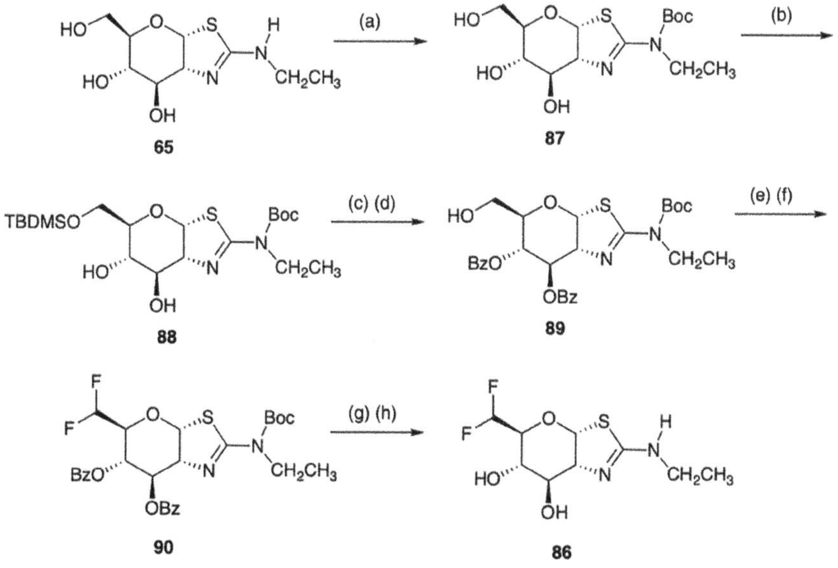

Scheme 10.11 Synthesis of MK8719 [67]. Reagents/solvent and yield: (a) Boc_2O, iPr_2NEt, DMF/MeOH, 64%; (b)TBDMSCl, imidazole, 76%; (c) BzCl, DMAP, 71%; (d) AcCl, MeOH, 90%; (e) Dess-Martin periodinane reagent (DMP), pyridine, DCM, 0 °C – rt; (f) DAST, DCM, 40%; (g) K_2CO_3, MeOH, 92%; (h) TFA, DCM, 95%.

regioselective protection of the primary alcohol by reaction with *terc*-butyldimethylsilyl chloride (TBDMSCl). Benzoylation of the free hydroxy groups of **88** followed by cleavage of the silyl group afforded compound **89**, whose primary alcohol was oxidized to aldehyde with Dess–Martin periodinane, further reacting with DAST to give compound **90** in 40% yield over the two steps. Deprotection with potassium carbonate to generate the hydroxy groups and with TFA for Boc cleavage afforded MK8719 (**86**) in 10.9% overall yield from Thiamet-G [67].

The hypothesis whereby increasing *O*-GlcNAc levels in Tau hinders Tau aggregation was recently investigated by Wang et al. [69] by applying MK8719 *in vivo* in a rTg4510 mouse model of tauopathy. As O-GlcNAcylation levels are reduced in AD brains, and the decrease of Tau *O*-GlcNAc levels is correlated with increased tau hyperphosphorylation, resulting in the formation of insoluble Tau aggregates, inhibition of OGA by MK8719 was envisioned as a strategy to attenuate Tau aggregation. It was found that oral administration of MK8719 to a rTg4510 mouse model of human tauopathy results in a significant increase of brain *O*-GlcNAc levels and reduction of pathological Tau, accompanied by attenuation of brain atrophy with reduction of forebrain volume loss as observed by volumetric magnetic resonance imaging. Although these are encouraging results for the usefulness of OGA inhibitors in reducing Tau pathology, it is crucial to understand the physiological and toxicological consequences of *O*-GlcNAc elevation *in vivo*, as there are many *O*-GlcNAcylated proteins that may also be influenced by OGA inhibition [69].

10.3 GalNAc in Neurodegeneration

In 2017, the relationship between AD and O-GalNAcylation was reported in several studies. Akasaka-Manya et al. [34] used real-time PCR to analyze the expression of human brain GalNAc-transferases (GalNAc-Ts), which transfer GalNAc from UDP-GalNAc to Ser or Thr residues, and showed that the expression of several GalNAc-Ts was altered with sporadic AD progression. To evaluate the impact of GalNAc-Ts overexpression on Aβ production, GalNAc-T1, GalNAc-T4, and GalNAc-T6 were transfected into HEK293T cells. Briefly, while GalNAc-T1 and GalNAc-T4 reduced Aβ1–40 formation, GalNAc-T6 reduced the production of both Aβ1–40 and Aβ1–42. From all three GalNAc-Ts, GalNAc transferase activity on APP was more relevant for GalNAc-T6, suggesting that enhanced O-glycosylation on APP by GalNAc-T6 inhibits Aβ production. GalNAc-T1, GalNAc-T4, and GalNAc-T6 were transfected into HEK293T cells to overexpress them and determine their effect on Aβ production. Transfection of GalNAc-T6 significantly reduced the generation of both Aβ1–40 and Aβ1–42, while GalNAc-T1 and GalNAc-T4 only reduced Aβ1–40 formation. Although the three GalNAc-Ts showed enzymatic activity on soluble APP, the activity of GalNAc-T6 on APP was the most prominent one. The expression of α-secretase and BACE1 was slightly altered in the transfected cells, but the activity of both secretases was not significantly altered. Their data suggested that excess O-glycosylation on APP by GalNAc-T6 inhibits Aβ production [34]. In 2020 Akasaka-Manya and Manya [70] reported that the reduced Aβ generation by coexpression of APP and GalNAc-T6 resulted from a decrease of β-cleavage, while the expression level of membrane-bound APP did not change. As possible explanations for the β-cleavage decrease, they suggest that either glycosylated APP is not transported to the location for β-cleavage, or BACE1 cannot approach glycosylated APP resulting from conformational changes. However, the upregulation of GalNAc-T6 observed in AD brains and the protective effect of GalNAc-T6 acting against Aβ generation remain contradictory, urging for further research to understand these results [70].

An interesting study highlighting the role of O-GalNAcylation on the amyloidogenic property of glycopeptides was carried out by Lin et al. in 2014 [14]. The authors showed that GalNAc α-*O*-linked to Ser135 residue of the prion peptide PrP(108–144), whose structure suffers a coil-to-β conversion associating into amyloid fibrils, had a prominent effect on the conformation of the polypeptide chain and inhibited amyloidogenesis. They suggested that the acetamido group in the equatorial position of GalNAc carbon 2 is important in the interaction between sugar and peptide, because the anti-amyloidogenic effect was not found when the sugar α-linked to Ser135 was galactose.

Another study by Liu et al. in 2017 [71] has shown that by reducing GalNAc-T2 activity with its competitive inhibitor luteolin, a tetrahydroxyflavone, the generation of total Aβ, Aβ1–40, and Aβ1–42 was reduced in a dose-dependent manner in cells coexpressing APP and GalNAc-T2. GalNAc-T1, GalNAc-T3, and GalNAc-T13 transfer GalNAc to APP as well but only GalNAc-T3 was also inhibited by luteolin. They found that both sAPPα and sAPPβ were reduced, suggesting that the process occurs prior to the intervention of secretases, and demonstrated that reduction of Aβ production resulted from the decrease of APP O-GalNAcylation. In addition, they have shown that the inhibitor luteolin also reduced Aβ production in the brain of APP/PS1 transgenic mice.

GalNAc is a residue present in complex molecules, the GAGs and the PGs, which together with proteins compose the cerebral extracellular matrix. Their accumulation surrounding certain neurons forms the perineuronal nets (PNNs) [72–75], whose components are synthesized by neurons, astrocytes, and oligodendrocytes, and form a unique environment. The major PNNs component is hyaluronan, a polysaccharide composed of disaccharide repeating units of glucuronic acid and GlcNAc. It is the only PNN GAG that is not bound to sulfate nor to a core protein. Other components are chondroitin sulfate proteoglycans (CSPGs), where GAGs are linked to a core protein and contain repeating units of disaccharides composed of an amino sugar (GalNAc or GlcNAc) and galactose or a uronic acid, namely glucuronic or iduronic acid. PNNs also contain hyaluronan and PGs binding proteins [73]. PNNs function as a neuron physical barrier, also forming a polyanionic microenvironment around neurons. They participate in brain signaling pathways and in synaptic plasticity and change from physiological to pathological conditions [73]. Baig et al. [72] showed, in 2005, that there is substantial loss of GalNAc from PNN CSPGs from the frontal cortex in AD, indicating a degradation of PNN composition around neurons and certainly affecting PNN functions that depend on GalNAc containing CSPGs side chains [72]. The authors concluded that further studies are required to characterize PNN degradation processes and their effects on AD. Indeed, further investigation on PNN has demonstrated its importance in controlling plasticity, regulating axonal growth and regeneration, and in memory storage. Inspired by modulating PNNs, either by targeting interactions with PNN core components or through PNN digestion, new options emerge for drug development in AD therapy [75].

10.4 Chitosan and Derivatives in AD Brain

Chitin is a linear polymer of β-(1→4) linked GlcNAc residues found in crustacean's exoskeletons and fungi's cell walls. It is the second most abundant natural polymer on earth and has been widely used in the field of biomaterials due to its nontoxic nature, biocompatibility, and biodegradability [76]. From chitin deacetylation, CTS, a linear polymer of β-(1→4) linked β-D-glucosamine (GlcN), can be obtained. Chitosan oligosaccharides (COSs) are biodegradation products of CTS or chitin, with low molecular weight and high-water solubility, due to their shorter chain lengths and free amino groups, making them easily absorbed by the human intestine [77]. CTS, COS, and their derivatives also possess antioxidant, anti-HIV, anti-inflammatory and neuroprotective effects [77].

Neuroinflammation results in oxidative stress in the synapses and mitochondria, contributing to neuronal and vascular degeneration in AD brain. Water-soluble CTS was found to have beneficial effects against the inflammatory response associated with Aβ, which induces inflammation by the production of pro-inflammatory cytokines tumor necrosis factor-α (TNF-α) or interleukin-6 (IL-6) in the brain. The secretion of both cytokines in human astrocytoma cells was induced by Aβ25–35 and interleukin-1β (IL-1β), a critical neurotoxic component in AD, the levels of which are increased in AD patients [78]. Kim et al. showed that pretreatment with water-soluble CTS ($1\,\mu g\,ml^{-1}$ and $10\,\mu g\,ml^{-1}$) significantly inhibited secretion of both TNF-α and IL-6. In addition, considering that Aβ25–35 induces the production of nitric oxide by regulating nitric-oxide synthase in neuroglial cells, and higher levels of this enzyme surround Aβ and produce neurotoxicity in astrocytes, damaging AD brain, the authors [78] also showed that expression of nitric oxide synthase is partially inhibited by treatment with water-soluble CTS. These experiments confirmed that the water-soluble CTS used has, indeed, regulatory effects on human astrocytes.

Jiang et al. [79] evaluated the activities of *N*-acetyl COS (NA-COS) in neurodegeneration. The results of their work demonstrated that NA-COS significantly improved learning and memory of AD rats. Furthermore, hematoxylin and eosin (HE) staining of brain tissue sections showed that NA-COS was able to diminish hippocampal neurodegeneration induced by Aβ25–35, while the treatment of intrahippocampal with NA-COS decreased hippocampal AChE and malondialdehyde levels, with an increase in ACh concentrations. In addition, with NA-COS treatment, antioxidant enzyme levels (superoxide dismutase and glutathione peroxidase) in rat's hippocampus were also enhanced [79].

Recently, alendronate (ALN)-loaded CTS nanoparticles (CTS-ALN-NPs) for brain delivery by a noninvasive intranasal route were investigated. CTS-ALN-NPs reduced the peripheral side effects and released ALN directly into brain. The *in vitro* and *ex vivo* release profiles revealed a sustained drug release through CTS-ALN-NPs when compared to the pure drug solution. Furthermore, intranasal CTS-ALN-NPs were evaluated against intracerebroventricular-streptozotocin (ICV-STZ), which induces AD-like pathologies in mice [80]. The intranasal CTS-ALN-NP altered the ICV-STZ-induced neurobehavioral, neurochemical, and histopathological changes in mice. The treatment of these mice with CTS-ALN-NPs resulted in a reduction of the neuroinflammatory cytokines IL-6, IL-1β, and TNF-α levels in the hippocampus, which highlights the anti-inflammatory activity of ALN. The effect of ALN was found to be more pronounced when administered in the form of CTS-NPs intranasally, which allowed the direct release of ALN into the brain, hence avoiding passage through the gastrointestinal tract [80].

The cholinergic neurotransmission mediated by ACh has a well-known role in modulation of learning and memory [81]. Indeed, the attenuation of ACh levels due to the aggravating activity of AChE and BChE enzymes significantly contributes to cognitive disabilities [82]. AChE and BChE levels were increased in the hippocampus of mice that received ICV-STZ injections, accompanied by important reductions in memory and learning abilities. In contrast, a drop in the levels of both ChEs was noticed after treatment with CTS-ALN-NPs, revealing that ALN is capable of reversing deficits in cognitive and cholinergic functions [80]. Furthermore, the inhibition of STZ-induced BACE-1 overexpression was observed in the hippocampus of mice treated with a subdiabetogenic ICV-STZ followed by the intranasal administration of CTS-ALN-NPs for 15 days (0.0352 mg kg^{-1}), along with reductions in Aβ1–42 levels. Interestingly, CTS-ALN-NPs led to stronger reductions in BACE-1 and Aβ1–42 peptide levels when compared to an ALN pure solution. These results show that CTS-ALN-NPs are capable of attenuating pathological changes observed in AD [80].

10.5 Cholinesterase Inhibitors

As previously referred, AD is a multifactorial neurodegenerative brain disorder, and its exact pathophysiology is not yet entirely known. One of the variables that has a direct link to AD is the cholinergic system, which directly contributes to regulation and memory processes and, therefore, represents a target for AD drug design [83]. Early studies involving AD's patients found an altered cholinergic activity, which resulted in cognitive and functional symptoms [84]. The two major forms of ChEs, in mammalian tissues, are AChE EC 3.1.1.7 and BChE EC 3.1.1.8. Both belong to the group of Ser hydrolases and are responsible for the breakdown of the neurotransmitters ACh and butyrylcholine (BCh), respectively [85]. AChE is substrate specific in nature and is found in high concentrations in the brain, while BChE is nonspecific and is distributed throughout the body. Under normal conditions, ACh is dominantly decomposed by AChE. Both enzymes exhibit different kinetic characteristics, depending on ACh concentrations. When ACh concentration is low, AChE's activity is major, whereas BChE presents higher activity when ACh concentration is elevated. In progressed AD, AChE in brain declines to 55–67% of normal values while BChE increases to 120% of normal levels, indicating that BChE plays a critical role in ACh hydrolysis, at a late stage of AD [83]. In fact, it has been reported that the specific inhibition of BChE is important for raising ACh levels and improving cognition [86]. Hence, designing selective, potent, and well-tolerated inhibitors of each ChE became a challenge for the scientific community, in order to determine which enzyme needs to be targeted for maximum effect in treating AD. Currently, there are some approved ChE

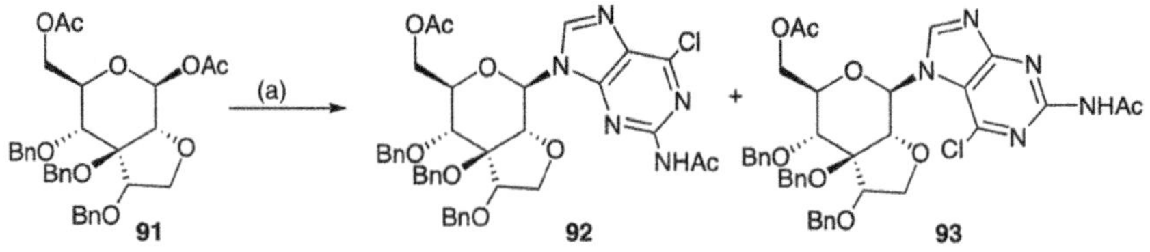

Scheme 10.12 Synthesis of the N^9 and N^7 nucleosides **92** and **93**, respectively. Reagents/solvent and yield: (a) BSA, 2-acetamido-6-chloropurine, TMSOTf, $(CH_2Cl)_2$, 85 °C, 25% for **92** and 37% for **93.** Source: Adapted from Marcelo et al. [88].

inhibitors, namely donepezil, rivastigmine, and galantamine, which help in controlling the symptoms of AD and do not treat the underlying disease or delay its progression [87]. In this perspective, Rauter and coworkers explored purine nucleosides as ChE inhibitors. Marcelo et al. [88], developed the first synthesis of 2-acetamidopurine nucleosides **92** and **93** starting from compound **91** incorporating a tetrahydrofuran ring fused to the pyranose (Scheme 10.12). The bicyclic sugar **94** was also modified to generate the elongated carboxylic ester appendage bearing the azido group at position 6 (**97a,b**, Scheme 10.13). For the purpose, Swern oxidation followed by a Grignard reaction afforded the allyl alcohols **95a,b**. Ozonolysis followed by sodium chlorite oxidation in *t*BuOH/H_2O/2-methylbut-2-ene (2 : 2 : 1) afforded the corresponding carboxylic acids, which were esterified to give the benzyl esters **96a,b**. Triflation and

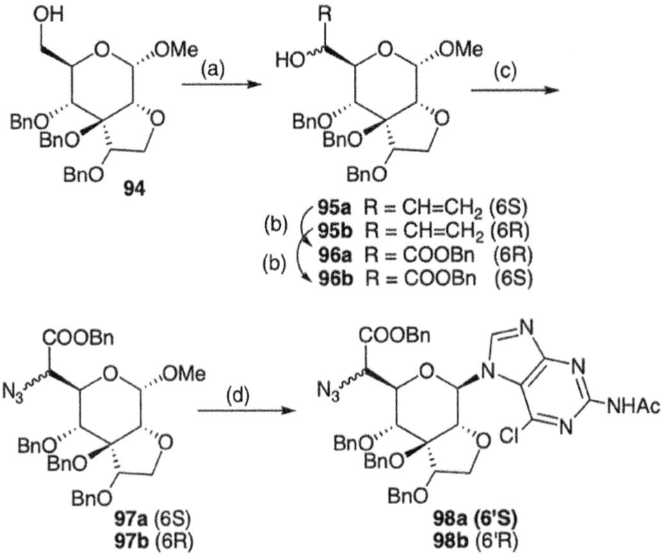

Scheme 10.13 Synthesis of the (6′*S*)- and (6′*R*)-configurated N^7 nucleosides **98a** and **98b**. Reagents/solvent and yield: (a) 1. $(COCl)_2$, DMSO, Et_3N; 2. CH_2=CHMgBr, THF, 65% over two steps; (b) 1. O_3, DMS, DCM; 2. $NaClO_2$, $NaH_2PO_4.H_2O$, *t*BuOH/H_2O/2-methylbut-2-ene (2 : 2 : 1); 3. BnBr, $KHCO_3$, Bu_4NI, DMF, 68% over three steps; (c) 1. Tf_2O, pyridine, DCM; 2. NaN_3, DMF, 72% for **97a,** 86% for **97b** both over two steps; (d) BSA, 2-acetamido-6-chloropurine, TMSOTf, CH_3CN, 60% for **98a** and 55% for **98b.** Source: Adapted from Marcelo et al. [88].

nucleophilic substitution with sodium azide afforded the diastereoisomers **97a** and **97b** in high yield. With these two bicyclic intermediates, coupling with silylated 2-acetamido-6-chloropurine catalyzed by TMSOTf afforded a mixture of N^9- and N^7-linked (**98a** and **98b**) nucleosides, which were screened for AChE and BChE inhibition. While none of the compounds tested inhibited AChE, the N^7 nucleosides showed potent inhibition toward BChE (Table 10.7). Nanomolar inhibition was obtained for

Table 10.7 Inhibition (%) of BChE for different concentrations of the compounds tested and IC_{50} values.

Compound nr	Concentration ($\mu g\,ml^{-1}$)	BChE inhibition (%)	BChE IC_{50} ± SEM[a]
92	100.00	48***	-----
	10.00	11	
93	100.00	84***	0.76 ± 0.05
	10.00	85***	
	1.00	70***	
	0.10	20***	
97a	100.00	91***	4.20 ± 0.05
	10.00	64***	
	1.00	36***	
	0.10	15*	
97b	100.00	98***	13.40 ± 0.80
	10.00	73***	
	1.00	14**	
98a	100.00	81***	22.00 ± 1.60
	10.00	34***	
	1.00	9	
98b	100	91***	0.14 ± 0.01
	10	79***	
	1	77***	
	0.1	63***	
	0.01	10	
Rivastigmine[b]	100	100***	0.17 ± 0.01
	10	100***	
	1	100***	
	0.1	76***	
	0.01	NI	

a) IC_{50} compound concentration inhibiting 50% of enzyme activity; values are expressed as mean ± standard error of the mean (SEM).
b) Rivastigmine is a drug to treat AD patients and was used as positive control.
***$P < 0.001$, **$P < 0.01$, *$P < 0.05$, NI: no inhibition. Source: Adapted from Marcelo et al. [88].

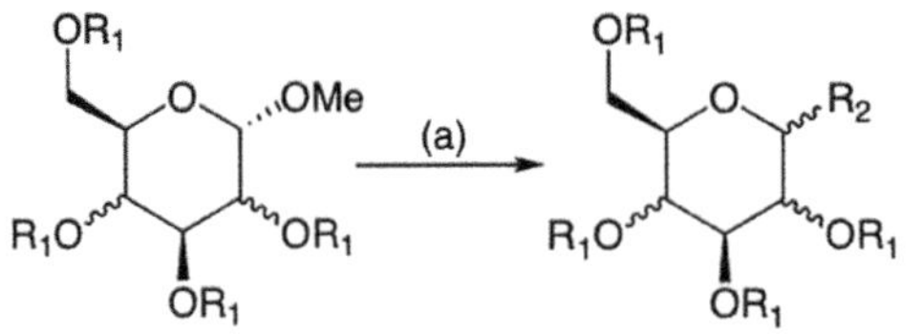

Scheme 10.14 Synthesis of the N[9] and N[7] nucleosides **99** to **122** as reported by Schwarz et al. [89] Reagents/solvent and yield: (a) BSA, 6-chloropurine (CP) or 2-acetamido-6-chloropurine (ACP), TMSOTf, CH_3CN, microwave irradiation (150 W); **99**: D-Glc, R_1=Ac, R_2=βN[7]CP, 20%; **100**: D-Glc, R_1=Ac, R_2=βN[9]CP, 41%; **101**: D-Glc, R_1=Bn, R_2=αN[7]CP, 6%; **102**: D-Glc, R_1=Bn, R_2=βN[7]CP, 17%; **103**: D-Glc, R_1=Bn, R_2=αN[9]CP, 10%; **104**: D-Glc, R_1=Bn, R_2=βN[9]CP, 45%; **105**: D-Gal, R_1=Bn, R_2=β-N[7]CP, 12%; **106**: D-Gal, R_1=Bn, R_2=αN[9]CP, 7%; **107**: D-Gal, R_1=Bn, R_2=β-N[9]CP, 43%; **108**: D-Man, R_1=Bn, R_2=αN[7]CP, 17%; **109**: D-Man, R_1=Bn, R_2=βN[7]CP, 14%; **110**: D-Man, R_1=Bn, R_2=αN[9]CP, 27%; **111**: D-Man, R_1=Bn, R_2=βN[9]CP, 10%; **112**: D-Glc, R_1=Bn, R_2=αN[7]ACP, 7%; **113**: D-Glc, R_1=Bn, R_2=βN[7]ACP, 34%; **114**: D-Glc, R_1=Bn, R_2=βN[9]ACP, 29%; **115**: D-Gal, R_1=Bn, R_2=αN[7]ACP, 3%; **116**: D-Gal, R_1=Bn, R_2=βN[7]ACP, 44%; **117**: D-Gal, R_1=Bn, R_2=αN[9]ACP, 10%; **118**: D-Gal, R_1=Bn, R_2=βN[9]ACP, 41%; **119**: D-Man, R_1=Bn, R_2=αN[7]ACP, 8%; **120**: D-Man, R_1=Bn, R_2=βN[7]ACP, 28%; **121**: D-Man, R_1=Bn, R_2=αN[9]ACP, 24%; **122**: D-Man, R_1=Bn, R_2=βN[9]ACP, 3%. Source: Adapted from Schwarz et al. [89].

compound **98b** competing well with rivastigmine, a drug currently in use for the treatment of AD [88]. Experimental results showed that the presence of benzyl groups on the carbohydrate scaffold and the N[7]-linked purine nucleobase were required for the strong BChE inhibition [84]. The preliminary evaluation of the acute cytotoxicity of the elongated bicyclic sugar precursors and nucleosides was also performed indicating low values, in the same order of magnitude as those of rivastigmine [88].

In 2014, Rauter and coworkers [89] presented a novel microwave-assisted synthesis and anticholinesterase activity of a series of 24 new purine nucleosides incorporating 6-chloropurine or 2-acetamido-6-chloropurine linked to D-glucosyl, D-galactosyl and D-mannosyl residues, with very interesting results compared to those previously obtained by Marcelo et al. [88]. Optimization of the reaction conditions adapted to formation of the new nucleoside structures was achieved by the use of microwave irradiation (150 W, 65 °C), reducing the reaction time from 2 hours to 15 minutes (Scheme 10.14) [89]. Compound structure was designed aiming at ChE (AChE and BChE) inhibition efficiency and selectivity by investigating sugar stereochemistry, purine structure, and N[7] or N[9]-linked purine ligation to the sugar (Table 10.8). The outcome of this work showed that α-anomers were the most active compounds. Regarding BChE inhibition, the nucleosides **110, 112, 119,** and **120** were noticeably more potent than the drug galantamine, and the most promising competitive and selective BChE inhibitor, the N[7]-linked 2-acetamido-α-D-mannosylpurine **119,** showed a K_i of 50 nM and a selectivity factor of 340-fold for BChE over AChE [89]. To the best of our knowledge, compound **119** shows the lowest K_i for BChE, being one of the best candidates toward the study of BChE selective inhibition against AD, when compared to other newly reported carbohydrate-based [90, 91] and noncarbohydrate-based compounds [82]. Interestingly, selectivity over AChE or BChE can be tuned by

Table 10.8 Inhibitory constant K_i for nucleosides **99–123** as determined by Elman's assay with BChE and AChE in comparison to galantamine hydrobromide.

Compound	BChE K_i (μM)	AChE *Ki* (μM)	Compound	BChE K_i (μM)	AChE K_i (μM)
99 β-Glc-N^7CP	>100	28.9 ± 2.1	**111** β-Man-N^9CP	>20	>20
100 β-Glc-N^9CP	>100	17.5 ± 2.6	**112** α-Glc-N^7ACP	2.5 ± 0.3	>10
101 α-Glc-N^7CP	>20	9.6 ± 2.3	**113** β-Glc-N^7ACP	>100	23.0 ± 1.9
102 β-Glc-N^7CP	>2	>2	**114** β-Glc-N^9ACP	>2	>2
103 α-Glc-N^9CP	>100	17.0 ± 2.0	**115** α-Gal-N^7ACP	50.0 ± 7.0	42.0 ± 9.2
104 β-Glc-N^9CP	>100	16.0 ± 2.7	**116** β-Gal-N^7ACP	>20	>20
105 β-Gal-N^7CP	>20	>20	**117** α-Gal-N^9ACP	>2	>2
106 α-Gal-N^9CP	>10	5.6 ± 0.8	**118** β-Gal-N^9ACP	>2	>2
107 β-Glc-N^9CP	>100	25.6 ± 2.1	**119** α-Man-N^7ACP	0.05 ± 0.01	18.1 ± 4.8
108 α-Man-N^7CP	30.9 ± 2.2	23.4 ± 3.6	**120** β-Man-N^7ACP	1.4 ± 0.1	3.0 ± 0.3
109 β-Man-N^7CP	9.6 ± 0.7	28.5 ± 6.5	**121** α-Man-N^9ACP	10.3 ± 2.6	17.1 ± 2.5
110 α-Man-N^9CP	2.8 ± 0.3	2.4 ± 0.3	**122** β-Man-N^9ACP	>10	>10
Galantamine hydrobromide	9.4 ± 0.7	0.5 ± 0.0	**Galantamine hydrobromide**	9.4 ± 0.7	0.5 ± 0.0

Source: Adapted from Schwarz et al. [89].

structural features. Indeed, glucosyl and galactosyl linkage to chloropurine favor the formation of selective AChE inhibitors, while α-glucosyl and α-mannosyl N^7 ligation to 2-acetamido-6-chloropurine favors selective BChE inhibitors.

Noteworthy, in 2021, Jasiecki et al. [92] revised and discussed the relationship between BChE levels and iron concentration in the brain, highlighting that production of BChE by glial cells is iron-dependent. Moreover, AD patients exhibit higher iron brain levels than healthy individuals, which contributes to the overexpression of BChE and subsequent decrease in ACh levels [89].

Several reports indicate that increased Aβ plaque formation is observed when Aβ is expressed in a context of increased AChE levels [93–95], with the neurotoxicity induced by AChE-Aβ complexes being greater than when promoted by Aβ alone [96]. Regarding BChE and Aβ interactions, however, much remains unclear. Although BChE is found in amyloid-β plaques and NFTs, its role in the amyloid hypothesis of AD is yet not fully understood. Some studies imply that BChE has a catalytic role in Aβ aggregation, while others point to a different scenario, where BChE is actually able to inhibit Aβ aggregation [97, 98]. In 2012, the results of a study published by Darvesh et al. suggested that BChE is associated with a subpopulation of Aβ deposits and could take part in AD plaque maturation [99]. On the other hand, in 2016, Kumar et al. reported that low levels of BChE (0.02–0.004 μM – cerebrospinal fluid concentrations in AD patients) increased the formation of Aβ42 fibrils by 20–30%; yet, higher concentrations (0.2 μM) of BChE attenuated the kinetics of amyloid-β fibrillization process [100]. Furthermore, when present simultaneously with apolipoprotein E4 (ApoE4), highly stable and soluble complexes between Aβ and BChE are formed (BAβACs), leading to a reduction in Aβ amyloid aggregation [100]. Also, it seems that Aβ interacts with a presumed activation site in the catalytic tunnel of BChE, which results in an increased ACh influx to the catalytic site, with a consequent increase in the rate of ACh hydrolysis [100]. Altogether, these data suggest that further studies are required, aiming to clarify the role of BChE in Aβ aggregation processes.

10.6 Fyn Kinase Inhibitors

Fyn kinase, a member of the Sec family of tyrosine kinases, is increasingly recognized as a promising disease-modifying therapeutic target against AD [11]. Over the past decade, evidence has shown that soluble amyloid β oligomers (Aβo) bind to the PrP^C at the neuronal surface of the postsynaptic neuron, resulting in the activation of the metabotropic glutamate receptor 5 (mGluR5), with subsequent Fyn kinase phosphorylation, followed by phosphorylation of downstream substrates, including the tau protein. Chronic exposure to soluble Aβo hence contributes to the accumulation of hyperphosphorylated Tau, which composes the NFTs responsible for synaptic dysfunction in the brain of AD patients. What's more, Fyn kinase can also phosphorylate tau directly. By taking part in this signaling pathway, Fyn arises as a unique linker of the two major hallmarks of AD pathology [11] (Figure 10.8).

APP is sequentially cleaved by BACE-1, and γ-gamma secretase, to produce Aβ monomers, including Aβ1–40 and Aβ1–42, at the surface of the postsynaptic neuronal membrane. These monomers aggregate into oligomers, which in turn bind to

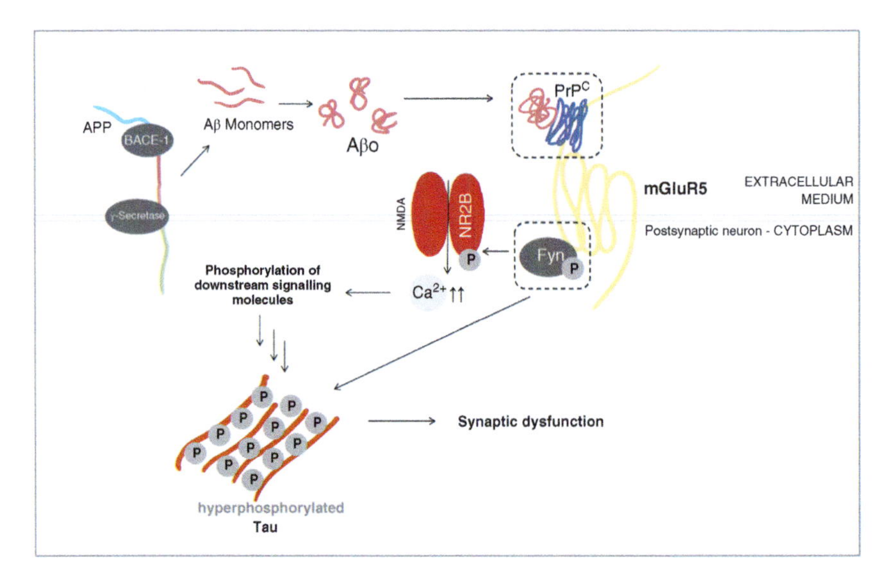

Figure 10.8 Simplified illustration of Aβo-induced Fyn activation leading to Tau hyperphosphorylation.

their high-affinity partner, the PrP^C. Once in complex with Aβo, PrP^C interacts physically with the mGluR5 and, in the cytoplasm, Fyn kinase binds to mGluR5 and is phosphorylated. This activated complex then results in the phosphorylation of the NR2B subunit of the *N*-methyl-D-aspartate receptor (NMDA), which then leads to a calcium influx that triggers the phosphorylation of downstream signaling molecules. With chronic exposure to Aβo at the neuronal surface, this cascade of events ultimately culminates in Tau hyperphosphorylation, which is responsible for synaptic dysfunction and cognitive decline in AD patients.

Not many compounds have been identified to this point as Fyn kinase inhibitors with proven concomitant downstream effects. In 2015, Kaufman et al. [101] investigated the brain-permeable Src family kinase inhibitor AZD0530, also known as saracatinib (**123**, Scheme 10.15), as a compound with potential against AD.

Saracatinib is a non-glycosylated C-5-substituted anilinoquinazoline and was originally discovered as an anticancer agent by Hennequin et al. [103] in 2006, along with a series of analogs investigated in structure–activity relationship (SAR) studies. Biological assays showed that, when given at $5\,mg\,kg^{-1}$ per day, this Fyn kinase inhibitor prevented Aβo-induced signaling, Tau phosphorylation, and deposition, while being able to rescue special memory and synaptic depletion in AD transgenic mice [98]. In another study by Tang et al. [104], later in 2020, the same dose of saracatinib promoted similar improvements in phosphorylated Tau accumulation in the

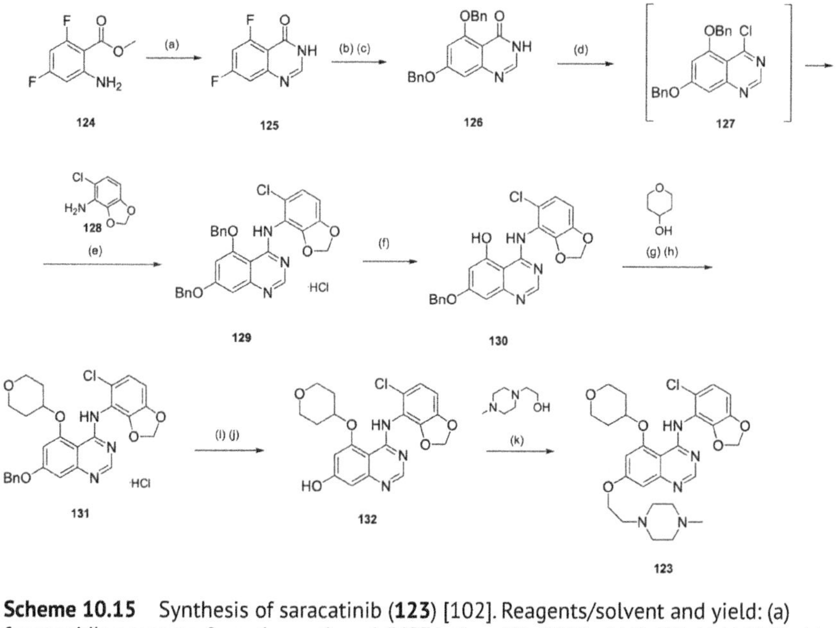

Scheme 10.15 Synthesis of saracatinib (**123**) [102]. Reagents/solvent and yield: (a) formamidine acetate, 2-methoxyethanol, DIPEA, then IPA (52%); (b) BnOH, NaH, DMA; (c) HCl, 98%; (d) anhydrous toluene, $POCl_3$, DIPEA; (e) anhydrous toluene, 69%; (f) pyridine, HCl, 92%; (g), Ph_3P, DTAD, tetrahydropyran-4-ol, toluene; (h) HCl, dioxane, then IPA, 80%; (i) TFA; (j) distillation, NaOH, 89%; (k) 2-(4-methylpiperazin-1-yl)ethanol, Ph_3P, DTAD, THF, then IPA, 86%.

hippocampus, accompanied by memory improvements in mice with transgenic and traumatic tauopathy.

The original 100 mg-scale synthesis of saracatinib [103] included 17 linear steps, 7 of which were protection/deprotection reactions. With some practical concerns raised for a potential scale-up synthesis, four years later, an optimized route for the kilogram manufacture of saracatinib was developed [102]. This route started from the difluoro quinazolinone **125**, accessed in 52% yield from ester **124** by using formamidine acetate in 2-methoxyethanol (Scheme 10.15), followed by addition of isopropanol and crystallization. Then, after a double S_NAr reaction to allow the replacement of both fluorine groups with benzyloxy moieties by reaction with benzyl alcohol in the presence of sodium hydride and dimethylacetamide (DMA), intermediate **126** was chlorinated in the presence of phosphorous oxychloride and DIPEA but **127** was not isolated and reacted with **128** to give anilinoquinazoline **129**, which was obtained as a hydrochloride salt in 69% yield. Selective debenzylation gave the 5-hydroxy intermediate **130** in 92% yield. Compound **131** was then accessed as a hydrochloride salt in 80% yield by reaction with tetrahydropyran-4-ol under Mitsunobu conditions, with subsequent addition of hydrochloric acid in 1,4-dioxane, followed by crystallization. After deprotection of the remaining benzyl group, followed by distillation to dryness and neutralization with sodium hydroxide, the final Mitsunobu coupling of intermediate **132** with 2-(4-methylpiperazin-1-yl)ethanol gave saracatinib **123** which, after crystallization, was obtained as a difumarate in 86% yield.

To the best of our knowledge, saracatinib (**123**) was the first Fyn kinase inhibitor with proven downstream effects in an AD model and with adequate pharmacokinetic properties. However, the very first sugar-based Fyn kinase inhibitors were not published until 2020 [103]. Indeed, Rauter and coworkers identified a new family of nature-inspired glucosylpolyphenols (**121–124,** Figure 10.9) that were shown to distinctively interfere with the Aβo-Fyn-Tau neuronal signaling cascade, with the advantage of having a much easier, straightforward, and efficient synthesis compared to the route described for saracatinib.

Synthesis of the three per-O-methylated *C*-glucosyl polyphenols **133–135** was conducted by Matos et al. [105] in only two reaction steps, in good to very good overall yield (Scheme 10.16). Briefly, after full methylation of the commercially available methyl α-D-glucopyranoside (**137**) with sodium hydride and methyl iodide, the TMSOTf-promoted *C*-glucosylation reaction of each polyphenol in the presence of drierite gave the desired final compound. This final reaction step afforded compound **133** in 53% yield, compound **134** in 37% yield, and compound **135** in 45% yield.

Synthesis of compound **136** required a few additional reaction steps (Scheme 10.17) but is still simpler than that of saracatinib (**123**). Firstly, the *C*-glycosylation reaction carried out as previously described afforded the intermediate **139** in 63% yield. It is interesting to note that this reaction occurred via a Friedel–Crafts-type mechanism, being the first exception to the Fries-type rearrangement ever reported in the *C*-glycosylation of unprotected phenols. Compound **139** was then submitted to a benzoylation reaction with benzoyl chloride, imidazole, and dichloromethane as a solvent to afford its derivative **140** in 88% yield, which was deprotected to give the

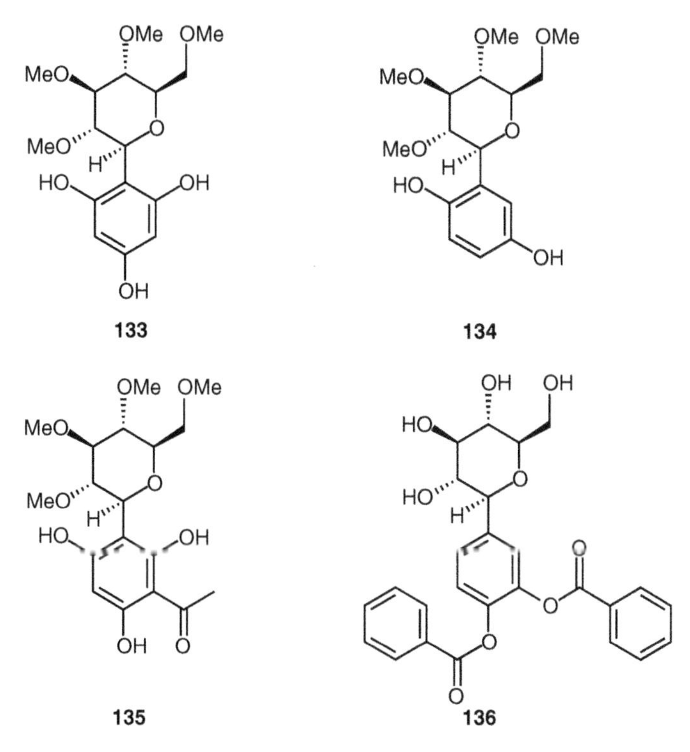

Figure 10.9 Structure of Fyn kinase inhibitor **134** and of the inhibitors of Aβo induced Fyn activation **133, 135,** and **136**.

desired compound **136** in 8% yield. The low yield of this final step was due to the hydrolysis of the ester groups and could potentially be further optimized.

In terms of their bioactivity, the per-O-methylated *C*-glucosyl phloroglucinol **133** and the per-O-methylated *C*-glucosyl hydroquinone **134** (Figure 10.9) were found to significantly inhibit Fyn kinase activity at 10 μM ($p < 0.01$ vs. untreated controls) in hiPSC-derived neural progenitor cells. On the other hand, compounds **133, 135,** and **136** (Figure 10.9) were able to inhibit Aβo-induced Fyn activation at the same

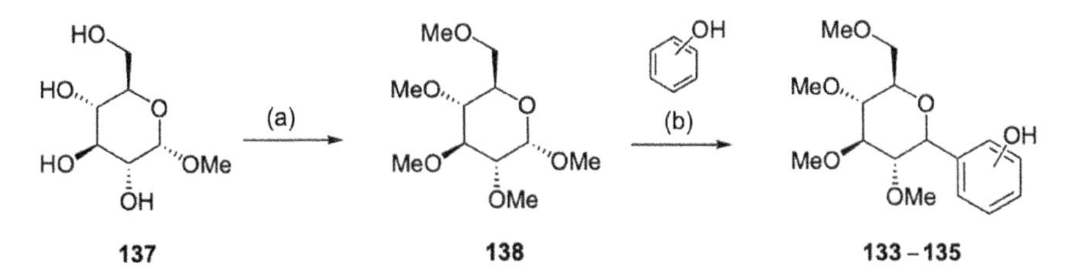

Scheme 10.16 Synthesis of per-O-methylated *C*-glucosyl polyphenols **133–135** [105]]. Reagents/solvent and yield: (a) DMF, NaH, MeI, 91%; (b) dry CH_3CN, hydroquinone, phloroglucinol or acetophloroglucinol, drierite, TMSOTf. Source: Adapted from Matos et al. [105].

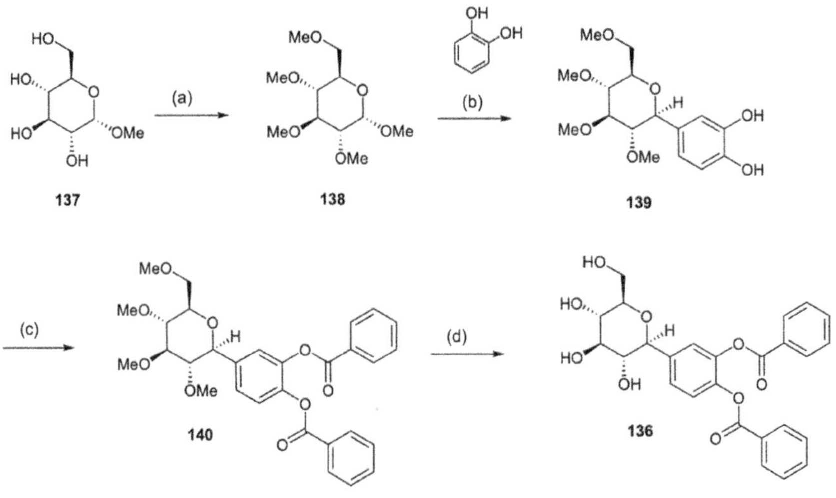

Scheme 10.17 Synthesis of per-O-methylated C-glucosyl polyphenols [103]. Reagents/solvent and yield: (a) DMF, NaH, MeI, 91%; (b) dry CH_3CN, drierite, TMSOTf, 63%; (c) DCM, imidazole, BzCl, 88%; (d) DCM, $BBr_3{\cdot}SMe_2$, 8%.

concentration (p <0.001 vs. untreated Aβo controls). Importantly, all four compounds (**133–136**) reduced Aβo-induced Tau hyperphosphorylation (at 10 μM) to levels below controls, which once more supports Fyn kinase as a therapeutic target for AD.

Even though other analog molecules also exhibited bioactivity in the SAR study conducted by the authors, this group of glycoconjugates displayed the adequate physicochemical features (optimal log D and effective permeability in a PAMPA assay) without any relevant cytotoxic effects at 50 μM [103], which makes them candidates for further development as CNS-active therapeutic agents.

10.7 Amyloid Protein–Protein Interaction Inhibitors

As previously discussed, the interactions between Aβo and its high-affinity partner PrP^C at the neuronal cell surface are crucial in triggering Fyn activation and subsequent events leading to tau hyperphosphorylation in AD (Figure 10.10). Hence, when thinking of protein–protein interaction inhibition strategies aiming to directly disrupt the involvement of Aβ in the development of this signaling cascade, two main approaches arise: (i) tackling the aggregation process that turns Aβ monomers (Aβm) into Aβo and (ii) inhibiting the binding between Aβo and the PrP^C (Figure 10.10).

In the context of approach (i), a group of flavonoids was investigated for their ability to inhibit Aβ aggregation [106]. The authors highlighted the ability of catechol-type flavonoids, such as the natural compound quercetin (**141,** Figure 10.11), to completely abolish Aβ aggregation by acting as Michael acceptors once oxidized into the corresponding *o*-quinones. This process is characteristic of Pan Assay

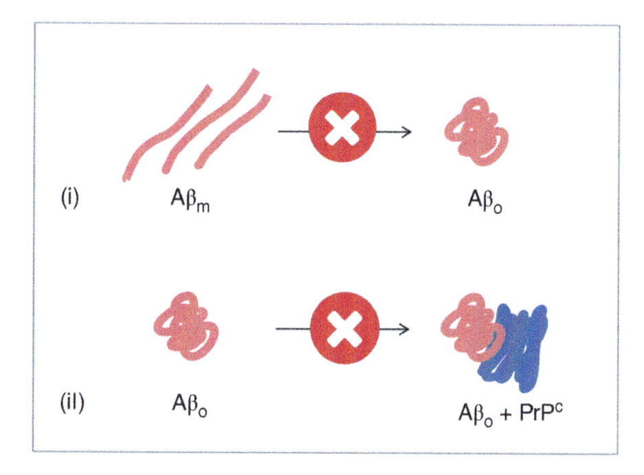

Figure 10.10 Two major processes to be targeted by amyloid protein–protein interaction inhibitors in the context of AD.

Interference Compounds (PAINS) with the catechol motif, which are able to indiscriminately interfere with many other molecular targets. Hence, highly promiscuous compounds such as quercetin (**141**) should not be regarded as ideal candidates for further optimization. Yet, they ultimately uncovered the potential of chrysin (**142**, Figure 10.11), another natural compound, to act as an anti-aggregation compound. Even though ThT fluorescence assays did not reveal a significant change in the presence of amyloid aggregates by the effects of this compound (tested in concentrations from 0.5 to 100 μM), Atomic Force Microscopy (AFM) experiments showed that, compared with structurally similar compounds, chrysin preferably

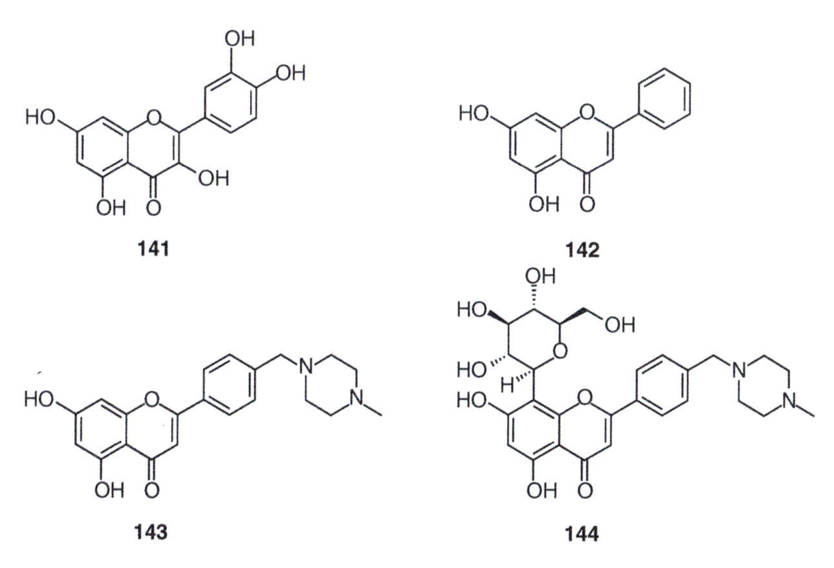

Figure 10.11 Amyloid anti-aggregation agents (**141, 142**) and Aβo-PrP^C disrupting agents (**143, 144**).

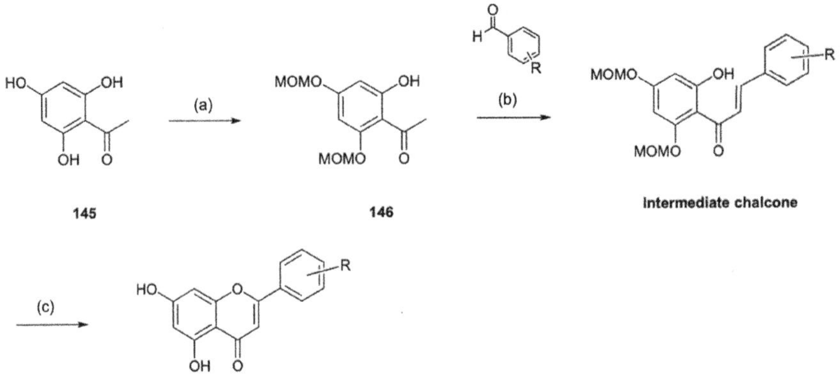

Scheme 10.18 Flavone synthesis as described by Marta et al. [106–108] Reagents/solvent and yield: (a) acetone, K_2CO_3, $ClCH_2OCH_3$, 52%; (b) 1,4-dioxane, aq. NaOH 50% (w/v), aromatic aldehyde; (c) pyridine, I_2, then TsOH, MeOH.

yields bigger Aβ1–42 aggregates instead of small, toxic oligomers. These results suggest a remodeling effect by the action of chrysin (**142**) and related molecules, serving as the basis for the generation of glycosylated flavones with neuroprotective potential, which shall be discussed in Section 10.8.

Importantly, chrysin was synthesized by Matos et al. [106] by a new methodology also used for the generation of similar flavones thereafter (Scheme 10.18) [107, 108]. Selective protection of two hydroxyl groups of acetophloroglucinol (**145**) with methoxymethyl ether (MOM), sodium hydroxide-promoted Claisen–Schmidt aldol condensation of **146** with the appropriate aromatic aldehyde gave the intermediate chalcone. In the synthesis of chrysin (**142**, R=H), the corresponding chalcone was obtained by reaction with benzaldehyde in 95% yield. Then, iodine and pyridine under reflux conditions were used for the oxidative cyclization step for the first time in the literature, affording the monoprotected flavone (as detected by LCMS). After work-up, acid-promoted deprotection of the MOM groups led to the synthesis of the desired product. This final one-pot reaction step afforded chrysin **142** in 81% yield. The individual synthesis of other flavones according to this general method will be mentioned throughout this chapter, along with the reaction yields obtained for the aldehyde used in each particular case.

In the context of strategy (ii) (Figure 10.10), a small library of nature-inspired flavones and their *C*-glucosyl derivatives was synthesized [108], from which the pair of *N*-methyl piperazinyl analogs **143** and **144** (Figure 10.11) showed therapeutic potential as protein–protein interaction inhibitors. In the initial Saturation Transfer Difference (STD) NMR screening assays that aimed to assess the binding potential of all synthesized molecules against Aβo, compound **143** stood out as the one presenting the highest potential for further investigation, with visible interactions with Aβo at 2 μM. Later on, both **143** and **144** inhibited Aβo-PrP^C binding in HEK cells, where the *C*-glucosyl derivative **144** decreased the binding between these two proteins by 41% ($p < 0.001$ vs. Aβo-PrP^C controls, with dose–dependent effects), and **143** by 26% ($p < 0.01$ vs. Aβo-PrP^C controls). Interestingly, the *C*-glucosyl moiety was found to enhance the extent of Aβo-PrP^C disruption, thus highlighting the importance of the C-linked sugar moiety for this type of therapeutic application ($p < 0.05$ for compound **143** vs. compound **144**) [108]. With no relevant cytotoxic effects observed *in vitro*, these compounds are therefore promising candidates for further studies, namely BBB permeation assays and *in vivo* tests aiming to prove the physiological effects of tackling the interaction between Aβo-PrP^C in AD animal models.

Interestingly, compound **133** (2,3,4,6-tetra-*O*-methylglucosyl)phloroglucinol (Figure 10.9) was able to inhibit the binding between Aβo and PrP^C by 26%, while its analogs **134–136** (Figure 10.9) inhibited by 16%, 11%, and 17%, respectively [102]. As earlier discussed, some of these compounds are Fyn kinase inhibitors (**134**), inhibit Aβo-induced Fyn activation (**133, 135,** and **136**), and all inhibited Aβo-induced Tau hyperphosphorylation to a significant extent.

The synthesis of **143** [108] was conducted on the basis of the previously described method for the synthesis of chrysin (**142**) (Scheme 10.18), with a slight variation regarding the phenol-protecting groups. Indeed, ethoxymethyl ether (EOM) was used instead of MOM ether, allowing the optimization of the protection step, which afforded the corresponding EOM-diprotected acetophloroglucinol derivative in 91% yield. Aldol condensation was carried out with 4-[(4-methylpiperazin-1-yl)methyl]benzaldehyde to give the intermediate chalcone in 99% yield. Oxidative cyclization and deprotection were conducted as earlier described [106] (Scheme 10.18), giving **143** in 90% yield. A similar synthetic route was employed for the preparation of the *C*-glucosyl analog **144** (Scheme 10.19) and all other *C*-glucosyl flavones

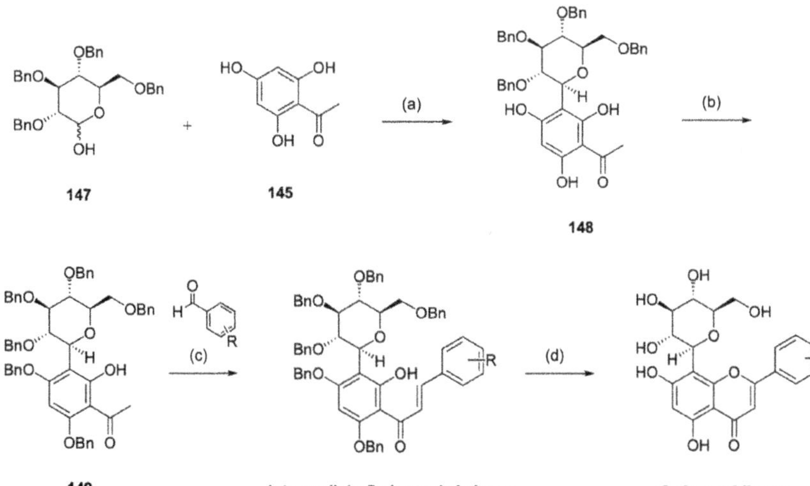

Scheme 10.19 Synthesis of *C*-glucosyl flavones as described by Matos et al. [107, 108]. Reagents/solvent and yield: (a) CH_3CN/) DCM, drierite, TMSOTf, 57%; (b) DMF, K_2CO_3, BnBr, 64%; (c) 1,4-dioxane, aq. NaOH 50% (w/v), aromatic aldehyde; (d) pyridine, I_2; then BBr_3, DCM.

described in this chapter. Succinctly, the TMSOTf-promoted *C*-glucosylation of acetophloroglucinol **145** with the commercially available glucosyl donor **147** at low temperature afforded intermediate **148** in 57% yield. Subsequently, after selective benzylation of two hydroxy groups, the base-catalyzed aldol condensation with the appropriate aldehyde, (4-[(4-methylpiperazin-1-yl)methyl]benzaldehyde in the case of compound **144**), led to the generation of the intermediate *C*-glucosyl chalcone which, in the synthesis of **144**, was accomplished in 58% yield. Finally, oxidative cyclization promoted by iodine in pyridine, followed by deprotection of all benzyl groups gave the desired *C*-glucosyl flavone. This final one-pot reaction step afforded **144** (R=4-methylpiperazin-1-yl) in 84% yield.

10.8 Inhibitors of Aβo and/or Oxidative Stress-Induced Neurotoxicity

In 2019, Rauter's group synthesized resveratrol deoxyglycosides **150** and **151** (Scheme 10.20) [109], and compared their effects with the corresponding aglycones regarding their ability to inhibit hydrogen peroxide-induced neurotoxicity in

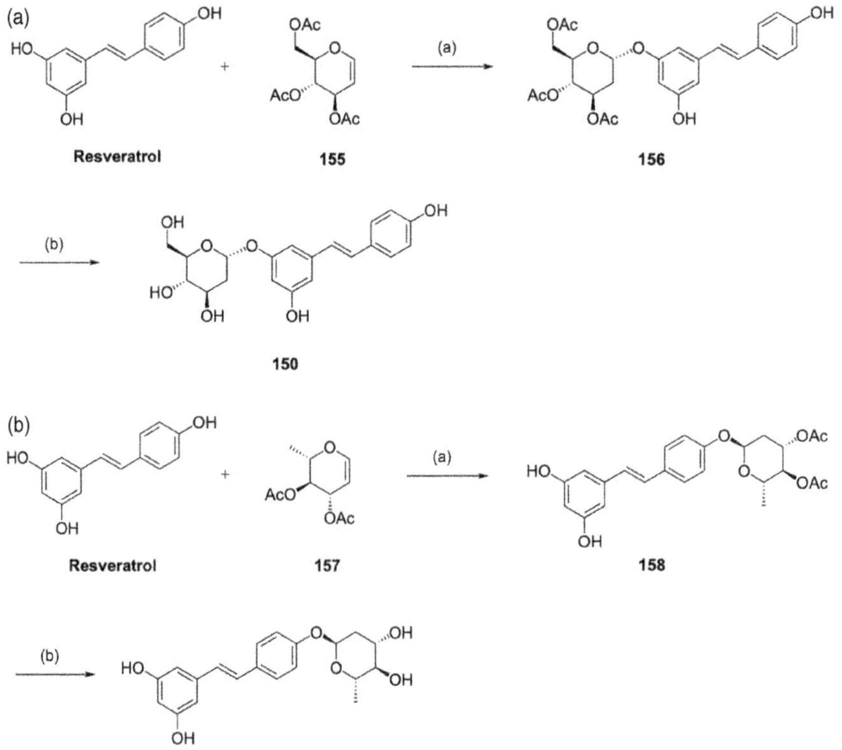

Scheme 10.20 Synthesis of resveratrol 2-deoxyglycosides with neuroprotective activity [109]. Reagents/solvent and yield: (a) TPHB, THF, **156**, 21%; **158**, 8%; (b) NaOMe, MeOH, **150**, 92%; **151**, 100%.

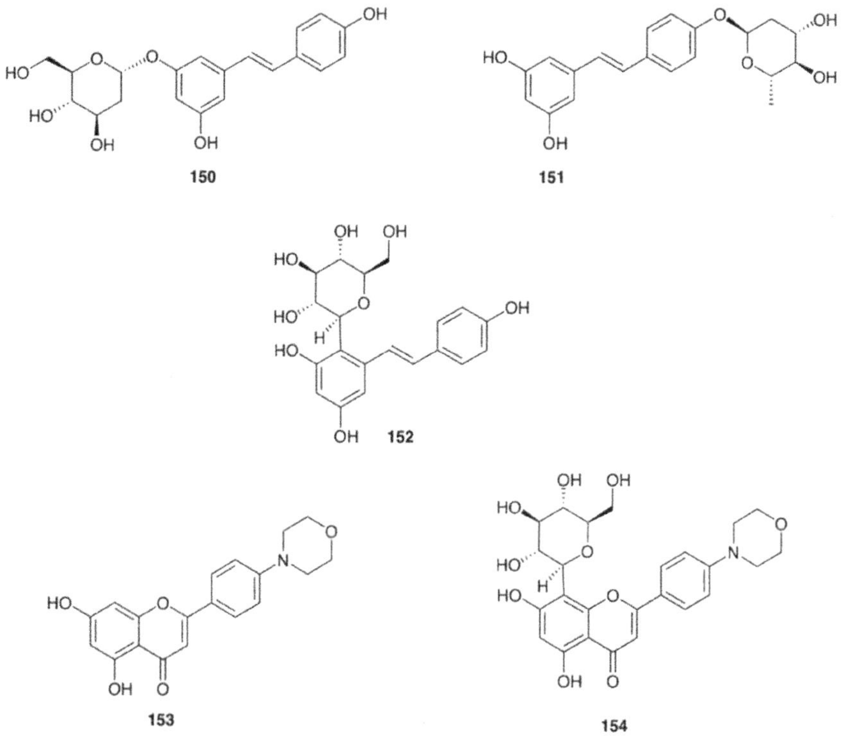

Figure 10.12 Polyphenol structures: resveratrol glycosides **150, 151**, glucosyl resveratrol **152**, and flavones **153, 154,** two promising compounds able to inhibit Aβo and/or oxidative stress-induced neurotoxicity.

SHSY-5Y cells. The resveratrol 2-deoxy-α-D-*arabino*-hexopyranoside **150** and the 2,6-dideoxy-α-L-*arabino*-hexopyranoside **151** were generated in a simple two-step process (Scheme 10.20a,b), by reaction of resveratrol with the adequate acetyl-protected glycals **155** or **157** in the presence of triphenylphosphane hydrobromide, followed by deprotection with sodium methoxide [109].

Both compounds were able to significantly reduce the damage caused by hydrogen peroxide when added to the neuronal cells at 50 μM, as observed in an MTT cell viability assay ($p < 0.05$ vs. hydrogen peroxide controls) [109]. Resveratrol was also able to show some effect ($p < 0.05$); however, it is important to note that resveratrol is a well-known Pan-Assay INterference compound (PAIN), and capable of interfering with biological membranes, altering their structure, fluidity, and the function of transmembrane proteins that are at the very beginning of cell signaling pathways [110, 111]. Thus, even though resveratrol is a powerful antioxidant, it is not the ideal candidate for optimization due to nonspecific effects that can be triggered by its action.

With this in mind, the potential of *C*-glycosylation as a tool for preventing the PAINS-type behavior exhibited by some polyphenolic compounds found in nature, including resveratrol, was further investigated [112]. It should be noted that though

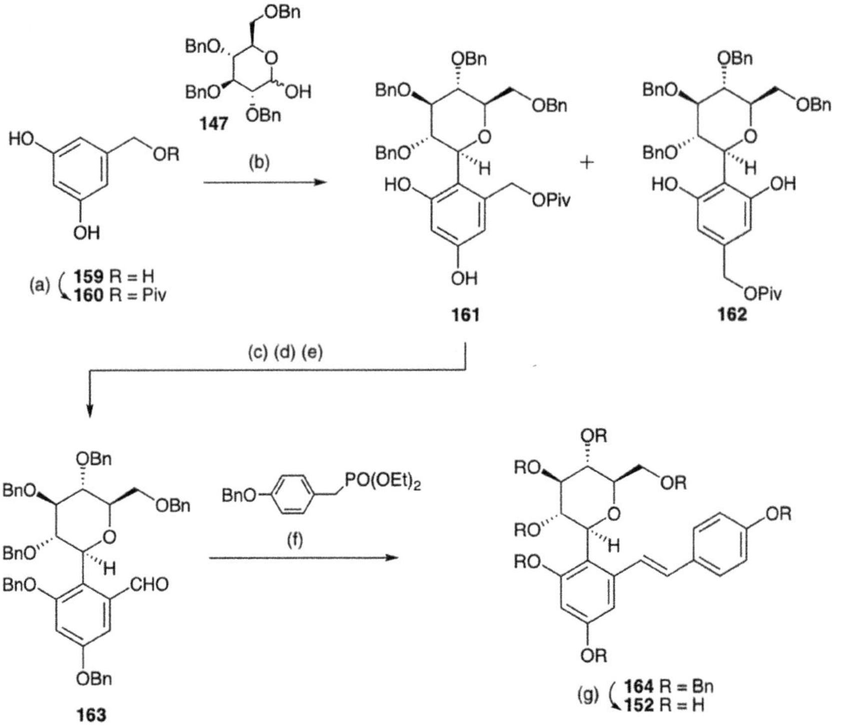

Scheme 10.21 The first synthesis of glucosylresveratrol **152** [112]. Reagents/solvent and yield: (a) PivCl, pyridine, 62% yield; (b) TMSOTf, drierite, DCM/CH_3CN, **162**, 24% isolated yield; (c) K_2CO_3, BnBr, DMF, 85% yield; (d) LiOH, MeOH/H_2O, 68% yield; (e) PCC, DCM, 81% yield; (f) *tert*-BuOK, DMF, 42% isolated yield for the (*E*)-isomer **164**; (g) DCM, BCl_3, 35% yield.

glycosides, such as **150** and **151** are easily synthesized, they are highly susceptible to hydrolysis in the gut, and therefore may reach their therapeutic targets as aglycones when administered orally. The same does not occur with the chemically and enzymatically stable *C*-glucosyl polyphenol derivatives, which became the focus of our research. The first synthesis of *C*-glucosyl resveratrol **152** (Scheme 10.21) was carried out starting by TMSOTf-catalyzed Fries-type reaction of benzylated glucopyranose **147** with the pivaloyl protected 5-(hydroxymethyl)benzene-1,3-diol **160** to afford the *C*-glucoside **161**, chromatographically separated from the secondary product **162** and isolated in 24% yield. Benzylation of the free hydroxy groups was followed by depivaloylation with lithium hydroxide in methanol/water and subsequent oxidation with pyridinium chlorochromate, obtaining the aldehyde **163** in 46.8% overall yield from **161**. Horner–Wadsworth–Emmons olefination with diethyl (4-benzyloxyphenyl)methylphosphonate in the presence of *tert*-BuOK in DMF afforded the protected (*E*)-isomer **164**, isolated in 55% yield. Full debenzylation was possible with BCl_3 to give purified (*E*)-glucosyl resveratrol **152** in 35% yield.

Compound **152** was then compared to its aglycone resveratrol concerning the ability to alter membrane dipole potential. Indeed, resveratrol and other polyphenols, typically regarded in the literature as PAINS, were found to significantly alter membrane dipole potential ($p < 0.0001$ vs. untreated controls), which could be a mechanism underpinning the broad bioactivities they have been described to exert when studied in cell models of disease. In contrast, the corresponding *C*-glucosyl derivatives, including compound **152**, did not exert the same effects (not statistically different vs. untreated controls), suggesting that *C*-glucosylation may be considered as a valuable approach to prevent the promiscuous membrane-disrupting effects of lipophilic, planar compounds, such as resveratrol.

It is important to note that the addition of the *C*-glucosyl moiety does not mean loss of bioactivity in the case of polyphenolic compounds. In fact, maintenance or even an increase in activity is possible with *C*-glucosylation, and this is well illustrated in this review by the results reported for compounds **143** and **144** (Figure 10.11) as well as those given in the literature by Rauter's group for other glucosylpolyphenols [113, 114]. Moreover, many other bioactive *C*-glycosides here are presented to confirm the usefulness of *C*-glycosylation as a means to generate bioactive molecular entities with therapeutic potential against neurodegenerative processes.

Another great example is here given by the 4′-morpholinylflavone derivatives **153** and **154** (Figure 10.12). Synthesized by Matos et al. [108] in 2019, these two compounds used at 50 μM concentrations were able to normalize cell viability of SHSY-5Y cells exposed to oxidative stress induced by H_2O_2, with significant differences relative to H_2O_2-treated controls ($p < 0.0001$ for compound **154** and $p < 0.05$ for **153**, vs. H_2O_2-treated controls, respectively). Furthermore, both compounds significantly reduced Aβ-induced neurotoxicity in the same cell model with a similar efficacy pattern ($p < 0.0001$ vs. Aβ-treated controls for **154**; $p < 0.05$ vs. Aβ-treated controls for **153**). Flavone **154**, the best in this study, showed adequate log D and effective permeability in a PAMPA assay, and was not cytotoxic in concentrations up to 100 μM, being, therefore, a promising lead for future investigation against AD [109].

10.9 Carbohydrate–Protein Interactions as Potential Therapeutic Targets Against AD

10.9.1 Lipid-Raft Gangliosides as Membrane Accumulation Sites for Toxic Aβ Aggregates

Carbohydrate–protein and carbohydrate–carbohydrate interactions are key in many physiological and pathological processes, and AD is no exception. Indeed, increasing evidence points toward membrane clusters enriched in ganglioside GM1, the most common brain ganglioside, as seeding locations for Aβ peptides [115]. Gangliosides are glycosphingolipids carrying one or more sialic acid units in their glycans and are usually located in the outer leaflet of the plasma membrane in the so-called lipid rafts, which are essential in signal transduction processes [116]. In these microdomains, also containing sphingomyelin and cholesterol, GM1 forms a string-like cluster with a hydrophobic sugar–lipid interface that is greatly prone to interact with and accommodate Aβ peptides, thereby restricting their spatial rearrangements and promoting the formation of cross-β-sheets [117]. The binding affinity between ganglioside clusters and Aβ peptides seems to increase with the number of glycolipid-decorating sugar moieties, with reports of hydrogen bond interactions between Aβ and sugar hydroxyl groups, or hydrophobic CH-π interactions between aromatic side chains of Aβ amino acid residues and the CH groups of the glycan backbone as the major driving force of binding of Aβ to lipid bilayers [118].

GM1-bound Aβ peptides exhibit an extremely elevated potential to accelerate Aβ aggregation [117]. Furthermore, contrarily to the earlier mentioned nontoxic Aβ fibrils formed in solution, membrane-anchored Aβ fibrils formed in such a hydrophobic context are, in fact, cytotoxic and display unique physicochemical properties [116]. Over time, such Aβ fibrils with potent toxicity accumulate in the form of amyloid plaques, causing lipid raft disruptions that affect cellular processes relying on the normal function of these microdomains [119]. Hence, compounds targeting the interactions between GM1 sialic acid units and Aβ side chains may be seen as promising therapeutic targets against AD, namely those able to suppress β-sheet formation by stabilizing α-type helical secondary structures of Aβ on the ganglioside clusters [117].

It is interesting to note that the formation of GM1 clusters in the first place is highly dependent on cholesterol concentrations, as the establishment of hydrogen bonds with GM1 and sphingomyelin assures the required cohesion between lipids for the clustering process to occur [116]. Notably, high levels of serum cholesterol positively correlate with an increased risk of dementia, with reports showing a decreased prevalence of AD in subjects prescribed with cholesterol-lowering drugs [120]. Moreover, the presence of the cholesterol transporter ApoE4 variant, linked to abnormal cholesterol metabolism in the brain, is strongly associated with the occurrence of late-onset AD [121]. The crucial role of cholesterol in ganglioside-containing lipid rafts may explain why that is the case. Therefore, tackling cholesterol metabolism and/or the assembly of GM1 clusters in the brain is also an appealing therapeutic target for drug discovery against AD.

10.9.2 The Role of Microglial Cells in Aβ Brain Clearance

Senile plaques, the extracellular deposits of Aβ aggregates typically found in the brain of patients with AD, are enriched in a variety of other components besides the Aβ peptide, namely sialylated glycoproteins and gangliosides. As previously described, gangliosides are able to bind Aβ and initiate their aggregation into toxic fibrils, thus playing a key role in the formation of these plaques [113]. It would be expected that microglial cells, the resident macrophages of the brain, would be able to effectively clear the amyloid aggregates observed in senile plaques by phagocytosis, but this is not the case [122, 123]. These cells express several recognition receptors at the cellular surface, including sialic acid-binding, immunoglobulin-like lectin receptors known as Siglecs. Specifically, microglia express Siglec-11, which belongs to the subfamily of CD33 proteins, known to suppress immune cell activation. CD33 are overexpressed in postmortem brain samples of AD patients and seem to inhibit the activity of microglia by specifically binding to sialylated conjugates, such as gangliosides, which might explain why senile plaques are not efficiently eliminated by microglial cells through phagocytosis [123, 124].

In 2019, a study was published by Dukhinova and coworkers [125] with brain-ganglioside deficient 5XFAD mice lacking the gene that encodes for α2,3-sialyltransferase, which is required for the synthesis of all major brain gangliosides, including GM1. This animal model for AD overexpresses three mutant human amyloid proteins and two presenilin *PS1* genes. These animals were found to have a significantly lower level of amyloid plaques compared to wild-type 5XFAD animals of the same age ($p < 0.0001$), without neuronal loss and with a comparable cognitive function to 5XFAD mice non-lacking the α2,3-sialyltransferase gene [125]. Interestingly, animals unable to produce sialylated brain gangliosides had lower microglia activation levels ($p < 0.05$), although these cells underwent important morphological changes of activation making them capable of phagocyting Aβ aggregates. Yet, the authors also reported that treatment of wild-type 5XFAD mice with a sialic acid-binding lectin targeting sialic acid units of major brain gangliosides successfully decreased amyloid deposition, inhibited neuroinflammation, boosted the expression of synaptic markers, and improved the cognitive function of these animals ($p < 0.05$ vs. controls) [125]. Moreover, in a recent study by Griciuc et al. [124], APP/PS1 mice, another animal model for the study of AD, treated with microRNA targeting CD33 (miRCD33) at an early stage, effectively displayed reduced CD33 microglial expression, and resulted in a significantly decreased Aβ plaque burden in the brain of these animals (24.4% decrease in Aβ plaque area, $p = 0.035$ vs. controls). However, it should be noted that in an *in vitro* study, where a CD33 knockout was performed in human macrophages and microglia, despite a higher phagocytic activity toward Aβ, CD33 deletion also resulted in oxidative damage and inflammation, contrarily to CD33-expressing microglia [126]. Together, this evidence validates the potential of ganglioside sialic acid masking as a therapeutic strategy to decrease the amyloid burden in AD.

10.10 Conclusion

With more than 50 million people affected globally, AD is undoubtedly one of the major public health concerns of the modern society for which effective disease-modifying therapies are urgently needed. Yet, the complex and multifactorial nature of this pathology makes the quest for new drug candidates a difficult challenge. As a matter of fact, reports on so far unexplored molecular players, cell–cell interactions, and key pathophysiological pathways in neurodegeneration are added to the literature almost everyday. While aggregation of Aβ is widely accepted as one of the causes of AD, other therapeutic targets are increasingly seen as alternative or additional choices when it comes to drug discovery and development.

As in many other pathological processes, carbohydrates play an important role in the processes leading to synaptic dysfunction and neuronal death, namely imbalances in posttranslational protein modification, ganglioside-promoted amyloid aggregation, recognition of amyloid aggregates by microglial cells via protein–carbohydrate interactions, among others. In this context, this critical review covered several different sugar-related therapeutic targets for drug development, as well as the most recently published carbohydrate-based drug candidates that act by mimicking natural substrates and transition states of glycosylation reactions. Both the synthesis of these compounds and the most important biological activity data available in the literature are here discussed. What's more, we have also covered a group of very recent nature-inspired sugar-linked polyphenols with remarkably promising results concerning Fyn inhibition, amyloid protein–protein interaction inhibition, and prevention of amyloid-oxidative stress-induced neuronal death. As here exemplified, in some cases the synthesis of such compounds can be much more straightforward and efficient compared to other known small molecules intended for the same use. Importantly, the sugar moiety can increase the desired bioactivity when linked to an aglycone with therapeutic potential, while mitigating undesirable PAINS-type behavior toward cell membranes.

The pool of data covered in this chapter illustrates the versatility of carbohydrates and carbohydrate-linked molecules in terms of their synthesis and bioactivity, while supporting their use as molecular scaffolds against neurodegenerative disorders, particularly AD.

List of Abbreviations

Aβ	amyloid β
Aβ1–40	40-Residue Aβ
Aβ1–42	42-Residue Aβ
Aβm	Aβ monomers
Aβo	Aβ oligomers
AcCl	acetyl chloride
ACh	acetylcholine
AChE	acetylcholinesterase
ACP	2-acetamido-6-chloropurine
AD	Alzheimer's disease
AFM	atomic force microscopy
AICD	APP intracellular domain
ALN	alendronate
ApoE4	apolipoprotein E4
APP	Aβ precursor protein

APP-CTFα	membrane-bound C-terminal fragment of APP generated by α-secretase
APP-CTFβ	membrane-bound C-terminal fragment of APP generated by β-secretase
AUC	area under the curve
BAβACs	soluble complexes between Aβ and BChE in the presence of ApoE4
BACE1	APP β-secretase
BChE	butyrylcholinesterase
Bn	benzyl
Boc	*tert*-butyloxycarbonyl
BSA	bis(trimethylsilyl)acetamide
BzCl	benzoyl chloride
CD33	sialic acid binding Ig-like lectin 3
ChEs	cholinesterases
CNS	central nervous system
COSs	chitosan oligosaccharides
CP	6-Chloropurine
CSA	camphorsulfonic acid
CSPGs	chondroitin sulfate proteoglycans
CTS	chitosan
CTS-ALN-NPs	ALN-loaded CTS nanoparticles
Cys	cysteine
DAST	diethylaminosulfur trifluoride
DBU	1,8-diazabicyclo(5.4.0)undec-7-ene
DCM	dichloromethane
DDQ	2,3-dichloro-5,6-dicyano-1,4-benzoquinone
DFT	density functional theory
D-Glc	D-glucose
DIPEA	N,N-diisopropylethylamine
DMA	dimethylacetamide
D-Man	D-mannose
DMAP	4-dimethylaminopyridine
DMF	dimethylformamide
DMS	dimethyl sulfide
DMSO	dimethylsulfoxide
DON	6-Diazo-5-oxonrleucine
DPPA	diphenylphosphoryl azide
DTAD	di-*tert*-butyl azodicarboxylate
DTT	dithiothreitol
EC50	half maximal effective concentration
EOM	ethoxymethyl ether
Fmoc	fluorenylmethyloxycarbonyl

GalNAc	*N*-acetyl glucosamine
GAGs	glycosaminoglycans
GalNAc-Ts	GalNAc-transferases
GlcN	glucosamine
HE	hematoxylin and eosin
HEXA	gene encoding for hexosaminidase subunit alpha
HEXB	gene encoding for hexosaminidase subunit beta
HIV	human immunodeficiency virus
hOGA	human P-GlcNAcase
ICV	intracerebroventricular
IL	interleukin
IPA	isopropyl alcohol
K_i	inhibition constant
LCMS	liquid chromatography–mass spectrometry
log *D*	pH-dependent distribution constant
m-GluR5	metabotropic glutamate receptor 5
miRCD33	microRNA targeting CD33
microRNA	micro ribonucleic acid
MOM	methoxymethyl ether
MTT	(3-(4,5-Dimethylthiazol-2-yl)-2,5-diphenyl-2*H*-tetrazolium bromide)
NA-COS	*N*-acetyl COS
NCS	*N*-chlorosuccinimide
NFTs	neurofibrillary tangles
NIS	*N*-iodosuccinimide
NMDA	*N*-methyl-D-aspartate receptor
NMR	nuclear magnetic resonance
OGA	*O*-GlcNAcase
OGT	*O*-GlcNAc transferase
PAIN	pan-assay interference compound
PAMPA	parallel artificial membrane permeability assay
PCC	pyridinium chlorochromate
PCR	polymerase chain reaction
PET	positron emission tomography
PGs	proteoglycans
PivCl	pivaloyl chloride
pK_a	$-\log_{10}$ of the acid dissociation constant, K_a
PMBCl	4-methoxybenzyl chloride
PNNs	perineural nets
PPTS	pyridinium *p*-toluenesulfonate
PrP	prion protein
PrP^C	cellular prion protein
PrP^{Sc}	Scrapie isoform of the prion protein
PUGNAc	2-Acetamido-2-deoxy-D-glucono-1,5-lactone *O*-(phenylcarbamoyl)oxime

Py	pyridine
PyBOP	benzotriazol-1-yloxytripyrrolidinophosphonium hexafluorophosphate
SAR	structure–activity relationships
Ser	serine
S_NAr	nucleophilic aromatic substitution
STD	saturation transfer difference
STZ	streptozotocin
TBDMSCl	*tert*-butyldimethylsilyl chloride
TBSCl	*tert*-butyldimethylsilyl chloride
TBSOTf	*tert*-butyldimethylsilyl trifluoromethanesulfonate
TFA	trifluoroacetic acid
THF	tetrahydrofuran
THP	tetrahydropyran
Thr	threonine
TIPSCl	triisopropylsilyl chloride
TMSOTf	trimethylsilyl trifluoromethanesulfonate
TNF	tumor necrosis factor
TPHB	triphenylphosphine hydrobromide
TPSA	topological polar surface area
UDP	uridine diphosphate

Acknowledgments

The authors wish to thank Fundação para a Ciência e a Tecnologia, Portugal, for supporting Centro de Química Estrutural (projects UIDB/00100/2020 e UIDP/00100/2020) and the Institute of Molecular Sciences (project LA/P/0056/2020). Ana Marta de Matos wishes to thank FCT for funding through the Individual Call for Scientific Employment Stimulus (2022.07037.CEECIND).

References

1 Varki, A. (2017). *Glycobiology* 27: 3–49.
2 Lin, B., Qing, X., Liao, J., and Zhuo, K. (2020). *Cell* 9: 1022.
3 Kappler, K. and Hennet, T. (2020). *Genes and Immunity* 21: 224–239.
4 Schnaar, R.L. (2015). *The Journal of Allergy and Clinical Immunology* 135: 609–615.
5 Sun, L., Middleton, D.R., Wantuch, P.L. et al. (2016). *Glycobiology* 26: 1029–1040.
6 Wani, W.Y., Chatham, J.C., Darley-Usmar, V. et al. (2017). *Brain Research Bulletin* 133: 80–87.
7 Estevez, A., Zhu, D., Blankenship, C., and Jiang, J. (2020). *Chemistry - A European Journal* 26: 12086–12100.

8 Alzheimer's Disease International. Numbers of people with dementia around the world, 2020. Available at: https://www.alzint.org/u/numbers-people-with-dementia-2017.pdf (accessed on June 1 2023).

9 Demetrius, L.A., Magistretti, P.J., and Pellerin, L. (2015). *Frontiers in Physiology* 5: 522.

10 Karran, E., Mercken, M., and De Stropper, B. (2011). *Nature Reviews Drug Discovery* 10: 698–712.

11 Nygaard, H.B. (2018). *Biological Psychiatry* 83: 369–376.

12 Um, J.W. and Strittmatter, S.M. (2013). *Prion* 7: 37–41.

13 Smith, L.M., Zhu, R., and Strittmatter, S.M. (2018). *Neuropharmacology* 130: 54–61.

14 Lin, C., Chen, E.H.-L., Lee, L.Y.-L. et al. (2014). *Carbohydrate Research* 387: 46–53.

15 Dias, C. and Rauter, A.P. (2015). *RSC Drug Discovery Series* 43: 180–208.

16 Bajad, N.G., Swetha, R., Gutti, G. et al. (2021). *Future Medicinal Chemistry* 13 (19): 1695–1711.

17 Ryan, P., Patel, B., Makwana, V. et al. (2018). *ACS Chemical Neuroscience* 9: 1530–1551.

18 Nordberg, A. and Svensson, A.-L. (1998). *Drug Safety* 19: 465–479.

19 Hitzeman, N. (2006). *American Family Physician* 74 (5): 747–749.

20 Lombard, V., Golaconda Ramulu, H., Drula, E. et al. (2014). *Nucleic Acids Research* 42: D490–D495.

21 Rexach, J.E., Clark, P.M., Mason, D.E. et al. (2012). *Nature Chemical Biology* 8: 253–261.

22 Francisco, H., Kollins, K., Varghis, N. et al. (2009). *Developmental Neurobiology* 69: 162–173.

23 Lagerlöf, O., Hart, G.W., and Huganir, R.L. (2017). *Proceedings. of the National Academy of Sciences of the United States of America* 114: 1684–1689.

24 Wang, A.C., Jensen, E.H., Rexach, J.E. et al. (2016). *Proceedings of the National Academy of Sciences of the United States of America* 113: 15120–15125.

25 Yuzwa, S.A. and Vocadlo, D.J. (2014). *Chemical Society Reviews* 43: 6839–6858.

26 Kim, C., Nam, D.W., Park, S.Y. et al. (2013). *Neurobiology of Aging* 34: 275–285.

27 Ryan, P., Xu, M., Davey, A.K. et al. (2019). *ACS Chemical Neuroscience* 10: 2209–2221.

28 Arnold, C.S., Johnson, G.V.W., Cole, R.N. et al. (1996). *The Journal of Biological Chemistry* 271: 28741–28744.

29 Di Domenico, F., Chiara Lanzillotta, C., and Tramutola, A. (2019). *Expert Review of Neurotherapeutics* 19: 1–3.

30 Drzezga, A., Lautenschlager, N., Siebner, H. et al. (2003). *European Journal of Nuclear Medicine and Molecular Imaging* 30: 1104–1113.

31 Sancheti, H., Akopian, G., Yin, F. et al. (2013). *PLoS One* 8: e69830.

32 Sancheti, H., Kanamori, K., Patil, I. et al. (2014). *Journal of Cerebral Blood Flow and Metabolism* 34: 288–296.

33 Chow, V.W., Mattson, M.P., Wong, P.C., and Gleichmann, M. (2010). *Neuromolecular Medicine* 12 (1): 1–12.

34 Akasaka-Manya, K., Kawamura, M., Tsumoto, H. et al. (2017). *Journal of Biochemistry* 161 (1): 99–111.
35 Gong, C.X., Liu, F., and Iqbal, K. (2016). *Alzheimer's & Dementia* 12: 1078–1089.
36 Liu, F., Shi, J., Tanimukai, H. et al. (2009). *Brain* 132: 1820–1832.
37 Frenkel-Pinter, M., Shmueli, M.D., Raz, C. et al. (2017). *Science Advances* 3: e1601576.
38 Hurtado-Guerrero, R., Dorfmueller, H.C., and van Aalten, D.M.F. (2008). *Current Opinion in Structural Biology* 18: 551–557.
39 Macauley, M.S., Whitworth, G.E., Debowski, A.W. et al. (2005). *The Journal of Biological Chemistry* 280: 25313–25322.
40 Cekic, N., Heinonen, J.E., Stubbs, K.A. et al. (2016). *Journal of Chemical Sciences* 7: 3742–3750.
41 Li, B., Li, H., Hu, C.-W., and Jiang, J. (2017). *Nature Communications* 8: 666.
42 Xiong, J. and Xu, D. (2020). *The Journal of Physical Chemistry B* 124: 9310–9322.
43 Liu, T., Duan, Y., and Yang, Q. (2018). *Biotechnology Advances* 36: 1127–1138.
44 Hattie, M., Cekic, N., Debowski, A.W. et al. (2016). *Organic & Biomolecular Chemistry* 14: 3193–3197.
45 Beer, D., Maloisel, J.-L., Rast, D.M., and Vasella, A. (1990). *Helvetica Chimica Acta* 73: 1918–1922.
46 Beer, D. and Vasella, A. (1985). *Helvetica Chimica Acta* 68: 2254–2274.
47 Horsch, M., Hoesch, L., Vasella, A., and Rast, D.M. (1991). *European Journal of Biochemistry* 197: 815–818.
48 Mohan, H. and Vasella, A. (2000). *Helvetica Chimica Acta* 83: 114–118.
49 Stubbs, K.A., Zhang, N., and Vocadlo, D.J. (2006). *Organic & Biomolecular Chemistry* 4: 839–845.
50 Goddard-Borger, E.D. and Stubbs, K.A. (2010). *The Journal of Organic Chemistry* 75: 3931–3934.
51 Stubbs, K.A., Macauley, M.S., and Vocadlo, D.J. (2009). *Angewandte Chemie* 121: 1326–1329.
52 Dorfmueller, H.C., Borodkin, V.S., Schimpl, M. et al. (2006). *Journal of the American Chemical Society* 128: 16484–16485.
53 Dorfmueller, H.C., Borodkin, V.S., Schimpl, M., and van Aalten, D.M.F. (2009). *The Biochemical Journal* 420: 221–227.
54 Dorfmueller, H.C., Borodkin, V.S., Schimpl, M. et al. (2010). *Chemistry & Biology* 17 (11): 1250–1255.
55 Aoyagi, T., Suda, H., Uotani, K. et al. (1992). *The Journal of Antibiotics* 45: 1404–1408.
56 Heightman, T.D. and Vasella, A.T. (1999). *Angewandte Chemie International Edition* 38: 750–770.
57 Perreira, M., Kim, J.E., Thomas, C.J., and Hanover, J.A. (2006). *Bioorganic & Medicinal Chemistry* 14: 837–846.
58 Rao, F.V., Dorfmueller, H.C., Villa, F. et al. (2006). *The EMBO Journal* 25 (7): 1569–1578.
59 Borodkin, V.S. and van Aalten, D.M.F. (2010). *Tetrahedron* 66: 7838–7849.

60 Gao, Y., Wells, L., Comer, F.I. et al. (2001). *The Journal of Biological Chemistry* 276: 9838–9845.
61 Yuzwa, S.A., Macauley, M.S., Heinonen, J.E. et al. (2008). *Nature Chemical Biology* 4: 483–490.
62 Yuzwa, S.A., Shan, X., Jones, B.A. et al. (2014). *Molecular Neurodegeneration* 9: 42.
63 Morrison, J.F. (1969). *Biochimica et Biophysica Acta* 185: 269–286.
64 Copeland, R.A. (2005). *Evaluation of Enzyme Inhibitors in Drug Discovery: A Guide for Medicinal Chemists and Pharmacologists*. Hoboken, New Jersey: John Wiley and Sons Inc.
65 Kuzmic, P., Elrod, K.C., Cregar, L.M. et al. (2000). *Analytical Biochemistry* 286: 45–50.
66 Perrin, L.C. and Fabian, M.A. (1996). *Analytical Chemistry* 68: 2127–2134.
67 Selnick, H.G., Hess, J.F., Tang, C. et al. (2019). *Journal of Medicinal Chemistry* 62: 10062–10097.
68 Sandhu, P., Lee, J., Ballard, J. et al. (2016). *Alzheimer's & Dementia* 12: P1028.
69 Wang, X., Li, W., Marcus, J. et al. (2020). *The Journal of Pharmacology and Experimental Therapeutics* 374: 252–263.
70 Akasaka-Manya, K. and Manya, H. (2020). *Biomolecules* 10: 1569.
71 Liu, F., Xu, K., Xu, Z. et al. (2017). *The Journal of Biological Chemistry* 292: 21304–21319.
72 Baig, S., Wilcock, G.K., and Love, S. (2005). *Acta Neuropathologica* 110: 393–401.
73 Testa, D., Prochiantz, A., and Di Nardo, A.A. (2019). *Seminars in Cell & Developmental Biology* 89: 125–135.
74 Duce, J.A., Zhu, X., Jacobson, L.H., and Beart, P.M. (2019). *British Journal of Pharmacology* 176: 3409–3412.
75 Duncan, J.A., Foster, R., and Kwok, J.C.F. (2019). *British Journal of Pharmacology* 176: 3611–3621.
76 Fonseca-Santos, B. and Chorilli, M. (2017). *Materials Science & Engineering. C, Materials for Biological Applications* 77: 1349–1362.
77 Naveed, M., Lucas Phil, L., Sohail, M. et al. (2019). *International Journal of Biological Macromolecules* 129: 827–843.
78 Kim, M.-S., Sung, M.-J., Seo, S.-B. et al. (2002). *Neuroscience Letters* 321: 105–109.
79 Jiang, Z., Liu, G., Yang, Y. et al. (2019). *Process Biochemistry* 84: 161–171.
80 Zameer, S., Ali, J., Vohora, D. et al. (2021). *Journal of Drug Targeting* 29: 199–216.
81 More, S.V., Kumar, H., Cho, D.-Y. et al. (2016). *International Journal of Molecular Sciences* 17: 1447.
82 Kumar, A., Pintus, F., Di Petrillo, A. et al. (2018). *Scientific Reports* 8: 4424.
83 Mushtaq, G., Greig, N.H., Khan, J.A., and Kamal, M.A. (2014). *CNS & Neurological Disorders Drug Targets* 13: 1432–1439.
84 Bartus, R.T., Dean, R.L., Beer, B., and Lippa, A.S. (1982). *Science* 217: 408–414.
85 Silver, A. (1974). *The Biology of Cholinesterases*. New York: American Elsevier Pub. Co.
86 Greig, N.H., Utsuki, T., Ingram, D.K. et al. (2005). *Proceedings of the National Academy of Sciences of the United States of America* 102: 17213–17218.

87 Sugimoto, H., Yamanishi, Y., Iimura, Y., and Kawakami, Y. (2000). *Current Medicinal Chemistry* 7: 303–339.

88 Marcelo, F., Silva, F.V., Goulart, M. et al. (2009). *Bioorganic & Medicinal Chemistry* 17: 5106–5116.

89 Schwarz, S., Csuk, R., and Rauter, A.P. (2014). *Organic & Biomolecular Chemistry* 12: 2446–2456.

90 Xavier, N.M., Schwarz, S., Vaz, P.D. et al. (2014). *European Journal of Organic Chemistry* 13: 2770–2779.

91 Batista, D., Schwarz, S., Loesche, A. et al. (2016). *Pure and Applied Chemistry* 2016 (88): 363–379.

92 Jasiecki, J., Targonska, M., and Wasag, B. (2021). *International Journal of Molecular Sciences* 22: 2033.

93 Inestrosa, N.C., Alvarez, A., Perez, C.A. et al. (1996). *Neuron* 16: 881–889.

94 Inestrosa, N.C., Dinamarca, M.C., and Alvarez, A. (2008). *The FEBS Journal* 275: 625–632.

95 Rees, T.M. and Brimijoin, S. (2003). *Drugs Today* 39: 75–83.

96 Garcia-Ayllón, M.S., Small, D.H., and Sáez-Valero, J. (2011). *Frontiers in Molecular Neuroscience* 4: 22.

97 Podoly, E., Bruck, T., Diamant, S. et al. (2008). *Neurodegenerative Diseases* 5: 232–236.

98 Podoly, E., Hanin, G., and Soreq, H. (2010). *Chemico-Biological Interactions* 197: 64–71.

99 Darvesh, S., Cash, M.K., Reid, G.A. et al. (2012). *Journal of Neuropathology and Experimental Neurology* 71: 2–14.

100 Kumar, R., Nordberg, A., and Darreh-Shori, T. (2016). *Brain* 139: 174–192.

101 Kaufman, A.C., Salazar, S.V., Haas, L.T. et al. (2015). *Annals of Neurology* 77: 953–971.

102 Ford, J.G., Pointon, S.M., Powell, L. et al. (2010). *Organic Process Research and Development* 14: 1078–1087.

103 Hennequin, L.F., Allen, J., Breed, J. et al. (2006). *Journal of Medicinal Chemistry* 49: 6465–6488.

104 Tang, S.J., Fesharaki-Zadeh, A., Takahashi, H. et al. (2020). *Acta Neuropathologica Communications* 8: 96.

105 Matos, A.M., Blázquez-Sánchez, M.T., Bento-Oliveira, A. et al. (2020). *Journal of Medicinal Chemistry* 63: 11663–11690.

106 Matos, A.M., Cristóvão, J.S., Yashunshy, D.V. et al. (2017). *Pure and Applied Chemistry* 89: 1305–1320.

107 Matos, A.M., Martins, A., Man, T. et al. (2019). *Pharmaceuticals* 12: 98.

108 Matos, A.M., Man, T., Idrissi, T. et al. (2019). *Pure and Applied Chemistry* 91: 1107–1136.

109 Dias, C., Matos, A.M., Blásquez-Sanchez, M.T. et al. (2019). *Pure and Applied Chemistry* 91: 1209–1221.

110 Neves, A.R., Nunes, C., Amenitsch, H., and Reis, S. (2016). *Soft Matter* 12: 2118–2126.

111 Baell, J.B. (2016). *Journal of Natural Products* 79: 616–628.

112 Matos, A.M., Blázquez-Sánchez, M.T., Sousa, C. et al. (2021). *Scientific Reports* 11: 4443.

113 Jesus, A.R., Vila-Viçosa, D., Machuqueiro, M. et al. (2017). *Journal of Medicinal Chemistry* 60: 568–579.

114 Jesus, R., Dias, C., Matos, A.M. et al. (2014). *Journal of Medicinal Chemistry* 57: 9463–9472.

115 Rudajev, V. and Novotny, J. (2020). *Membranes* 10: 226.

116 Matsuzaki, K. (2014). *Accounts of Chemical Research* 47: 2397–2404.

117 Kamiya, Y., Yagi-Utsumi, M., Yagi, H., and Kato, K. (2011). *Current Pharmaceutical Design* 17: 1672–1684.

118 Ikeda, K. and Matsuzaki, K. (2008). *Biochemical and Biophysical Research Communications* 370: 525–529.

119 Molander-Melin, M., Blennow, K., Bogdanovic, N. et al. (2005). *Journal of Neurochemistry* 92: 171–182.

120 Loera-Valencia, R., Goikolea, J., Parrando-Fernandez, C. et al. (2019). *The Journal of Steroid Biochemistry and Molecular Biology* 190: 104–114.

121 Jeong, W., Lee, H., Cho, S., and Seo, J. (2019). *Molecules and Cells* 42: 739–746.

122 Linnartz, B., Bodea, L.-G., and Neumann, H. (2012). *Cell and Tissue Research* 349: 215–227.

123 Salminen, A. and Kaarniranta, K. (2009). *Journal of Molecular Medicine* 87: 697–701.

124 Griciuc, A., Federico, A.N., Natasan, J. et al. (2020). *Human Molecular Genetics* 29: 2920–2935.

125 Dukhinova, M., Veremeyko, T., Yung, A.W.Y. et al. (2019). *Neurobiology of Aging* 77: 128–143.

126 Wißfeld, J., Nozaki, I., Mathews, M. et al. (2021). *Glia* 69: 1393–1412.

11

Carbohydrate-Based Antithrombotics

Antonella Bisio, Marco Guerrini, and Annamaria Naggi

Istituto di Ricerche Chimiche e Biochimiche G. Ronzoni, V. G. Colombo 81, Milan, 20133, Italy

11.1 Introduction

The history of carbohydrate-based antithrombotics began in Toronto about a century ago, with the chance discovery of heparin by Jay McLean and William Henry Howell [1, 2]. They had actually been looking for a procoagulant substance and suspected that their fat-soluble anticoagulant tissue extract might be a phospholipid. Despite their initial misinterpretation, the commercial potential of heparin was recognized immediately, and the first pharmaceutical heparin product appeared in the US market in 1939, produced by Roche Organon with the trade name Liquaemin. It was only in the early 1960s though, following the publication of a landmark clinical trial based on the use of heparin for the treatment of pulmonary embolism (PE) [3], that heparin was recognized as a powerful therapeutic drug for the prevention and treatment of venous thrombosis by intravenous administration. In the early 1970s, with the publication of a seminal paper by the group Kakkar, introducing the concept of subcutaneous administration of low-dose heparin for the prevention of postoperative deep vein thrombosis (DVT) [4], heparin entered routine clinical use as an anticoagulant drug.

In parallel with clinical studies, extensive chemical and biochemical investigations have been undertaken, contributing to the elucidation of the molecular basis of the anticoagulant and antithrombotic activities of heparin and heparin-like glycosaminoglycans (GAGs), such as heparan sulfate (HS) and dermatan sulfate (DS), in particular regarding their role in the coagulation system. The mechanism of the interaction of heparin and DS with two key proteins of the coagulation process, antithrombin (AT) and heparin cofactor II (HCII), and the subsequent enhancement of inactivation of the two key coagulation enzymes, factor Xa (fXa) and thrombin (also named fIIa) has been unraveled at the molecular level.

The wealth of knowledge acquired, especially from the mid-1970s onward, has led to the development of low molecular weight heparins (LMWHs) derived from

Carbohydrate-Based Therapeutics, First Edition. Edited by Roberto Adamo and Luigi Lay.

unfractionated heparin (UFH) through distinct chemical or enzymatic depolymerization processes and to a unique pentasaccharide obtained by chemical synthesis, as distinct drugs.

11.2 Antithrombotic Drugs

Under physiological conditions, blood flows smoothly and efficiently in the arteries and veins, but if a clot or thrombus breaks free from the vessel wall and obstructs blood flow, the result – termed thrombosis – can have serious, even lethal, consequences. Thrombosis can occur both in arteries, which carry blood from the heart to the rest of the body, as well as in veins, which carry blood from the body back to the heart. Arterial thrombosis can result in a heart attack when it occurs in coronary arteries, or in stroke when it occurs in blood vessels in the brain. Venous thrombosis can lead to DVT, often in the legs, groin or arms, and later to PE: DVT and PE are known collectively as venous thromboembolism (VTE). Both arterial and venous thromboses are major healthcare concerns, causing an estimated 18 million deaths worldwide each year [5].

The most important components of a thrombus are platelets and fibrin, the latter being the final product of the coagulation cascade and is a protein that forms a mesh encapsulating a high concentration of red blood cells. Both platelets and fibrin stabilize the thrombus and prevent it from breaking down, whereas fibrin predominates in venous thrombi, and platelets are the main component of arterial clots. Accordingly, two classes of antithrombotic drugs have been developed; antiplatelet drugs and anticoagulants. Whereas the former prevents platelets from clumping, anticoagulants slow down clotting by reducing fibrin production and preventing the formation and growth of clots. All carbohydrate-based antithrombotic agents of a GAG nature, such as heparin, DS, LMWHs, and fondaparinux, together with drugs based on GAG mixtures, share this latter mode of action. Defibrotide (DF) and pentosan polysulfate (PPS), the only non-GAG antithrombotic drugs considered here, have a broader and more complex spectrum of action.

11.3 Heparin

Heparin originates in mammalian mast cell granules as a large (about 80 kDa) polymeric component of the proteoglycan serglycin [6]. Its biosynthesis begins with the formation of a tetrasaccharide linkage region, which is synthesized from the proteoglycan core protein and continues with the alternating addition of 1→4 linked α-*N*-acetyl-D-glucosamine (GlcNAc) and β-D-glucuronic acid (GlcA), which leads to linear polysaccharide chain elongation. Immediately after polymerization, the sequential and coordinated action of a series of enzymes gives rise to important enzymatic modifications to the sequence. First, *N*-deacetylase–*N*-sulfotransferase partially removes the *N*-acetyl group from GlcNAc by simultaneously adding a sulfate group (GlcNS), and then uronyl C-5 epimerase converts some GlcA into α-L-iduronic acid (IdoA) and various *O*-sulfotransferases catalyze the addition of sulfate groups at position 2 of IdoA, 6 of glucosamine or, less frequently, position 3 of glucosamine. The final result is the formation of structural domains with different substitution patterns and sulfation degrees [7]. Such an array of structural features can generate potentially a total of 48 different disaccharide combinations, but due to the restrictions of the biosynthetic route, only 23 disaccharides have been identified to date in heparin [8]. Based on the analysis of HS/heparin biosynthetic enzymes, a biosynthetic scheme structured in two branches has been proposed, the major one containing commonly occurring IdoA–GlcNS disaccharides, and the minor one involving less represented structures, such as IdoA–GlcNAc [9]. The nodal aspect of this scheme arises from the different efficiency with which epimerase converts GlcA–GlcNS into IdoA–GlcNS and GlcA–GlcNAc into IdoA–GlcNAc.

Following mast cell activation and consequent degranulation, heparin is released in the extracellular matrix of endothelium and partially hydrolyzed by heparanase, an endo-β-glucuronidase into fragments ranging in mass from 5 to 30 kDa [10], resulting in a mixture of highly heterogenous polymeric chains arising from both the different molecular weight as well as the number and position of sulfate groups of the constituent disaccharides. The main repeating disaccharide structure of the resulting polymer is the trisulfated unit [-4)-α-L-IdoA2S (1→4) α-D-GlcNS,6S-(1-], which constitutes the regular polymer regions, accounting for more than 70% of heparin chains and resulting in an overall degree of sulfation of about 2.4 per disaccharide [11, 12]. The remaining 30% of the heparin polymer, constituting the so-called irregular regions, has a more complex composition, including both less O-sulfated and N-acetylated domains, and an important pentasaccharide sequence, containing a 3-O-sulfated glucosamine, which interestingly, is usually preceded by an un-sulfated IdoA residue, IdoA–[GlcNAc,6S–GlcA–GlcNS,3S,6S–IdoA2S–GlcNS,6S–] [13, 14]. Such a pentasaccharide, also called AGA*IA, with the asterisk indicating the peculiar 3-O sulfation, is endowed with a high affinity for the plasma protein AT. The structures of the basic disaccharide units, together with the AT-binding pentasaccharide are shown in Figure 11.1. Despite the observation that irregular regions do not appear to be distributed in an orderly manner, a low-sulfated domain resides close to the original heparin core protein [15]. Moreover, the 3-O-sulfated pentasaccharide turns out to be enriched toward the nonreducing terminus of the heparin chain [16–19], in agreement with the finding that NMR signals typically associated with the "linkage region" [20] are missing in heparin fractions with high affinity for AT [21]. In contrast, other studies have suggested a random distribution [22].

The heterogeneity of heparin can be related to the species and organs of origin and to the process of production. Heparins from different tissues and/or animal sources appear to have different IdoA2S and IdoA content. In particular, the trisulfated disaccharide comprises up to 90% of bovine lung heparin extracts, an organ that was formerly a major source of heparin, whereas it represents about 75% of porcine mucosa, which has largely replaced bovine lung as the principal source of clinical heparin. Heparins from various animal origins also exhibit distinct levels of the antithrombin binding pentasaccharide sequence (ATBPS). Heparins extracted

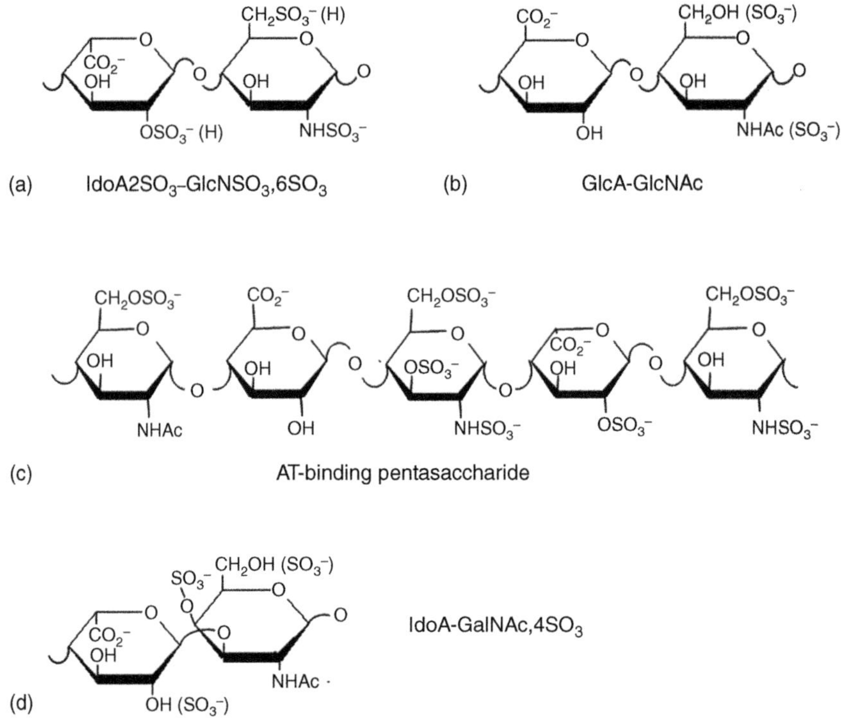

(a) IdoA2SO_3-GlcNSO_3,6SO_3 (b) GlcA-GlcNAc

(c) AT-binding pentasaccharide

(d) IdoA-GalNAc,4SO_3

Figure 11.1 Structures of the prevailing (a) and minor (b) heparin disaccharide units, of the peculiar AT-binding pentasaccharide (c), and of dermatan sulfate repeating disaccharide unit (d). The forms in parenthesis occur less frequently.

from clams are surprisingly rich in ATBPS and exhibit very potent anticoagulant properties [23]. Pharmaceutical-grade UFH of porcine origin consists of chains ranging from under 5000 Da to over 60 000 Da, with a mean molecular weight (M_W) from 15 to 20 kDa, which also depends on the method of preparation [24, 25].

A further level of complexity of heparin structure, besides its disaccharide composition, sequence of disaccharide components and clustering of similar disaccharides to constitute different domains, is provided by the IdoA conformation, which exists in equilibrium mostly between the chair $^1C_4$ and the skew $^2S_0$ conformation. The relative population of these forms in solution depends on the presence of the 2-*O*-sulfate group and on the nature and structure of the neighboring units [26]. Whereas the overall geometry of the heparin chains remains similar for the two IdoA conformers, the spatial arrangement of the sulfate groups is significantly different for the two local conformations. While in the $^1C_4$ conformation, the sulfate groups are more dispersed, in the $^2S_0$ form, sulfate groups are arranged in symmetrically oriented clusters (Figure 11.2) [27]. The "plasticity" of IdoA residues, modulated through particular sequences, can generate local distortions in the secondary structure of the polysaccharide chains, facilitating the most effective docking of the anionic groups of the GAG to the basic amino acids of proteins. In this regard, it is noteworthy that iduronate residues of the pentasaccharide sequence adopt a purely

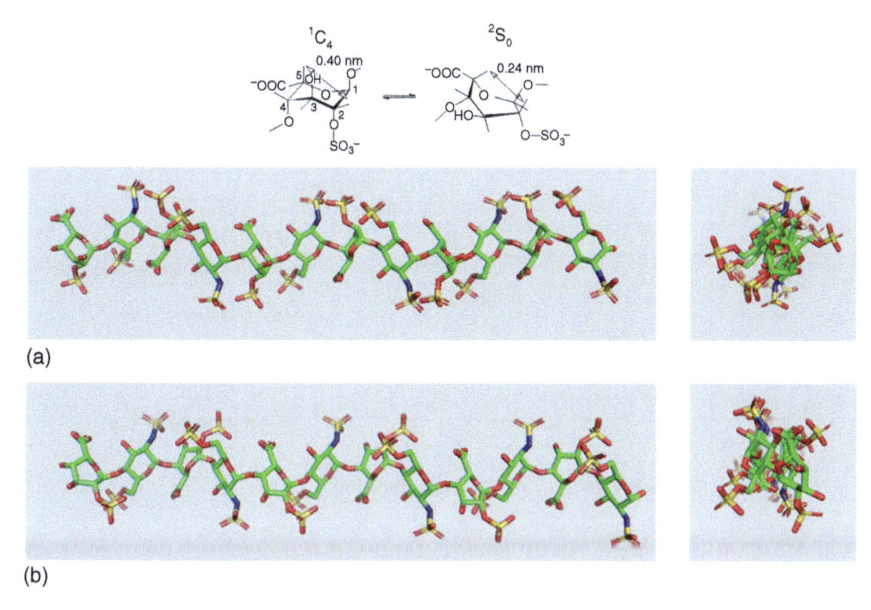

Figure 11.2 Structural differences associated with different conformations of iduronate residues. Structure of heparin sequences of the NS region along and across the chain for the [1]C_4 (a) and the [2]S_0 conformation (b). Source: Adapted from Mulloy et al. [27].

$^2S_0$ conformation when bound to AT, independent of its sulfation, or the structure of the neighboring units [28–30]. Some of the challenges of studying conformational aspects of this class of molecules and the application of density functional theory as a solution have been highlighted in a recent study of heparin tetrasaccharides [31].

11.4 Mechanism of Interaction with Coagulation Factors

11.4.1 Antithrombin-Mediated Activity

Heparin exerts its antithrombotic/anticoagulant action in several ways, the most important of which is through the enhancement of the protein cofactor AT, a plasma serine protease inhibitor. Heparin catalyzes the inhibitory activity of AT principally against two key coagulation enzymes, fXa and thrombin, by irreversibly blocking their procoagulant activity through a proposed combined template-based/allosteric mechanism. Despite heparin acting at many levels of the coagulation cascade by accelerating the inhibition of several factors, such as VIIa, IXa, XIa, XIIa, and kallikrein by AT, all of these activities combined are far less significant than the inhibition of fXa and thrombin [32, 33]. (Figure 11.3).

Heparin promotes the interaction between AT and thrombin or fXa by binding with the ATBPS AGA*IA and inducing in the protein a conformational change, which is mainly driven by specific electrostatic interactions between sulfate groups

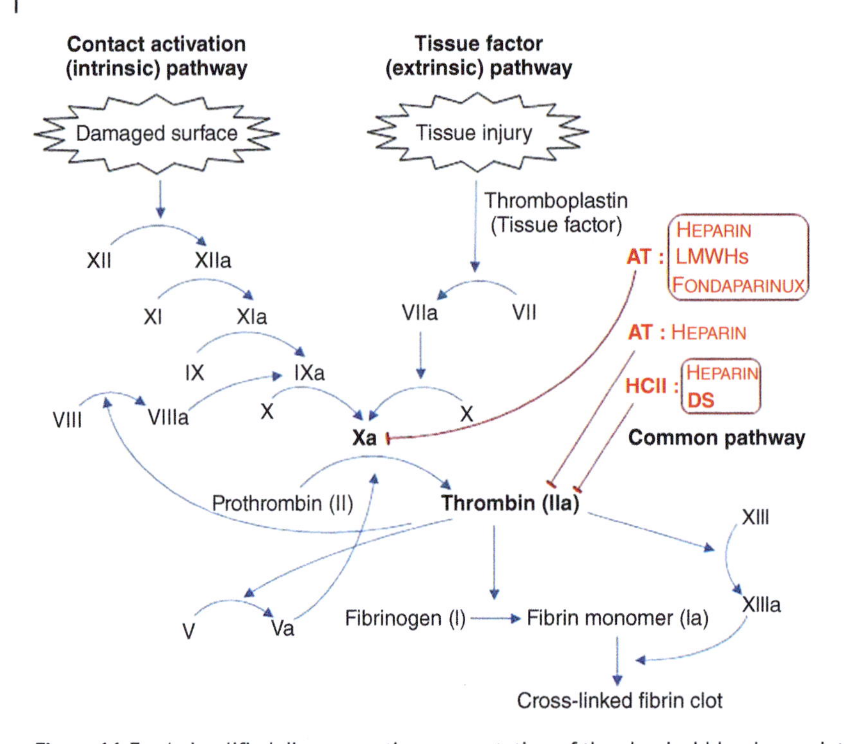

Figure 11.3 A simplified diagrammatic representation of the classical blood coagulation cascade, with the conventional division into three pathways; the intrinsic, extrinsic, and common pathways. The intrinsic pathway is activated by internal damage to endothelial cells of the vessel walls, whereas the extrinsic pathway can be activated in several ways, including tissue damage outside of the blood vessel, hypoxia, sepsis, malignancy, and inflammation. The two pathways converge in a final common pathway that ultimately converts fibrinogen into a fibrin clot. The function of blood coagulation is to maintain hemostasis, an intricate process, which is achieved by a series of clotting factors (indicated by Roman numerals) that are transformed in sequence from an inactive to an activated form (a), each one catalyzing the following reaction in a cascade. Most of these factors are serine proteases (except for glycoproteins fV, fVIII, and for fXIII, a transglutaminase), which act by hydrolyzing downstream proteins. Blue arrows represent the conversion of pro-enzymes into their active forms; red lines indicate the inhibition of coagulation factors fXa and fIIa, by AT and HCII, accelerated by complexation with heparin or structurally related molecules. The AT:heparin complex also inhibits factors fVIIa, fIXa, fXIa, and fXIIa.

and positively charged amino acids. The molecular basis of the high affinity AT: pentasaccharide interaction has been studied extensively [34–36]. Interestingly, a two-step mechanism was proposed, consisting of an activating step induced by AGA* trisaccharide followed by a stabilizing step promoted by IA disaccharide, which should be facilitated by the conformational flexibility of iduronate residues [37]. Subsequently, Lima et al. [38] suggested that the 3-*O*-sulfate group in the central glucosamine could play a stabilizing role of AT rather than an activating one. More recently, a three-state model for heparin allosteric activation of AT was proposed,which explains both the need for a core conformational change and the role of the peculiar reactive-center-loop hinge insertion [39]. Moreover, structural variants of the ATBPS have been also isolated from porcine and bovine heparins, which appear to contain N-sulfated instead of N-acetylated glucosamine at the non-reducing end and/or a 6-O-desulfated central 3-O-sulfated glucosamine [29, 40].

Extensive details of the mechanism of thrombin and fXa interactions with the AT-heparin complex have been reported [41, 42]. Importantly, there are substantial differences between the way thrombin and fXa interact with heparin-activated AT. Whereas fXa establishes a direct protein–protein interaction, which is favored by, but not strictly dependent on, direct binding with heparin [36], thrombin must bind to the same heparin chain that is bound to AT. This additional interaction requires that the heparin chain is sufficiently long to reach both proteins (Figure 11.4). In particular, a chain extension of at least 13 monosaccharide units toward the nonreducing end of the ATBPS is required. Such an octa-decasaccharide minimum motif for thrombin inhibition through AT called the C-domain [44], has been isolated [45] and has a molecular weight of 5400 Da.

For fXa inactivation, the dependence on the molecular weight of heparin is much less stringent. Interaction with both heparin and AT, while occurring, is not an essential requirement for fXa inhibition. Accordingly, heparin chains below 5400 Da containing the ATBPS, can still inhibit fXa [46]. This finding constitutes the biochemical rationale for the development of low molecular-weight heparins. For UFH the ratio between anti-Xa and anti-IIa activity is 1 but, for low molecular-weight heparins, is higher [47]. Importantly, by inactivating thrombin, heparin also

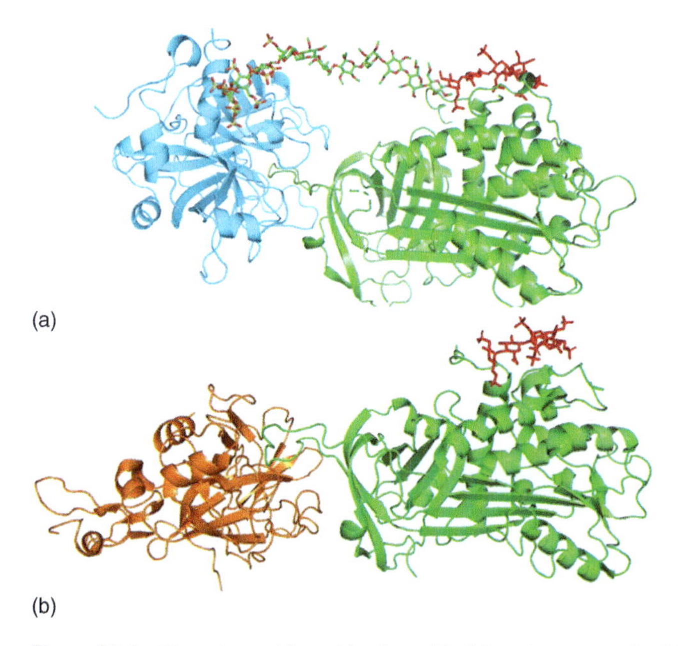

Figure 11.4 Heparin-antithrombin-thrombin (a) and pentasaccharide-antithrombin-FXa (b) ternary complexes. Antithrombin, thrombin, and fXa are shown in green, cyan, and orange, respectively. The pentasaccharide sequence is colored in red. Source: Adapted from Johnson and coworkers [36, 43].

inhibits fibrin formation and additionally prevents thrombin-induced activation of factor V and factor VIII [48, 49]. The anticoagulant activity of heparin is heterogenous because only approximately one-third of UFH chains contain the ATBPS and hence are capable of binding with high affinity to AT (high-affinity heparin). Heparin chains lacking this specific sequence (termed low-affinity heparin) have minimal anticoagulant activity at therapeutic concentrations. They can activate AT *in vitro*, but only at very much higher concentrations [50].

11.4.2 Heparin Cofactor II Mediated Activity

The anticoagulant activity of heparin is also mediated by a second plasma cofactor, HCII, another glycoprotein belonging to the serine protease inhibitor family, with structural similarities to AT. HCII selectively inhibits thrombin by forming a stoichiometric 1 : 1 covalent bimolecular complex, which is favored by the presence of heparin [51, 52], however, the potentiation of thrombin inhibition by HCII is an order of magnitude smaller than the action of the heparin-AT complex on thrombin [53]. Moreover, HCII binds heparin in a nonspecific manner, the apparent binding affinity being affected by ionic strength and chain length. The minimum heparin chain length interacting with the HCII binding site comprises 13 monosaccharide units. Interestingly, HCII and AT have the same affinity for low-affinity heparin (Kd$\sim$25 μM) [54]. HCII can also be activated by other GAGs and, in the vascular system, it appears to have specificity for DS, indicating that it may represent a potential alternative to heparin for the prevention and treatment of thrombosis [55] (see Section 11.6.1).

11.4.3 Additional Factors

As mentioned previously, heparin antithrombotic activity is not limited to its anticoagulant effect. It is also capable of enhancing the electronegative potential and hence the antithrombogenic property of vessel walls [56]. Heparin can also act by enhancing the activity of other two serin protease inhibitors, such as protein C inhibitor (PCI) and tissue factor pathway inhibitor (TFPI), the former promoting thrombin generation, the latter inhibiting factors Xa and VIIa. Both PCI and TFPI together participate in the balancing mechanism between procoagulant and anticoagulant forces that maintain hemostasis [42].

11.4.4 Adverse Effects of Heparin

The limitations of heparin use are caused by AT-independent mechanisms attributable to its charge-dependent binding properties with numerous proteins, including plasma proteins released from platelets and endothelial cells [57, 58]. The most important side effect of heparin, as with many other antithrombotic agents, is bleeding [59]. Owing to the relatively short plasma half-life of heparin, this effect is relatively transient and can be reversed by protamine, a highly basic protein composed predominantly of arginine residues. Protamine and heparin polyanionic chains interact in a straightforward manner, forming stable complexes, which are totally devoid of anticoagulant activity and are removed rapidly from the circulation.

11.4.4.1 Heparin-Induced Thrombocytopenia

A further well-described and potentially life-threatening side effect is the immune-mediated heparin-induced thrombocytopenia (HIT), which occurs typically after 10–14 days of heparin therapy in about 5% of treated patients [60]. HIT is a prothrombotic disorder caused by the formation of antigenic complexes of heparin and platelet factor 4 (PF4), a positively charged protein released by platelets following their activation. It is usually associated with a fall in platelet count of more than 50% and with the onset of serious vascular endothelial injury, such as thrombi formation and disseminated intravascular coagulation [61]. Thus, HIT appears paradoxical because, despite thrombocytopenia inducing an anticoagulant effect, the major clinical consequence is a worsening of the venous or arterial thrombotic risk.

11.4.4.2 Osteoporosis

The other common serious side effect of long-term heparin use is osteoporosis, with an incidence of 2.2–5%. Despite the mechanism by which heparin affects bone metabolism is unclear, it has been hypothesized that osteoporosis may be caused by the interaction of heparin with osteoblasts, resulting in the release of factors that activate osteoclasts [62]. All of these effects are dependent upon heparin chain length and are significantly reduced by the administration of LMWH.

11.5 Low Molecular Weight Heparins

First introduced into clinical practice some 30 years ago, LMWHs have gradually replaced heparin for the prophylaxis and treatment of DVT. They constitute a class of depolymerized heparin derivatives endowed with distinct biochemical and pharmacological profiles, which are determined by their composition [63, 64]. Developed with the intention of minimizing the adverse side effects of heparin, LMWHs should in principle differ from their parent heparins only in terms of their chain length, which is approximately one-third of that of the original UFH. They differ not only in molecular weight but also in monosaccharide composition and disaccharide sequences, depending on the method used for their manufacture. Different chemical, enzymatic, and physical strategies have been employed to prepare LMWHs, each one exhibiting its own cleavage selectivity of the heparin chains, and each one introducing distinct chemical groups, thereby providing process-related fingerprints. An overview of the methods for preparing LMWHs has been reported [65].

The four most used LMWHs in US and Europe are enoxaparin, dalteparin, tinzaparin, and nadroparin each of which is characterized by unique structural modifications at the reducing and/or nonreducing ends of the cleaved heparin chains [66, 67]. Enoxaparin derives from chemical β-eliminative cleavage by alkaline treatment of heparin benzyl esters that introduces Δ4,5 unsaturation into uronic acid residues at the nonreducing end. An additional characteristic is the presence of 1,6-anhydro amino sugars at the reducing end [68]. Tinzaparin, produced by enzymatic β-eliminative cleavage of UFH by heparinase-I, is also characterized by the presence of an unsaturated uronate residue at the nonreducing chain end. Dalteparin and nadroparin are both obtained by deaminative cleavage with nitrous acid followed by reduction, which results in the formation of an anhydromannitol ring at the reducing end. They differ in the ionic form, dalteparin being a sodium salt, such as enoxaparin and tinzaparin, nadroparin a calcium salt. All of these new, non-native, residues increase the structural complexity of LMWHs compared to UFH. Moreover, owing to the domain structure, the fragmentation of heparin chains based on the distinct preferential cleavage points of the various methods generates oligosaccharides, which differ significantly in their disaccharide sequences. Furthermore, the important ATPBS, when preserved by the depolymerization method, can be located differently within the oligosaccharide chain, with consequently different binding abilities toward AT. Both the reducing and nonreducing extensions of the AGA*IA sequence, together with possible structural modifications inside the pentasaccharide itself, influence the AT-binding properties of heparin oligosaccharides in a manner dependent on the structures of those additional residues [14].

Despite the undeniable importance of all these structural alterations, the most relevant structural differences among LMWHs, directly affecting their pharmacological activity are, nevertheless, still significantly dependent on the average molecular weight and oligosaccharide distribution [69, 70]. *In vitro* functional assays assessing plasma protein binding and AT-mediated antiprotease activity identify wide variations among the commercially available LMWHs [71] (Table 11.1). Most of the anticoagulant and pharmacokinetic differences between LMWHs and UFH, such as reduced bleeding and lowered risk of thrombocytopenia, sustained antithrombotic activity, better bioavailability, and longer half-life, can be explained

Table 11.1 Average molecular weight (M_w) and polydispersity degree (D), anti-Xa, anti-Xa/anti-IIa ratio (Xa/IIa) activities, elimination half-life ($t½$), and percentage content of GA* disaccharide (the structural marker of ATBPS determined by NMR) of LMWHs in comparison with fondaparinux and a typical UFH.

LMWH	M_w (Da)	D	Anti-Xa ($IU\,mg^{-1}$)	Anti-IIa ($IU\,mg^{-1}$)	Xa/IIa	$t½$ (h)	GA*
UFH	17 500 [25]	1.14 [25]	193 [72]	193 [72]	1 [72]	0.5–2.5 [73] dose-dependent	2.6 [74]
Enoxaparin	5400 [70]	1.34 [70]	104 [71]	32 [71]	3.3 [71]	4.5 [73]	3.2 [69]
Dalteparin	6900 [70]	1.22 [70]	122 [71]	60 [71]	2.0 [71]	2.0–2.3 [73]	4.1 [71]
Tinzaparin	8300 [70]	1.40 [70]	90 [71]	50 [71]	1.8 [71]	3.4 [73]	2.1 [71]
Nadroparin	4500 [71]	na	94 [71]	31 [71]	3.0 [71]	3.5 [73]	nd
Bemiparin	3600 [75]	na	80–120 [75]	5–20 [75]	8.0 [75]	5.2–5.4 [75]	nd
Fondaparinux	1728°	1.0	930 [75]	0	—	17 [73]	—

na: not available; nd: not determined; °molecular mass of sodium salt.

by their lower affinity for several plasma and tissue proteins resulting from their reduced chain length [76–78]. Importantly, owing to their lower Mw, they have reduced ability to inhibit thrombin compared to UFH. The specific activity of LMWHs in anticoagulant assays ranges from 30 to 60 U mg^{-1} for anti-IIa, and 80 to 130 U mg^{-1} for anti-Xa [71]. The numerous advantages of LMWHs over UFH have limited the current use of the latter mainly to the prevention of clotting during procedures involving the extracorporeal circulation of blood (e.g. cardiopulmonary bypass and dialysis), owing to its ability to be neutralized by protamine [79].

11.5.1 Ultralow Molecular Weight Heparins

Ultralow molecular weight heparins (ULMWHs) represent the recent evolution of depolymerized heparin derivatives, a new "second generation" LMWH, designed with the aim of minimizing the risk of bleeding and HIT [73]. With an average molecular weight of 36 kDa, an anti-fXa/anti-fIIa activity ratio of 8 : 1 and a half-life of about five hours (Table 11.1), bemiparin was the first ULMWH that was approved in Spain in 1998, obtained by chemical β-elimination and fractionation of porcine mucosal UFH [75]. Currently marketed in 58 countries except for the USA, it is approved for once-daily subcutaneous use in the treatment and prophylaxis of VTE, and for the prevention of clotting in the extracorporeal circuit during hemodialysis.

The development of two other ULMWH compounds, semuloparin and RO-14, the latter being a bemiparin derivative [73, 80, 81], were suspended by their respective manufacturers, despite promising initial clinical results, because they were considered to be commercially unattractive following the introduction of fondaparinux into clinical use (see paragraph 8).

11.6 Drugs Based on Natural GAG Mixtures

Sulodexide, danaparoid, and mesoglycan are the active components of the most clinically studied drugs based on natural GAG mixtures. These are obtained from porcine intestinal mucosa following heparin extraction, comprising heparin/HS and DS in different proportions and, in the case of danaparoid and mesoglycan, also by a minor amount of chondroitin sulfate (CS). Owing to their overall lower sulfation degree with respect to heparin, they have higher bioavailability, a longer half-life and a reduced effect on systemic clotting and bleeding. Moreover, these heparinoids are often used as alternatives to heparin therapy in patients who have developed HIT [82–86]. Their efficacy derives from the dual action of catalysing thrombin inhibition through the simultaneous activation of AT and HCII, exerted by both heparin/HS and DS components [87, 88].

The role of DS in these compounds deserves particular attention. Despite it not being the exclusive constituent of any drug currently available, the contribution of DS to the regulation of the coagulation cascade has been extensively described, together with its potential therapeutic role [89–91].

11.6.1 The Role of Dermatan Sulfate

Dermatan sulfate, a constituent of the extracellular matrix of several tissues also present on the cell surface of vasculature endothelium, is an IdoA containing GAG, like heparin. It differs significantly, however, from heparin in its lower sulfation degree, as the IdoA residues are mostly unsulfated, as well as in the position and configuration of glycosidic bonds. DS is largely composed of monosulfated disaccharide repeating unit [-4)-β-L-IdoA-1→3-α-D-GalNAc,4S-(1-], where *α*-D-GalNAc,4S is an N-acetylated galactosamine residue, O-sulfated at position 4. Dermatan sulfate is also intrinsically heterogenous; some of its uronic acids being β-D-GlcA, and some of its disaccharide units bearing additional sulfation, either at position 2 of the IdoA residue or, at position 6 of the GalNAc4S residue (Figure 11.1). These over-sulfated sequences are deemed responsible for the antithrombotic properties of this GAG [92]. Commercial DS preparations are polydisperse mixtures of molecules with M_W values up to 60 kDa. On average, the M_W of purified DS extracts varies from approximately 24–47 kDa [25].

Actually, despite HCII not requiring specific structural sequences in heparin, it does need specific IdoA2SO_3-GalNAc, 4SO_3 disulfated sequences in DS [93]. The minimum structural motif for binding HCII should be composed of at least three disulfated disaccharides, but the enhancement of HCII-mediated thrombin inhibition requires at least three to four additional disaccharide units [94]. More precisely, the longer the disulfated disaccharide sequence, the higher the HCII-mediated inhibitory activity [95]. Dermatan sulfate polymers from marine invertebrate, mainly composed of IdoA2SO_3-GalNAc4SO_3 activate HCII at low concentrations, whereas polymers mainly composed of IdoA2SO_3-GalNAc6SO_3 exhibit 1000-times lower activity [96, 97].

Dermatan sulfate has been employed for prophylaxis of VTE in patients undergoing surgery for cancer, despite its reduced antithrombotic activity compared to heparin, DS is a safer drug for intramuscular administration, because of reduced bleeding complications [98]. Dermatan sulfate has also been employed successfully as anticoagulant therapy in patients who developed renal failure following major cardiovascular surgery [99]. More recently it has been proposed as an anticoagulant in regular dialysis treatment instead of the standard use of heparin, providing an alternative approach in case of thrombocytopenia or other adverse effects of UFH [100]. As the high molecular weight of DS chains prevents its absorption following subcutaneous administration, low molecular weight DS derivatives, such as Desmin 370, have also been proposed [101]. Despite a reduced *in vitro* potency, they exhibit better pharmacokinetic properties, including improved bioavailability and longer duration of action [102, 103].

11.6.2 Sulodexide

Sulodexide, present on the European and US markets mainly with the trade names of Vessel and Sulonex, respectively, is a highly purified GAG composed by 80% of fast-moving heparin (Fm-Hep) fraction, defined on the basis of its electrophoretic mobility under specific conditions, and 20% DS [104, 105]. According to a recent detailed structural characterization of sulodexide, the Fm-Hep component has a molecular weight of about 10–11 kDa and an overall sulfation degree (number of sulfate groups per disaccharide) of 2.13, both lower than typical current UFH, 15–20 kDa, and 2.45, respectively. The DS component also has a significantly lower molecular weight than typical DS samples, about 20–21 kDa vs. 30–50 kDa [25, 105]. Sulodexide pharmacological properties, being mediated by both AT and HCII, are comparable to UFH, as is the release of TFPI, which further contributes to its antithrombotic effect. Owing to the structural features of its GAG components, however, it differs from heparin in its extensive absorption by the vascular endothelium both following parenteral and oral routes and in its longer half-life [106, 107]. Several clinical studies demonstrated that prolonged sulodexide administration decreases the incidence of recurrences of thromboembolic events without detectable risk of bleeding [108]. Currently, sulodexide is widely accepted in many countries as an effective and safe endothelial-protecting agent [106].

11.6.3 Danaparoid

Danaparoid is a low molecular weight heparinoid composed of a mixture of HS, DS, and CS, and constitutes the active principle of Orgaran. From a recent in-depth structural study of danaparoid, the quantitative relationship among the three components determined by 2D-NMR is in the following ranges: 77.6–88.4% for HS, 9.0–16.8% for DS, and 2.3–8.0% for CS [109]. Regarding the molecular weight evaluation, the weight average (M_W) of the whole mixture ranges between 4.2 and 4.6 kDa, whereas the two fractions comprising DS/CS and HS vary in the range 6.2–7.0 kDa and 3.3–3.5 kDa, respectively. The prevalence of HS (77.6–88.4%) compared to relatively low levels of DS in the sample explains the relatively low average molecular weight of the sample. Together with an overall sulfation degree of 1.19–1.32, the same study describes the presence of oxidized gluco/galactosamine at the reducing end as process signatures [109]. Owing to its peculiar structural features, danaparoid has high bioavailability and inactivates fXa more than thrombin by a ratio greater than 20 : 1 [110]. Moreover, the elimination half-life is in the range 19.2–24.5 hours during anti-Xa activity and in the range 1.8–4.3 hours during anti-IIa activity [111]. Renal excretion is the main route of elimination, accounting for approximately 40–50% of the total clearance of anti-fXa activity following intravenous administration [111]. Danaparoid does not differ from UFH in terms of efficacy in the treatment of existing DVT and it has been proven effective in the prophylaxis of DVT following ischaemic stroke, in the treatment of patients with renal failure or dependent on renal replacement therapy, and in patients who developed HIT [83, 112].

11.6.4 Mesoglycan

Mesoglycan is marketed in many countries with the trade name Prisma. Despite its structural characterization never having been reported, mesoglycan is commonly described as being composed by 47.5–52% of HS, 8–8.5% of slow-moving heparin

(Sm-hep), 35–36% of DS, and 5–8.5% of CS [113], where Sm-hep is a heparin fraction endowed with higher sulfation degree than the complementary Fm-hep, based on electrophoretic mobility [104]. More recently, a broader composition was reported; 25–60% of DS, 3–15% of CS, and HS as complement to 100% [84]. Mesoglycan is approved both for oral and intramuscular administration routes. An extensive description of the positive results of a treatment with mesoglycan in the management of vascular diseases was recently reported, ranging from profibrinolytic activity, to decreasing neurologic deficits, such as efficacy in preventing thrombotic recurrence in patients affected by DVT, to name but a few [114]. More recently, it has also been reported to be beneficial in inner ear disease of a vascular origin [115].

11.7 Defibrotide

Isolated for the first time in 1968 as a phosphorus containing fraction derived from bovine lung and later recognized as a fragment of DNA [116], DF differs from the other antithrombotic heparinoids in its non-GAG structure. It is currently derived from the controlled depolymerization of porcine intestinal mucosal DNA, consisting in a polydisperse mixture of single-stranded and double-stranded phosphodiester oligonucleotides, approximately in a 9 : 1 ratio, with oligonucleotide length of 9–80 mer (average 50 mer) and a mean molecular weight of 15–30 kDa [117, 118]. Like heparin and natural heparin-related GAGs, DF is a multitarget drug with considerable structural complexity and a controversial and, as yet, not well-understood mechanism of action [119]. Initially considered primarily an antithrombotic and profibrinolytic agent, due to the ability of some aptamers to inhibit thrombin [120] and to the dual ability to increase the activity of plasma tissue plasminogen activator (tPA, a serine protease involved in the breakdown of blood clots) and to decrease the activity of its inhibitor (PAI-1) [119], it has also been shown to inhibit platelet activation and aggregation, with no significant systemic anticoagulant effect in pharmacological studies [121]. As a polyanion, it interacts with several heparin-binding proteins, predominantly through charge–charge interactions, including vascular endothelial growth factor (VEGF) and basic fibroblast growth factor (FGF2), thereby promoting endothelial cell proliferation and microvessel formation [119]. These activities together with other multiple pharmacological properties, including anti-inflammatory, anti-atherosclerotic, and angio-protective effects, are reported to promote an anticoagulant phenotype specific to the endothelium [121].

Marketed in Europe and US under the trade name of Defitelio, DF is currently approved specifically for the treatment of hepatic veno-occlusive disease, also known as sinusoidal obstruction syndrome. It is the only successful therapy available for this potentially life-threatening complication, the most severe forms resulting in multiorgan dysfunction, that can occur both in adult and pediatric patients, particularly after bone marrow transplantation [121]. Recently, DF has been suggested as a prophylactic as well as a preventive treatment in patients at high risk of developing hepatic syndromes [122].

11.8 Pentosan Polysulfate

PPS is a heparinoid of plant origin, being manufactured from beechwood hemicellulose, through a semisynthetic process of sulfate esterification, which confers upon the molecule a highly polyanionic property [123] resulting in a complex mixture of heterogneous polysaccharides that structurally resemble GAGs. The backbone is made up of repeating β-D-xylopyranose (Xyl) units linked 1–4 to each other with some interspersed 4-*O*-methylglucuronic acid units that are α(1–2) linked to the D-xylose sugar. The 2-*O* and 3-*O* hydroxyl positions of the Xyl residues can also be acetylated [124]. Additional structural peculiarities have been reported, including the average M_W: a value of about 6.6 kDa, with 50% of chains in the range 3–8 kDa makes PPS similar to LMWH [125]. Despite it displaying a broad spectrum of pharmacological activities, from anticoagulant and fibrinolytic to anti-inflammatory and antiallergic, the mechanism of action is not completely understood [126–129]. It has been reported to be a good activator of HCII and an AT-independent inhibitor of fXa and fIIa, despite the degree of inhibition remaining a controversial matter [130–132]. Initially intended in the 1950s as an oral anticoagulant, a potential substitute of parenteral UFH, PPS was found to be too weak to be clinically useful [133]. By virtue of its profibrinolytic and moderate anticoagulant properties, however, it is currently marketed under the trade name Fibrase for the prophylaxis and therapy of vascular pathology with thrombotic risk. As the active ingredient of Elmiron, since 1996, it is the only US FDA-approved oral drug for the effective treatment of bladder pain and discomfort in adults with interstitial cystitis [134].

11.9 Fondaparinux and Related Synthetic Oligosaccharides

Fondaparinux is a synthetic pentasaccharide, the first of a new class of antithrombotic agents, which inhibits thrombin generation through the selective AT-dependent inactivation of factor Xa. The pioneering idea of an *ab initio* synthetic project emerged in the early 1980s – in parallel with the development of LMWHs – following the structural elucidation of the unique pentasaccharide domain responsible for heparin anticoagulant activity, the culmination of efforts by many research groups [135]. A few years later, two independent research groups each synthesized the first pentasaccharide analog of the ATPBS, containing an *N*-sulfate group instead of the *N*-acetyl group in the nonreducing end glucosamine [136, 137] and also one stabilized by the introduction of an *O*-methyl group at the reducing-end anomeric position [138]. The resulting antithrombotic drug, marketed as Arixtra, entered clinical use both in the USA and Europe in 2002. Unlike heparin and LMWHs, which interact with several factors, fondaparinux targets one site only. It is reported not to interact with plasma proteins other than AT and catalyzes the inactivation of a single coagulation factor fXa [139]. Accordingly, it exhibits nearly complete bioavailability by the subcutaneous route, with a rapid onset of action, and a prolonged half-life of about 17 hours compared to 1 hour for heparin and 4 hours for LMWHs, which enables single daily administration [140, 141] (Table 11.1). Moreover, several clinical trials proved that the administration of 2.5 mg day^{-1} of fondaparinux in major orthopedic surgery led to a decrease of thrombotic risk of 50% compared to LMWH, administered 40 mg once daily or 30 mg twice daily [142]. Importantly, this higher efficacy is achievable without increasing the risk of clinically relevant bleeding [143]. Furthermore, with fondaparinux, there is no release of TFPI and no risk of HIT [144, 145]. Very recently, a potentially scalable, simplified synthesis of fondaparinux has been reported, based on the use of thioglycoside building blocks with well-defined reactivity [146]. By reducing the number of synthetic steps and eliminating multiple purification steps, it is proposed as a simplified and cost-effective procedure with improved synthetic efficiency. The chemical method originally developed for the synthesis of fondaparinux paved the way for heparin-related oligosaccharides generally. Among many, three oligosaccharides in particular, with varied, but promising pharmacological profiles, have been considered for clinical investigation. These are idraparinux and the biotynilated derivatives idrabiotaparinux and EP217609.

Idraparinux, a poly-O-methylated O-sulfated analogue of fondaparinux, is much easier to synthesize, has higher anti-Xa activity and possesses an elimination half-life about five to six times longer and even more for prolonged prophylaxis, with consequent severe bleeding complications [142, 147, 148].

Idrabiotaparinux, a derivative of idraparinux with a biotin molecule linked to the nonreducing hexosamime, was synthesized with the aim of obtaining a neutralizable antithrombitic. While maintaining the same potency and long-lasting anti-Xa effect as the parent compound, permitting single weekly administration, idrabiotaparinux offers the advantage of being neutralized efficiently by avidin, an egg-derived protein with low antigenicity [149].

EP217609 is a more complex antithrombotic compound, combining in one molecule an indirect anti-fXa inhibitor, a direct anti-fIIa inhibitor, separated by an inert spacer, and a biotin molecule that can be neutralized quickly by an injection of avidin [150]. In addition to possessing high AT-mediated anti-Xa and direct anti-IIa potencies that do not dissociate *in vivo*, in a first clinical study, it was well-tolerated and exhibited a half-life of about 20 hours [151]. From the above, it is clear that a drawback in the use of fondaparinux is that no effective antidote is available to neutralize its effect. Currently, complete reversal is only possible for UFH, through use of protamine sulfate, capable of forming a stable complex with the negatively charged heparin chains. Protamine only partially neutralizes the anticoagulant effect of LMWHs, however, due to their reduced binding capacity [152, 153]. Two antidotes have been proposed, which efficiently reverse the *in vitro* and *in vivo* anticoagulant activity of both fondaparinux and LMWHs, an AT variant (AT-N135Q-Pro394) and a recombinant form of fXa (PRT064445) [154, 155]. The former has significantly increased affinity for both drugs, and no anticoagulant activity, while the latter is catalytically inactive but, retains the ability to bind activated AT. Both compounds have the potential to be used as universal reversal agents for a broad range of direct or indirect fXa inhibitors.

11.10 Chemoenzymatic Synthesis of Oligosaccharides

The chemoenzymatic synthesis of heparin oligosaccharides deserves to be mentioned for its promising potential to produce safe and effective antithrombotic therapeutic agents. It is proposed as an alternative to a synthetic approach, since it generates authentic heparin molecules, instead of mimetics, involving a much shorter synthetic route [156]. The synthesis of a series of size-defined, predominantly N-sulfated oligosaccharides with up to 21 saccharide residues, capable of anti-Xa activity with or without anti-IIa activity depending on the length, has been accomplished using a set of bacterial biosynthetic enzymes, such as glycosyltransferases, sulfotransferases and C_5-epimerase, and various defined precursors [157]. Moreover, in a library of 3-O-sulfated oligosaccharides recently produced, the octasaccharide [GlcNS6S-GlcA-GlcNS3S6S-IdoA2S-GlcNS6S-IdoA2S-GlcNS6S-GlcA-pNP] stood out for its high anti-Xa activity and faster clearance than fondaparinux in a rat model, making this compound a potential short-acting anticoagulant drug candidate [158]. The chemoenzymatic approach provides a general tool for preparing tailor-made oligosaccharides with different therapeutic targets. The limit seems to be represented only by our detailed knowledge of the precise heparin/HS saccharide sequences involved in critical protein interactions.

11.11 Conclusions and Perspectives

Despite heparin owing its fame and success as a drug to the chance discovery of one of its many properties, which may not even be its principal role *in vivo*, during its century of life, heparin has paved the way for several generations of derivatives, including LMWHs and, more recently, the synthetic drug fondaparinux, and these have replaced it almost completely on account of their improved safety and ease of use. While many of the most recent studies of heparin are dedicated to investigating its non-anticoagulant properties in-depth, the search for new alternative heparinoids, is under continuous development, in the quest for increasingly safe, efficacious, and cost-effective weapons against thrombosis. There is every chance that the number and breadth of antithrombotic applications of these versatile materials will continue to grow in the future. The shortage of the raw materials of animal origin, which is presently a threat for the market of heparin derivatives, makes synthetic strategies the most promising path for the development of new antithrombotic drugs.

Acknowledgment

The authors thank prof. Edwin A. Yates (University of Liverpool) for fruitful discussion.

References

1 McLean, J. (1916). The thromboplastic action of cephalin. *The American Journal of Physiology* 41: 250–257.

2 Howell, W.H. and Holt, E. (1918). Two new factors in blood coagulation – heparin and pro-antithrombin. *The American Journal of Physiology* 47: 328–341.

3 Barritt, D. and Jordan, S. (1960). Anticoagulant drugs in the treatment of pulmonary embolism: a controlled trial. *Lancet* 1: 1309–1312.

4 Kakkar, V.V., Corrigan, T., Spindler, J. et al. (1972). Efficacy of low doses of heparin in prevention of deep-vein thrombosis after major surgery. A double-blind, randomised trial. *Lancet* 2 (7768): 101–106.

5 Roth, G.A., Forouzanfar, M.H., Moran, A.E. et al. (2015). Demographic and epidemiologic drivers of global cardiovascular mortality. *The New England Journal of Medicine* 372: 1333–1341.

6 Kolset, S.O. and Tveit, H. (2008). Serglycin – structure and biology. *Cellular and Molecular Life Sciences* 65: 1073–1085.

7 Carlsson, P. and Kjellen, L. (2012). Heparin biosynthesis. In: *Heparin – A Century of Progress* (ed. R. Lever, B. Mulloy, and C.P. Page), 23–41. Berlin/Heidelberg, Germany: Springer Verlag.

8 Esko, J.D. and Selleck, S.B. (2002). Order out of chaos: assembly of ligand binding site in heparan sulfate. *Annual Review of Biochemistry* 71: 435–471.

9 Rudd, T.R. and Yates, E.A. (2012). A highly efficient tree structure for the biosynthesis of heparan sulfate accounts for the commonly observed disaccharides and suggests a mechanism for domain synthesis. *Molecular BioSystems* 8: 1499–1506.

10 Wang, B., Jia, J., Zhang, X. et al. (2011). Heparanase affects secretory granules homeostasis of murine mast cells through degrading heparin. *The Journal of Allergy and Clinical Immunology* 128: 1310–1317.

11 Stringer, S., Kandola, B., Pye, D., and Gallagher, J. (2003). Heparin sequencing. *Glycobiology* 13: 97–107.

12 Mulloy, B. (2012). Structure and physicochemical characterization of heparin. In: *Heparin: A Century of Progress* (ed. R. Lever, B. Mullo, and C.P. Page), 77–98. Berlin/Heidelberg, Germany: Springer Verlag.

13 Lindahl, U., Bäckström, G., Thunberg, L., and Leder, I.G. (1980). Evidence for a 3-O-sulfated D-glucosamine residue in the antithrombin-binding sequence of heparin. *Proceedings of the National Academy of Sciences* 77: 6551–6555.

14 Guerrini, M., Guglieri, S., Casu, B. et al. (2008). Antithrombin-binding octasaccharides and role of extensions of the active pentasaccharide sequence in the specificity and strength of interaction. Evidence for very high affinity induced by an unusual glucuronic acid residue. *The Journal of Biological Chemistry* 283 (39): 26662–26675.

15 Sugahara, K., Tsuda, H., Yoshida, K. et al. (1995). Structure determination of the octa- and decasaccharide sequences isolated from the carbohydrate-protein linkage region of porcine intestinal heparin. *The Journal of Biological Chemistry* 270 (39): 22914–22923.

16 Rosenfeld, L. and Danishefsky, I. (1988). Location of specific oligosaccharides in heparin in terms of their distance from the protein linkage region in the native proteoglycan. *The Journal of Biological Chemistry* 263: 262–266.

17 Yamada, S., Yamane, Y., Tsuda, H. et al. (1998). A major common trisulfated hexasaccharide core sequence, hexuronic acid(2-sulfate)-glucosamine(*N*-sulfate)-iduronic acid-*N*-acetylglucosamine-glucuronic sequence, acid-glucosamine(*N*-sulfate), isolated from the low sulfated irregular region of porcine intestinal heparin. *The Journal of Biological Chemistry* 273: 1863–1871.

18 Radoff, S. and Danishefsky, I. (1984). Location on heparin of the oligosaccharide section essential for anticoagulant activity. *The Journal of Biological Chemistry* 259: 166–172.

19 Gong, F., Jemth, P., Escobar-Galvis, M.L. et al. (2003). Processing of macromolecular heparin by heparinase. *The Journal of Biological Chemistry* 278: 35152–35158.

20 Iacomini, M., Casu, B., Guerrini, M. et al. (1999). "Linkage region" sequences of heparins and heparan sulfates: detection and quantification by nuclear magnetic resonance spectroscopy. *Analytical Biochemistry* 274: 50–58.

21 Bisio, A., Guerrini, M., Yates, E.A. et al. (1997). NMR identification of structural environment for 2, 3 and 2, 3, 6 trisulfated glucosamine residues in heparin with high and no affinity for antithrombin. *Glycoconjugate Journal* 14: 89.

22 Oscarsson, L.-G., Pejler, G., and Lindahl, U. (1989). Location of the antithrombin-binding sequence in the heparin chain. *The Journal of Biological Chemistry* 264: 296–304.

23 Pejler, G., Danielsson, A., Björk, I. et al. (1987). Structure and antithrombin binding properties of heparin isolated from the clams Anomalocardia brasiliana and Tivela macrolides. *The Journal of Biological Chemistry* 262: 11413–11421.

24 Mulloy, B., Gray, E., and Barrowcliffe, T.W. (2000). Characterization of unfractionated heparin: comparison of materials from the last 50 years. *Thrombosis and Haemostasis* 84: 1052–1056.

25 Bertini, S., Bisio, A., Torri, G. et al. (2005). Molecular weight determination of heparin and dermatan sulfate by size exclusion chromatography with a triple detector array. *Biomacromolecules* 6: 168–173.

26 Ferro, D.R., Provasoli, A., Ragazzi, M. et al. (1986). Evidence for conformational equilibrium of the sulfated L-iduronate residue in heparin and in synthetic heparin mono- and oligo-saccharides: NMR and force-field studies. *Journal of the American Chemical Society* 108: 6773–6778.

27 Mulloy, B., Forster, M.J., Jones, C., and Davies, D.B. (1993). N.m.r. and molecular-modelling studies of the solution conformation of heparin. *The Biochemical Journal* 293: 849–858.

28 Das, S.K., Mallet, J.-M., Esnault, J. et al. (2001). Synthesis of conformationally locked carbohydrates: a skew-boat conformation of l-iduronic acid governs the antithrombotic activity of heparin. *Angewandte Chemie, International Edition* 40: 1670–1673.

29 Guerrini, M., Mourier, P.A.J., Torri, G., and Viskov, C. (2014). Antithrombin-binding oligosaccharides: structural diversities in a unique function? *Glycoconjugate Journal* 31 (6, 7): 409–416.

30 Stancanelli, E., Elli, S., Hsieh, P.-H. et al. (2018). Recognition and conformational properties of an alternative antithrombin binding sequence obtained by chemoenzymatic synthesis. *ChemBioChem* 19: 1178–1188.

31 Hricovini, M. and Hricovini, M. (2018). Solution conformation of heparin tetrasaccharide. DFT analysis of structure and spin–spin coupling constants. *Molecules* 23 (11): 3042–3054.

32 Rosenberg, R.D. and Bauer, K.A. (1994). The heparin-antithrombin system: a natural anticoagulant mechanism. In: *Hemostasis and Thrombosis: Basic Principles and Clinical Practice*, 3e (ed. R.W. Colman, J. Hirsh, V.J. Marder, and E.W. Salzman), 837–860. Philadelphia, PA: JB Lippincott Co.

33 Beurskens, D.M.H., Huckriede, J.P., Schrijver, R. et al. (2020). The anticoagulant and nonanticoagulant properties of heparin. *Thrombosis and Haemostasis* 120: 1371–1383.

34 van Boeckel, C.A.A. and Petitou, M. (1993). The unique antithrombin III binding domain of heparin: a lead to new synthetic antithrombotics. *Angewandte Chemie* 32: 1671–1690.

35 Hricovini, M., Guerrini, M., Bisio, A. et al. (2001). Conformation of heparin pentasaccharide bound to antithrombin III. *The Biochemical Journal* 359: 265–272.

36 Johnson, D.J.D., Li, W., Adams, T.E., and Huntington, J.A. (2006). Antithrombin–S195A factor Xa-heparin structure reveals the allosteric mechanism of antithrombin activation. *The EMBO Journal* 25: 2029–2037.

37 Desai, U.R., Petitou, M., Biörk, I., and Olson, S.T. (1988). Mechanism of heparin activation of antithrombin: evidence for an induced-fit model of allosteric activation involving two interaction subsites. *Biochemistry* 37: 13033–13041.

38 Lima, A.M., Hughes, A.J., Veraldi, N. et al. (2013). Antithrombin stabilisation by sulfated carbohydrates correlates with anticoagulant activity. *Medicinal Chemistry Communications* 4: 870–873.

39 Izaguirre, G., Swanson, R., Roth, R. et al. (2021). Paramount importance of core conformational changes for heparin allosteric activation of antithrombin. *Biochemistry* 60 (15): 1201–1213.

40 Loganathan, D.H., Wang, M., Mallis, L.M., and Linhardt, R.J. (1990). Structural variation in the antithrombin III binding site region and its occurrence in heparin from different sources. *Biochemistry* 29: 4362–4368.

41 Huntington, J.A. (2003). Mechanism of glycosaminoglycan activation of the serpins in hemostasis. *Journal of Thrombosis and Haemostasis* 1: 1535–1549.

42 Gray, E., Hogwood, J., and Mulloy, B. (2012). The anticoagulant and antithrombotic mechanism of heparin. In: *Heparin – A Century of Progress* (ed. R. Lever, B. Mulloy, and C.P. Page), 43–61. Berlin/Heidelberg, Germany: Springer Verlag.

43 Li, W., Johnson, D.J., Esmon, C.T., and Huntington, J.A. (2004). Structure of the antithrombin-thrombin-heparin ternary complex reveals the antithrombotic mechanism of heparin. *Nature Structural & Molecular Biology* 11: 857–862.

44 Al Dieri, R., Wagenvoord, R., van Dedem, G.W. et al. (2003). The inhibition of blood coagulation by heparins of different molecular weight is caused by a common functional motif – the C-domain. *Journal of Thrombosis and Haemostasis* 1: 907–914.

45 Mourier, P.A.J., Guichard, O.Y., Herman, F. et al. (2017). New insights in thrombin inhibition structure–activity relationships by characterization of octadecasaccharides from low molecular weight heparin. *Molecules* 22 (3): 428.

46 Wagenvoord, R., Al, D.R., Van, D.G. et al. (2008). Linear diffusion of thrombin and factor Xa along the heparin molecule explains the effects of extended heparin chain lengths. *Thrombosis Research* 122: 237–245.

47 Gray, E. (2012). Standardisation of low-molecular-weight heparin. In: *Heparin – A Century of Progress* (ed. R. Lever, B. Mulloy, and C.P. Page), 65–76. Berlin/Heidelberg, Germany: Springer Verlag.

48 Ofosu, F.A., Sie, P., Modi, G.J. et al. (1987). The inhibition of thrombin-dependent feedback reactions is critical to the expression of anticoagulant effects of heparin. *The Biochemical Journal* 243: 579–588.

49 Beguin, S., Lindhout, T., and Hemker, H.C. (1988). The mode of action of heparin in plasma. *Thrombosis and Haemostasis* 60: 457–462.

50 Streusand, V.J., Bjork, I., Gettins, P.G. et al. (1995). Mechanism of acceleration of antithrombin-proteinase reactions by low affinity heparin. Role of the antithrombin binding pentasaccharide in heparin rate enhancement. *The Journal of Biological Chemistry* 270: 9043–9051.

51 Parker, K.A. and Tollefsen, D.M. (1985). The protease specificity of heparin cofactor II. Inhibition of thrombin generated during coagulation. *The Journal of Biological Chemistry* 260: 3501–3505.

52 Griffith, M.J. (1983). Heparin-catalyzed inhibitor/protease reactions: kinetic evidence for a common mechanism of action of heparin. *Proceedings of the National Academy of Sciences United States of America* 80: 5460–5464.

53 Tollefsen, D.M. and Blank, M.K. (1981). Detection of a new heparin-dependent inhibitor of thrombin in human plasma. *The Journal of Clinical Investigation* 68: 589–596.

54 O'Keeffe, D., Olson, S.T., Gasiunas, N. et al. (2004). The heparin binding properties of heparin cofactor II suggest an antithrombin-like activation mechanism. *The Journal of Biological Chemistry* 279 (48): 50267–50273.

55 McGuire, E.A. and Tollefsen, D.M. (1987). Activation of heparin cofactor II by fibroblast and vascular smooth cells. *The Journal of Biological Chemistry* 262: 169–175.

56 Jaques, L.B. (1985). Ein neues Konzept seiner Natur und seiner Wirkung. Teil III. Die klinische Wirkung des Heparins. *Hämostaseologie* 5: 121–126.

57 Young, E., Wells, P., Holloway, S. et al. (1994). Ex-vivo and in-vitro evidence that low molecular weight heparins exhibit less binding to plasma proteins than unfractionated heparin. *Thrombosis and Haemostasis* 71 (03): 300–304.

58 Lane, D.A. (1989). Heparin binding and neutralizing protein. In: *Heparin: Chemical and Biological Properties. Clinical Applications* (ed. D.A. Lane and U. Lindahl), 363–374. London, UK: Edward Arnold.

59 Schulman, S., Beyth, R.J., Kearon, C., and Levine, M.N. (2008). Hemorrhagic complications of anticoagulant and thrombolytic treatment: American College of

Chest Physicians Evidence-Based Clinical Practice Guidelines (8th Edition). *Chest* 133: 257S–298S.

60 Ahmed, I., Majeed, A., and Powell, R. (2007). Heparin induced thrombocytopenia: diagnosis and management update. *Postgraduate Medical Journal* 83 (983): 575–582.

61 Alban, S., Krauel, K., and Greinacher, A. (2012). The role of sulfated polysaccharides in the pathogenesis of heparin-induced thrombocytopenia. In: *Heparin-Induced Thrombocytopenia, CRC Press*, 5e (ed. T.E. Warkentin and A. Greinacher), 181–208. Boca Raton, FL: Taylor and Francis Group.

62 Shaughnessy, S.G., Young, E., Deschamps, P., and Hirsh, J. (1995). The effects of low molecular weight and standard heparin on calcium loss from fetal rat calvaria. *Blood* 86: 1368–1373.

63 Merli, G., Vanscoy, G.J., Rihn, T.L. et al. (2001). Applying scientific criteria to therapeutic interchange: a balanced analysis of low-molecular-weight heparins. *Journal of Thrombosis* 11: 247–259.

64 Fareed, J., Fu, K., Yang, L.H., and Hoppensteadt, D.A. (1999). Pharmacokinetics of low molecular weight heparins in animal models. *Seminars in Thrombosis and Hemostasis* 25 (3): 51–55.

65 Guerrini, M. and Bisio, A. (2012). Low-molecular-weight heparins: differential characterization/physical characterization. In: *Heparin – A Century of Progress* (ed. R. Lever, B. Mulloy, and C.P. Page), 127–157. Berlin/Heidelberg, Germany: Springer Verlag.

66 Linhardt, R.J. and Gunay, N.S. (1999). Production and chemical processing of low molecular weight heparins. *Seminars in Thrombosis and Hemostasis* 25 (3): 5–16.

67 Guerrini, M., Guglieri, S., Naggi, A. et al. (2007). Low molecular weight heparins: structural differentiation by bidimensional nuclear magnetic resonance spectroscopy. *Seminars in Thrombosis and Hemostasis* 33: 478–487.

68 Mascellani, G., Guerrini, M., Torri, G. et al. (2007). Characterization of di- and monosulfated, unsaturated heparin disaccharides with terminal N-sulfated 1,6-anhydro-β-d-glucosamine or N-sulfated 1,6-anhydro-β-d-mannosamine residues. *Carbohydrate Research* 342: 835–842.

69 Bisio, A., Vecchietti, D., Citterio, L. et al. (2009). Structural features of low-molecular-weight heparins affecting their affinity to antithrombin. *Thrombosis and Haemostasis* 102 (5): 865–873.

70 Bisio, A., Mantegazza, A., Vecchietti, D. et al. (2015). Determination of the molecular weight of low-molecular-weight heparins by using high-pressure size exclusion chromatography on-line with a triple detector array and conventional methods. *Molecules* 20: 5085–5098.

71 Fareed, J., Ma, Q., Florian, M. et al. (2003). Unfractionated and low-molecular-weight heparins, basic mechanism of action and pharmacology. *Seminars in Cardiothoracic and Vascular Anesthesia* 7 (4): 357–377.

72 Gerotziafas, G.T., Petropoulou, A.D., Verdy, E. et al. (2007). Effect of the anti-factor Xa and anti-factor IIa activities of low-molecular-weight heparins upon the phase of thrombin generation. *Journal of Thrombosis and Haemostasis* 5: 955–962.

73 Walenga, J. and Lyman, G.H. (2013). Evolution of heparin anticoagulants to ultra-low-molecular-weight heparins: a review of pharmacologic and clinical differences and applications in patients with cancer. *Critical Reviews in Oncology/Hematology* 88: 1–18.

74 Naggi, A., Gardini, C., Pedrinola, G. et al. (2016). Structural peculiarity and antithrombin binding region profile ofmucosal bovine and porcine heparins. *Journal of Pharmaceutical and Biomedical Analysis* 118: 52–63.

75 Martinez-Gonzalez, J., Vila, L., and Rodriguez, C. (2008). Bemiparin: second generation, low-molecular-weight heparin for treatment and prophylaxis of venous thromboembolism. *Expert Reviews in Cardiovascular Therapy* 6 (6): 793–802.

76 Nader, H.B., Walenga, J.M., Berkowitz, S.D. et al. (1999). Preclinical differentiation of low molecular weight heparins. *Seminars in Thrombosis and Hemostasis* 25 (3): 63–72.

77 Fareed, J., Ma, Q., Florian, M. et al. (2004). Differentiation of low molecular-weight heparins: impact on the future of the management of thrombosis. *Seminars in Thrombosis and Hemostasis* 30 (1): 89–104.

78 Bertini, S., Fareed, J., Madaschi, L. et al. (2017). Characterization of PF4-heparin complexes by photon correlation spectroscopy and zeta potential. *Clinical and Applied Thrombosis/Hemostasis* 23 (7): 725–734.

79 Boer, C., Meesters, M.I., Veerhoek, D., and Vonk, A.B.A. (2018). Anticoagulant and side effects of protamine in cardiac surgery: a narrative review. *British Journal of Anaesthesia* 120 (5): 914–927.

80 Viskov, C., Just, M., Laux, V. et al. (2009). Description of the chemical and pharmacological characteristics of a new hemisynthetic ultra-low-molecular-weight heparin, AVE5026. *Journal of Thrombosis and Haemostasis* 7: 1143–1151.

81 Rico, S., Antonijoan, R.M., Gich, I. et al. (2011). Safety assessment and pharmacodynamics of a novel ultra low molecular weight heparin (RO-14) in healthy volunteers – A first-time-in-human single ascending dose study. *Thrombosis Research* 127: 292–298.

82 Acostamadiedo, J.M., Iyer, U.G., and Owen, J. (2000). Danaparoid sodium. *Expert Opinion on Pharmacotherapy* 1 (4): 803–814.

83 Ibbotson, T. and Perry, C.M. (2002). Danaparoid: a review of its use in thromboembolic and coagulation disorders. *Drugs* 62 (15): 2283–2314.

84 Davenport, A. (2012). Alternatives to standard unfractionated heparin for pediatric hemodialysis treatments. *Pediatric Nephrology* 27 (10): 1869–1879.

85 Borawski, J., Zbroch, E., Rydzewska-Rosolowska, A. et al. (2007). Sulodexide for hemodialysis anticoagulation in heparin-induced thrombocytopenia type II. *Journal of Nephrology* 20 (3): 370–372.

86 Adiguzel, C., Iqbal, O., Hoppensteadt, D. et al. (2009). Comparative anticoagulant and platelet modulatory effects of enoxaparin and sulodexide. *Clinical and Applied Thrombosis/Hemostasis* 15 (5): 501–511.

87 Lauver, D.A. and Lucchesi, B.R. (2006). Sulodexide: a renewed interest in this glycosaminoglycan. *Cardiovascular Drug Reviews* 24 (3, 4): 214–226.

88 Cosmi, B., Cini, M., Legnani, C. et al. (2003). Additive thrombin inhibition by fast moving heparin and dermatan sulfate explains the anticoagulant effect of sulodexide, a natural mixture of glycosaminoglycans. *Thrombosis Research* 109 (5, 6): 333–339.

89 Harenberg, J., Jescheck, M., Acker, M. et al. (1996). Effects of low-molecular-weight dermatan sulfate on coagulation, fibrinolysis and tissue factor pathway inhibitor in healthy volunteers. *Blood Coagulation & Fibrinolysis* 7 (1): 49–56.

90 Tovar, A.M.F., de Mattos, D.A., Stelling, M.P. et al. (2005). Dermatan sulfate is the predominant antothrombotic glycosaminoglycan in vessel walls: implications for a possible physiological function of heparin cofactor II. *Biochimica et Biophysica Acta* 1740: 45–53.

91 Linhardt, R.J. and Hileman, R.E. (1995). Dermatan sulfate as a potential therapeutic agent. *General Pharmacology* 26 (3): 443–451.

92 Casu, B., Guerrini, M., and Torri, G. (2004). Structural and conformational aspects of the anticoagulant and antithrombotic activity of heparin and dermatan sulfate. *Current Pharmaceutical Design* 10: 939–949.

93 Maimone, M.M. and Tollefsen, D.M. (1991). Structure of a dermatan sulfate hexasaccharide that binds to heparin cofactor II with high affinity. *The Journal of Biological Chemistry* 119: 653–661.

94 Mascellani, G., Liverani, L., Bianchini, P. et al. (1993). Structure and contribution to the heparin cofactor Il-mediated inhibition of thrombin of naturally oversulphated sequences of dermatan sulphate. *The Biochemical Journal* 296: 639–648.

95 Mascellani, G., Liverani, L., Prete, A. et al. (1994). Quantitation of dermatansulfate active site for heparin cofactor II by 1H nuclear magnetic resonance spectroscopy. *Analytical Biochemistry* 223: 135–141.

96 Pavão, M.S.G., Aiello, K.R., Werneck, C.C. et al. (1998). Highly sulfated dermatan sulfates from Ascidians. Structure versus anticoagulant activity of these glycosaminoglycans. *The Journal of Biological Chemistry* 273: 2748–27857.

97 Pavão, M.S.G., Mourão, P.A.S., Mulloy, B., and Tollefsen, D.M. (1995). A unique dermatan sulfate- like glycosaminoglycan from ascidian. Its structure and the effect of its unusual sulfation pattern on anticoagulant activity. *The Journal of Biological Chemistry* 270: 31027–31036.

98 Di Carlo, V., Agnelli, G., Prandoni, P. et al. (1999). Dermatan sulphate for the prevention of postoperative venous thromboembolism in patients with cancer. DOS (Dermatan sulphate in Oncologic Surgery) Study Group. *Thrombosis Haemostasis* 82 (1): 30–34.

99 Vitale, C., Verdecchia, C., Bagnis, C. et al. (2008). Effects of dermatan sulfate for anticoagulation in continuous renal replacement therapy. *Journal of Nephrology* 21 (2): 205–212.

100 Vitale, C., Berutti, S., Bagnis, C. et al. (2013). Dermatan sulfate: an alternative to unfractionated heparin for anticoagulation in hemodialysis patients. *Journal of Nephrology* 26 (1): 158–163.

101 Sié, P., Dupouy, D., Caranobe, C. et al. (1993). Antithrombotic properties of a dermatan sulfate hexadecasaccharide fractionated by affinity for heparin cofactor II. *Blood* 81: 1771–1777.

102 Dol, F., Petitou, M., Lormeau, J.C. et al. (1990). Pharmacologic properties of a low molecular weight dermatan sulfate: comparison with unfractionated dermatan sulfate. *The Journal of Laboratory and Clinical Medicine* 115 (1): 43–51.

103 Barbanti, M., Calanni, F., Milani, M.R. et al. (1993). Therapeutic effect of a low molecular weight dermatan sulphate (Desmin 370) in rat venous thrombosis – evidence for an anticoagulant-independent mechanism. *Thrombosis and Haemostasis* 69: 147–151.

104 Volpi, N. (1993). "Fast moving" and "slow moving" heparins, dermatan sulfate, and chondroitin sulfate: qualitative and quantitative analysis by agarose-gel electrophoresis. *Carbohydrate Research* 247: 263–278.

105 Veraldi, N., Guerrini, M., Urso, E. et al. (2018). Fine structural characterization of sulodexide. *Journal of Pharmaceutical and Biomedical Analysis* 156: 67–79.

106 Coccheri, S. and Mannello, F. (2014). Development and use of sulodexide in vascular diseases: implications for treatment. *Drug Design, Development and Therapy* 4 (8): 49–65.

107 Hoppensteadt, D.A. and Fareed, J. (2014). Pharmacological profile of sulodexide. *International Angiology* 33: 229–235.

108 Andreozzi, G.M., Bignamini, A.A., Davì, G. et al. (2015). Sulodexide for the prevention of recurrent venous thromboembolism. The sulodexide in secondary prevention of recurrent deep vein thrombosis (SURVET) study: a multicenter, randomized, double-blind, placebo-controlled trial. *Circulation* 132: 1891–1897.

109 Gardini, C., Urso, E., Guerrini, M. et al. (2017). Characterization of danaparoid complex extractive drug by an orthogonal analytical approach. *Molecules* 22: 1116–1141.

110 Meuleman, D.G. (1992). Orgaran (Org 10172): its pharmacological profile in experimental models. *Haemostasis* 22 (2): 59–65.

111 Wilde, M.I. and Danaparoid, M.A. (1997). A review of its pharmacology and clinical use in the management of heparin-induced thrombocytopenia. *Drugs* 54 (6): 903–924.

112 De Pont, A.-C.J.M., Hofstra, J.-J.H., Pik, D.R. et al. (2007). Pharmacokinetics and pharmacodynamics of danaparoid during continuous venovenous hemofiltration: a pilot study. *Critical Care* 11: R102.

113 Vittoria, A., Messa, G.L., Frigerio, C. et al. (1988). Effect of a single dose of mesoglycan on the human fibrinolytic system, and the profibrinolytic action of nine daily doses. *International Journal of Tissue Reactions* 10 (4): 261–266.

114 Tufano, A., Arturo, C., Cimino, E. et al. (2010). Mesoglycan: clinical evidences for use in vascular diseases. *International Journal of Vascular Medicine* 2010 (4): 390643.

115 Neri, G., Marcelli, V., Califano, L., and Investigators, G. (2018). Assessment of the effect of mesoglycan in the treatment of audiovestibular disorders of vascular origin. *International Journal of Immunopathology and Pharmacology* 32: 1–5.

116 Niada, R., Mantovani, M., Prino, G. et al. (1981). Antithrombotic activity of a polydeoxyribonucleotidic substance extracted from mammalian organs: a possible link with prostacyclin. *Thrombosis Research* 23 (3): 233–246.

117 Pescador, R., Capuzzi, L., Mantovani, M. et al. (2013). Defibrotide: properties and clinical use of an old/new drug. *Vascular Pharmacology* 59 (1, 2): 1–10.

118 Baker, D.E. and Demaris, K. (2016). Defibrotide. *Hospital Pharmacy* 51 (10): 847–854.

119 Stein, C., Castanotto, D., Krishnan, A., and Nikolaenko, L. (2016). Defibrotide (Defitelio): a new addition to the stockpile of Food and Drug Administration-approved oligonucleotide drugs. *Molecular Therapy-Nucleic Acids* 5 (8): e346.

120 Bracht, F. and Schror, K. (1994). Isolation and identification of aptamers from defibrotide that acts as thrombin antagonists in vitro. *Biochemical and Biophysical Research Communications* 200: 933–937.

121 Richardson, P.G., Carreras, E., Iacobelli, M., and Nejadnik, B. (2018). The use of defibrotide in blood and marrow transplantation. *Blood Advances* 2 (12): 1495–1509.

122 Mohty, M., Malard, F., Abecasis, M. et al. (2020). Prophylactic, preemptive, and curative treatment for sinusoidal obstruction syndrome/veno-occlusive disease in adult patients: a position statement from an international expert group. *Bone Marrow Transplantation* 55: 485–495.

123 de Ferra, L., Naggi, A., Zenoni, M., and Pinto, B. (2014). Patent WO 2014114723 A1: method for the qualification of preparations of pentosan polysulfate, raw materials and production processes thereof.

124 Scheller, H. and Ulvskov, P. (2010). Hemicelluloses. *Annual Review of Plant Biology* 61: 263–289.

125 Alekseeva, A., Raman, R., Eisele, G. et al. (2020). In-depth structural characterization of pentosan polysulfate sodium complex drug using orthogonal analytical tools. *Carbohydrate Polymers* 234: 115913–115926.

126 Fischer, A.M., Merton, R.E., Marsh, N.A. et al. (1982). A comparison of pentosan polysulfate and heparin. II: effect of subcutaneous injection. *Thrombosis and Haemostasis* 30: 109–113.

127 Mulholland, S.G., Hanno, P., Parsons, C.L. et al. (1990). Pentosan polysulfate sodium for therapy of interstitial cystitis: a double-blind placebo-controlled clinical study. *Urology* 35 (6): 552–558.

128 Parson, C.L., Benson, G., Childs, S.J. et al. (1994). A Quantitatively controlled method to study prospectively interstitial cystitis and demonstrate the efficacy of pentosanpolysulfate. *The Journal of Urology* 150: 845–848.

129 Sandem, C., Mori, M., Jogdand, P. et al. (2017). Broad Th2 neutralization and anti-inflammatory action of pentosan polysulfate sodium in experimental allergic rhinitis. *Immunity, Inflammation and Disease* 5 (3): 300–309.

130 Dunn, F., Soria, J., Soria, C. et al. (1983). Fibrinogen binding on human platelets: influence of different heparins and pentosan polysulphate. *Thrombosis Research* 29: 141–148.

131 Soria, C., Soria, J., Rijckewaart, J.J. et al. (1980). Anticoagulant activities of a pentosan polysulphate: comparison with standard heparin and a fraction of low molecular heparin. *Thrombosis Research* 19: 455–463.
132 Ofosu, F.A., Blajchman, M.A., Modi, G.J. et al. (1985). The importance of thrombin inhibition for the expression of the anticoagulant activities of heparin, dermatan sulphate, low molecular weight heparin and pentosan polysulphate. *British Journal of Haematology* 60: 695–704.
133 Teichman, J.M. (2002). The role of pentosan polysulfate in treatment approaches for interstitial cystitis. *Revista de Urologia* 1 (1): S21–S27.
134 Anderson, V.A. and Perry, C.M. (2006). Pentosan polysulfate: a review of its use in the relief of bladder pain or discomfort in interstitial cystitis. *Drugs* 66 (6): 821–835.
135 Petitou, M., Casu, B., and Lindahl, U. (2003). 1976–1983, a critical period in the history of heparin: the discovery of the antithrombin binding site. *Biochimie* 85 (1–2): 83–89.
136 Sinaÿ, P., Jaquinet, J.-C., Petitou, M. et al. (1984). Total synthesis of a heparin pentasaccharide fragment having high affinity for antithrombin III. *Carbohydrate Research* 132 (2): c5–c9.
137 van Boeckel, C.A.A., Beetz, T., Vos, N.J. et al. (1985). Synthesis of a pentasaccharide corresponding to the antithrombin III binding fragment of heparin. *Journal of Carbohydrate Chemistry* 4 (3): 293–321.
138 Petitou, M., Duchaussoy, P., Lederman, I. et al. (1987). Synthesis of heparin fragments: a α-methyl pentaoside with high affinity for antithrombin III. *Carbohydrate Research* 167: 67–75.
139 Paolucci, F., Claviés, M.-C., François, D., and Necciari, J. (2012). Fondaparinux sodium mechanism of action: identification of specific binding to purified and human plasma-derived proteins. *Clinical Pharmacokinetics* 41: 11–18.
140 Bauer, K.A., Hawkins, D.W., Peters, P.C. et al. (2002). Fondaparinux, a synthetic pentasaccharide: the first in a new class of antithrombotic agents – the selective factor Xa inhibitors. *Cardiovascular Drug Reviews* 20 (1): 37–52.
141 Nutescu, E.A., Burnett, A., Fanikos, J. et al. (2016). Pharmacology of anticoagulants used in the treatment of venous thromboembolism. *Journal of Thrombosis and Thrombolysis* 41: 15–31.
142 Petitou, M. and van Boeckel, C.A.A. (2004). A synthetic antithrombin III binding pentasaccharide is now a drug! What comes next? *Angewandte Chemie* 43: 3118–3133.
143 Petitou, M., Duchaussoy, P., Herbert, J.-M. et al. (2002). The synthetic pentasaccharide fondaparinux: first in the class of antithrombotic agents that selectively inhibit coagulation factor Xa. *Seminars in Thrombosis Research* 28 (4): 393–402.
144 Walenga, J., Jeske, W., Samama, M.M. et al. (2002). Fondaparinux: a synthetic heparin pentasaccharide as a new antithrombotic agent. *Expert Opinion on Investigational Drugs* 11 (3): 397–407.

145 Rauova, L., Poncz, M., McKenzie, S.E. et al. (2005). Ultralarge complexes of PF4 and heparin are central to the pathogenesis of heparin-induced thrombocytopenia. *Blood* 105 (1): 131–138.

146 Dey, S., Lo, H.-J., and Wong, C.-H. (2020). Programmable one-pot synthesis of heparin pentasaccharide fondaparinux. *Organic Letters* 22 (12): 4638–4642.

147 Chen, C. and Yu, B. (2009). Efficient synthesis of Idraparinux, the anticoagulant pentasaccharide. *Bioorganic & Medicinal Chemistry* 19 (14): 3875–3879.

148 Harenberg, J. (2010). Idraparinux and Idrabiotaparinux. *Expert Review of Clinical Pharmacology* 3 (1): 9–16.

149 Savi, P., Herault, J.P., Duchaussoy, P. et al. (2008). Reversible biotinylated oligosaccharides: a new approach for a better management of anticoagulant therapy. *Journal of Thrombosis and Haemostasis* 6: 1697–1706.

150 Petitou, M., Nancy-Portebois, V., Dubreucq, G. et al. (2009). From heparin to EP217609: The long way to a new pentasaccharide-based neutralizable anticoagulant with an unprecedented pharmacological profile. *Thrombosis and Haemostasis* 102: 804–810.

151 Gueret, P., Combe, S., Krezel, C. et al. (2016). First in man study of EP217609, a new long-acting, neutralisable parenteral antithrombotic with a dual mechanism of action. *European Journal of Clinical Pharmacology* 72 (9): 1041–1050.

152 Hirsh, J., Bauer, K.A., Donati, M.B. et al. (2008). Parenteral anticoagulants: American College of Chest Physicians Evidence-Based Clinical Practice Guidelines (8th Edition). *Chest* 133 (6 Suppl): 141S–159S.

153 Crowther, M.A., Berry, L.R., Monagle, P.T., and Chan, A.K. (2002). Mechanisms responsible for the failure of protamine to inactivate low-molecular-weight heparin. *British Journal of Haematology* 116 (1): 178–186.

154 Bianchini, E.P., Fazavana, J., Picard, V., and Borgel, D. (2011). Development of a recombinant antithrombin variant as a potent antidote to fondaparinux and other heparin derivatives. *Blood* 117 (6): 2054–2060.

155 Lu, G., DeGuzman, F.R., Hollenbach, S.J. et al. (2013). A specific antidote for reversal of anticoagulation by direct and indirect inhibitors of coagulation factor Xa. *Nature Medicine* 19 (4): 446–451.

156 Xu, Y., Masuko, S., Takieddin, M. et al. (2011). Chemoenzymatic synthesis of homogeneous ultra low molecular weight heparins. *Science* 334: 498–501.

157 Xu, Y., Pempe, E.H., and Liu, J. (2012). Chemoenzymatic synthesis of heparin oligosaccharides with both anti-factor Xa and anti-factor IIa activities. *The Journal of Biological Chemistry* 287 (34): 29054–29061.

158 Wang, Z., Hsieh, P.-H., Xu, Y. et al. (2017). Synthesis of 3-O-sulfated oligosaccharides to understand the relationship between structures and functions of heparan sulfate. *Journal of the American Chemical Society* 139 (14): 5249–5256.

Index

Carbohydrate-Based Therapeutics, First Edition. Edited by Roberto Adamo and Luigi Lay.

h

i

k